INTRODUCTION TO FLIGHT

INTRODUCTION
TO
FLIGHT

Second Edition

John D. Anderson, Jr.
Professor of Aerospace Engineering
University of Maryland

McGraw-Hill Book Company

New York St. Louis San Francisco Auckland Bogotá Hamburg
London Madrid Mexico Montreal New Delhi
Panama Paris São Paulo Singapore Sydney Tokyo Toronto

This book was set in Times Roman by Science Typographers, Inc.
The editors were Anne Murphy and Jo Satloff;
the production supervisor was Charles Hess.
The cover was designed by Nicholas Krenitsky.
New drawings were done by ECL Art.

Cover Photograph Credits
The Wright Brothers—The Bettmann Archive
The Space Shuttle Columbia—NASA

INTRODUCTION TO FLIGHT

567890 DODO 897

ISBN 0-07-001639-9

Library of Congress Cataloging in Publication Data

Anderson, John David.
    Introduction to flight.

    Includes bibliographies and index.
    1. Aerodynamics.   2. Airplanes—Design and construction.
3. Space flight.   I. Title.
TL570.A68   1985   629.1   84-21813
ISBN 0-07-001639-9

CONTENTS

PREFACE TO THE
SECOND EDITION

The purpose of the second edition is the same as that of the first: to present the basic fundamentals of aerospace engineering at the introductory level in the clearest, simplest, and most motivating style possible. It is meant to be *understood* and *enjoyed* by the reader. I have made every effort to talk to the reader in a fashion that is readable and completely understandable. The choice and organization of the subject matter, the sequence in which ideas are introduced, and the way of explaining these ideas has been very carefully constructed with the uninitiated reader in mind. It is intended to be a self-learning, self-contained text at the first- and second-year levels. Tedious detail and massive "handbook" data are avoided; rather, fundamental concepts are introduced and discussed in as straightforward and clear-cut a manner as possible.

The response to the first edition from students, faculty, and practicing professionals throughout the country has been overwhelmingly favorable. Therefore, for the second edition most of the content of the first edition has been carried over virtually intact, with minor corrections, modernizations, and extensions made to the existing sections as appropriate. Moreover, the second edition contains much added material as follows:

1. Chapter 6 has been expanded to include takeoff and landing performance, turning flight, load factor, the V-n diagram, and energy concepts for accelerated rate of climb.
2. Sections on the neutral point, static margin, longitudinal control, elevator trim angle and hinge moments, and stick-free longitudinal stability have been added to Chap. 7.
3. Chapter 8 has been expanded to include the basic concepts dealing with atmospheric reentry, particularly ballistic trajectories and aerodynamic heating.
4. In the first edition, the sections on the history of aeronautical engineering have been particularly popular. Therefore, in the second edition, eight new

history sections have been added, including discussions of the contributions of Osborne Reynolds and Ernst Mach. Moreover, the interrelationships among the Wright brothers, Samuel Langley, and Glenn Curtiss are discussed at length in light of their massive influence on the early development of aeronautics in the United States. The connections between these men form "the aeronautical triangle," a new interpretation of early U.S. aeronautical history introduced for the first time in this book. Also, several of the original history sections have been greatly expanded, especially those dealing with the first manned supersonic flight and the advent of space flight.

5. Many new illustrations have been added, and a number of the existing figures have been improved to reflect actual experimental and theoretical data rather than generic variations.

6. A concise summary section has been added at the end of each chapter to help the reader consolidate his or her understanding of the material.

7. Homework problems are included at the end of all chapters (except Chap. 1). For the first edition, both the problems and solutions appeared in a separate Problems and Solutions Manual available only to instructors. A selection of these problems have been carried over to the second edition, along with many new problems.

Hence, the present book is longer and more complete than the first edition. However, the original philosophy as described in the preface to the first edition is totally preserved.

What constitutes a proper introduction to aerospace engineering? The answer is as varied as the number of people addressing the question. I have made choices based on ten years of experience with students at the University of Maryland. These choices have also been influenced by numerous conversations with university faculty and practicing professionals throughout the United States. Special thanks are due to the faculty of the Department of Aeronautics at the U.S. Air Force Academy, and in particular to the head of the department, Colonel Daniel H. Daley. I have been privileged to participate in the annual aerodynamics workshop at the academy since its inception in 1979. During these workshops, I have had many stimulating discussions with the excellent academy faculty and students, and this interaction has served to mold and influence parts of this second edition.

At the University of Maryland, this book is used in a year-long introductory course for sophomore students in aerospace engineering. It leads directly into a second book by the author, *Fundamentals of Aerodynamics* (McGraw-Hill, 1984), which is used in a two-semester junior-senior aerodynamics course. This in turn feeds into a third book, *Modern Compressible Flow: With Historical Perspective* (McGraw-Hill, 1982), used in a course for advanced undergraduates and first-year graduate students. The complete triad is intended by the author to give the students a reasonable technical, historical, and philosophical perspective on aerospace engineering in general, and on aerodynamics and gas dynamics in particular.

I am very grateful to Mrs. Susan Cunningham, who did such an excellent job typing the manuscript for this second edition. I am fortunate to have such dedicated and professional help from one of the best administrative assistants in the world.

I would also like to express my thanks for the many useful comments and suggestions provided by colleagues who reviewed this text during the course of its development, especially to Dale F. Moses, San Diego State University; John Sherfesee, United States Air Force Academy; V. V. Utgoff, United States Naval Academy; and Robert C. Winn, United States Air Force Academy.

Finally, it is my firm conviction that the study and mastery of the profession of aerospace engineering is one of the most gratifying of human endeavors. This book is designed to set the student on such a course, and to give him or her a sense of enthusiasm, dedication, and genuine love of the profession. To the reader, I simply say—have fun.

John D. Anderson, Jr.

PREFACE TO THE FIRST EDITION

This book is an introduction to aerospace engineering from both the technological and historical points of view. It is written to appeal to several groups of people: (1) students of aerospace engineering in their freshman or sophomore years in college who are looking for a comprehensive introduction to their profession, (2) advanced high school seniors who simply want to learn what aerospace engineering is all about, (3) both college undergraduate and graduate students who want to obtain a wider perspective on the glories, the intellectual demands, and the technical maturity of aerospace engineering, and (4) working engineers who simply want to obtain a firmer grasp on the fundamental concepts and historical traditions that underlie their profession.

As an introduction to aerospace engineering, this book is unique in at least three ways. First, the vast majority of aerospace engineering professionals and students have little knowledge or appreciation of the historical traditions and background associated with the technology that they use almost everyday. To fill this vacuum, the present book attempts to marble some history of aerospace engineering into the parallel technical discussions. For example, such questions as who was Bernoulli, where did the Pitot tube originate, how did wind tunnels evolve, who were the first true aeronautical engineers, and how did wings and airfoils develop are answered. The present author feels strongly that such material should be an integral part of the background of all aerospace engineers.

Second, this book incorporates both the SI and English engineering system of units. Modern students of aerospace engineering must be bilingual—on one hand, they must fully understand and feel comfortable with the SI units, because most modern and all future literature will deal with the SI system; on the other hand, they must be able to read and feel comfortable with the vast bulk of existing literature, which is predominantly in engineering units. In this book, the SI system is emphasized, but an honest effort is made to give the reader a feeling for and understanding of both systems. To this end some example problems are worked out in the SI system and others in the English system.

Third, the author feels that technical books do not have to be dry and sterile in their presentation. Instead, the present book is written in a rather informal style. It attempts to *talk* to the reader. Indeed, it is intended to be almost a self-teaching, self-pacing vehicle that the reader can use to obtain a fundamental understanding of aerospace engineering.

This book is a product of several years of teaching the introductory course in aerospace engineering at the University of Maryland. Over these years, students have constantly encouraged the author to write a book on the subject, and their repeated encouragement could not be denied. The present book is dedicated in part to these students.

Writing a book of this magnitude is a total commitment of time and effort for a longer time than the author likes to remember. In this light, this book is dedicated to my wife, Sarah-Allen, and my two daughters, Katherine and Elizabeth, who relinquished untold amounts of time with their husband and father so that these pages could be created. To them I say thank you, and hello again. Also, hidden between the lines, but ever-so-much present is Edna Brothers, who typed the manuscript in such a dedicated fashion. In addition, the author wishes to thank Dr. Richard Hallion and Dr. Thomas Crouch, curators of the National Air and Space Museum of the Smithsonian Institution, for their helpful comments on the historical sections of this manuscript, and especially Dick Hallion, for opening the vast archives of the museum for the author's historical research. Also, many thanks are due to the reviewers of this manuscript, Professor J. J. Azar of the University of Tulsa, Dr. R. F. Brodsky of Iowa State University, Dr. David Caughey of Sibley School of Mechanical and Aerospace Engineering, and Professor Francis J. Hale of North Carolina State University; their comments have been most constructive, especially those of Dr. Caughey and Professor Hale. Finally, the author wishes to thank his many colleagues in the profession for stimulating discussions about what constitutes an introduction to aerospace engineering. Hopefully, this book is a reasonable answer.

John D. Anderson, Jr.

THE FIRST AERONAUTICAL ENGINEERS

Nobody will fly for a thousand years!

Wilbur Wright, 1901, in a fit of despair

SUCCESS FOUR FLIGHTS THURSDAY MORNING ALL AGAINST TWENTY ONE MILE WIND STARTED FROM LEVEL WITH ENGINE POWER ALONE AVERAGE SPEED THROUGH AIR THIRTY ONE MILES LONGEST 57 SECONDS INFORM PRESS HOME CHRISTMAS.

OREVELLE WRIGHT
A telegram, with the original misprints, from Orville
Wright to his father, December 17, 1903

1.1 INTRODUCTION

The scene: Wind-swept sand dunes of Kill Devil Hills, 4 mi south of Kitty Hawk, North Carolina. *The time:* About 10:35 A.M. on Thursday, December 17, 1903. *The characters:* Orville and Wilbur Wright and five local witnesses. *The action:* Poised, ready to make history, is a flimsy, odd-looking machine, made from spruce and cloth in the form of two wings, one placed above the other, a horizontal elevator mounted on struts in front of the wings, and a double vertical rudder behind the wings (see Figure 1.1). A 12-hp engine is mounted on the top surface of the bottom wing, slightly right of center. To the left of this engine lies a man—Orville Wright—prone on the bottom wing, facing into the brisk and cold December wind. Behind him rotate two ungainly looking airscrews (propellers), driven by two chain and pulley arrangements connected to the same engine. The machine begins to move along a 60-ft launching rail on level ground. Wilbur Wright runs along the right side of the machine, supporting the wingtip so that it will not drag the sand. Near the end of the starting rail, the machine lifts into the air; at this moment, John Daniels of the Kill Devil Life Saving Station takes a photograph which preserves for all time the most historic moment in aviation history (see Figure 1.2). The machine flies unevenly, rising suddenly to about 10 ft, then ducking quickly toward the ground. This type of erratic flight continues for 12 s, when the machine darts to the sand, 120 ft from the point where it lifted from the starting rail. Thus ends a flight which, in Orville Wright's own words, was "the first in the history of the world in which a machine carrying a man had raised itself by its own power into the air in full flight, had sailed forward without reduction of speed, and had finally landed at a point as high as that from which it started."

The machine was the Wright Flyer I, which is shown in Figures 1.1 and 1.2 and which is now preserved for posterity in the Air and Space Museum of the

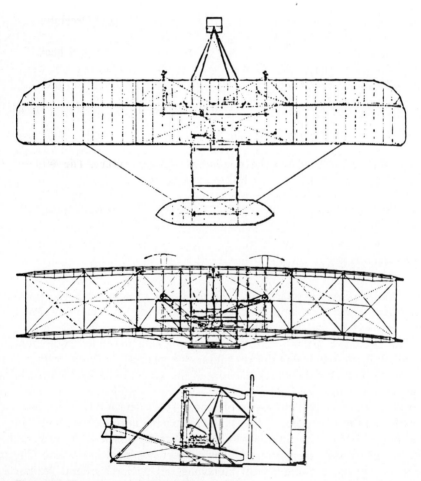

Figure 1.1 Three views of the Wright Flyer I, 1903.

Figure 1.2 The first heavier-than-air flight in history: the Wright Flyer I with Orville Wright at the controls, December 17, 1903. *(National Air and Space Museum.)*

Smithsonian Institution in Washington, D.C. The flight on that cold December 17 was momentous: it brought to a realization the dreams of centuries, and it gave birth to a new way of life. It was the first genuine powered flight of a heavier-than-air machine. With it, and with the further successes to come over the next five years, came the Wright brothers' clear right to be considered the premier aeronautical engineers of history.

However, contrary to some popular belief, the Wright brothers did not truly *invent* the airplane; rather, they represent the fruition of a century's worth of prior aeronautical research and development. The time was ripe for the attainment of powered flight at the beginning of the twentieth century. The Wright brothers' ingenuity, dedication, and persistence earned them the distinction of being first. The purpose of this chapter is to look back over the years which led up to successful powered flight and to single out an important few of those inventors and thinkers who can rightfully claim to be the first aeronautical engineers. In this manner, some of the traditions and heritage that underlie modern aerospace engineering will be more appreciated when we develop the technical concepts of flight in subsequent chapters.

1.2 VERY EARLY DEVELOPMENTS

Since the dawn of human intelligence, the idea of flying in the same realm as birds has possessed human minds. Witness the early Greek myth of Daedalus and his son Icarus. Imprisoned on the island of Crete in the Mediterranean Sea, Daedalus is said to have made wings fastened with wax. With these wings, they both escaped by flying through the air. However, Icarus, against his father's warnings, flew too close to the sun; the wax melted, and Icarus fell to his death in the sea.

All early thinking of human flight centered on the imitation of birds. Various unsung ancient and medieval people fashioned wings and met with sometimes disastrous and always unsuccessful consequences in leaping from towers or roofs, flapping vigorously. In time, the idea of strapping a pair of wings to arms fell out of favor. It was replaced by the concept of wings flapped up and down by various mechanical mechanisms, powered by some type of human arm, leg, or body movement. These machines are called *ornithopters*. Recent historical research has uncovered that Leonardo da Vinci himself was possessed by the idea of human flight and that he designed vast numbers of ornithopters toward the end of the fifteenth century. In his surviving manuscripts, over 35,000 words and 500 sketches deal with flight. One of his ornithopter designs is shown in Figure 1.3, which is an original da Vinci sketch made sometime between 1486 and 1490. It is not known whether da Vinci ever built or tested any of his designs. However, human-powered flight by flapping wings was always doomed to failure. In this sense, da Vinci's efforts did not make important contributions to the technical advancement of flight.

Human efforts to fly literally got off the ground on November 21, 1783, when a balloon carrying Pilatre de Rozier and the Marquis d'Arlandes ascended into the air and drifted 5 mi across Paris. The balloon was inflated and buoyed up by hot air from an open fire burning in a large wicker basket underneath. The design and construction of the balloon were due to the Montgolfier brothers, Joseph and Etienne. In 1782, Joseph Montgolfier, gazing into his fireplace, conceived the idea

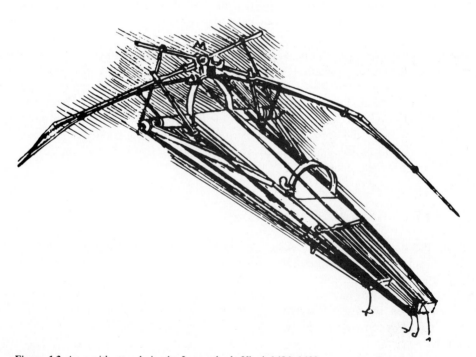

Figure 1.3 An ornithopter design by Leonardo da Vinci, 1486–1490.

of using the "lifting power" of hot air rising from a flame to lift a person from the surface of the earth. The brothers instantly set to work, experimenting with bags made of paper and linen, in which hot air from a fire was trapped. After several public demonstrations of flight without human passengers, including the 8-min voyage of a balloon carrying a cage containing a sheep, a rooster, and a duck, the Montgolfiers were ready for the big step. At 1:54 P.M. on November 21, 1783, the first flight with human passengers rose majestically into the air and lasted for 25 min (see Figure 1.4). It was the first time in history that a human being had been lifted off the ground for a sustained period of time. Very quickly after this, the noted French physicist J. A. C. Charles (of Charles' gas law in physics) built and flew a hydrogen-filled balloon from the Tuileries Gardens in Paris on December 1, 1783.

So people were finally off the ground! Balloons, or "aerostatic machines" as called by the Montgolfiers, made no real technical contributions to human heavier-than-air flight. However, they served a major purpose in triggering the public's interest in flight through the air. They were living proof that people could really leave the ground and sample the environs heretofore exclusively reserved for birds. Moreover, they were the only means of human flight for almost 100 years.

Figure 1.4 The first aerial voyage in history: the Montgolfier hot-air balloon lifts from the ground near Paris, November 21, 1783.

1.3 SIR GEORGE CAYLEY (1773–1857)—THE TRUE INVENTOR OF THE AIRPLANE

The modern airplane has its origin in a design set forth by George Cayley in 1799. It was the first concept to include a *fixed* wing for generating lift, another *separate* mechanism for propulsion (Cayley envisioned paddles), and a combined horizontal and vertical (cruciform) tail for stability. Cayley inscribed his idea on a silver disc (presumably for permanence), shown in Figure 1.5. On the reverse side of the disc is a diagram of the lift and drag forces on an inclined plane (the wing). The disc is now preserved in the Science Museum in London. Before this time, thoughts of mechanical flight had been oriented towards the flapping wings of ornithopters, where the flapping motion was supposed to provide both lift and propulsion. (Da Vinci designed his ornithopter wings to flap simultaneously downward and backward for lift and propulsion.) However, Cayley is responsible for breaking this unsuccessful line of thought; he separated the concept of lift from propulsion and, in so doing, set into motion a century of aeronautical development that culminated in the Wright brothers' success in 1903. George Cayley is a giant in aeronautical history: he is the parent of modern aviation and is the first true aeronautical engineer. Let us look at him more closely.

Cayley was born at Scarborough in Yorkshire, England, on December 27, 1773. He was educated at York and Nottingham and later studied chemistry and electricity under several noted tutors. He was a scholarly man of some rank, a baronet who spent much of his time on the family estate called Brompton. A portrait of Cayley is shown in Figure 1.6. He was a well-preserved person, of extreme intellect and open mind, active in many pursuits over a long life of 84 years. In 1825, he invented the caterpillar tractor, forerunner of all modern

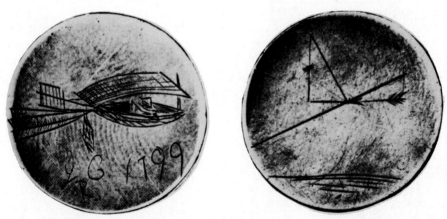

Figure 1.5 The silver disc on which Cayley engraved his concept for a fixed-wing aircraft, the first in history, in 1799. The reverse side of the disc shows the resultant aerodynamic force on a wing resolved into lift and drag components, indicating Cayley's full understanding of the function of a fixed wing. The disc is presently in the Science Museum in London.

Figure 1.6 A portrait of Sir George Cayley, painted by Henry Perronet Briggs in 1841. The portrait now hangs in the National Portrait Gallery in London.

tracked vehicles. In addition, he was chairman of the Whig Club of York, founded the Yorkshire Philosophical Society (1821), cofounded the British Association for the Advancement of Science (1831), was a member of Parliament, was a leading authority on land drainage, and published papers dealing with optics and railroad safety devices. Moreover, he had a social conscience: he appealed for, and donated to, the relief of industrial distress in Yorkshire.

However, by far his major and lasting contribution to humanity was in aeronautics. After experimenting with model helicopters beginning in 1796, Cayley engraved his revolutionary fixed-wing concept on the silver disc in 1799 (see Figure 1.5). This was followed by an intensive 10-year period of aerodynamic investigation and development. In 1804, he built a whirling arm apparatus, shown in Figure 1.7, for testing airfoils; this was simply a lifting surface (airfoil) mounted on the end of a long rod, which was rotated at some speed to generate a flow of air over the airfoil. In modern aerospace engineering, wind tunnels now serve this function, but in Cayley's time the whirling arm was an important development, which allowed the measurement of aerodynamic forces and the center of pressure on a lifting surface. Of course, these measurements were not very accurate, because after a number of revolutions of the arm, the surrounding air would begin to rotate with the device. Nevertheless, it was a first step in aerodynamic testing. Also in 1804, Cayley designed, built, and flew the small

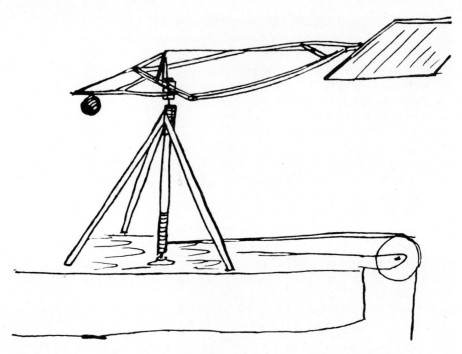

Figure 1.7 George Cayley's whirling arm apparatus for testing airfoils.

model glider shown in Figure 1.8; this may seem trivial today, something that you may have done as a child, *but in 1804 it represented the first modern-configuration airplane of history*, with a fixed wing, and a horizontal and vertical tail that could be adjusted. (Cayley generally flew his glider with the tail at a positive angle of incidence, as shown in his sketch in Figure 1.8.) A full-scale replica of this glider is on display at the Science Museum in London—the model is only about 1 m long.

Cayley's first outpouring of aeronautical results was documented in his momentous triple paper of 1809–1810. Entitled "On Aerial Navigation," and published in the November 1809, February 1810, and March 1810 issues of Nicholson's *Journal of Natural Philosophy*, this document ranks as one of the most important aeronautical works in history. (Note that the words "natural philosophy" in history are synonymous with physical science.) Cayley was prompted to write his triple paper after hearing reports that Jacob Degen had

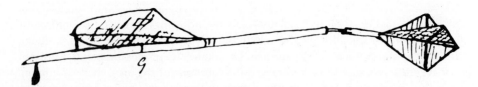

Figure 1.8 The first modern configuration airplane in history: Cayley's model glider, 1804.

recently flown in a mechanical machine in Vienna. In reality, Degen flew in a contraption which was lifted by a balloon. It was of no significance, but Cayley did not know the details. In an effort to let people know of his activities, Cayley documented many aspects of aerodynamics in his triple paper. It was the first treatise on theoretical and applied aerodynamics in history to be published. In it, Cayley elaborates on his principle of separation of lift and propulsion and his use of a fixed wing to generate lift. He states that the basic aspect of a flying machine is "to make a surface support a given weight by the application of power to the resistance of air." He notes that a surface inclined at some angle to the direction of motion will generate lift and that a cambered (curved) surface will do this more efficiently than a flat surface. He also states for the first time in history that lift is generated by a region of low pressure on the upper surface of the wing. The modern technical aspects of these phenomena will be developed and explained in Chaps. 4 and 5; however, stated by Cayley in 1809–1810, these phenomena were new and unique. His triple paper also addressed the matter of flight control and was the first document to discuss the role of the horizontal and vertical tail planes in airplane stability. Interestingly enough, Cayley goes off on a tangent in discussing the use of flappers for propulsion. Note that on the silver disc (see Figure 1.5) Cayley shows some paddles just behind the wing. From 1799 until his death in 1857, Cayley was obsessed with such flappers for aeronautical propulsion. He gave little attention to the propeller (airscrew); indeed, he seemed to have an aversion to rotating machinery of any type. However, this should not detract from his numerous positive contributions. Also in his triple paper, Cayley tells us of the first successful full-size glider of history, built and flown without passengers by him at Brompton in 1809. However, there is no clue as to its configuration.

Curiously, the period from 1810 to 1843 was a lull in Cayley's life in regard to aeronautics. Presumably, he was busy with his myriad other interests and activities. During this period, he showed interest in airships (controlled balloons), as opposed to heavier-than-air machines. He made the prophetic statement that "balloon aerial navigation can be done readily, and will probably, in the order of things, come into use before mechanical flight can be rendered sufficiently safe and efficient for ordinary use." He was correct; the first successful airship, propelled by a steam engine, was built and flown by the French engineer Henri Giffard in Paris in 1852, 51 years before the first successful airplane.

Cayley's second outpouring of aeronautical results occurred in the period from 1848 to 1854. In 1849, he built and tested a full-size airplane. During some of the flight tests, a 10-year-old boy was carried along and was lifted several meters off the ground while gliding down a hill. Cayley's own sketch of this machine, called the *boy carrier*, is shown in Figure 1.9. Note that it is a triplane (three wings mounted on top of each other). Cayley was the first to suggest such multiplanes (i.e., biplanes and triplanes), mainly because he was concerned with the possible structural failure of a single large wing (a monoplane). Stacking smaller, more compact, wings on top of each other made more sense to him, and his concept perpetuated far into the twentieth century. It was not until the late

1930s that the monoplane became the dominant airplane configuration. Also note from Figure 1.9 that, strictly speaking, this was a "powered" airplane, i.e., it was equipped with propulsive flappers.

One of Cayley's most important papers was published in *Mechanics Magazine* for September 25, 1852. By this time he was 79 years old! The article was entitled "Sir George Cayley's Governable Parachutes." It gave a full description of a large human-carrying glider which incorporated almost all the features of the modern airplane. This design is shown in Figure 1.10, which is a facsimile of the illustration which appeared in the original issue of *Mechanics Magazine*. This airplane had (1) a main wing at an angle of incidence for lift, with a dihedral for lateral stability, (2) an adjustable cruciform tail for longitudinal and directional stability, (3) a pilot-operated elevator and rudder, (4) a fuselage in the form of a car, with a pilot's seat and three-wheel undercarriage, and (5) a tubular beam and box beam construction. These combined features were not to be seen again until the Wright brothers' designs at the beginning of the twentieth century. Incredibly, this 1852 paper by Cayley went virtually unnoticed, even though *Mechanics Magazine* had a large circulation. It was recently rediscovered by the eminent British aviation historian Charles H. Gibbs-Smith in 1960 and republished by him in the June 13, 1960, issue of *The Times*.

Sometime in 1853—the precise date is unknown—George Cayley built and flew the world's first human-carrying glider. Its configuration is not known, but Gibbs-Smith states that it was most likely a triplane on the order of the earlier boy carrier (see Figure 1.9) and that the planform (top view) of the wings was

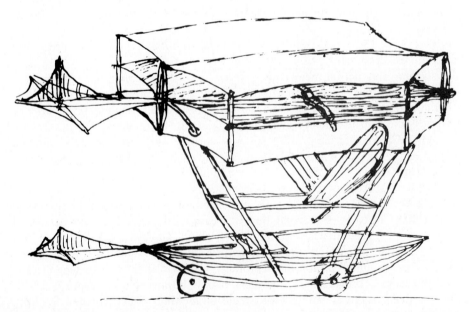

Figure 1.9 Cayley's triplane from 1849—the boy carrier. Note the vertical and horizontal tail surfaces and the flapperlike propulsive mechanism.

𝔐echanics' 𝔐agazine,

MUSEUM, REGISTER, JOURNAL, AND GAZETTE.

No. 1520.]   SATURDAY, SEPTEMBER 25, 1852.   [Price 3*d*., Stamped 4*d*.

Edited by J. C. Robertson, 166, Fleet-street.

SIR GEORGE CAYLEY'S GOVERNABLE PARACHUTES.

Fig. 2.

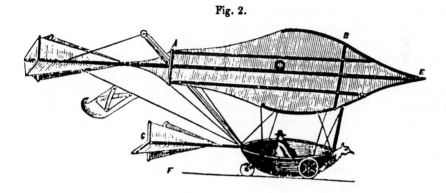

Fig. 1.

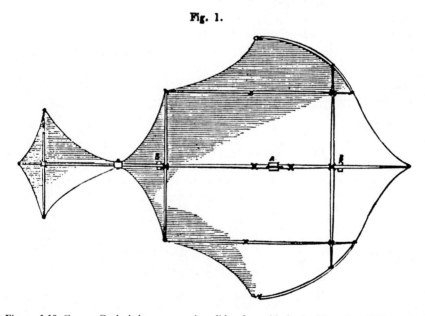

Figure 1.10 George Cayley's human-carrying glider, from *Mechanics Magazine*, 1852.

probably shaped much like the glider in Figure 1.10. According to several eyewitness accounts, a gliding flight of several hundred yards was made across a dale at Brompton with Cayley's coachman aboard. The glider landed rather abruptly, and after struggling clear of the vehicle, the shaken coachman is quoted as saying: "Please, Sir George, I wish to give notice. I was hired to drive, and not to fly." Very recently, this flight of Cayley's coachman was reenacted for the public in a special British Broadcasting Corporation television show on Cayley's life. While visiting the Science Museum in London in August of 1975, the present author was impressed to find the television replica of Cayley's glider (minus the coachman) hanging in the entranceway.

George Cayley died at Brompton on December 15, 1857. During his almost 84 years of life, he laid the basis for all practical aviation. He was called the "father of aerial navigation" by William Samuel Henson in 1846. However, for reasons that are not clear, the name of George Cayley retreated to the background soon after his death. His works became obscure to virtually all later aviation enthusiasts in the latter half of the nineteenth century. This is incredible, indeed unforgivable, considering that his published papers were available in known journals. Obviously, many subsequent inventors did not make the effort to examine the literature before forging ahead on their own ideas. (This is certainly a problem for engineers today, with the virtual explosion of written technical papers since World War II.) However, Cayley's work has been brought to light by the research of several modern historians in the twentieth century. Notable among them is C. H. Gibbs-Smith, from whose book entitled *Sir George Cayley's Aeronautics* (1962) much of the above material has been gleaned. Gibbs-Smith states that had Cayley's work been extended directly by other aviation pioneers, and had they digested ideas espoused in his triple paper of 1809–1810 and in his 1852 paper, successful powered flight would have most likely occurred in the 1890s. Probably so!

As a final tribute to George Cayley, we note that the French aviation historian Charles Dollfus said the following in 1923:

> The aeroplane is a British invention: it was conceived in all essentials by George Cayley, the great English engineer who worked in the first half of last century. The name of Cayley is little known, even in his own country, and there are very few who know the work of this admirable man, the greatest genius of aviation. A study of his publications fills one with absolute admiration both for his inventiveness, and for his logic and common sense. This great engineer, during the Second Empire, did in fact not only invent the aeroplane entire, as it now exists, but he realized that the problem of aviation had to be divided between theoretical research—Cayley made the first aerodynamic experiments for aeronautical purposes—and practical tests, equally in the case of the glider as of the powered aeroplane.

1.4 THE INTERREGNUM—FROM 1853 TO 1891

For the next 50 years after Cayley's success with the coachman-carrying glider, there were no major advances in aeronautical technology comparable to the previous 50 years. Indeed, as stated above, much of Cayley's work became obscure to all but a few dedicated investigators. However, there was considerable activity, with numerous people striking out (sometimes blindly) in various uncoordinated directions to conquer the air. Some of these efforts are noted below, just to establish the flavor of the period.

William Samuel Henson (1812–1888) was a contemporary of Cayley. In April 1843, he published in England a design for a fixed-wing airplane powered by a steam engine driving two propellers. Called the *aerial steam carriage*, this design received wide publicity throughout the nineteenth century, due mainly to a series of illustrative engravings which were reproduced and sold around the world. This was a publicity campaign of which Madison Avenue would have been proud; one of these pictures is shown in Figure 1.11. Note some of the qualities of modern aircraft in Figure 1.11: the engine inside a closed fuselage, driving two propellers; tricycle landing gear; and a single rectangular-shaped wing of relatively high aspect ratio. (We will discuss the aerodynamic characteristics of such wings in Chap. 5.) Henson's design was a direct product of George Cayley's ideas and research in aeronautics. The aerial steam carriage was never built, but the design, along with its widely published pictures, served to engrave George Cayley's fixed-wing concept on the minds of virtually all subsequent workers. Thus, even though Cayley's published papers fell into obscurity after his death, his major concepts were partly absorbed and perpetuated by following generations of inventors, even though most of these inventors did not know the true source of the ideas. In this manner, Henson's aerial steam carriage was one of the most influential airplanes in history, even though it never flew!

John Stringfellow, a friend of Henson, made several efforts to bring Henson's design to fruition. Stringfellow built several small steam engines and attempted to power some model monoplanes off the ground. He was close, but unsuccessful.

Figure 1.11 Henson's aerial steam carriage, 1842–1843. *(National Air and Space Museum.)*

Figure 1.12 Stringfellow's model triplane exhibited at the first aeronautical exhibition in London, 1868.

However, his most recognized work appeared in the form of a steam-powered triplane, a model of which was shown at the 1868 aeronautical exhibition sponsored by the Aeronautical Society at the Crystal Palace in London. A photograph of Stringfellow's triplane is shown in Figure 1.12. This airplane was also unsuccessful, but again it was extremely influential because of worldwide publicity. Illustrations of this triplane appeared throughout the end of the nineteenth century. Gibbs-Smith, in his book *Aviation: An Historical Survey from its Origins to the End of World War II* (1970), states that these illustrations were later a strong influence on Octave Chanute, and through him the Wright brothers, and strengthened the concept of superimposed wings. Stringfellow's triplane was the main bridge between George Cayley's aeronautics and the modern biplane.

During this period, the first powered airplanes actually hopped off the ground, but for only hops. In 1857–1858, the French naval officer and engineer Felix Du Temple flew the first successful powered model airplane in history; it was a monoplane with swept-forward wings and was powered by clockwork! Then, in 1874, Du Temple achieved the world's first powered takeoff by a piloted, full-size airplane. Again, the airplane had swept-forward wings, but this time it was powered by some type of hot-air engine (the precise type is unknown). A sketch of Du Temple's full-size airplane is shown in Figure 1.13. The machine, piloted by a young sailor, was launched down an inclined plane at Brest, France; it left the ground for a moment but did not come close to anything resembling sustained flight. In the same vein, the second powered airplane with a pilot left the ground near Leningrad (then St. Petersburg), Russia, in July 1884. Designed by Alexander F. Mozhaiski, this machine was a steam-powered monoplane, shown in Figure 1.14. Mozhaiski's design was a direct descendant from Henson's aerial steam carriage—it was even powered by an English steam engine! With I. N. Golubev as pilot, this airplane was launched down a ski ramp and flew for a

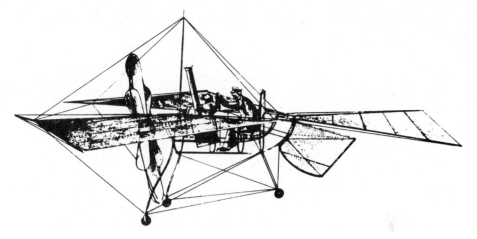

Figure 1.13 Du Temple's airplane: the first aircraft to make a powered but assisted takeoff, 1874.

few seconds. As with Du Temple's airplane, no sustained flight was achieved. At various times, the Russians have credited Mozhaiski with the first powered flight in history, but of course it did not satisfy the necessary criteria to be called such. Du Temple and Mozhaiski achieved the first and second *assisted* powered takeoffs, respectively, in history, but neither experienced sustained flight. In his book *The World's First Aeroplane Flights* (1965), C. H. Gibbs-Smith states the

Figure 1.14 The second airplane to make an assisted takeoff: Mozhaiski's aircraft, Russia, 1884.

following criteria used by aviation historians to judge a successful powered flight:

In order to qualify for having made a simple powered and sustained flight, a conventional aeroplane should have sustained itself freely in a horizontal or rising flight path—without loss of airspeed—beyond a point where it could be influenced by any momentum built up before it left the ground: otherwise its performance can only be rated as a powered leap, i.e., it will not have made a fully self-propelled flight, but will only have followed a ballistic trajectory modified by the thrust of its propeller and by the aerodynamic forces acting upon its aerofoils. Furthermore, it must be shown that the machine can be kept in satisfactory equilibrium. Simple sustained flight obviously need not include full controllability, but the maintenance of adequate equilibrium in flight is part and parcel of sustentation.

Under these criteria, there is no doubt in the mind of any major aviation historian that the first powered flight was made by the Wright brothers in 1903. However, the assisted "hops" described above put two more rungs in the ladder of aeronautical development in the nineteenth century.

Of particular note during this period is the creation in London in 1866 of the Aeronautical Society of Great Britain. Before this time, work on "aerial navigation" (a phrase coined by George Cayley) was looked upon with some disdain by many scientists and engineers. It was too out of the ordinary and was not to be taken seriously. However, the Aeronautical Society soon attracted scientists of stature and vision, people who shouldered the task of solving the problems of mechanical flight in a more orderly and logical fashion. In turn, aeronautics took on a more serious and meaningful atmosphere. The society, through its regular meetings and technical journals, provided a cohesive scientific outlet for the presentation and digestion of aeronautical engineering results. The society is still flourishing today in the form of the highly respected Royal Aeronautical Society. Moreover, it served as a model for the creation of both the American Rocket Society and the Institute of Aeronautical Sciences in the United States in this century; both of these societies merged in 1964 to form the American Institute of Aeronautics and Astronautics (AIAA), one of the most influential channels for aerospace engineering information exchange today.

In conjunction with the Aeronautical Society of Great Britain, at its first meeting on June 27, 1866, Francis H. Wenham read a paper entitled "Aerial Locomotion," one of the classics in aeronautical engineering literature. Wenham was a marine engineer who later was to play a prominent role in the society and who later designed and built the first wind tunnel in history (see Chap. 4). His paper, which was also published in the first annual report of the society, was the first to point out that most of the lift of a wing was obtained from the portion near the leading edge. He also established that a wing with high aspect ratio was the most efficient for producing lift. (We will see why in Chap. 5.)

As noted in our previous discussion about Stringfellow, the Aeronautical Society started out in style: When it was only two years old, in 1868, it put on the first aeronautical exhibition in history at the Crystal Palace. It attracted an assortment of machines and balloons and for the first time offered the general public a first-hand overview of the efforts being made to conquer the air.

Stringfellow's triplane (discussed earlier) was of particular interest. Zipping over the heads of the enthralled onlookers, the triplane moved through the air along an inclined cable strung below the roof of the exhibition hall (see Figure 1.12). However, it did not achieve sustained flight on its own. In fact, the 1868 exhibition did nothing to advance the technical aspects of aviation, but it was a masterstroke of good public relations.

1.5 OTTO LILIENTHAL (1848–1896)—THE GLIDER MAN

With all the efforts that had taken place in the past, it was still not until 1891 that a human literally jumped into the air and flew with wings in any type of controlled fashion. This person was Otto Lilienthal, one of the giants in aeronautical engineering (and in aviation in general). Lilienthal designed and flew the first successful controlled gliders in history. He was a man of aeronautical stature comparable to Cayley and the Wright brothers. Let us examine the man and his contributions more closely.

Lilienthal was born on May 23, 1848, at Anklam, Prussia (Germany). He obtained a good technical education at trade schools in Potsdam and Berlin, the latter at the Berlin Technical Academy, graduating with a degree in mechanical engineering in 1870. After a one-year stint in the army during the Franco-Prussian War, Lilienthal went to work designing machinery in his own factory. However, from early childhood he was interested in flight and performed some youthful experiments on ornithopters of his own design. Toward the late 1880s, his work and interests took a more mature turn, which ultimately led to fixed-wing gliders.

In 1889, Lilienthal published a book entitled *Der Vogelflug als Grundlage der Fliegekunst* (Bird Flight as the Basis of Aviation). This is another of the early classics in aeronautical engineering, because not only did he study the structure and types of birds' wings, but he also applied the resulting aerodynamic information to the design of mechanical flight. Lilienthal's book contained some of the most detailed aerodynamic data available at that time. Translated sections were later read by the Wright brothers, who incorporated some of his data in their first glider designs in 1900 and 1901. (However, the Wright brothers finally found it necessary to correct some of Lilienthal's aerodynamic data, as will be discussed in a subsequent section.)

By 1889, Lilienthal had also come to a philosophical conclusion which was to have a major impact on the next two decades of aeronautical development. He concluded that to learn practical aerodynamics, he had to get up in the air and experience it himself. In his own words,

One can get a proper insight into the practice of flying only by actual flying experiments.... The manner in which we have to meet the irregularities of the wind, when soaring in the air, can only be learnt by being in the air itself.... The only way which leads us to a quick development in human flight is a systematic and energetic practice in actual flying experiments.

Figure 1.15 A monoplane hang glider by Lilienthal, 1894.

To put this philosophy into practice, Lilienthal designed a glider in 1889, and another in 1890—both were unsuccessful. However, in 1891, Lilienthal's first successful glider flew from a natural hill at Derwitz, Germany. (Later, he was to build an artificial hill about 50 ft high near Lichterfelde, a suburb of Berlin; this conically shaped hill allowed glider flights to be made into the wind, no matter what the direction.) The general configuration of his monoplane gliders is shown in Figure 1.15, which is a photograph showing Lilienthal as the pilot. Note the rather birdlike planform of the wing. Lilienthal used cambered (curved) airfoil shapes on the wing and incorporated vertical and horizontal tail planes in the back for stability. These machines were hang gliders, the grandparents of the sporting vehicles of today. Flight control was exercised by one's shifting one's center of gravity under the glider.

Contrast Lilienthal's flying philosophy with those of previous would-be aviators before him. During most of the nineteenth century, powered flight was looked upon in a brute-force manner: build an engine strong enough to drive an airplane, slap it on an airframe strong enough to withstand the forces and generate the lift, and presumably you could get into the air. What would happen *after* you got into the air would be just a simple matter of steering the airplane around the sky like a carriage or automobile on the ground—at least this was the general feeling. Gibbs-Smith called the people taking this approach the "chauffeurs." In contrast were the "airmen"—Lilienthal was the first—who recognized the need to get up in the air, fly around in gliders, and obtain the "feel" of an airplane *before* an engine was used for powered flight. The chauffeurs were mainly interested in thrust and lift, whereas the airmen were firstly concerned with flight control in the air. The airmen's philosophy ultimately led to successful powered flight; the chauffeurs were singularly unsuccessful.

Lilienthal made over 2500 successful glider flights. The aerodynamic data he obtained were published in papers circulated throughout the world. In fact, his work was timed perfectly with the rise of photography and the printing industry. In 1871 the dry-plate negative was invented, which by 1890 could "freeze" a moving object without a blur. Also, the successful halftone method of printing photographs in books and journals had been developed. As a result, photos of Lilienthal's flights were widely distributed; indeed, Lilienthal was the first human to be photographed in an airplane (see, for example, Figure 1.15). Such widespread dissemination of his results inspired other pioneers in aviation. The Wright brothers' interest in flight did not crystalize until Wilbur first read some of Lilienthal's papers about 1894.

On Sunday, August 9, 1896, Lilienthal was gliding from the Gollenberg hill near Stollen in Germany. It was a fine summer's day. However, a temporary gust of wind brought Lilienthal's monoplane glider to a standstill; he stalled and crashed to the ground. Only the wing was crumpled; the rest of the glider was undamaged. However, Lilienthal was carried away with a broken spine. He died the next day in the Bergmann Clinic in Berlin. During the course of his life, Lilienthal remarked several times that "sacrifices must be made." This epitaph is carved on his gravestone in Lichterfelde cemetery.

There is some feeling that had Lilienthal lived, he would have beaten the Wright brothers to the punch. In 1893 he built a powered machine; however, the prime mover was a carbonic acid gas motor which twisted six slats at each wingtip, obviously an ornithopter-type idea to mimic the natural mode of propulsion for birds. In the spring of 1895 he built a second, but larger, powered machine of the same type. Neither one of these airplanes was ever flown with the engine operating. It seems to this author that this mode of propulsion was doomed to failure. If Lilienthal had lived, would he have turned to the gasoline engine driving a propeller and thus achieved powered flight before 1903? It is a good question for conversation.

1.6 PERCY PILCHER (1867–1899)—EXTENDING THE GLIDER TRADITION

In June of 1895, Otto Lilienthal received a relatively young and very enthusiastic visitor in Berlin—Percy Pilcher, a Scot who lived in Glasgow and who had already built his first glider. Under Lilienthal's guidance Pilcher made several glides from the artificial hill. This visit added extra fuel to Pilcher's interest in aviation; he returned to the British Isles and over the next four years built a series of successful gliders. His most noted machine was the *Hawk*, built in 1896 (see Figure 1.16). Pilcher's experiments with his hang gliders made him the most distinguished British aeronautical engineer since George Cayley. Pilcher was an "airman," and along with Lilienthal he underscored the importance of learning the practical nature of flight in the air before lashing an engine to the machine.

However, Pilcher's sights were firmly set on powered flight. In 1897, he calculated that an engine of 4 hp weighing no more than 40 lb, driving a 5-ft-diameter propeller, would be necessary to power his *Hawk* off the ground. Since no such engine was available commercially, Pilcher (who was a marine engineer by training) spent most of 1898 designing and constructing one. It was

Figure 1.16 Pilcher's hang glider the *Hawk*, 1896.

completed and bench-tested by the middle of 1899. Then, in one of those quirks of fate that dot many aspects of history, Pilcher was killed while demonstrating his *Hawk* glider at the estate of Lord Braye in Leicestershire, England. The weather was bad, and on his first flight the glider was thoroughly water-soaked. On his second flight, the heavily sodden tail assembly collapsed, and Pilcher crashed to the ground. Like Lilienthal, Pilcher died one day after this disaster. Hence, England and the world also lost the only man other than Lilienthal who might have achieved successful powered flight before the Wright brothers.

1.7 AERONAUTICS COMES TO AMERICA

Look at the geographical distribution of the early developments in aeronautics as portrayed in the previous sections. After the advent of ballooning, due to the Montgolfiers' success in France, progress in heavier-than-air machines was focused in England until the 1850s: witness the contributions of Cayley, Henson, and Stringfellow. This is entirely consistent with the fact that England also gave birth to the industrial revolution during this time. Then the spotlight moved to the European continent with Du Temple, Mozhaiski, Lilienthal, and others. There were some brief flashes again in Britain, such as those due to Wenham and the Aeronautical Society. In contrast, throughout this time virtually no important progress was being made in the United States. The fledgling nation was busy consolidating a new government and expanding its frontiers. There was not much interest or time for serious aeronautical endeavors.

However, this vacuum was broken by Octave Chanute (1832–1910), a French-born naturalized citizen who lived in Chicago. Chanute was a civil engineer who became interested in mechanical flight about 1875. For the next 35 years, he collected, absorbed, and assimilated every piece of aeronautical information he could find. This culminated in 1894 with the publishing of his book entitled *Progress in Flying Machines*, a work that ranks with Lilienthal's *Der Vogelflug* as one of the great classics in aeronautics. Chanute's book summarized all the important progress in aviation to that date; in this sense, he was the first serious aviation historian. In addition, Chanute made positive suggestions as to the future directions necessary to achieve success in powered flight. The Wright brothers avidly read *Progress in Flying Machines* and subsequently sought out Chanute in 1900. A close relationship and interchange of ideas developed between them. A friendship developed which was to last in various degrees until Chanute's death in 1910.

Chanute was an "airman." Following this position, he began to design hang gliders, in the manner of Lilienthal, in 1896. His major specific contribution to aviation was the successful biplane glider shown in Figure 1.17, which introduced the effective Pratt truss method of structural rigging. The Wright brothers were directly influenced by this biplane glider, and in this sense Chanute provided the natural bridge between Stringfellow's triplane (1868) and the first successful powered flight (1903).

About 500 miles to the east, in Washington, D.C., the United States' second noted pre-Wright aeronautical engineer was hard at work. Samuel Pierpont Langley (1834–1906), secretary of the Smithsonian Institution, was tirelessly designing and building a series of powered aircraft, which finally culminated in two attempted piloted flights, both in 1903, just weeks before the Wrights' success on December 17.

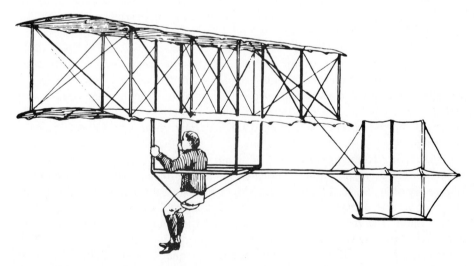

Figure 1.17 Chanute's hang glider, 1896. *(National Air and Space Museum.)*

Langley was born at Roxbury, Massachusetts, on August 22, 1834. He received no formal education beyond high school, but his childhood interest in astronomy spurred him to a lifelong program of self-education. Early in his career, he worked 13 years as an engineer and architect. Then, after making a tour of European observatories, Langley became an assistant at the Harvard Observatory in 1865. He went on to become a mathematics professor at the U.S. Naval Academy, a physics and astronomy professor at the University of Pittsburgh, and the director of the Allegheny Observatory at Pittsburgh. By virtue of his many scientific accomplishments, Langley was appointed secretary of the Smithsonian Institution in 1887.

In this same year, Langley, who was by now a scientist of international reputation, began his studies of powered flight. Following the example of Cayley, he built a large whirling arm, powered by a steam engine, with which he made force tests on airfoils. He then built nearly 100 different types of rubber-band-powered model airplanes, graduating to steam-powered models in 1892. However, it was not until 1896 that Langley achieved any success with his powered models; on May 6 one of his aircraft made a free flight of 3300 ft, and on November 28 another flew for over $\frac{3}{4}$ mi. These Aerodromes (a term due to Langley) were tandem-winged vehicles, driven by two propellers between the wings, powered by a 1-hp steam engine of Langley's own design. (However, Langley was influenced by one of John Stringfellow's small aerosteam engines, which was presented to the Smithsonian in 1889. After studying this historic piece of machinery, Langley set out to design a better engine.)

Langley was somewhat satisfied with his success in 1896. Recognizing that further work toward a piloted aircraft would be expensive in both time and money, he "made the firm resolution not to undertake the construction of a large man-carrying machine." (Note that it was this year that the Wright brothers became interested in powered flight, another example of the flow and continuity of ideas and developments in physical science and engineering. Indeed, Wilbur and Orville were directly influenced and encouraged by Langley's success with powered aircraft. After all, here was a well-respected scientist who believed in the eventual attainment of mechanical flight and who was doing something about it.)

Consequently, there was a lull in Langley's aeronautical work until December 1898. Then, motivated by the Spanish-American War, the War Department, with the personal backing of President McKinley himself, invited Langley to build a machine for passengers. They backed up their invitation with $50,000. Langley accepted.

Departing from his earlier use of steam, Langley correctly decided that the gasoline-fueled engine was the proper prime mover for aircraft. He first commissioned Stephan Balzer of New York to produce such an engine, but dissatisfied with the results, Langley eventually had his assistant Charles Manly redesign the power plant. The resulting engine produced 52.4 hp and yet weighed only 208 lb, a spectacular achievement for that time. Using a smaller, 3.2-hp, gasoline-fueled engine, Langley made a successful flight with a quarter-scale model aircraft in August of 1903.

Figure 1.18 Langley's full-size Aerodrome on the houseboat launching catapult, 1903. *(National Air and Space Museum.)*

Encouraged by this success, Langley stepped directly to the full-size airplane. He mounted this tandem-winged aircraft on a catapult in order to provide an assisted takeoff. In turn, the airplane and catapult were placed on top of a houseboat on the Potomac River (see Figure 1.18). On October 7, 1903, with Manly at the controls, the airplane was ready for its first attempt. The launching was given wide advance publicity, and the press was present to watch what might be the first successful powered flight in history. Here is the resulting report from the *Washington Post* the next day:

> A few yards from the houseboat were the boats of the reporters, who for three months had been stationed at Widewater. The newspapermen waved their hands. Manly looked down and smiled. Then his face hardened as he braced himself for the flight, which might have in store for him fame or death. The propeller wheels, a foot from his head, whirred around him one thousand times to the minute. A man forward fired two skyrockets. There came an answering "toot, toot," from the tugs. A mechanic stooped, cut the cable holding the catapult; there was a roaring, grinding noise—and the Langley airship tumbled over the edge of the houseboat and disappeared in the river, sixteen feet below. It simply slid into the water like a handful of mortar....

Manly was unhurt. Langley believed the airplane was fouled by the launching mechanism, and he tried again on December 8, 1903. Again, the Aerodrome fell into the river, and again Manly was fished out, unhurt. It is not entirely certain what happened this time; again the fouling of the catapult was blamed, but some experts maintain that the tail boom cracked due to structural weakness. At any rate, that was the end of Langley's attempts. The War Department gave up, stating that "we are still far from the ultimate goal (of human flight)." Members of Congress and the press leveled vicious and unjustified attacks on Langley (human flight was still looked upon with much derision by most people). In the face of this ridicule, Langley retired from the aeronautical scene. He died on February 27, 1906, a man in despair.

In contrast to Chanute and the Wright brothers, Langley was a "chauffeur." Most modern experts feel that his Aerodrome would not have been capable of sustained, equilibrium flight had it been successfully launched. Langley made no experiments with gliders with passengers to get the feel of the air. He ignored completely the important aspects of flight control. He attempted to launch Manly into the air on a powered machine without Manly having one second of flight experience. Nevertheless, Langley's aeronautical work was of some importance because he lent the power of his respected technical reputation to the cause of mechanical flight, and his Aerodromes were to provide encouragement to others.

Nine days after Langley's second failure, the Wright Flyer I rose from the sands of Kill Devil Hills.

1.8 WILBUR (1867–1912) AND ORVILLE (1871–1948) WRIGHT—INVENTORS OF THE FIRST PRACTICAL AIRPLANE

The scene now shifts to the Wright brothers, the premier aeronautical engineers of history. Only George Cayley may be considered comparable. In Sec. 1.1, it was stated that the time was ripe for the attainment of powered flight at the beginning of the twentieth century. The ensuing sections then provided numerous historical brushstrokes to emphasize this statement. Thus, the Wright brothers drew on an existing heritage that is part of every aerospace engineer today.

Wilbur Wright was born on April 16, 1867 (2 years after the Civil War) on a small farm at Millville, Indiana. Four years later, Orville was born on August 19, 1871, at Dayton, Ohio. The Wrights were descendants of an old Massachusetts family, and their father was a bishop of the United Brethren Church. The two brothers grew up in Dayton and benefited greatly from the intellectual atmosphere of their family. Their father had some mechanical talent. He invented an early form of the typewriter. Their mother had a college degree in mathematics; she often lent her kitchen to her sons for experiments. Interestingly enough, neither Wilbur nor Orville officially received a high school diploma; Wilbur did not bother to go to the commencement services, and Orville took a special series of courses in his senior year that did not lead to a prescribed degree. Afterward, the brothers immediately sampled the business world. In 1889, they first published a weekly four-page newspaper on a printing press of their own design. However, Orville had talent as a prize-winning cyclist, and this prompted the brothers to set up a bicycle sales and repair shop in Dayton in 1892. Three years later they began to manufacture their own bicycle designs, using homemade tools. These enterprises were profitable and helped to provide the financial resources for their later work in aeronautics.

In 1896, Otto Lilienthal was accidently killed during a glider flight (see Sec. 1.5). In the wake of the publicity, the Wright brothers' interest in aviation, which had been apparent since childhood, was given much impetus. Wilbur and Orville had been following Lilienthal's progress intently; recall that Lilienthal's gliders were shown in flight by photographs distributed around the world. In fact, an article on Lilienthal in an issue of *McClure's Magazine* in 1894 was apparently the first to trigger Wilbur's mature interest; but it was not until 1896 that Wilbur really became a serious thinker about human flight.

Like several pioneers before him, Wilbur took up the study of bird flight as a guide on the path toward mechanical flight. This led him to conclude in 1899 that birds "regain their lateral balance when partly overturned by a gust of wind, by a torsion of the tips of the wings." Thus emerged one of the most important developments in aviation history: the use of wing twist to control airplanes in lateral (rolling) motion. Ailerons are used on modern airplanes for this purpose, but the idea is the same. (The aerodynamic fundamentals associated with wing twist or ailerons are discussed in Chaps. 5 and 7). In 1903, Chanute, in describing the work of the Wright brothers, coined the term "wing warping" for this idea, a term that was to become accepted but which was to cause some legal confusion later.

Anxious to pursue and experiment with the concept of wing warping, Wilbur wrote to the Smithsonian Institution in May 1899 for papers and books on aeronautics; in turn he received a brief bibliography of flying, including works by Chanute and Langley. Most important among these was Chanute's *Progress in Flying Machines* (see Sec. 1.7). Also at this time, Orville became as enthusiastic as his brother, and they both digested all the aeronautical literature they could find. This led to their first aircraft, a biplane kite with a wingspan of 5 ft, in August of 1899. This machine was designed to test the concept of wing warping, which was accomplished by means of four controlling strings from the ground. The concept worked!

Encouraged by this success, Wilbur wrote to Chanute in 1900, informing him of their initial, but fruitful, progress. This letter began a close friendship between the Wright brothers and Chanute, a friendship which was to benefit both parties in the future. Also, following the true "airman" philosophy, the Wrights were convinced they had to gain experience in the air before applying power to an aircraft. By writing to the U.S. Weather Bureau, they found an ideal spot for glider experiments, the area around Kitty Hawk, North Carolina, where there were strong and constant winds. A full-size biplane glider was ready by September 1900 and was flown in October of that year at Kitty Hawk. Figure 1.19 shows a photograph of the Wright's no. 1 glider. It had a 17-ft wingspan and a horizontal elevator in front of the wings and was usually flown on strings from the ground; only a few brief piloted flights were made.

With some success behind them, Wilbur and Orville proceeded to build their no. 2 glider (see Figure 1.20). Moving their base of operations to Kill Devil Hills, 4 mi south of Kitty Hawk, they tested no. 2 during July and August of 1901. These were mostly manned flights, with Wilbur or Orville lying prone on the

Figure 1.19 The Wright brothers' no. 1 glider at Kitty Hawk, North Carolina, 1900. *(National Air and Space Museum.)*

bottom wing, facing into the wind, as shown in Figure 1.20. This new glider was somewhat larger, with a 22-ft wingspan. As with all Wright machines, it had a horizontal elevator in front of the wings. The Wrights felt that a forward elevator would, among other functions, protect them from the type of fatal nosedive that killed Lilienthal.

During these July and August test flights Octave Chanute visited the Wrights' camp. He was much impressed by what he saw. This led to Chanute's invitation to Wilbur to give a lecture in Chicago. In giving this paper on September 18, 1901, Wilbur laid bare their experiences, including the design of their gliders and the concept of wing warping. Chanute described Wilbur's presentation as "a devilish good paper which will be extensively quoted." Chanute, as usual, was

Figure 1.20 The Wright brothers' no. 2 glider at Kill Devil Hills, 1901. *(National Air and Space Museum.)*

serving his very useful function as a collector and disseminator of aeronautical data.

However, the Wrights were not close to being satisfied with their results. When they returned to Dayton after their 1901 tests with the no. 2 glider, both brothers began to suspect the existing data which appeared in the aeronautical literature. To this date, they had faithfully relied upon detailed aerodynamic information generated by Lilienthal and Langley. Now they wondered about its accuracy. Wilbur wrote that "having set out with absolute faith in the existing scientific data, we were driven to doubt one thing after another, until finally, after two years of experiment, we cast it all aside, and decided to rely entirely upon our own investigations." And investigate they did! Between September 1901 and August 1902, the Wrights undertook a major program of aeronautical research. They built a wind tunnel (see Chap. 4) in their bicycle shop at Dayton and tested over 200 different airfoil shapes. They designed a force balance to measure accurately lift and drag. This period of research was a high-water mark in early aviation development. The Wrights learned, and with them ultimately did the world. This sense of learning and achievement by the brothers is apparent simply from reading through *The Papers of Wilbur and Orville Wright* (1953), edited by Marvin W. McFarland. The aeronautical research carried out during this period ultimately led to their no. 3 glider, which was flown in 1902. It was so successful that Orville wrote "that our tables of air pressure which we made in our wind tunnel would enable us to calculate in advance the performance of a machine." Here is the first example in history of the major impact of wind-tunnel testing on the flight development of a given machine, an impact that has been repeated for all major airplanes of the twentieth century.

The no. 3 glider was a classic. It was constructed during August and September of 1902. It first flew at Kill Devil Hills on September 20, 1902. It was a biplane glider with a 32-ft 1-in wingspan, the largest of the Wright gliders to date. This no. 3 glider is shown in Figure 1.21. Note that, after several modifications, the Wrights added a vertical rudder behind the wings. This rudder was movable, and when connected to move in unison with the wing warping, it enabled the no. 3 glider to make a smooth, banked turn. This combined use of rudder with wing warping (or later, ailerons) was another major contribution of the Wright brothers to flight control in particular, and aeronautics in general.

So the Wrights now had the most practical and successful glider in history. During 1902, they made over 1000 perfect flights. They set a distance record of 622.5 ft and a duration record of 26 s. In the process, both Wilbur and Orville became highly skilled and proficient pilots, something that would later be envied worldwide.

Powered flight was now just at their fingertips, and the Wrights knew it! Flushed with success, they returned to Dayton to face the last remaining problem: propulsion. As with Langley before them, they could find no commercial engine that was suitable. So they designed and built their own during the winter months of 1903. It produced 12 hp and weighed about 200 lb. Moreover, they conducted their own research, which allowed them to design an effective propeller. These

Figure 1.21 The Wright brothers' no. 3 glider, 1902. *(National Air and Space Museum.)*

accomplishments, which had eluded people for a century, gushed forth from the Wright brothers like natural spring water.

With all the major obstacles behind them, Wilbur and Orville built their Wright Flyer I from scratch during the summer of 1903. It closely resembled the no. 3 glider but had a wingspan of 40 ft 4 in and used a double rudder behind the wings and a double elevator in front of the wings. And of course, there was the spectacular gasoline-fueled Wright engine, driving two pusher propellers by means of bicycle-type chains. A three-view diagram and photograph of the Flyer I are shown in Figures 1.1 and 1.2, respectively.

From September 23 to 25, the machine was transported to Kill Devil Hills, where the Wrights found their camp in some state of disrepair. Moreover, their no. 3 glider had been damaged over the winter months. They set about to make repairs and afterward spent many weeks of practice with their no. 3 glider. Finally, on December 12, everything was in readiness. However, this time the elements interfered: bad weather postponed the first test of the Flyer I until December 14. On that day, the Wrights called witnesses to the camp and then flipped a coin to see who would be the first pilot. Wilbur won. The Flyer began to move along the launching rail under its own power, picking up flight speed. It lifted off the rail properly but suddenly went into a steep climb, stalled, and thumped back to the ground. It was the first recorded case of pilot error in powered flight: Wilbur admitted that he put on too much elevator and brought the nose too high.

With minor repairs made, and with the weather again favorable, the Flyer was again ready for flight on December 17. This time it was Orville's turn at the controls. The launching rail was again laid on the level sand. A camera was adjusted to take a picture of the machine as it reached the end of the rail. The engine was put on full throttle, the holding rope was released, and the machine began to move. The rest is history, as portrayed in the opening paragraphs of this chapter.

One cannot read nor write of this epoch-making event without experiencing some of the excitement of the time. Wilbur Wright was 36 years old; Orville was 32. Between them, they had done what no one before them had accomplished. By their persistent efforts, their detailed research, and their superb engineering, the Wrights had made the world's first successful heavier-than-air flight, satisfying all the necessary criteria laid down by responsible aviation historians. After Orville's first flight on that December 17, three more flights were made during the morning, the last covering 852 ft and remaining in the air for 59 s. The world of flight—and along with it the world of successful aeronautical engineering—had been born!

It is interesting to note that, even though the press was informed of these events via Orville's telegram to his father (see the introduction to this chapter), virtually no notice appeared before the public; even the Dayton newspapers did not herald the story. This is a testimonial to the widespread cynicism and disbelief among the general public about flying. Recall that just nine days before, Langley had failed dismally in full view of the public. In fact, it was not until Amos I. Root observed the Wrights flying in 1904 and published his inspired account in a journal of which he was the editor, *Gleanings in Bee Culture* (January 1, 1905, issue), that the public had its first detailed account of the Wrights' success. However, the article had no impact.

The Wright brothers did not stop with their Flyer I. In May of 1904, their second powered machine, the Wright Flyer II, was ready. This aircraft had a smaller wing camber (airfoil curvature) and a more powerful and efficient engine. In outward appearance, it was essentially like the 1903 machine. During 1904, over 80 brief flights were made with the Flyer II, all at a 90-acre field called Huffman Prairie, 8 mi east of Dayton. (Huffman Prairie still exists today; it is on the huge Wright-Patterson Air Force Base, a massive aerospace development center named in honor of the Wrights.) These tests included the first circular flight —made by Wilbur on September 20. The longest flight lasted 5 min and 4 s, traversing over $2\frac{3}{4}$ mi.

More progress was made in 1905. Their Flyer III was ready by June. The wing area was slightly smaller than the Flyer II, the airfoil camber was increased back to what it was in 1903, the biplane elevator was made larger and was placed farther in front of the wings, and the double rudder was also larger and placed farther back behind the wings. New, improved propellers were used. This machine, the Flyer III, was the first *practical* airplane in history. It made over 40 flights during 1905, the longest being 38 min and 3 s, covering 24 mi. These flights were generally terminated only after gas was used up. C. H. Gibbs-Smith writes about the Flyer III: "The description of this machine as the world's first practical powered aeroplane is justified by the sturdiness of its structure, which withstood constant takeoffs and landings; its ability to bank, turn, and perform figures of eight; and its reliability in remaining airborne (with no trouble) for over half an hour."

Then the Wright brothers, who heretofore had been completely open about their work, became secretive. They were not making any progress in convincing the U.S. Government to buy their airplane, but at the same time various people and companies were beginning to make noises about copying the Wright design. A patent applied for by the Wrights in 1902 to cover their ideas of wing warping combined with rudder action was not granted until 1906. So, between October 16, 1905 and May 6, 1908, neither Wilbur nor Orville flew, nor did they allow anybody to view their machines. However, their aeronautical engineering did not stop. During this period, they built at least six new engines. They also designed a new flying machine which was to become the standard Wright type A, shown in Figure 1.22. This airplane was similar to the Wright Flyer III, but it had a 40-hp engine and provided for two people seated upright between the wings. It also represented the progressive improvement of a basically successful design, a concept of airplane design carried out to the present day.

The public and the Wright brothers finally had their meeting, and in a big way, in 1908. The Wrights signed contracts with the U.S. Army in February 1908, and with a French company in March of the same year. After that, the wraps were off. Wilbur traveled to France in May, picked up a crated type A which had been waiting at Le Havre since July of 1907, and completed the assembly in a friend's factory at Le Mans. With supreme confidence, he announced his first public flight in advance—to take place on August 8, 1908. Aviation pioneers from all over Europe, who had heard rumors about the Wrights' successes since 1903, the press, and the general public all flocked to a small race course at Hunaudieres, 5 mi south of Le Mans. On the appointed day, Wilbur took off, made an impressive, circling flight for almost 2 min, and landed. It was like a revolution. Aeronautics, which had been languishing in Europe since Lilienthal's death in 1896, was suddenly alive. The Frenchman Louis Bleriot, soon to become famous for being first to fly across the English Channel, exclaimed: "For us in France and everywhere, a new era in mechanical flight has commenced—it is marvelous." The French press, after being skeptical for years of the Wrights' supposed accomplishments, called Wilbur's flight "one of the most exciting spectacles ever presented in the history of applied science." More deeply, echoing the despair of many would-be French aviators who were in a race with the Wrights to be first with powered flight, Leon Delagrange said: "Well, we are beaten. We just don't exist." Subsequently, Wilbur made 104 flights in France before the end of 1908. The acclaim and honor due the Wright brothers since 1903 had finally arrived.

Orville was experiencing similar success in the United States. On September 3, 1908, he began a series of demonstrations for the Army at Fort Myer, near Washington, D.C. Flying a type A machine, he made 10 flights, the longest for 1 h and 14 min, before September 17. On that day, Orville experienced a propeller failure that ultimately caused the machine to crash, seriously injuring himself and killing his passenger, Lt. Thomas E. Selfridge. This was the first crash of a

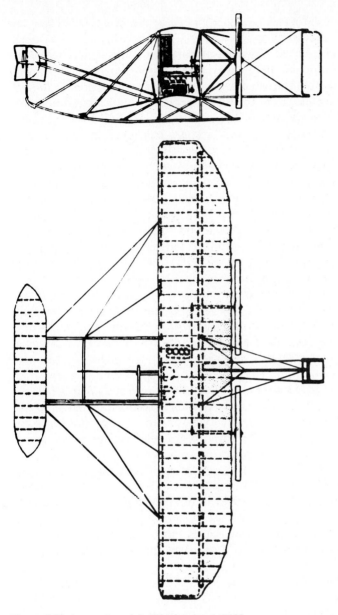

Figure 1.22 A two-view of the Wright type A, 1908.

powered aircraft, but it did not deter either Orville or the Army. Orville made a fast recovery and was back to flying in 1909; and the Army bought the airplane.

The accomplishments of the Wright brothers were monumental. Their zenith was during the years 1908 to 1910; after that European aeronautics quickly caught up and went ahead in the technological race. The main reason for this was that all the Wright machines, from the first gliders, were statically unstable (see Chap. 7). This meant that the Wright airplanes would not fly "by themselves"; rather, they had to be constantly, every instant, controlled by the pilot. In contrast, European inventors believed in inherently stable aircraft. After their lessons in flight control from Wilbur in 1908, workers in France and England moved quickly to develop controllable, but stable, airplanes. These were basically safer, and easier to fly. The concept of static stability has carried over to virtually all airplane designs through the present century. (It is interesting to note that the new designs for lightweight fighters, such as the General Dynamics F-16, are statically *unstable*, which represents a return to the Wrights' design philosophy. However, unlike the Wright Flyers, these new aircraft are flown constantly, every moment, by electrical means, by the new "fly-by-wire" concept.)

To round out the story of the Wright brothers, Wilbur died in an untimely fashion of typhoid fever on May 30, 1912. In a fitting epitaph, his father said: "This morning, at 3:15 Wilbur passed away, aged 45 years, 1 month, and 14 days. A short life full of consequences. An unfailing intellect, imperturbable temper, great self-reliance and as great modesty. Seeing the right clearly, pursuing it steadily, he lived and died."

Orville lived on until January 30, 1948. During World War I, he was commissioned a major in the Signal Corps Aviation Service. Although he sold all his interest in the Wright company and "retired" in 1915, he afterward performed research in his own shop. In 1920, he invented the split flap for wings, and he continued to be productive for many years.

As a final footnote to this story of two great men, there occurred a dispute between Orville and the Smithsonian Institution concerning the proper historical claims on powered flight. As a result, Orville sent the historic Wright Flyer I, the original, to the Science Museum in London in 1928. It resided there, through the bombs of World War II, until 1948, when the museum sent it to the Smithsonian. It is now part of the National Air and Space Museum and occupies a central position in the gallery.

1.9 THE AERONAUTICAL TRIANGLE—LANGLEY, THE WRIGHTS, AND GLENN CURTISS

In 1903—a milestone year for the Wright brothers, with their first successful powered flight—Orville and Wilbur faced serious competition from Samuel P. Langley. As portrayed in Sec. 1.7, Langley was the secretary of the Smithsonian Institution and was one of the most respected scientists in the United States at that time. Beginning in 1886, Langley mounted an intensive aerodynamic research and development program, bringing to bear the resources of the Smithsonian, and later the War Department. He carried out this program with a dedicated zeal that matched the fervor that the Wrights themselves demonstrated later. Langley's efforts culminated in the full-scale Aerodrome shown in Figure 1.18. In October 1903, this Aerodrome was ready for its first attempted flight, in the full glare of publicity in the national press.

The Wright brothers were fully aware of Langley's progress. During their preparations with the Wright Flyer at Kill Devil Hills in the summer and fall of 1903, Orville and Wilbur kept in touch with Langley's progress via the newspapers. They felt this competition keenly, and the correspondence of the Wright brothers at this time indicates an uneasiness that Langley might become the first to successfully achieve powered flight, before they would have a chance to test the Wright Flyer. In contrast, Langley felt no competition at all from the Wrights. Although the aeronautical activity of the Wright brothers was generally known throughout the small circle of aviation enthusiasts in the United States and Europe—thanks mainly to reports on their work by Octave Chanute—this activity was not taken seriously. At the time of Langley's first attempted flight on October 7, 1903, there is no recorded evidence that Langley was even aware of the Wrights' powered machine sitting on the sand dunes of Kill Devil Hills, and certainly no appreciation by Langley of the degree of aeronautical sophistication achieved by the Wrights. As it turned out, as was related in Sec. 1.7, Langley's attempts at manned powered flight, first on October 7 and again on December 8, resulted in total failure. A photograph of Langley's Aerodrome, lying severely

Figure 1.23 Langley's Aerodrome resting in the Potomac River after its first unsuccessful flight on October 7, 1903. Charles Manly, the pilot, was fished out of the river, fortunately unhurt.

damaged in the Potomac River on October 7, is shown in Figure 1.23. In hindsight, the Wrights had nothing to fear in competition with Langley.

Such was not the case in their competition with another aviation pioneer—Glenn H. Curtiss—beginning five years later. In 1908—another milestone year for the Wrights, with their glorious first public flights in France and the United States—Orville and Wilbur faced a serious challenge and competition from Curtiss, which was to lead to acrimony and a flurry of lawsuits that left a smudge on the Wrights' image and resulted in a general inhibition of the development of early aviation in the United States. By 1910, the name of Glenn Curtiss was as well known throughout the world as Orville and Wilbur Wright, and indeed, Curtiss-built airplanes were more popular and easier to fly than those produced by the Wrights. How did these circumstances arise? Who was Glenn Curtiss, and what was his relationship with the Wrights? What impact did Curtiss have on the early development of aviation, and how did his work compare and intermesh with that of Langley and the Wrights? Indeed, the historical development of aviation in the United States can be compared to a triangle, with the Wrights at one apex, Langley at another, and Curtiss at the third. This "aeronautical triangle" is shown in Figure 1.24. What was the nature of this triangular relationship? These and other questions are addressed in this section. They make a fitting conclusion to the overall early historical development of aeronautical engineering as portrayed in this chapter.

Let us first look at Glenn Curtiss, the man. Curtiss was born in Hammondsport, New York, on May 21, 1878. Hammondsport at that time was a small town —population less than 1000—bordering on Keuka Lake, one of the Finger Lakes in upstate New York. (Later, Curtiss was to make good use of Keuka Lake for the development of amphibian aircraft—one of his hallmarks.) The son of a harness maker who died when Curtiss was 5 years old, Curtiss was raised by his mother and grandmother. Their modest financial support came from a small vineyard which grew in their front yard. His formal education ceased with the eighth grade, after which he moved to Rochester, where he went to work for the Eastman Dry Plate and Film Company (later to become Kodak), stenciling numbers on the paper backing of film. In 1900, he returned to Hammondsport, where he took over a bicycle repair shop (shades of the Wright brothers). At this time, Glenn Curtiss began to show a passion that would consume him for his lifetime—a passion for *speed*. He became active in bicycle racing and quickly earned a reputation as a winner. In 1901 he incorporated an engine on his bicycles and became an avid motorcycle racer. By 1902, his fame was spreading, and he was receiving numerous orders for motorcycles with engines of his own design. By 1903 Curtiss had established a motorcycle factory at Hammondsport, and he was designing and building the best (highest horsepower-to-weight ratio) engines available anywhere. In January 1904, at Ormond Beach, Florida, Curtiss established a new world's speed record for a ground vehicle—67 mi/h over a 10-mi straightaway—a record that was to stand for 7 years.

Curtiss "backed into" aviation. In the summer of 1904, he received an order from Thomas Baldwin, a California balloonist, for a two-cylinder engine. Baldwin

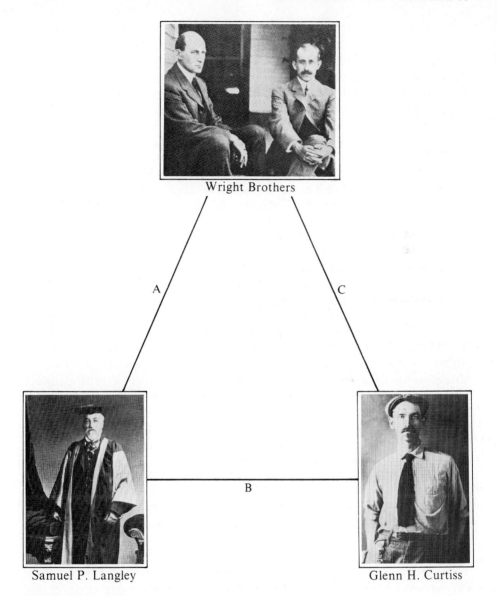

Figure 1.24 The "aeronautical triangle," a relationship that dominated the early development of aeronautics in the United States during the period 1886–1916.

was developing a powered balloon—a dirigible. The Baldwin dirigibles, with the highly successful Curtiss engines, soon became famous around the country. In 1906 Baldwin moved his manufacturing facilities to Hammondsport, to be next to the source of his engines. A lifelong friendship and cooperation developed between Baldwin and Curtiss and provided Curtiss with his first experience in aviation—as a pilot of some of Baldwin's powered balloons.

In August 1906, Baldwin traveled to the Dayton Fair in Ohio for a week of dirigible flight demonstrations; he brought Curtiss along to personally maintain the engines. The Wright brothers also attended the fair—specifically to watch Thomas Baldwin perform. They even lent a hand in retrieving the dirigible when it strayed too far afield. This was the first face-to-face encounter between Curtiss and the Wrights. During that week, Baldwin and Curtiss visited the Wrights at the brothers' bicycle shop and entered upon long discussions on powered flight. Recall from Sec. 1.8 that the Wrights had discontinued flying one year earlier, and at the time of their meeting with Curtiss, Orville and Wilbur were actively trying to interest the United States, as well as England and France, in buying their airplane. The Wrights had become very secretive about their airplane and allowed no one to view it. Curtiss and Baldwin were no exceptions. However, that week in Dayton, the Wrights were relatively free with Curtiss, giving him information and technical suggestions about powered flight. Years later, these conversations became the crux of the Wrights' claim that Curtiss had stolen some of their ideas and used them for his own gain.

This claim was probably not entirely unjustified, for by that time Curtiss had a vested interest in powered flight; a few months earlier he had supplied Alexander Graham Bell with a 15-hp motor to be used in propeller experiments, looking toward eventual application to a manned, heavier-than-air, powered aircraft. The connection between Bell and Curtiss is important. Bell, renowned as inventor of the telephone, had an intense interest in powered flight. He was a close personal friend of Samuel Langley and, indeed, was present for Langley's successful unmanned Aerodrome flights in 1896. By the time Langley died in 1906, Bell was actively carrying out kite experiments and was testing air propellers on a catamaran at his Nova Scotia coastal home. In the summer of 1907, Bell formed the Aerial Experiment Association, a group of five men whose officially avowed purpose was simply "to get into the air." The Aerial Experiment Association (A.E.A.) consisted of Bell himself, Douglas McCurdy (son of Bell's personal secretary, photographer, and very close family friend), Frederick W. Baldwin (a freshly graduated mechanical engineer from Toronto and close friend to McCurdy), Thomas E. Selfridge (an Army lieutenant with an extensive engineering knowledge of aeronautics), and Glenn Curtiss. The importance of Curtiss to the A.E.A. is attested by the stipends that Bell paid to each member of the association—Curtiss was paid five times more than the others. Bell had asked Curtiss to join the association because of Curtiss's excellent engine design and superb mechanical ability. Curtiss was soon playing a role much wider than just designing engines. The plan of the A.E.A. was to conduct intensive research and development on powered flight and to build five airplanes—one for each member. The first aircraft, the *Red Wing*, was constructed by the A.E.A. with Selfridge as the chief designer. On March 12, 1908, the *Red Wing* was flown at Hammondsport for the first time, with Baldwin at the controls. It covered a distance of 318 ft and was billed as "the first public flight" in the United States.

Recall that the tremendous success of the Wright brothers from 1903 to 1905 was not known by the general public, mainly because of indifference in the press, as well as the Wrights' growing tendency to be secretive about their airplane design until they could sell an airplane to the U.S. Government. However, the Wrights' growing apprehension about the publicized activities of the A.E.A. is reflected in a letter from Wilbur to the editor of the *Scientific American* after the flight of the *Red Wing*. In this letter, Wilbur states:

> In 1904 and 1905, we were flying every few days in a field alongside the main wagon road and electric trolley line from Dayton to Springfield, and hundreds of travelers and inhabitants saw the machine in flight. Anyone who wished could look. We merely did not advertise the flights in the newspapers.

On March 17, 1908, the second flight of the *Red Wing* resulted in a crash which severely damaged the aircraft. Work on the *Red Wing* was subsequently abandoned in lieu of a new design of the A.E.A., the *White Wing*, with Baldwin as the chief designer. The *White Wing* was equipped with ailerons—the first in the United States. Members of the A.E.A. were acutely aware of the Wrights' patent on wing warping for lateral control, and Bell was particularly sensitive to making certain that his association did not infringe upon this patent. Therefore, instead of using wing warping, the *White Wing* utilized triangular-shaped, movable ailerons at the wingtips of both wings of the biplane. Beginning on May 18, 1908, the *White Wing* successfully made a series of flights piloted by various members of the A.E.A. One of these flights, with Glenn Curtiss at the controls, was reported by Selfridge to the Associated Press as follows:

> G. H. Curtiss of the Curtiss Manufacturing Company made a flight of 339 yards in two jumps in Baldwin's White Wing this afternoon at 6:47PM. In the first jump he covered 205 yards then touched, rose immediately and flew 134 yards further when the flight ended on the edge of a ploughed field. The machine was in perfect control at all times and was steered first to the right and then to the left before landing. The 339 yards was covered in 19 seconds or 37 miles per hour.

Two days later, with an inexperienced McCurdy at the controls, the *White Wing* crashed and was never flown again.

However, by this time, the Wright brothers' apprehension about the A.E.A. was growing into bitterness toward its members. Wilbur and Orville genuinely felt that the A.E.A. had pirated their ideas and was going to use them for commercial gain. For example, on June 7, 1908, Orville wrote to Wilbur (who was in France preparing for his spectacular first public flights that summer at Le Mans—see Sec. 1.8): "I see by one of the papers that the Bell outfit is offering Red Wings for sale at $5,000 each. They have some nerve." On June 28, he related to Wilbur: "Curtiss et al. are using our patents, I understand, and are now offering machines for sale at $5,000 each, according to the *Scientific American*. They have got good cheek."

The relations between the Wrights and the A.E.A.—particularly Curtiss—were exacerbated on July 4, 1908, when the A.E.A. achieved their crowning success. A new airplane had been constructed—the *June Bug*—with

Figure 1.25 Glenn Curtiss flying *June Bug* on July 4, 1908, on his way to the *Scientific American* prize for the first public flight of over 1 km.

Glenn Curtiss as the chief designer. The previous year, the *Scientific American* had offered a trophy, through the Aero Club of America, worth more than $3000 to the first aviator making a straight flight of 1 km (3281 ft). On Independence Day in 1908, at Hammondsport, New York, Glenn Curtiss at the controls of his *June Bug* was ready for an attempt at the trophy. A delegation of 22 members of the Aero Club were present, and the official starter was none other than Charles Manly, Langley's dedicated assistant and pilot of the ill-fated Aerodrome (see Sec. 1.7 and Figure 1.23). Late in the day, at 7:30 P.M., Curtiss took off and in 1 min and 40 s had covered a distance of more than 1 mi, easily winning the *Scientific American* prize. A photograph of the *June Bug* during this historic flight is shown in Figure 1.25.

The Wright brothers could have easily won the *Scientific American* prize long before Curtiss; they simply chose not to. Indeed, the publisher of the *Scientific American*, Charles A. Munn, wrote to Orville on June 4 inviting him to make the first attempt at the trophy, offering to delay Curtiss's request for an attempt. On June 30, the Wrights responded negatively—they were too involved with preparations for their upcoming flight trials in France and at Fort Myer in the United States. However, Curtiss's success galvanized the Wrights' opposition. Remembering their earlier conversations with Curtiss in 1906, Orville wrote to Wilbur on July 19:

> I had been thinking of writing to Curtiss. I also intended to call attention of the *Scientific American* to the fact that the Curtiss machine was a poor copy of ours; that we had furnished

them the information as to how our older machines were constructed, and that they had followed this construction very closely, but have failed to mention the fact in any of their writings.

Curtiss's publicity in July was totally eclipsed by the stunning success of Wilbur during his public flights in France beginning August 8, 1908, and by Orville's Army trials at Fort Myer beginning on September 3, 1908. During the trials at Fort Myer, the relationship between the Wrights and the A.E.A. took an ironic twist. One member of the evaluation board assigned by the Army to observe Orville's flights was Lt. Thomas Selfridge. Selfridge had been officially detailed to the A.E.A. by the Army for a year and was now back at his duties of being the Army's main aeronautical expert. As part of the official evaluation, Orville was required to take Selfridge on a flight as a passenger. During this flight, on September 17, one propeller blade cracked and changed its shape, thus losing thrust. This imbalanced the second propeller, which cut a control cable to the tail. The cable subsequently wrapped around the propeller and snapped it off. The Wright type A went out of control and crashed. Selfridge was killed, and Orville was severely injured; he was in the hospital for $1\frac{1}{2}$ months. For the rest of his life, Orville would walk with a limp as a result of this accident. Badly shaken by Selfridge's death, and somewhat overtaken by the rapid growth of aviation after the events of 1908, the Aerial Experiment Association dissolved itself on March 31, 1909. In the written words of Alexander Graham Bell, "The A.E.A. is now a thing of the past. It has made its mark upon the history of aviation and *its work will live*."

After this, Glenn Curtiss struck out in the aviation world on his own. Forming an aircraft factory at Hammondsport, Curtiss designed and built a new airplane, improved over the *June Bug* and named the *Golden Flyer*. In August 1909, a massive air show was held at Reims, France, attracting huge crowds and the crown princes of Europe. For the first time in history, the Gordon Bennett trophy was offered for the fastest flight. Glenn Curtiss won this trophy with his *Golden Flyer*, averaging a speed of 47.09 mi/h over a 20-km course and defeating a number of pilots flying Wright airplanes. This launched Curtiss on a meteoric career as a daredevil pilot and a successful airplane manufacturer. His motorcycle factory at Hammondsport was converted entirely to the manufacture of airplanes. His airplanes were popular with other pilots of that day because they were statically stable and hence easier and safer to fly than the Wright airplanes, which had been intentionally designed by the Wright brothers to be statically unstable (see Chap. 7). By 1910, aviation circles and the general public held Curtiss and the Wrights in essentially equal esteem. At the lower right of Figure 1.24 is a photograph of Curtiss at this time; the propeller ornament in his cap was a good luck charm which he took on his flights. By 1911, a Curtiss airplane had taken off from and landed on a ship. Also in that year, Curtiss had developed the first successful seaplanes and had forged a lasting relationship with the U.S. Navy. In June 1911, the Aero Club of America issued its first official pilot's license to Curtiss in view of the fact that he had made the first public flight in America, an honor which otherwise would have gone to the Wrights.

In September 1909, the Wright brothers filed suit against Curtiss for patent infringements. They argued that their wing warping patent of 1906, liberally interpreted, covered all forms of lateral control, including the ailerons used by Curtiss. This triggered five years of intensive legal maneuvering, which dissipated much of the energies of all the parties. Curtiss was not alone in this regard. The Wrights brought suit against a number of fledgling airplane designers during this period, both in the United States and in Europe. Such litigation consumed Wilbur's attention, in particular, and effectively removed him from being a productive worker toward technical aeronautical improvements. It is generally agreed by aviation historians that this was not the Wrights' finest hour. Their legal actions not only hurt their own design and manufacturing efforts, they also effectively discouraged the early development of aeronautics by others, particularly in the United States. (It is quite clear that when World War I began in 1914, the United States—birthplace of aviation—was far behind Europe in aviation technology.) Finally, in January 1914, the courts ruled in favor of the Wrights, and Curtiss was forced to pay royalties to the Wright family. (By this time, Wilbur was dead, having succumbed to typhoid fever in 1912.)

In defense of the Wright brothers, their actions against Curtiss grew from a genuine belief on their part that Curtiss had wronged them and had consciously stolen their ideas, which Curtiss had subsequently parlayed into massive economic gains. This went strongly against the grain of the Wrights' staunchly ethical upbringing. In contrast, Curtiss bent over backward to avoid infringing on the letter of the Wrights' patent, and there is numerous evidence that Curtiss was consistently trying to mend relations with the Wrights. It is this author's opinion that both sides became entangled in a complicated course of events that followed those heady days after 1908, when aviation burst on the world scene, and that neither Curtiss nor the Wrights should be totally faulted for their actions. These events simply go down in history as a less-than-glorious, but nevertheless important, chapter in the early development of aviation.

An important postscript should be added here regarding the triangular relationship between Langley, the Wrights, and Curtiss, as shown in Figure 1.24. In Secs. 1.7 and 1.8, we have already seen the relationship between Langley and the Wrights and the circumstances leading up to the race for the first flight in 1903. This constitutes side *A* in Figure 1.24. In the present section, we have seen the strong connection between Curtiss and the work of Langley, via Alexander Graham Bell—close friend and follower of Langley and creator of the Aerial Experiment Association, which gave Curtiss a start in aviation. We have even noted that Charles Manly, Langley's assistant, was the official starter for Curtiss's successful competition for the *Scientific American* trophy. Such relationships form side *B* of the triangle in Figure 1.24. Finally, we have seen the relationship, although somewhat acrimonious, between the Wrights and Curtiss, which forms side *C* in Figure 1.24.

In 1914 an event occurred which simultaneously involved all three sides of the triangle in Figure 1.24. When the Langley Aerodrome failed for the second time in 1903 (see Figure 1.23), the wreckage was simply stored away in an unused room in the back of the Smithsonian Institution. When Langley died in 1906, he was replaced as secretary of the Smithsonian by Dr. Charles D. Walcott. Over the ensuing years, Secretary Walcott felt that the Langley Aerodrome should be given a third chance. Finally, in 1914, the Smithsonian awarded a grant of $2000 for the repair and flight of the Langley Aerodrome to none other than Glenn Curtiss. The Aerodrome was shipped to Curtiss's factory in Hammondsport, where it was not only repaired, but 93 separate technical modifications were made, aerodynamically, structurally, and to the engine. For help during this restoration and modification, Curtiss hired Charles Manly. Curtiss added pontoons to the Langley Aerodrome and on May 28, 1914, personally flew the modified aircraft for a distance of 150 ft over Keuka Lake. Figure 1.26 shows a photograph of the Langley Aerodrome in graceful flight over the waters of the lake. Later, the Aerodrome was shipped back to the Smithsonian, where it was carefully restored to its original configuration and in 1918 was placed on display in the old Arts and Industries Building. Underneath the Aerodrome was placed a plaque reading: "Original Langley flying machine, 1903. The first man-carrying aeroplane in the history of the world capable of sustained free flight." The plaque did *not* mention that the Aerodrome demonstrated its sustained flight capability only after the 93 modifications made by Curtiss in 1914. It is no surprise that Orville Wright was deeply upset by this state of affairs, and this is the principal reason why the original 1903 Wright Flyer was not given to the Smithsonian until 1948, the year of Orville's death. Instead, from 1928 to 1948, the Flyer resided in the Science Museum in London.

Figure 1.26 The modified Langley Aerodrome in flight over Keuka Lake in 1914.

This section ends on two ironies. In 1915, Orville sold the Wright Aeronautical Corporation to a group of New York businesspeople. During the 1920s, this corporation became a losing competitor in aviation. Finally, on June 26, 1929, in a New York office, the Wright Aeronautical Corporation was officially merged with the successful Curtiss Aeroplane and Motor Corporation, forming the Curtiss-Wright Corporation. Thus, ironically, the names of Curtiss and Wright finally came together after all those earlier turbulent years. The Curtiss-Wright Corporation went on to produce numerous famous aircraft, perhaps the most notable being the P-40 of World War II fame. Unfortunately, the company could not survive the lean years immediately after World War II, and its aircraft development and manufacturing ceased in 1948. This leads to the second irony. Although the very foundations of powered flight rest on the work of Orville and Wilbur Wright and Glenn Curtiss, there is not an airplane either produced or in standard operation today that bears the name of either Wright or Curtiss.

1.10 THE PROBLEM OF PROPULSION

During the nineteenth century, numerous visionaries predicted that manned heavier-than-air flight was inevitable once a suitable power plant could be developed to lift the aircraft off the ground. It was just a matter of developing an engine having enough horsepower while at the same time not weighing too much, i.e., an engine with a high horsepower-to-weight ratio. This indeed was the main stumbling block to such people as Stringfellow, Du Temple, and Mozhaiski—the steam engine simply did not fit the bill. Then, in 1860, the Frenchman Jean Joseph Etienne Lenoir built the first practical gas engine. It was a single-cylinder engine, burning ordinary street-lighting gas for fuel. By 1865, 400 of Lenoir's engines were doing odd jobs around Paris. Further improvements in such internal combustion engines came rapidly. In 1876, N. A. Otto and E. Langen of Germany developed the four-cycle engine (the ancestor of all modern automobile engines), which also used gas as a fuel. This led to the simultaneous but separate development in 1885 of the four-cycle gasoline-burning engine by Gottlieb Daimler and Karl Benz, both in Germany. Both Benz and Daimler put their engines in motor cars, and the automobile industry was quickly born. After these "horseless carriages" were given legal freedom of the roads in 1896 in France and Britain, the auto industry expanded rapidly. Later, this industry was to provide much of the technology and many of the trained mechanics for the future development of aviation.

This development of the gasoline-fueled internal combustion engine was a godsend to aeronautics, which was beginning to gain momentum in the 1890s. In the final analysis, it was the Wright brothers' custom-designed and constructed gasoline engine that was responsible for lifting their Flyer I off the sands of Kill Devil Hills that fateful day in December 1903. A proper aeronautical propulsion device had finally been found.

It is interesting to note that the brotherhood between the automobile and the aircraft industries persists to the present day. For example, in June 1926, Ford introduced a very successful three-engine, high-wing, transport airplane—the Ford 4-AT Trimotor. During World War II, virtually all the major automobile companies built airplane engines and airframes. General Motors still maintains an airplane engine division—the Allison Division in Indianapolis, Indiana—noted for its turboprop designs. More recently, automobile designers are turning to aerodynamic streamlining and wind-tunnel testing to reduce drag, hence increase fuel economy. Thus, the parallel development of the airplane and the automobile over the past 100 years has been mutually beneficial.

It can be argued that propulsion has paced every major advancement in the speed of airplanes. Certainly, the advent of the gasoline engine opened the doors to the first successful flight. Then, as the power of these engines increased from the 12-hp, Wright-designed engine of 1903 to the 2200-hp, radial engines of 1945, airplane speeds correspondingly increased from 28 to over 500 mi/h. Finally, jet and rocket engines today provide enough thrust to propel aircraft at thousands of miles per hour—many times the speed of sound. So throughout the history of manned flight, propulsion has been the key that has opened the doors to flying faster and higher.

1.11 FASTER AND HIGHER

The development of aeronautics in general, and aeronautical engineering in particular, was exponential after the Wrights' major public demonstrations in 1908 and has continued to be so to the present day. It is beyond the scope of this book to go into all the details. However, marbled into the engineering text in the following chapters are various historical highlights of technical importance. It is hoped that the following parallel presentations of the fundamentals of aerospace engineering and some of their historical origins will be synergistic and that, in combination with the present chapter, they will give the reader a certain appreciation for the heritage of this profession.

As a final note, the driving philosophy of many advancements in aeronautics since 1903 has been to fly *faster and higher*. This is dramatically evident from Figure 1.27, which gives the flight speeds for typical aircraft as a function of chronological time. Note the continued push for increased speed over the years, and the particular increase in recent years made possible by the jet engine. Singled out in Figure 1.27 are the winners of the Schneider Cup races between 1913 and 1931 (with a moratorium during World War I). The Schneider Cup races were started in 1913 by Jacques Schneider of France as a stimulus to the development of high-speed float planes. They prompted some early but advanced development of high-speed aircraft. The winners are shown by the dashed line in Figure 1.27, for comparison with standard aircraft of the day. Indeed, the winner of the last Schneider race in 1931 was the Supermarine S.6B, a forerunner of the

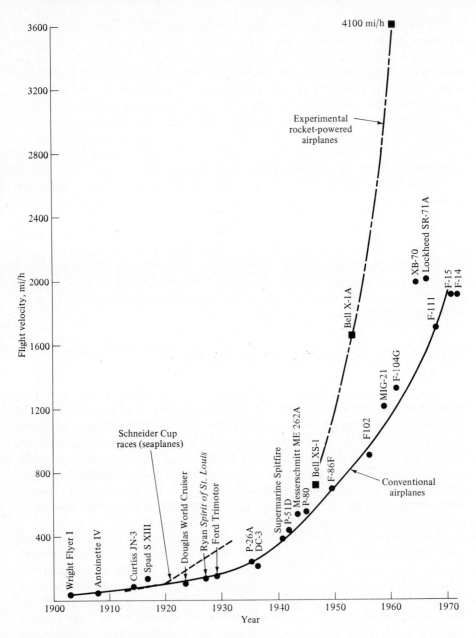

Figure 1.27 Typical flight velocities over the years.

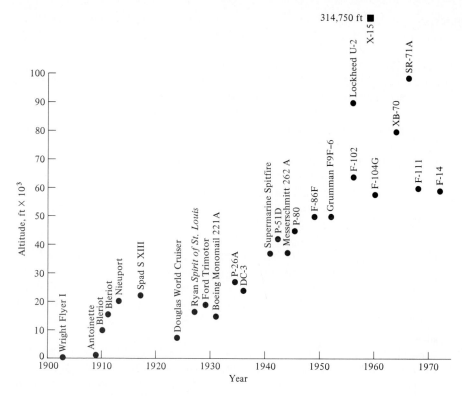

Figure 1.28 Typical flight altitudes over the years.

famous Spitfire of World War II. Of course, today the maximum speed of flight has been pushed to the extreme value of 36,000 ft/s, escape velocity from the earth, by the Apollo lunar spacecraft.

As a companion to speed, the maximum altitudes of typical manned aircraft are shown in Figure 1.28 as a function of chronological time. The same push to higher values is evident; so far, the record is the moon in 1969.

In the present chapter, we have only been able to note briefly several important events and people in the historical development of aeronautics. Moreover, there are many other places, people, and accomplishments which we simply could not mention in the interest of brevity. Therefore, the reader is urged to consult the short bibliography at the end of this chapter for additional modern reading on the history of aeronautics.

1.12 CHAPTER SUMMARY

You are about to embark on a study of aerospace engineering. This chapter has presented a short historical sketch of some of the heritage behind modern aerospace engineering. The major stepping-stones to controlled, manned, heavier-than-air, powered flight are summarized as follows:

1. Leonardo da Vinci conceives the ornithopter and leaves over 500 sketches of his design, drawn during 1486 to 1490. However, this approach to manned flight proves to be unsuccessful over the ensuing centuries.
2. The Montgolfier hot-air balloon floats over Paris on November 21, 1783. For the first time in history, a human being is lifted and carried through the air for a sustained period of time.
3. A red-letter date in the progress of aeronautics is 1799. In that year, Sir George Cayley in England engraves on a silver disk his concept of a fuselage, a fixed wing, and horizontal and vertical tails. He is the first person to propose separate mechanisms for the generation of lift and propulsion. He is the grandparent of the concept of the modern airplane.
4. The first two powered hops in history are achieved by the Frenchman Felix Du Temple in 1874 and the Russian Alexander F. Mozhaiski in 1884. However, they do not represent truly controlled, sustained flight.
5. Otto Lilienthal designs the first fully successful gliders in history. During the period 1891–1896, he achieves over 2500 successful glider flights. If he had not been killed in a glider crash in 1896, Lilienthal might have achieved powered flight before the Wright brothers.
6. Samuel Pierpont Langley, secretary of the Smithsonian Institution, achieves the first sustained heavier-than-air, *unmanned*, powered flight in history with his small-scale Aerodrome in 1896. However, his attempts at manned flight are unsuccessful, the last one failing on December 8, 1903—just nine days before the Wright brothers' stunning success.
7. Another red-letter date in the history of aeronautics, indeed in the history of humanity, is December 17, 1903. On that day, at Kill Devil Hills in North Carolina, Orville and Wilbur Wright achieve the first controlled, sustained, powered, heavier-than-air, manned flight in history. This flight is to revolutionize life during the twentieth century.
8. The development of aeronautics takes off exponentially after the Wright brothers' public demonstrations in Europe and the United States in 1908. The ongoing work of Glenn Curtiss and the Wrights and the continued influence of Langley's early work form an important "aeronautical triangle" in the development of aeronautics before World War I.

Throughout the remainder of this book, various historical notes will appear, to continue to describe the heritage of aerospace engineering as its technology advances over the twentieth century. It is hoped that such historical notes will add a new dimension to your developing understanding of this technology.

BIBLIOGRAPHY

Angelucci, E., *Airplanes from the Dawn of Flight to the Present Day*, McGraw-Hill, New York, 1973.
Combs, H., *Kill Devil Hill*, Houghton Mifflin, Boston, 1979.

Crouch, T. D., *A Dream of Wings*, W. W. Norton, New York, 1981.

Gibbs-Smith, C. H., *Sir George Cayley's Aeronautics 1796–1855*, Her Majesty's Stationery Office, London, 1962.

_____, *The Invention of the Aeroplane (1799–1909)*, Faber, London, 1966.

_____, *Aviation: An Historical Survey from its Origins to the End of World War II*, Her Majesty's Stationery Office, London, 1970.

_____, *Flight Through the Ages*, Crowell, New York, 1974.

The following are a series of small booklets prepared for the British Science Museum by C. H. Gibbs-Smith, published by Her Majesty's Stationery Office, London:

The Wrights Brothers, 1963

The World's First Aeroplane Flights, 1965

Leonardo da Vinci's Aeronautics, 1967

A Brief History of Flying, 1967

Sir George Cayley, 1968

Josephy, A. M., and Gordon, A., *The American Heritage History of Flight*, Simon and Schuster, New York, 1962.

McFarland, Marvin W. (ed.), *The Papers of Wilbur and Orville Wright*, McGraw-Hill, New York, 1953.

Roseberry, C. R., *Glenn Curtiss: Pioneer of Flight*, Doubleday, Garden City, NY, 1972.

Taylor, J. W. R., and Munson, K., *History of Aviation*, Crown, New York, 1972.

TWO

FUNDAMENTAL THOUGHTS

Engineering: "The application of scientific principles to practical ends." From the Latin word "ingenium," meaning inborn talent and skill, ingenious.

The American Heritage Dictionary of the English Language, 1969

The language of engineering and physical science is a logical collection and assimilation of symbols, definitions, formulas, and concepts. To the average person in the street, this language is frequently esoteric and incomprehensible. In fact, when you become a practicing engineer, do not expect to converse with your spouse across the dinner table about your great technical accomplishments of the day. Chances are that he or she will not understand what you are talking about. The language is intended to convey physical thoughts. It is our way of describing the phenomena of nature as observed in the world around us. It is a language that has evolved over at least 2500 years. For example, in 400 B.C., the Greek philosopher Democritus introduced the word and concept of the "atom," the smallest bit of matter that could not be cut. The purpose of this chapter is to introduce some of the everyday language used by aerospace engineers; in turn, this language will be extended and applied throughout the remainder of this book.

2.1 FUNDAMENTAL PHYSICAL QUANTITIES OF A FLOWING GAS

As you read through this book, you will soon begin to appreciate that the flow of air over the surface of an airplane is the basic source of the lifting or sustaining force that allows a heavier-than-air machine to fly. In fact, the shape of an airplane is designed to encourage the airflow over the surface to produce a lifting force in the most efficient manner possible. (You will also begin to appreciate that the design of an airplane is in reality a *compromise* between many different requirements, the production of aerodynamic lift being just one.) The science that deals with the flow of air (or for that matter, the flow of any gas) is called *aerodynamics*, and the person who practices this science is called an *aerodynamicist*. The study of the flow of gases is important in many other aerospace applications, e.g., the design of rocket and jet engines, propellers, vehicles entering planetary atmospheres from space, wind tunnels, and rocket and projectile configurations. Even the motion of the global atmosphere, and the flow of effluents through smokestacks, fall within the realm of aerodynamics. The applications are almost limitless.

Four fundamental quantities in the language of aerodynamics are pressure, density, temperature, and velocity. Let us look at each one.

A Pressure

When you hold your hand outside the window of a moving automobile, with your palm perpendicular to the incoming airstream, you can feel the air pressure exerting a force and tending to push your hand rearward, in the direction of the airflow. The *force per unit area* on your palm is defined as the *pressure*. The pressure exists basically because air molecules (oxygen and nitrogen molecules) are striking the surface of your hand and transferring some of their *momentum* to the surface. More precisely:

> *Pressure is the* normal *force per unit area exerted on a surface due to the time rate of change of momentum of the gas molecules impacting on that surface.*

It is important to note that, even though pressure is defined as force per unit area, e.g., newton/meter^2 or pound/foot^2, you do not need a surface that is actually 1 m^2 or 1 ft^2 to talk about pressure. In fact, pressure is usually defined at a point in the gas or a point on a surface and can vary from one point to another. We can use the language of differential calculus to see this more clearly. Referring to Figure 2.1, consider a point B in a volume of gas. Let

dA = an incremental area around B

dF = force on one side of dA due to pressure

Then, the pressure p at point B in the gas is defined as

$$p = \lim\left(\frac{dF}{dA}\right) \qquad dA \to 0 \qquad (2.1)$$

Equation (2.1) says that, in reality, the pressure p is the limiting form of the force per unit area where the area of interest has shrunk to zero around the point B. In

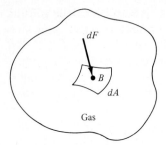

Figure 2.1 Definition of pressure.

this formalism, it is easy to see that p is a point property and can have a different value from one point to another in the gas.

Pressure is one of the most fundamental and important variables in aerodynamics, as we shall soon see. Common units of pressure are newtons/meter^2, dynes/centimeter^2, pounds/foot^2, and atmospheres. Abbreviations for these quantities are N/m^2, dyn/cm^2, lb/ft^2, and atm, respectively. See Appendix C for a list of common abbreviations for physical units.

B Density

The density of a substance (including a gas) is the mass of that substance per unit volume.

Density will be designated by the symbol ρ. For example, consider air in a room that has a volume of 250 m^3. If the mass of the air in the room is 306.25 kg and is evenly distributed throughout the space, then $\rho = 306.25$ kg/250 m^3 = 1.225 kg/m^3 and is the same at every point in the room.

Analogous to the previous discussion on pressure, the definition of density does not require an actual volume of 1 m^3 or 1 ft^3. Rather, ρ is a point property and can be defined as follows. Referring to Figure 2.2, consider point B inside a volume of gas. Let

$$dv = \text{an elemental volume around the point } B$$
$$dm = \text{the mass of gas inside } dv$$

Then, ρ at point B is

$$\rho = \lim\left(\frac{dm}{dv}\right) \qquad dv \to 0 \tag{2.2}$$

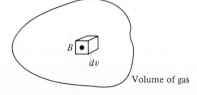

Volume of gas   **Figure 2.2** Definition of density.

Therefore, ρ is the mass per unit volume where the volume of interest has shrunk to zero around point B. The value of ρ can vary from point to point in the gas. Common units of density are kg/m^3, $slug/ft^3$, gm/cm^3, and lb_m/ft^3. (The pound mass, lb_m, will be discussed in a subsequent section.)

C Temperature

Consider a gas as a collection of molecules and atoms. These particles are in constant motion, moving through space and occasionally colliding with one another. Since each particle has motion, it also has kinetic energy. If we watch the motion of a single particle over a long period of time during which it experiences numerous collisions with its neighboring particles, then we can meaningfully define the average kinetic energy of the particle over this long duration. If the particle is moving rapidly, it has a higher average kinetic energy than if it were moving slowly. The temperature T of the gas is directly proportional to the average molecular kinetic energy. In fact, we can define T as follows:

Temperature is a measure of the average kinetic energy of the particles in the gas. If KE is the mean molecular kinetic energy, then temperature is given by $KE = \frac{3}{2}kT$, *where k is the Boltzmann constant.*

The value of k is 1.38×10^{-23} J/K, where J is an abbreviation for joules.

Hence, we can qualitatively visualize a high-temperature gas as one in which the particles are randomly rattling about at high speeds, whereas in a low-temperature gas, the random motion of the particles is relatively slow. Temperature is an important quantity in dealing with the aerodynamics of supersonic and hypersonic flight, as we shall soon see. Common units of temperature are the kelvin (K), degree Celsius (°C), degree Rankine (°R), and degree Fahrenheit (°F).

D Flow Velocity and Streamlines

The concept of speed is commonplace: it represents the distance traveled by some object per unit time. For example, we all know what is meant by traveling at a speed of 55 mi/h down the highway. However, the concept of the velocity of a flowing gas is somewhat more subtle. First of all, velocity connotes *direction* as well as speed. The automobile is moving at a velocity of 55 mi/h *due north in a horizontal plane*. To designate velocity, we must quote both speed and direction. For a flowing gas, we must further recognize that each region of the gas does not necessarily have the same velocity; i.e., the speed and direction of the gas may vary from point to point in the flow. Hence, flow velocity, along with p, ρ, and T, is a point property.

To see this more clearly, consider the flow of air over an airfoil or the flow of combustion gases through a rocket engine, as sketched in Figure 2.3. To orient yourself, lock your eyes on a specific, infinitesimally small element of mass in the gas, and watch this element move with time. Both the speed and direction of this

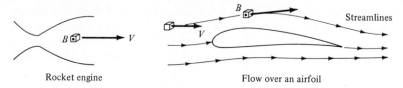

Rocket engine   Flow over an airfoil

Figure 2.3 Flow velocity and streamlines.

element (usually called a fluid element) can vary as it moves from point to point in the gas. Now, fix your eyes on a specific fixed point in the gas flow, say point *B* in Figure 2.3. We can now define flow velocity as follows:

> *The velocity at any fixed point B in a flowing gas is the velocity of an infinitesimally small fluid element as it sweeps through B.*

Again, emphasis is made that velocity is a point property and can vary from point to point in the flow.

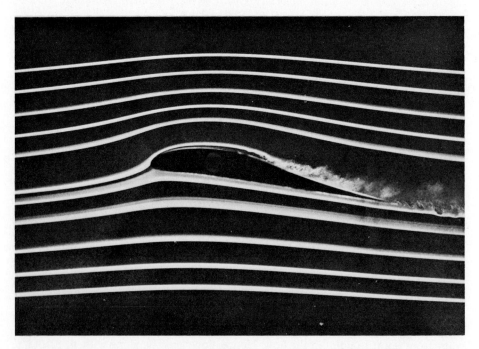

Figure 2.4 Smoke photograph of the low-speed flow over a Lissaman 7769 airfoil at 10° angle of attack. The Reynolds number based on chord is 150,000. This is the airfoil used on the Gossamer Condor man-powered aircraft. *(The photograph was taken in one of the Notre Dame University smoke tunnels by Dr. T. J. Mueller, Professor of Aerospace Engineering at Notre Dame, and is shown here through his courtesy.)*

Referring again to Figure 2.3, we note that as long as the flow is steady (as long as it does not fluctuate with time), a moving fluid element is seen to trace out a fixed *path* in space. This path taken by a moving fluid element is called a *streamline* of the flow. Drawing the streamlines of the flow field is an important way of visualizing the motion of the gas; we will frequently be sketching the streamlines of the flow about various objects. For example, the streamlines of the flow about an airfoil are sketched in Figure 2.3 and clearly show the direction of motion of the gas. Figure 2.4 is an actual photograph of streamlines over an airfoil model in a low-speed subsonic wind tunnel. The streamlines are made visible by injection of filaments of smoke upstream of the model; these smoke filaments follow the streamlines in the flow. Using another flow field visualization technique, Figure 2.5 shows a photograph of a flow where the surface streamlines are made visible by coating the model with a mixture of white pigment in mineral oil. Clearly, the visualization of flow streamlines is a useful aid in the study of aerodynamics.

Figure 2.5 An oil streak photograph showing the surface streamline pattern for a fin mounted on a flat plate in supersonic flow. The parabolic curve in front of the fin is due to the bow shock wave and flow separation ahead of the fin. Note how clearly the streamlines can be seen in this complex flow pattern. Flow is from right to left. The Mach number is 5 and the Reynolds number is 6.7×10^6. *(Courtesy of Allen E. Winkelmann, University of Maryland, and the Naval Surface Weapons Center.)*

2.2 THE SOURCE OF ALL AERODYNAMIC FORCES

We have just discussed the four basic aerodynamic flow quantities: p, ρ, T, and **V**, where **V** is velocity which has both magnitude and direction; i.e., velocity is a vector quantity. A knowledge of p, ρ, T, and **V** at each point of a flow fully defines the *flow field*. For example, if we were concerned with the flow about a sharp-pointed cone as shown in Figure 2.6, we could imagine a cartesian *xyz* three-dimensional space, where the velocity far ahead of the cone, V_∞, is in the x direction and the cone axis is also along the x direction. The specification of the following quantities then fully defines the *flow field*:

$$p = p(x, y, z)$$
$$\rho = \rho(x, y, z)$$
$$T = T(x, y, z)$$
$$\mathbf{V} = \mathbf{V}(x, y, z)$$

(In practice, the flow field about a right-circular cone is more conveniently described in terms of cylindrical coordinates, but we are concerned only with the general ideas here.)

Theoretical and experimental aerodynamicists labor to calculate and measure flow fields of many types. But why? What practical information does knowledge of the flow field yield with regard to airplane design or to the shape of a rocket engine? A substantial part of the first five chapters of this book endeavors to answer these questions. However, the roots of the answers lie in the following discussion.

Probably the most practical consequence of the flow of air over an object is that the object experiences a force, an aerodynamic force, such as your hand feels outside the open window of a moving car. Subsequent chapters will discuss the nature and consequences of such aerodynamic forces. The purpose here is to state that the aerodynamic force exerted by the airflow on the surface of an airplane,

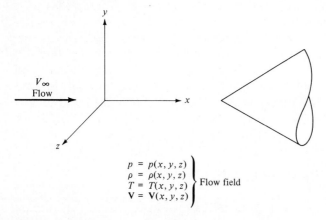

Figure 2.6 Specification of a flow field.

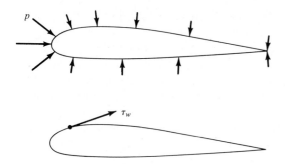

Figure 2.7 Pressure and shear-stress distributions.

missile, etc., stems from only two simple natural sources:

1. Pressure distribution on the surface
2. Shear stress (friction) on the surface

We have already discussed pressure. Referring to Figure 2.7, we see pressure exerted by the gas on the solid surface of an object always acts *normal* to the surface, as shown by the directions of the arrows. The lengths of the arrows denote the magnitude of the pressure at each local point on the surface. Note that the surface pressure varies with location. The net *unbalance* of the varying pressure distribution over the surface creates a force, an aerodynamic force. The second source, shear stress acting on the surface, is due to the frictional effect of the flow "rubbing" against the surface as it moves around the body. The shear stress τ_w is defined as the force per unit area acting *tangentially* on the surface due to friction, as shown in Figure 2.7. It is also a point property; it also varies along the surface; and the net unbalance of the surface shear stress distribution also creates an aerodynamic force on the body. *No matter how complex the flow field, and no matter how complex the shape of the body, the only way nature has of communicating an aerodynamic force to a solid object or surface is through the pressure and shear stress distributions which exist on the surface.* These are the basic fundamental sources of all aerodynamic forces.

Finally, we can state that a primary function of theoretical and experimental aerodynamics is to predict and measure the aerodynamic forces on a body. In many cases, this implies prediction and measurement of p and τ_w along a given surface. Furthermore, a prediction of p and τ_w on the surface frequently requires knowledge of the complete flow field around the body. This helps to answer our earlier question as to what practical information is yielded by knowledge of the flow field.

2.3 EQUATION OF STATE FOR A PERFECT GAS

Air under normal conditions of temperature and pressure, such as that encountered in subsonic and supersonic flight through the atmosphere, behaves very much like a *perfect gas*. The definition of a perfect gas can best be seen by returning to the molecular picture. A gas is a collection of particles (molecules, atoms, electrons, etc.) in random motion, where each particle is, on the average, a long distance away from its neighboring particles. Each molecule has an *intermolecular force field* about it, a ramification of the complex interaction of the electromagnetic properties of the electrons and nucleus. The intermolecular force field of a given particle extends a comparatively long distance through space and changes from a strong repulsive force at close range to a weak attractive force at long range. The intermolecular force field of a given particle reaches out and is felt by the neighboring particles. If the neighboring particles are far away, they feel only the tail of the weak attractive force; hence the motion of the neighboring particles is only negligibly affected. On the other hand, if they are close, their motion can be greatly affected by the intermolecular force field. Since the pressure and temperature of a gas are tangible quantities derived from the motion of the particles, then p and T are directly influenced by intermolecular forces, especially when the molecules are packed closely together (i.e., at high densities). This leads to the definition of a perfect gas:

A perfect gas is one in which intermolecular forces are negligible.

Clearly, from the above discussion, a gas in nature where the particles are widely separated (low densities) approaches the definition of a perfect gas. The air in the room about you is one such case; each particle is separated, on the average, by more than 10 molecular diameters from any other. Hence, air at standard conditions can be readily approximated by a perfect gas. Such is also the case for the flow of air about ordinary flight vehicles at subsonic and supersonic speeds. Therefore, in this book, we will always deal with a perfect gas for our aerodynamic calculations.

The relation among p, ρ, and T for a gas is called the *equation of state*. For a perfect gas, the equation of state is

$$\boxed{p = \rho R T} \tag{2.3}$$

where R is the specific gas constant, the value of which varies from one type of gas to another. For normal air we have

$$R = 287 \frac{\text{J}}{(\text{kg})(\text{K})} = 1716 \frac{\text{ft} \cdot \text{lb}}{(\text{slug})(^\circ\text{R})}$$

It is interesting that the *deviation* of an actual gas in nature from perfect gas behavior can be expressed approximately by the modified Berthelot equation of state:

$$\frac{p}{\rho R T} = 1 + \frac{ap}{T} - \frac{bp}{T^3}$$

where a and b are constants of the gas. Thus, the deviations become smaller as p

decreases and T increases. This makes sense, because if p is high, the molecules are packed closely together, intermolecular forces become important, and the gas behaves less like a perfect gas. On the other hand, if T is high, the molecules move faster. Thus their average separation is larger, intermolecular forces become less significant in comparison to the inertia forces of each molecule, and the gas behaves more like a perfect gas.

Also, it should be noted that when the air in the room around you is heated to temperatures above 2500 K, the oxygen molecules begin to dissociate (tear apart) into oxygen atoms; at temperatures above 4000 K, the nitrogen begins to dissociate. For these temperatures, air becomes a *chemically reacting gas*, where its chemical composition becomes a function of both p and T; i.e., it is no longer normal air. As a result, R in Eq. (2.3) becomes a variable, $R = R(p, T)$, simply because the gas composition is changing. The perfect gas equation of state, Eq. (2.3), is still valid for such a case, except that R is no longer a constant. This situation is encountered in very high speed flight, e.g., the atmospheric entry of the Apollo capsule, in which case the temperatures in some regions of the flow field reach 11,000 K.

Again, in this book, we will always assume that air is a perfect gas, obeying Eq. (2.3), with a constant $R = 287$ J/(kg)(K) or 1716 ft $\cdot$ lb/(slug)($°$R).

2.4 DISCUSSION ON UNITS

Physical units are vital to the language of engineering. In the final analysis, the end result of most engineering calculations or measurements is a number which represents some physical quantity, e.g., pressure, velocity, or force. The number is given in terms of combinations of units, for example, 10^5 N/m^2, 300 m/s, or 5 N, where the newton, meter, and second are examples of *units*. (See Appendix C.)

Historically, various branches of engineering have evolved and favored systems of units which seemed to most conveniently fit their needs. These various sets of "engineering" units usually differ among themselves and are different from the metric system preferred for years by physicists and chemists. In the modern world of technology, where science and engineering interface on almost all fronts, such duplicity and variety of units has become an unnecessary burden. Metric units are now the accepted norm in both science and engineering in most countries outside the United States. More importantly, in 1960 the Eleventh General Conference on Weights and Measures defined and officially established the *Système International d'Unites*—the SI system of units—which was adopted as the preferred system of units by 36 participating countries, including the United States. Since then, the United States has made progress toward the voluntary implementation of SI units in engineering. For example, several NASA (National Aeronautics and Space Administration) laboratories have made SI units virtually mandatory for all results reported in technical reports, although engineering units can be shown as a duplicate set. The AIAA (American Institute of Aeronautics and Astronautics) has made a policy of encouraging SI units for all papers reported in their technical journals. At the time of this writing, the United States Congress is studying legislation to require total conversion to SI units for all uses of weights and measures. It is apparent that in a few decades, the United States, along with the rest of the world, will be using SI units almost exclusively.

For these reasons, students who prepare themselves for practicing engineering in the last quarter of the twentieth century must do "double duty" with regard to familiarization with units. They must be familiar with the old engineering units in order to use the vast bulk of existing technical literature quoted in such units. At the same time they must be intimately familiar with the SI system for all future work and publications; i.e., our next generation of engineers must be bilingual with regard to units.

In order to promote fluency in both the engineering and SI units, this book will incorporate both sets. However, you should adopt the habit of working all your problems using the SI units from the start, and when results are requested in engineering units, *convert* your final SI results to the desired units. This will give you every chance to develop a "feel" for SI units. It is important that you develop a natural feeling for SI units; e.g., you should feel as at home with pressures in N/m^2 as you probably already do with lb/in^2 (psi). A mark of successful experienced engineers is their feel for correct magnitudes of physical quantities in familiar units. Make the SI your familiar units.

For all practical purposes, the SI system is a metric system based on the meter, kilogram, second, and kelvin as basic units of length, mass, time, and temperature. It is a *coherent*, or *consistent*, set of units. Such consistent sets of units allow physical relationships to be written without the need for "conversion factors" in the basic formulas. For example, in a consistent set of units, Newton's second law can be written

$$F = m \times a$$

Force = mass × acceleration

In the SI system,

$$F = ma$$

$$(1 \text{ newton}) = (1 \text{ kilogram})(1 \text{ meter/second}^2) \tag{2.4}$$

The newton is a force defined such that it accelerates a mass of 1 kilogram by 1 meter/per second per second.

The English engineering system of units is another consistent set of units. Here the basic units of mass, length, time, and temperature are the slug, foot, second, and Rankine, respectively. In this system,

$$F = ma$$

$$(1 \text{ pound}) = (1 \text{ slug})(1 \text{ foot/second}^2) \tag{2.5}$$

The pound is a force defined such that it accelerates a mass of 1 slug by 1 foot/second^2. Note that in both systems, Newton's second law is written simply as $F = ma$, with no conversion factor on the right-hand side.

In contrast, a nonconsistent set of units defines force and mass such that Newton's second law must be written with a conversion factor, or constant, as

$$F = (1/g_c) \quad \times \quad m \quad \times \quad a$$

$$\uparrow \qquad\quad \uparrow \qquad\qquad \uparrow \qquad \uparrow$$

Force   Conversion   Mass   Acceleration

factor

A nonconsistent set of units frequently used in the past by mechanical engineers includes the pound or pound force, pound mass, foot, and second, whereby

$$g_c = 32.2 \quad (lb_m)(ft)/(s^2)(lb_f)$$

$$F = (1/g_c) \quad m \quad \times \quad a \qquad\qquad (2.6)$$

$$\uparrow \qquad \uparrow \qquad \uparrow \qquad\quad \uparrow$$

$$lb_f \quad (1/32.2) \quad lb_m \qquad ft/s^2$$

In this system, the unit of mass is the pound mass lb_m. By comparing Eqs. (2.5) and (2.6), we see that 1 slug = 32.2 lb_m. A slug is a large hunk of mass, whereas the lb_m is considerably smaller, by a factor of 32.2. This is illustrated in Figure 2.8.

Already, you can sense how confusing the various sets of units can become, especially nonconsistent units. Confusion on the use of Eq. (2.6) with the g_c factor has resulted in many heat exchangers and boilers being designed 32.2 times either too large or too small. On the other hand, the use of a consistent set of units, such as in Eqs. (2.4) and (2.5), which have no conversion factor, eliminates such confusion. This is one of the beauties of the SI system.

For this reason, we will always deal with a consistent set of units in this book. We will use both the SI units from Eq. (2.4) and the English engineering units from Eq. (2.5). As stated before, you will frequently encounter the engineering units in the existing literature, whereas you will be seeing the SI units with increasing frequency in the future literature; i.e., you must become bilingual. To summarize, we will deal with the English engineering system units—lb, slug, ft, s, °R—and the Système International (SI) units—N, kg, m, s, K.

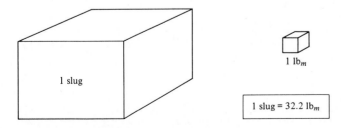

Figure 2.8 Comparison between the slug and pound mass.

Therefore, returning to the equation of state, Eq. (2.3), where $p = \rho R T$, the units are

	English Engineering System	SI
p	lb/ft^2	N/m^2
ρ	slugs/ft^3	kg/m^3
T	°R	K
R (for air)	1716 ft · lb/(slug)(°R)	287 J/(kg)(K)

There are two final points about units that you should note. First, the units of a physical quantity can frequently be expressed in more than one combination simply by appealing to Newton's second law. From Newton's law, the relation between N, kg, m, and s is

$$F = ma$$

$$N = kg \cdot m/s^2$$

Thus, a quantity such as $R = 287$ J/(kg)(K) can also be expressed in the equivalent way as

$$R = 287\frac{J}{(kg)(K)} = 287\frac{N \cdot m}{(kg)(K)} = 287\frac{kg \cdot m}{s^2}\frac{m}{(kg)(K)} = 287\frac{m^2}{(s^2)(K)}$$

Thus, R can also be expressed in the equivalent terms of velocity squared divided by temperature. In the same vein,

$$R = 1716\frac{ft \cdot lb}{(slug)(°R)} = 1716\frac{ft^2}{(s^2)(°R)}$$

Secondly, in the equation of state, Eq. (2.3), T is always the *absolute* temperature, where zero degrees is the absolutely lowest temperature possible. Both K and °R are absolute temperature scales, where 0 °R = 0 K = temperature at which almost all molecular translational motion theoretically stops. On the other hand, the familiar Fahrenheit (F) and Celsius (C) scales are *not* absolute scales. Indeed,

$$0°F = 460\ °R$$

$$0°C = 273\ K = 32°F$$

For example, 90°F is the same as $460 + 90 = 550°R$,

and   10°C is the same as $273 + 10 = 283$ K.

Please remember: T in Eq. (2.3) *must* be the absolute temperature, either Kelvin or Rankine.

2.5 SPECIFIC VOLUME

Density ρ is the *mass per unit volume*. The inverse of this quantity is also frequently used in aerodynamics. It is called the specific volume v and is defined as the *volume per unit mass*. By definition,

$$v = \frac{1}{\rho}$$

Hence, from the equation of state

$$p = \rho RT = \frac{1}{v} RT$$

we also have

$$\boxed{pv = RT} \tag{2.7}$$

Units for v are m^3/kg and ft^3/slug.

Example 2.1 The air pressure and density at a point on the wing of a Boeing 747 are 1.10×10^5 N/m^2 and 1.20 kg/m^3, respectively. What is the temperature at that point?

SOLUTION From Eq. (2.3), $p = \rho RT$, hence $T = p/\rho R$

$$T = \frac{1.10 \times 10^5 \text{ N/m}^2}{(1.20 \text{ kg/m}^3)[287 \text{ J/(kg)(K)}]} = \boxed{319 \text{ K}}$$

Example 2.2 The high-pressure air storage tank for a supersonic wind tunnel has a volume of 1000 ft^3. If air is stored at a pressure of 30 atm and a temperature of 530 °R, what is the mass of gas stored in the tank in slugs? In lb$_m$?

SOLUTION

$$1 \text{ atm} = 2116 \text{ lb/ft}^2$$

Hence, $p = (30)(2116)$ lb/ft$^2 = 6.348 \times 10^4$ lb/ft^2. Also, from Eq. (2.3), $p = \rho RT$, hence $\rho = p/RT$.

$$\rho = \frac{6.348 \times 10^4 \text{ lb/ft}^2}{[1716 \text{ ft·lb/(slug)(°R)}](530 \text{ } R)} = 6.98 \times 10^{-2} \text{ slug/ft}^3$$

This is the density, which is mass *per unit volume*. The total mass M in the tank of volume V is

$$M = \rho V = (6.98 \times 10^{-2} \text{ slug/ft}^3)(1000 \text{ ft}^3) = \boxed{69.8 \text{ slugs}}$$

Recall that 1 slug = 32.2 lb$_m$.

Hence

$$M = (69.8)(32.2) = \boxed{2248 \text{ lb}_m}$$

Example 2.3 Air flowing at high speed in a wind tunnel has pressure and temperature equal to 0.3 atm and -100°C, respectively. What is the air density? What is the specific volume?

SOLUTION

$$1 \text{ atm} = 1.01 \times 10^5 \text{ N/m}^2$$

Hence
$$p = (0.3)(1.01 \times 10^5) = 0.303 \times 10^5 \text{ N/m}^2$$

Note that $T = -100°\text{C}$ is *not* an absolute temperature.

Hence
$$T = -100 + 273 = 173 \text{ K}$$

From Eq. (2.3), $p = \rho RT$, hence $\rho = p/RT$.

$$\rho = \frac{0.303 \times 10^5 \text{ N/m}^2}{[287 \text{ J/(kg)(K)}](173 \text{ K})} = \boxed{0.610 \text{ kg/m}^3}$$

$$v = \frac{1}{\rho} = \frac{1}{0.610} = \boxed{1.64 \text{ m}^3/\text{kg}}$$

Note: It is worthwhile to remember that

$$\boxed{\begin{array}{l} 1 \text{ atm} = 2116 \text{ lb/ft}^2 \\ 1 \text{ atm} = 1.01 \times 10^5 \text{ N/m}^2 \end{array}}$$

2.6 HISTORICAL NOTE: THE NACA AND NASA

Let us pick up the string of aeronautical engineering history from Chap. 1. After Orville and Wilbur Wright's dramatic public demonstrations in the United States and Europe in 1908, there was a virtual explosion in aviation developments. In turn, this rapid progress had to be fed by new technical research in aerodynamics, propulsion, structures, and flight control. It is important to realize that then, as well as today, aeronautical research was sometimes expensive, always demanding in terms of intellectual talent, and usually in need of large testing facilities. Such research in many cases was either beyond the financial resources of, or seemed too out of the ordinary for, private industry. Thus, the fundamental research so necessary to fertilize and pace the development of aeronautics in the twentieth century had to be established and nurtured by national governments. It is interesting to note that George Cayley himself (see Chap. 1) as far back as 1817 called for "public subscription" to underwrite the expense of the development of airships. Responding about 80 years later, the British Government set up a school for ballooning and military kite flying at Farnborough, England. By 1910, the Royal Aircraft Factory was in operation at Farnborough with the noted Geoffrey de Havilland as its first airplane designer and test pilot. This was the first major governmental aeronautical facility in history. This operation was soon to evolve into the Royal Aircraft Establishment (RAE), which today is still conducting viable aeronautical research for the British Government.

In the United States, aircraft development as well as aeronautical research languished after 1910. During the next decade, the United States embarrassingly fell far behind Europe in aeronautical progress. This set the stage for the U.S. Government to establish a formal mechanism for pulling itself out of its aeronautical "dark ages." On March 3, 1915, by an act of Congress, the National Advisory Committee for Aeronautics (NACA) was created, with an initial appropriation of $5000 per year for 5 years. This was at first a true committee, consisting of 12 distinguished members knowledgeable about aeronautics. Among the charter members in 1915 were Professor Joseph S. Ames of Johns Hopkins University (later to become President of Johns Hopkins) and Professor William F. Durand of Stanford University, both of whom were to make major impressions on aeronautical research in the first half century of powered flight. This advisory committee, the NACA, was originally to meet annually in Washington, D.C., on "the Thursday after the third Monday of October of each year," with any special meetings to be called by the chair. Its purpose was to advise the government on aeronautical research and development and to bring some cohesion to such activities in the United States.

The committee immediately noted that a single advisory group of 12 members was not sufficient to breathe life into U.S. aeronautics. Their insight is apparent in the letter of submittal for the first annual report of the NACA in 1915, which contained the following passage:

> There are many practical problems in aeronautics now in too indefinite a form to enable their solution to be undertaken. The committee is of the opinion that one of the first and most important steps to be taken in connection with the committee's work is the provision and equipment of a flying field together with aeroplanes and suitable testing gear for determining the forces acting on full-sized machines in constrained and in free flight, and to this end the estimates submitted contemplate the development of such a technical and operating staff, with the proper equipment for the conduct of full-sized experiments.
>
> It is evident that there will ultimately be required a well-equipped laboratory specially suited to the solving of those problems which are sure to develop, but since the equipment of such a laboratory as could be laid down at this time might well prove unsuited to the needs of the early future, it is believed that such provision should be the result of gradual development.

So the first action of this advisory committee was to call for major governmental facilities for aeronautical research and development. The clouds of war in Europe—World War I had already started a year earlier—made their recommendations even more imperative. In 1917, when the United States entered the conflict, actions followed the committee's words. We find the following entry in the third annual NACA report: "To carry on the highly scientific and special investigations contemplated in the act establishing the committee, and which have, since the outbreak of the war, assumed greater importance, and for which facilities do not already exist, or exist in only a limited degree, the committee has contracted for a research laboratory to be erected on the Signal Corps Experimental Station, Langley Field, Hampton, Virginia." The report goes on to describe a single, two-story laboratory building with physical, chemical, and structural testing laboratories. The building contract was for $80,900; actual construction began in 1917. Two wind tunnels and an engine test stand were contemplated "in the near future." The selection of a site 4 mi north of Hampton, Virginia, was based on general health conditions and the problems of accessibility to Washington and the larger industrial centers of the East, protection from naval attack, climatic conditions, and cost of the site.

Thus, the Langley Memorial Aeronautical Research Laboratory was born. It was to remain the only NACA laboratory and the only major U.S. aeronautical laboratory of any type for the next 20 years. Named after Samuel Pierpont Langley (see Sec. 1.7), it pioneered in wind-tunnel and flight research. Of particular note is the airfoil and wing research performed at Langley during the 1920s and 1930s. We will return to the subject of airfoils in Chap. 5, at which time the reader should note that the airfoil data included in Appendix D were obtained at Langley. With the work which poured out of the Langley laboratory, the United States took the lead in aeronautical development. High on the list of accomplishments, along with the systematic testing of airfoils, was the development of the NACA engine cowl (see Sec. 6.19), an aerodynamic fairing built around radial piston engines which dramatically reduced the aerodynamic drag of such engines.

In 1936, Dr. George Lewis, who was then NACA Director of Aeronautical Research (a position he held from 1924 to 1947), toured major European laboratories. He noted that the NACA's lead in aeronautical research was quickly disappearing, especially in light of advances being made in Germany. As World War II drew close, the NACA clearly recognized the need for two new laboratory operations: an advanced aerodynamics laboratory to probe into the mysteries of high-speed (even supersonic) flight, and a major engine-testing laboratory. These needs eventually led to the construction of the Ames Aeronautical Laboratory at Moffett Field, near Mountain View, California (authorized in 1939) and the Lewis Engine Research Laboratory at Cleveland, Ohio (authorized in 1941). Along with Langley, these two new NACA laboratories again helped to spearhead the United States to the forefront of aeronautical research and development in the 1940s and 1950s.

The dawn of the space age occurred on October 4, 1957, when Russia launched *Sputnik I*, the first artificial satellite to orbit the earth. Taking its somewhat embarrassed technical pride in hand, the United States moved quickly to compete in the race for space. On July 29, 1958, by another act of Congress (Public Law 85-568), the National Aeronautics and Space Administration (NASA) was born. At this same moment, the NACA came to an end. Its programs, people, and facilities were instantly transferred to NASA, lock, stock, and barrel. However, NASA was a larger organization than just the old NACA; it absorbed in addition numerous Air Force, Navy, and Army projects for space. Within two years of its birth, NASA was authorized four new major installations: an existing Army facility at Huntsville, Alabama, renamed the George C. Marshall Space Flight Center; the Goddard Space Flight Center at Greenbelt, Maryland; the Manned Spacecraft Center (now the Johnson Spacecraft Center) in Houston, Texas; and the Launch Operations Center (now the John F. Kennedy Space Center) at Cape Canaveral, Florida. These, in addition to the existing but slightly renamed Langley, Ames, and Lewis research centers, were the backbone of NASA. Thus, the aeronautical expertise of the NACA now formed the seeds for

NASA, shortly thereafter to become one of the world's most important forces in space technology.

This capsule summary of the roots of the NACA and NASA is included in the present chapter on fundamental thoughts because it is virtually impossible for a student or practitioner of aerospace engineering in the United States not to be influenced or guided by NACA or NASA data and results. The extended discussion on airfoils in Chap. 5 is a case in point. Thus, the NACA and NASA are "fundamental" to the discipline of aerospace engineering, and it is important to have some impression of the historical roots and tradition of these organizations. Hopefully, this short historical note provides such an impression. A much better impression can be obtained by taking a journey through the NACA and NASA technical reports in the library, going all the way back to the first NACA report in 1915. In so doing, a panorama of aeronautical and space research through the years will unfold in front of you.

2.7 CHAPTER SUMMARY

Some of the major ideas in this chapter are listed below.

1. The language of aerodynamics involves pressure, density, temperature, and velocity. In turn, the illustration of the velocity field can be enhanced by drawing streamlines for a given flow.
2. The source of all aerodynamic forces on a body is the pressure distribution and the shear stress distribution over the surface.
3. A perfect gas is one where intermolecular forces can be neglected. For a perfect gas, the equation of state which relates p, ρ, and T is

$$p = \rho RT \tag{2.3}$$

where R is the specific gas constant.
4. In order to avoid confusion, errors, and a number of unnecessary "conversion factors" in the basic equations, always use consistent units. In this book, the SI system (newton, kilogram, meter, second) and the English engineering system (pound, slug, foot, second) will be used.

1 nm = 6075 ft

BIBLIOGRAPHY

Gray, George W., *Frontiers of Flight*, Knopf, New York, 1948.

Hartman, E. P., *Adventures in Research: A History of Ames Research Center 1940–1965*, NASA SP-4302, 1970.

Mechtly, E. A. *The International System of Units*, NASA SP-7012, 1969.

PROBLEMS

2.1 Consider the low-speed flight of the space shuttle as it is nearing a landing. If the air pressure and temperature at the nose of the shuttle are 1.2 atm and 300 K, respectively, what are the density and specific volume?

2.2 If 1500 lb_m of air is pumped into a previously empty 900-ft^3 storage tank and the air temperature within the tank is uniformly 70°F, what is the air pressure within the tank in atmospheres?

2.3 In the above problem, assume the rate at which air is being pumped into the tank is 0.5 lb_m/s. Consider the instant in time at which there is 1000 lb_m of air in the tank. Assume the air temperature is uniformly 50°F at this instant and is increasing at the rate of 1°F/min. Calculate the rate of change of pressure at this instant.

2.4 Assume that, at a point on the wing of the Concorde supersonic transport, the air temperature is -10°C and the pressure is 1.7×10^4 N/m^2. Calculate the density at this point.

2.5 At a point in the test section of a supersonic wind tunnel, the air pressure and temperature are 0.5×10^5 N/m^2 and 240 K, respectively. Calculate the specific volume.

2.6 Consider a flat surface in an aerodynamic flow (say a flat side wall of a wind tunnel). The dimensions of this surface are 3 ft in the flow direction (the x direction) and 1 ft perpendicular to the flow direction (the y direction). Assume that the pressure distribution (in lb/ft^2) is given by $p = 2116 - 10x$ and is independent of y. Assume also that the shear-stress distribution (in lb/ft^2) is given by $\tau_w = 90/(x + 9)^{1/2}$ and is independent of y. In the above expressions, x is in feet, and $x = 0$ at the front of the surface. Calculate the magnitude and direction of the net aerodynamic force on the surface.

THREE

THE STANDARD ATMOSPHERE

Sometimes gentle, sometimes capricious, sometimes awful, never the same for two moments together; almost human in its passions, almost spiritual in its tenderness, almost divine in its infinity.

John Ruskin, The Sky

Aerospace vehicles can be divided into two basic categories: atmospheric vehicles such as airplanes and helicopters, which always fly within the sensible atmosphere, and space vehicles such as satellites, the Apollo lunar vehicle, and deep space probes, which operate outside the sensible atmosphere. However, space vehicles do encounter the earth's atmosphere during their blast-offs from the earth's surface and again during their reentries and recoveries after completion of their missions. If the vehicle is a planetary probe, then it may encounter the atmospheres of Venus, Mars, Jupiter, etc. Therefore, during the design and performance of any aerospace vehicle, the properties of the atmosphere must be taken into account.

The earth's atmosphere is a dynamically changing system, constantly in a state of flux. The pressure and temperature of the atmosphere depend on altitude, location on the globe (longitude and latitude), time of day, season, and even solar sunspot activity. To take all these variations into account when considering the design and performance of flight vehicles is impractical. Therefore, a *standard atmosphere* is defined in order to relate flight tests, wind-tunnel results, and general airplane design and performance to a common reference. The standard atmosphere gives mean values of pressure, temperature, density, and other properties as functions of altitude; these values are obtained from experimental balloon and sounding-rocket measurements combined with a mathematical model of the atmosphere. To a reasonable degree, the standard atmosphere reflects average atmospheric conditions, but this is not its main importance. Rather, its main function is to provide tables of common reference conditions that can be used in an organized fashion by aerospace engineers everywhere. The purpose of this chapter is to give you some feeling for what the standard atmosphere is all about and how it can be used for aerospace vehicle analyses.

It should be mentioned that several different standard atmospheres exist, compiled by different agencies at different times, each using slightly different experimental data in their models. For all practical purposes, the differences are insignificant below 30 km (100,000 ft), which is the domain of contemporary airplanes. A standard atmosphere in common use is the 1959 ARDC Model Atmosphere. (ARDC stands for the U.S. Air Force's previous Air Research and Development Command, which is now the Air Force Systems Command.) The atmospheric tables used in this book are taken from the 1959 ARDC Model Atmosphere.

3.1 DEFINITION OF ALTITUDE

Intuitively, we all know the meaning of altitude. We think of it as the distance above the ground. But like so many other general terms, it must be more precisely defined for quantitative use in engineering. In fact, in the following sections we will define and use six different altitudes: absolute, geometric, geopotential, pressure, temperature, and density altitudes.

First, imagine that we are at Daytona Beach, Florida, where the ground is sea level. If we would fly straight up in a helicopter and drop a tape measure to the ground, the measurement on the tape would be, by definition, the geometric altitude h_G, i.e., the geometric height above sea level.

Now, if we would bore a hole through the ground to the center of the earth and extend our tape measure until it hit the center, then the measurement on the tape would be, by definition, the *absolute altitude* h_a. If r is the radius of the earth, then $h_a = h_G + r$. This is illustrated in Figure 3.1.

The absolute altitude is important, especially for space flight, because the local acceleration of gravity, g, varies with h_a. From Newton's law of gravitation, g varies inversely as the square of the distance from the center of the earth. Letting g_0 be the gravitational acceleration at *sea level*, the local gravitational acceleration g at a given absolute altitude h_a is:

$$g = g_0 \left(\frac{r}{h_a} \right)^2 = g_0 \left(\frac{r}{r + h_G} \right)^2 \tag{3.1}$$

The variation of g with altitude must be taken into account when you are dealing with mathematical models of the atmosphere, as follows.

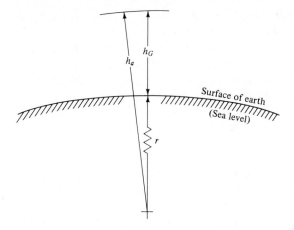

Figure 3.1 Definition of altitude.

3.2 THE HYDROSTATIC EQUATION

We will now begin to piece together a model which will allow us to calculate variations of p, ρ, and T as functions of altitude. The foundation of this model is the hydrostatic equation, which is nothing more than a force balance on an element of fluid at rest. Consider the small stationary fluid element of air shown in Figure 3.2. We take for convenience an element with rectangular faces, where the top and bottom faces have sides of unit length and the side faces have an infinitesimally small height dh_G. On the bottom face, the pressure p is felt, which gives rise to an upward force of $p \times 1 \times 1$ exerted on the fluid element. The top face is slightly higher in altitude (by the distance dh_G), and because pressure varies with altitude, the pressure on the top face will be slightly different from that on the bottom face, differing by the infinitesimally small value dp. Hence, on the top face, the pressure $p + dp$ is felt. It gives rise to a downward force of

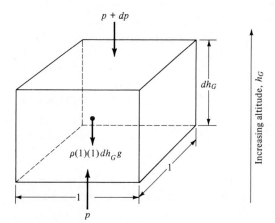

Figure 3.2 Force diagram for the hydrostatic equation.

$(p + dp)(1)(1)$ on the fluid element. Moreover, the volume of the fluid element is $(1)(1) dh_G = dh_G$, and since ρ is the mass per unit volume, then the mass of the fluid element is simply $\rho(1)(1) dh_G = \rho\, dh_G$. If the local acceleration of gravity is g, then the weight of the fluid element is $g\rho\, dh_G$, as shown in Figure 3.2. The three forces shown in Figure 3.2, pressure forces on the top and bottom and the weight must balance because the fluid element is not moving. Hence

$$p = (p + dp) + \rho g\, dh_G$$

Thus

$$dp = -\rho g\, dh_G \qquad (3.2)$$

Equation (3.2) is the *hydrostatic equation* and applies to any fluid of density ρ, e.g., water in the ocean as well as air in the atmosphere.

Strictly speaking, Eq. (3.2) is a differential equation; i.e., it relates an infinitesimally small change in pressure dp to a corresponding infinitesimally small change in altitude dh_G, where in the language of differential calculus, dp and dh_G are differentials. Also, note that g is a variable in Eq. (3.2); g depends on h_G as given by Eq. (3.1).

To be made useful, Eq. (3.2) should be integrated to give us what we want, namely, the variation of pressure with altitude, $p = p(h_G)$. To simplify the integration, we will make the *assumption* that g is constant through the atmosphere, equal to its value at sea level, g_0. This is something of a historical convention in aeronautics. At the altitudes encountered during the earlier development of human flight (less than 15 km or 50,000 ft), the variation of g is negligible. Hence, we can write Eq. (3.2) as

$$dp = -\rho g_0\, dh \qquad (3.3)$$

However, to make Eqs. (3.2) and (3.3) numerically identical, the altitude h in Eq. (3.3) must be slightly different from that of h_G in Eq. (3.2), to compensate for the fact that g is slightly different from g_0. Suddenly, we have defined a new altitude h, which is called the *geopotential altitude* and which differs from the geometric altitude. For the practical mind, geopotential altitude is a "fictitious" altitude, defined by Eq. (3.3) for ease of future calculations. However, many standard atmosphere tables quote their results in terms of geopotential altitude, and care must be taken to make the distinction. Again, geopotential altitude can be thought of as that fictitious altitude which is physically compatible with the assumption of $g = \text{const} = g_0$.

3.3 RELATION BETWEEN GEOPOTENTIAL AND GEOMETRIC ALTITUDES

We still seek the variation of p with geometric altitude, $p = p(h_G)$. However, our calculations using Eq. (3.3) will give, instead, $p = p(h)$. Therefore, we need to relate h to h_G, as follows. Dividing Eq. (3.3) by (3.2), we obtain

$$1 = \frac{g_0}{g} \frac{dh}{dh_G}$$

or

$$dh = \frac{g}{g_0} dh_G \tag{3.4}$$

Substitute Eq. (3.1) into (3.4):

$$dh = \frac{r^2}{(r + h_G)^2} dh_G \tag{3.5}$$

By convention, we set both h and h_G equal to zero at sea level. Now, consider a given point in the atmosphere. This point is at a certain geometric altitude h_G, and associated with it is a certain value of h (different from h_G). Integrating Eq. (3.5) between sea level and the given point, we have

$$\int_0^h dh = \int_0^{h_G} \frac{r^2}{(r + h_G)^2} dh_G = r^2 \int_0^{h_G} \frac{dh_G}{(r + h_G)^2}$$

$$h = r^2 \left(\frac{-1}{r + h_G} \right)_0^{h_G} = r^2 \left(\frac{-1}{r + h_G} + \frac{1}{r} \right)$$

$$= r^2 \left(\frac{-r + r + h_G}{(r + h_G)r} \right)$$

Thus

$$h = \left(\frac{r}{r + h_G} \right) h_G \tag{3.6}$$

where h is geopotential altitude and h_G is geometric altitude. This is the desired relation between the two altitudes. When we obtain relations such as $p = p(h)$, we can use Eq. (3.6) to subsequently relate p to h_G.

A quick calculation using Eq. (3.6) shows that there is little difference between h and h_G for low altitudes. For such a case, $h_G \ll r$, $r/(r + h_G) \cong 1$, hence $h \cong h_G$. Putting in numbers, $r = 6.356766 \times 10^6$ m (at a latitude of 45°), and if $h_G = 7$ km (about 23,000 ft), then the corresponding value of h is, from Eq. (3.6), $h = 6.9923$ km, about 0.1 of 1 percent difference! Only at altitudes above 65 km (213,000 ft) does the difference exceed 1 percent. (It should be noted that 65 km is an altitude where aerodynamic heating of NASA's space shuttle becomes important during reentry into the earth's atmosphere from space.)

3.4 DEFINITION OF THE STANDARD ATMOSPHERE

We are now in a position to obtain p, T, and ρ as functions of h for the standard atmosphere. The keystone of the standard atmosphere is a *defined* variation of T with altitude, based on experimental evidence. This variation is shown in Figure 3.3. Note that it consists of a series of straight lines, some vertical (called the constant-temperature, or *isothermal*, regions) and some inclined (called the *gradient* regions). Given $T = T(h)$ as *defined* by Figure 3.3, then $p = p(h)$ and $\rho = \rho(h)$ follow from the laws of physics, as shown below.

First, consider again Eq. (3.3):

$$dp = -\rho g_0\, dh$$

Divide by the equation of state, Eq. (2.3):

$$\frac{dp}{p} = -\frac{\rho g_0\, dh}{\rho RT} = -\frac{g_0}{RT}\, dh \tag{3.7}$$

Consider first the isothermal (constant-temperature) layers of the standard atmosphere, as given by the vertical lines in Figure 3.3 and sketched in Figure 3.4. The temperature, pressure, and density at the base of the isothermal layer shown in

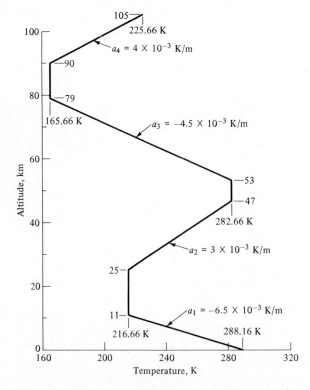

Figure 3.3 Temperature distribution in the standard atmosphere.

Figure 3.4 are T_1, p_1, and ρ_1, respectively. The base is located at a given geopotential altitude h_1. Now consider a given point in the isothermal layer above the base, where the altitude is h. The pressure p at h can be obtained by integrating Eq. (3.7) between h_1 and h.

$$\int_{p_1}^{p} \frac{dp}{p} = -\frac{g_0}{RT} \int_{h_1}^{h} dh \tag{3.8}$$

Note that g_0, R, and T are constants that can be taken outside the integral. (This clearly demonstrates the simplification obtained by assuming that $g = g_0 = $ const, and therefore dealing with geopotential altitude h in the analysis.) Performing the integration in Eq. (3.8), we obtain

$$\ln \frac{p}{p_1} = -\frac{g_0}{RT}(h - h_1)$$

or

$$\boxed{\frac{p}{p_1} = e^{-(g_0/RT)(h-h_1)}} \tag{3.9}$$

From the equation of state:

$$\frac{p}{p_1} = \frac{\rho T}{\rho_1 T_1} = \frac{\rho}{\rho_1}$$

Thus

$$\boxed{\frac{\rho}{\rho_1} = e^{-(g_0/RT)(h-h_1)}} \tag{3.10}$$

Equations (3.9) and (3.10) give the variation of p and ρ versus geopotential altitude for the isothermal layers of the standard atmosphere.

Considering the gradient layers, as sketched in Figure 3.5, we find the temperature variation is linear and is geometrically given as

$$\frac{T - T_1}{h - h_1} = \frac{dT}{dh} \equiv a$$

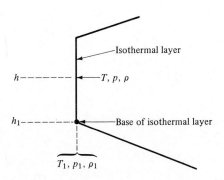

h ------- —T, p, ρ

Isothermal layer

h_1 ------- —Base of isothermal layer

T_1, p_1, ρ_1

Figure 3.4 Isothermal layer.

where a is a *specified* constant for each layer obtained from the defined temperature variation in Figure 3.3. The value of a is sometimes called the *lapse rate* for the gradient layers.

$$a \equiv \frac{dT}{dh}$$

Thus

$$dh = \frac{1}{a} dT$$

Substitute this result into Eq. (3.7):

$$\frac{dp}{p} = -\frac{g_0}{aR} \frac{dT}{T} \tag{3.11}$$

Integrating between the base of the gradient layer (shown in Figure 3.5) and some point at altitude h, also in the gradient layer, Eq. (3.11) yields

$$\int_{p_1}^{p} \frac{dp}{p} = -\frac{g_0}{aR} \int_{T_1}^{T} \frac{dT}{T}$$

$$\ln \frac{p}{p_1} = -\frac{g_0}{aR} \ln \frac{T}{T_1}$$

Thus

$$\boxed{\frac{p}{p_1} = \left(\frac{T}{T_1} \right)^{-g_0/aR}} \tag{3.12}$$

From the equation of state

$$\frac{p}{p_1} = \frac{\rho T}{\rho_1 T_1}$$

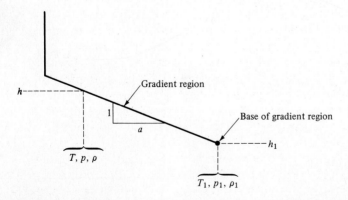

Figure 3.5 Gradient layer.

Hence, Eq. (3.12) becomes

$$\frac{\rho T}{\rho_1 T_1} = \left(\frac{T}{T_1}\right)^{-g_0/aR}$$

$$\frac{\rho}{\rho_1} = \left(\frac{T}{T_1}\right)^{-(g_0/aR)-1}$$

or

$$\frac{\rho}{\rho_1} = \left(\frac{T}{T_1}\right)^{-[(g_0/aR)+1]} \qquad (3.13)$$

Recall that the variation of T is linear with altitude and is given by the specified relation

$$T = T_1 + a(h - h_1) \qquad (3.14)$$

Equation (3.14) gives $T = T(h)$ for the gradient layers; when it is plugged into Eq. (3.12), we obtain $p = p(h)$; similarly from Eq. (3.13) we obtain $\rho = \rho(h)$.

Now we can see how the standard atmosphere is pieced together. Looking at Figure 3.3, start at sea level ($h = 0$), where standard sea level values of pressure, density, and temperature, p_s, ρ_s, and T_s, respectively, are

$$p_s = 1.01325 \times 10^5 \text{ N/m}^2 = 2116.2 \text{ lb/ft}^2$$

$$\rho_s = 1.2250 \text{ kg/m}^3 = 0.002377 \text{ slug/ft}^3$$

$$T_s = 288.16 \text{ K} = 518.69°\text{R}$$

These are the base values for the first gradient region. Use Eq. (3.14) to obtain values of T as a function of h until $T = 216.66$ K, which occurs at $h = 11.0$ km. With these values of T use Eqs. (3.12) and (3.13) to obtain the corresponding values of p and ρ in the first gradient layer. Next, starting at $h = 11.0$ km as the base of the first isothermal region (see Figure 3.3), use Eqs. (3.9) and (3.10) to calculate values of p and ρ versus h, until $h = 25$ km, which is the base of the next gradient region. In this manner, with Figure 3.3 and Eqs. (3.9), (3.10), and (3.12) to (3.14), a table of values for the standard atmosphere can be constructed.

Such a table is given in Appendix A for SI units and Appendix B for English engineering units. Look at these tables carefully and become familiar with them. They *are* the standard atmosphere. The first column gives the geometric altitude, and the second column gives the corresponding geopotential altitude obtained from Eq. (3.6). The third through fifth columns give the corresponding standard values of temperature, pressure, and density, respectively, for each altitude, obtained from the discussion above.

We emphasize again that the standard atmosphere is a reference atmosphere only and certainly does not predict the actual atmospheric properties that may exist at a given time and place. For example, Appendix A says that at an altitude (geometric) of 3 km, $p = 0.70121 \times 10^5$ N/m^2, $T = 268.67$ K, and $\rho = 0.90926$

kg/m³. In reality, situated where you are, if you would right now levitate yourself to 3 km above sea level, you would most likely feel a p, T, and ρ different from the above values obtained from Appendix A. The standard atmosphere allows us only to reduce test data and calculations to a convenient, agreed-upon reference, as will be seen in subsequent sections of this book.

Example 3.1 Calculate the standard atmosphere values of T, p, and ρ at a geopotential altitude of 14 km.

SOLUTION Remember that T is a *defined* variation for the standard atmosphere. Hence, we can immediately refer to Figure 3.3 and find that at $h = 14$ km,

$$T = 216.66 \text{ K}$$

To obtain p and ρ, we must use Eqs. (3.9) to (3.14), piecing together the different regions from sea level up to the given altitude with which we are concerned. Beginning at sea level, the first region (from Figure 3.3) is a gradient region from $h = 0$ to $h = 11.0$ km. The lapse rate is

$$a = \frac{dT}{dh} = \frac{216.66 - 288.16}{11.0 - 0} = -6.5 \text{ K/km}$$

or

$$a = -0.0065 \text{ K/m}$$

Therefore, using Eqs. (3.12) and (3.13), which are for a gradient region and where the base of the region is sea level (hence $p_1 = 1.01 \times 10^5$ N/m² and $\rho_1 = 1.23$ kg/m³), we find at $h = 11.0$ km,

$$p = p_1 \left(\frac{T}{T_1}\right)^{-g_0/aR} = (1.01 \times 10^5)\left(\frac{216.66}{288.16}\right)^{-9.8/-0.0065(287)}$$

where $g_0 = 9.8$ m/s² in the SI units. Hence, p (at $h = 11.0$ km) $= 2.26 \times 10^4$ N/m².

$$\rho = \rho_1 \left(\frac{T}{T_1}\right)^{-(g_0/aR + 1)}$$

$$\rho = (1.23)\left(\frac{216.66}{288.16}\right)^{-[9.8/-0.0065(287) + 1]}$$

$$\rho = 0.367 \text{ kg/m}^3 \qquad \text{at } h = 11.0 \text{ km}$$

The above values of p and ρ now form the *base* values for the first isothermal region (see Figure 3.3). The equations for the isothermal region are Eqs. (3.9) and (3.10), where now $p_1 = 2.26 \times 10^4$ N/m² and $\rho_1 = 0.367$ kg/m³. For $h = 14$ km, $h - h_1 = 14 - 11 = 3$ km $= 3000$ m. From Eq. (3.9),

$$p = p_1 e^{-(g_0/RT)(h - h_1)} = (2.26 \times 10^4) e^{-[9.8/287(216.66)](3000)}$$

$$\boxed{p = 1.41 \times 10^4 \text{ N/m}^2}$$

From Eq. (3.10),

$$\frac{\rho}{\rho_1} = \frac{p}{p_1}$$

Hence

$$\rho = \rho_1 \left(\frac{p}{p_1}\right) = 0.367 \frac{1.41 \times 10^4}{2.26 \times 10^4} = \boxed{0.23 \text{ kg/m}^3}$$

These values check, within round-off error, with the values given in Appendix A. Note: This example demonstrates how the numbers in Appendixes A and B are obtained!

3.5 PRESSURE, TEMPERATURE, AND DENSITY ALTITUDES

With the tables of Appendixes A and B in hand, we can now define three new "altitudes"—pressure, temperature, and density altitudes. This is best done by example. Imagine that you are in an airplane flying at some real, geometric altitude. The value of your actual altitude is immaterial for this discussion. However, at this altitude, you measure the actual outside air pressure to be 6.16×10^4 N/m². From Appendix A, you find that the standard altitude that corresponds to a pressure of 6.16×10^4 N/m² is 4 km. Therefore, by *definition*, you say that you are flying at a *pressure altitude* of 4 km. Simultaneously, you measure the actual outside air temperature to be 265.4 K. From Appendix A, you find that the standard altitude that corresponds to a temperature of 265.4 K is 3.5 km. Therefore, by definition, you say that you are flying at a *temperature altitude* of 3.5 km. Thus, you are simultaneously flying at a pressure altitude of 4 km and a temperature altitude of 3.5 km while your actual geometric altitude is yet a different value. The definition of *density altitude* is made in the same vein. These quantities—pressure, temperature, and density altitudes—are just convenient numbers that, via Appendix A or B, are related to the actual p, T, and ρ for the actual altitude at which you are flying.

Example 3.2 If an airplane is flying at an altitude where the actual pressure and temperature are 4.72×10^4 N/m² and 255.7 K, respectively, what are the pressure, temperature, and density altitudes?

SOLUTION For the pressure altitude, look in Appendix A for the standard altitude value corresponding to $p = 4.72 \times 10^4$ N/m². This is 6000 m. Hence

$$\boxed{\text{Pressure altitude} = 6000 \text{ m} = 6 \text{ km}}$$

For the temperature altitude, look in Appendix A for the standard altitude value corresponding to $T = 255.7$ K. This is 5000 m. Hence

$$\boxed{\text{Temperature altitude} = 5000 \text{ m} = 5 \text{ km}}$$

For the density altitude, we must first calculate ρ from the equation of state:

$$\rho = \frac{p}{RT} = \frac{4.72 \times 10^4}{287(255.7)} = 0.643 \text{ kg/m}^3$$

Looking in Appendix A and interpolating between 6.2 km and 6.3 km, we find that the standard altitude value corresponding to $\rho = 0.643$ kg/m³ is about 6.240 m. Hence

$$\boxed{\text{Density altitude} = 6240 \text{ m} = 6.24 \text{ km}}$$

It should be noted that temperature altitude is not a unique value. The above answer for temperature altitude could equally well be 5.0, 38.2, or 59.5 km because of the multivalued nature of the altitude-versus-temperature function. In this section, only the lowest value of temperature altitude is used.

3.6 HISTORICAL NOTE: THE STANDARD ATMOSPHERE

With the advent of ballooning in 1783 (see Chap. 1), people suddenly became interested in acquiring a greater understanding of the properties of the atmosphere above ground level. However, a compelling reason for such knowledge did not arise until the coming of heavier-than-air flight in the twentieth century. As we shall see in subsequent chapters, the flight performance of aircraft is dependent upon such properties as pressure and density of the air. Thus, a knowledge of these properties, or at least some agreed-upon standard for worldwide reference, is absolutely necessary for intelligent aeronautical engineering.

The situation in 1915 was summarized by C. F. Marvin, Chief of the U.S. Weather Bureau and chairman of an NACA subcommittee to investigate and report upon the existing status of atmospheric data and knowledge. In his "Preliminary Report on the Problem of the Atmosphere in Relation to Aeronautics," NACA Report No. 4, 1915, Marvin writes:

> The Weather Bureau is already in possession of an immense amount of data concerning atmospheric conditions, including wind movements at the earth's surface. This information is no doubt of distinct value to aeronautical operations, but it needs to be collected and put in form to meet the requirements of aviation.

The following four years saw such efforts to collect and organize atmospheric data for use by aeronautical engineers. In 1920, the Frenchman A. Toussaint, director of the Aerodynamic Laboratory at Saint-Cyr-l'Ecole, France, suggested the following formula for the temperature decrease with height:

$$T = 15 - 0.0065h$$

where T is in degrees Celsius and h is the geopotential altitude in meters. Toussaint's formula was formally adopted by France and Italy with the Draft of Inter-Allied Agreement on Law Adopted for the Decrease of Temperature with Increase of Altitude, issued by the Ministere de la Guerre, Aeronautique Militaire, Section Technique, in March 1920. One year later, England followed suit. The United States was close behind. Since Marvin's report in 1915, the U.S. Weather Bureau had compiled measurements of the temperature distribution and found Toussaint's formula to be a reasonable representation of the observed mean annual values. Therefore, at its executive committee meeting of December 17, 1921, the NACA adopted Toussaint's formula for airplane performance testing, with the statement: "The subcommittee on aerodynamics recommends that for the sake of uniform practice in different countries that Toussaint's formula be adopted in determining the standard atmosphere up to 10 km (33,000 ft)... ."

Much of the technical data base that supported Toussaint's formula was reported in NACA Report No. 147, "Standard Atmosphere," by Willis Ray Gregg in 1922. Based on free-flight tests at McCook Field in Dayton, Ohio, and at Langley Field in Hampton, Virginia, and on other flights at Washington, D.C., as well as artillery data from Aberdeen, Maryland, and Dahlgren, Virginia, and sounding-balloon observations at Fort Omaha, Nebraska, and at St. Louis, Missouri, Gregg was able to compile a table of mean annual atmospheric properties. An example of his results is as follows:

Altitude, m	Mean Annual Temperature in United States, K	Temperature from Toussaint's Formula, K
0	284.5	288
1,000	281.0	281.5
2,000	277.0	275.0
5,000	260.0	255.5
10,000	228.5	223.0

Clearly, Toussaint's formula provided a simple and reasonable representation of the mean annual results in the United States. This was the primary message in Gregg's report in 1922. However, the report neither gave extensive tables nor attempted to provide a document for engineering use.

Thus, it fell to Walter S. Diehl (who later became a well-known aerodynamicist and airplane designer as a captain in the Naval Bureau of Aeronautics) to provide the first practical tables for a standard atmosphere for aeronautical use. In 1925, in NACA Report No. TR 218, entitled (again) "Standard Atmosphere," Diehl presented extensive tables of standard atmospheric properties in both metric and English units. The tables were in increments of 50 m up to an altitude of 10 km and then in increments of 100 m up to 20 km. In English units, the tables were in increments of 100 ft up to 32,000 ft and then in increments of 200 ft up to a maximum altitude of 65,000 ft. Considering the aircraft of that day (see Figure 1.28), these tables were certainly sufficient. Moreover, starting from Toussaint's formula for T up to 10,769 m, then assuming $T = \text{const} = -55°C$ above 10,769 m, Diehl obtained p and ρ in precisely the same fashion as described in the previous sections of this chapter.

The 1940s saw the beginning of serious rocket flights, with the German V-2 and the initiation of sounding rockets. Moreover, airplanes were flying higher than ever. Then, with the advent of intercontinental ballistic missiles in the 1950s and space flight in the 1960s, altitudes became quoted in terms of hundreds of miles rather than feet. Therefore, new tables of the standard atmosphere were created, mainly extending the old tables to higher altitudes. Popular among the various tables is the ARDC 1959 Standard Atmosphere, which is used in this book and is given in Appendixes A and B. For all practical purposes, the old and new tables agree for altitudes of most interest. Indeed, it is interesting to compare values, as shown below:

Altitude, m	T from Diehl, 1925, K	T from ARDC, 1959, K
0	288	288.16
1,000	281.5	281.66
2,000	275.0	275.16
5,000	255.5	255.69
10,000	223.0	223.26
10,800	218.0	218.03
11,100	218.0	216.66
20,000	218.0	216.66

So Diehl's standard atmosphere from 1925, at least up to 20 km, is just as good as the values today.

3.7 CHAPTER SUMMARY

Some of the major ideas of this chapter are listed below.

1. The standard atmosphere is defined in order to relate flight tests, wind-tunnel results, and general airplane design and performance to a common reference.
2. The definitions of the standard atmospheric properties are based on a given temperature variation with altitude, representing a mean of experimental data. In turn, the pressure and density variations with altitude are obtained from this empirical temperature variation by using the laws of physics. One of these laws is the hydrostatic equation:

$$dp = -\rho g \, dh_G \qquad (3.2)$$

3. In the isothermal regions of the standard atmosphere, the pressure and density variations are given by

$$\frac{p}{p_1} = \frac{\rho}{\rho_1} = e^{-(g_0/RT)(h-h_1)} \qquad (3.9) \text{ and } (3.10)$$

4. In the gradient regions of the standard atmosphere, the pressure and density variations are given by

$$\frac{p}{p_1} = \left(\frac{T}{T_1}\right)^{-g_0/aR} \qquad (3.12)$$

$$\frac{\rho}{\rho_1} = \left(\frac{T}{T_1}\right)^{-[(g_0/aR)+1]} \qquad (3.13)$$

where $T = T_1 + a(h - h_1)$ and a is the given lapse rate.

5. The pressure altitude is that altitude in the standard atmosphere which corresponds to the actual ambient pressure encountered in flight or laboratory experiments. For example, if the ambient pressure of a flow, no matter where it is or what it is doing, is 393.12 lb/ft^2, the flow is said to correspond to a pressure altitude of 40,000 ft (see Appendix B). The same idea can be used to define density and temperature altitudes.

BIBLIOGRAPHY

Minzner, R. A., Champion, K. S. W., and Pond, H. L., *The ARDC Model Atmosphere, 1959*, Air Force Cambridge Research Center Report No. TR-59-267, U.S. Air Force, Bedford, MA, 1959.

PROBLEMS

3.1 At 12 km in the standard atmosphere, the pressure, density, and temperature are 1.9399×10^4 N/m^2, 3.1194×10^{-1} kg/m^3, and 216.66 K, respectively. Using these values, calculate the standard atmospheric values of pressure, density, and temperature at an altitude of 18 km, and check with the standard altitude tables.

3.2 Consider an airplane flying at some real altitude. The outside pressure and temperature are 2.65×10^4 N/m^2 and 220 K, respectively. What are the pressure and density altitudes?

3.3 Consider an airplane flying at a pressure altitude of 33,500 ft and a density altitude of 32,000 ft. Calculate the outside air temperature.

3.4 At what value of the geometric altitude is the difference $h - h_G$ equal to 2 percent of h?

3.5 Using Toussaint's formula, calculate the pressure at a geopotential altitude of 5 km.

FOUR

BASIC AERODYNAMICS

Mathematics up to the present day have been quite useless to us in regard to flying.

From the fourteenth Annual Report of
the Aeronautical Society of Great Britain, 1879

Mathematical theories from the happy hunting grounds of pure mathematicians are found suitable to describe the airflow produced by aircraft with such excellent accuracy that they can be applied directly to airplane design.

Theodore von Karman, 1954

Consider an airplane flying at an altitude of 3 km (9840 ft) at a velocity of 112 m/s (367 ft/s or 251 mi/h). At a given point on the wing, the pressure and airflow velocity are specific values, dictated by the laws of nature. One of the objectives of the science of aerodynamics is to decipher these laws and to give us methods to calculate the flow properties. In turn, such information allows us to calculate practical quantities, such as the lift and drag on the airplane. Another example is the flow through a rocket engine of a given size and shape. If this engine is sitting on the launch pad at Cape Canaveral and given amounts of fuel and oxidizer are ignited in the combustion chamber, the flow velocity and pressure at the nozzle exit are again specific values dictated by the laws of nature. The basic principles of aerodynamics allow us to calculate the exit flow velocity and pressure, which in turn allow us to calculate the thrust. For reasons such as these, the study of aerodynamics is vital to the overall understanding of flight. The purpose of this chapter is to provide an introduction to the basic laws and concepts of aerodynamics and to show how they are applied to solving practical problems.

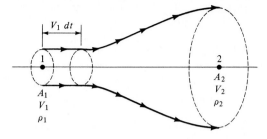

Figure 4.1 Stream tube with mass conservation.

4.1 THE CONTINUITY EQUATION

The laws of aerodynamics are formulated by applying to a flowing gas several basic principles from physics. For example,

Physical principle: *Mass can be neither created nor destroyed.**

To apply this principle to a flowing gas, consider an imaginary circle drawn perpendicular to the flow direction, as shown in Figure 4.1. Now look at all the streamlines that go through the circumference of the circle. These streamlines form a tube, called a *stream tube*. As we move along with the gas confined inside the stream tube, we see that the cross-sectional area of the tube may change, say in moving from point 1 to point 2 in Figure 4.1. However, as long as the flow is steady (invariant with time), the mass that flows through the cross section at point 1 must be the same as the mass that flows through the cross section at point 2, because by the definition of a streamline, there can be no flow across streamlines. The mass flowing through the stream tube is confined by the streamlines of the boundary, much as the flow of water through a flexible garden hose is confined by the wall of the hose.

Let A_1 be the cross-sectional area of the stream tube at point 1. Let V_1 be the flow velocity at point 1. Now, at a given instant in time, consider all the fluid elements that are momentarily in the plane of A_1. After a lapse of time dt, these same fluid elements all move a distance $V_1 dt$, as shown in Figure 4.1. In so doing, the elements have swept out a volume $(A_1 V_1 dt)$ downstream of point 1. The *mass* of gas, dm, in this volume is equal to the density times the volume, i.e.,

$$dm = \rho_1 (A_1 V_1 dt) \tag{4.1}$$

This is the mass of gas that has *swept through* area A_1 during the time interval dt.

Definition: *The **mass flow** $\dot{m}$ through area A is the mass crossing A per unit time.*

---

* Of course, Einstein has shown that $e = mc^2$ and hence mass is truly not conserved in situations where energy is released. However, for any noticeable change in mass to occur, the energy release must be tremendous, such as occurs in a nuclear reaction. We are generally not concerned with such a case in practical aerodynamics.

Therefore, from Eq. (4.1), for area A_1,

$$\text{Mass flow} = \frac{dm}{dt} \equiv \dot{m}_1 = \rho_1 A_1 V_1 \qquad \text{kg/s or slugs/s}$$

Also, the mass flow through A_2, bounded by the same streamlines that go through the circumference of A_1, is obtained in the same fashion, as

$$\dot{m}_2 = \rho_2 A_2 V_2$$

Since mass can be neither created nor destroyed, we have $\dot{m}_1 = \dot{m}_2$. Hence

$$\boxed{\rho_1 A_1 V_1 = \rho_2 A_2 V_2} \qquad (4.2)$$

This is the *continuity equation* for steady fluid flow. It is a simple algebraic equation which relates the values of density, velocity, and area at one section of the stream tube to the same quantities at any other section.

4.2 INCOMPRESSIBLE AND COMPRESSIBLE FLOW

Before we proceed any further, it is necessary to point out that all matter in real life is *compressible* to some greater or lesser extent. That is, if we take an element of matter and squeeze on it hard enough with some pressure, the volume of the element of matter will decrease. However, its mass will stay the same. This is shown schematically in Figure 4.2. As a result, the *density* ρ of the element changes as it is squeezed. The amount by which ρ changes depends on the nature of the material of the element and how hard we squeeze it, i.e., the magnitude of the pressure. If the material is solid, such as steel, the change in volume is insignificantly small and ρ is constant for all practical purposes. If the material is a liquid, such as water, the change in volume is also very small and again ρ is essentially constant. (Try pushing a tightly fitting lid into a container of liquid, and you will find out just how "solid" the liquid can be.) On the other hand, if the material is a gas, the volume can readily change and ρ can be a variable.

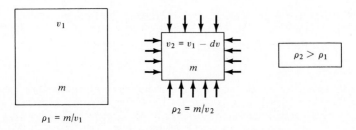

Figure 4.2 Illustration of compressibility.

The preceding discussion allows us to characterize two classes of aerodynamic flow: compressible flow and incompressible flow.

1. *Compressible flow*—flow in which the density of the fluid elements can change from point to point. Referring to Eq. (4.2), we see if the flow is compressible, $\rho_1 \neq \rho_2$. The variability of density in aerodynamic flows is particularly important at high speeds, such as for high-performance subsonic aircraft, all supersonic vehicles, or rocket engines. Indeed, all real-life flows, strictly speaking, are compressible. However, there are some circumstances in which the density changes only slightly. These circumstances lead to the second definition, as follows.

2. *Incompressible flow*—flow in which the density of the fluid elements is always constant. Referring to Eq. (4.2), we see if the flow is incompressible, $\rho_1 = \rho_2$, hence

$$A_1 V_1 = A_2 V_2 \qquad (4.3)$$

Incompressible flow is a myth. It can never actually occur in nature, as discussed above. However, for those flows in which the actual variation of ρ is negligibly small, it is convenient to make the *assumption* that ρ is constant in order to simplify our analysis. (Indeed, it is an everyday activity of engineering and physical science to make idealized assumptions about real physical systems in order to make such systems amenable to analysis. However, care must always be taken not to apply results obtained from such idealizations to those real problems where the assumptions are grossly inaccurate or inappropriate.) The assumption of incompressible flow is an excellent approximation for the flow of liquids, such as water or oil. Moreover, the low-speed flow of air, where $V < 100$ m/s (or $V < 225$ mi/h) can also be assumed to be incompressible to a close approximation. A glance at Figure 1.27 shows that such velocities were the domain of almost all airplanes from the Wright Flyer (1903) to the late 1930s. Hence, the early development of aerodynamics always dealt with incompressible flows, and for this reason there exists a huge body of incompressible flow literature with its attendant technology. At the end of this chapter, we will be able to prove *why* airflow at velocities less than 100 m/s can be safely assumed to be incompressible.

In solving and examining aerodynamic flows, you will constantly be faced with making distinctions between incompressible and compressible flows. It is important to start that habit now, because there are some striking quantitative and qualitative differences between the two types of flow.

As a parenthetical comment, for incompressible flow, Eq. (4.3) explains why all common garden hose nozzles are convergent shapes, such as shown in Figure 4.3. From Eq. (4.3),

$$V_2 = \left(\frac{A_1}{A_2} \right) V_1$$

If A_2 is less than A_1, then the velocity increases as the water flows through the nozzle, as desired. The same principle is used in the design of nozzles for subsonic

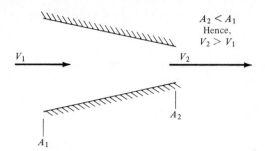

$A_2 < A_1$
Hence,
$V_2 > V_1$

Figure 4.3 Incompressible flow in a convergent duct.

wind tunnels built for aerodynamic testing, as will be discussed in a subsequent section.

Example 4.1 Consider a convergent duct with an inlet area $A_1 = 5$ m². Air enters this duct with a velocity $V_1 = 10$ m/s and leaves the duct exit with a velocity $V_2 = 30$ m/s. What is the area of the duct exit?

SOLUTION Since the flow velocities are less than 100 m/s, we can assume incompressible flow. From Eq. (4.3)

$$A_1 V_1 = A_2 V_2$$

$$A_2 = A_1 \frac{V_1}{V_2} = 5 \text{ m}^2 \frac{10}{30} = \boxed{1.67 \text{ m}^2}$$

4.3 THE MOMENTUM EQUATION

The continuity equation, Eq. (4.2), is only part of the story. For example, it says nothing about the pressure in the flow; yet, we know, just from intuition, that pressure is an important flow variable. Indeed, differences in pressure from one point to another in the flow create forces that act on the fluid elements and cause them to move. Hence, there must be some relation between pressure and velocity, and that relation will be derived in this section.

Again, we first state a fundamental law of physics, namely, Newton's second law.

Physical principle: *Force = mass × acceleration*

or $F = ma$ (4.4)

To apply this principle to a flowing gas, consider an infinitesimally small fluid element moving along a streamline with velocity V, as shown in Figure 4.4. At some given instant, the element is located at point P. The element is moving in the x direction, where the x axis is oriented parallel to the streamline at point P. The y and z axes are mutually perpendicular to x. The fluid element is very small, infinitesimally small. However, looking at it through a magnifying glass, we see the picture shown at the upper right of Figure 4.4. Question: What is the force

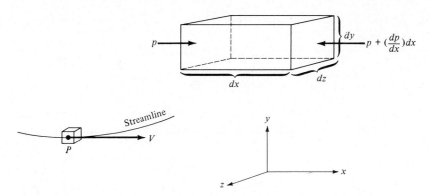

Figure 4.4 Force diagram for the momentum equation.

on this element? Physically, the force is a combination of three phenomena:

1. Pressure acting in a normal direction on all six faces of the element
2. Frictional shear acting tangentially on all six faces of the element
3. Gravity acting on the mass inside the element

For the time being, we will ignore the presence of frictional forces; moreover, the gravity force is generally a small contribution to the total force. Therefore, we will assume that the only source of a force on the fluid element is pressure.

To calculate this force, let the dimensions of the fluid element be dx, dy, and dz, as shown in Figure 4.4. Consider the left and right faces, which are perpendicular to the x axis. The pressure on the left face is p. The area of the left face is $dy\,dz$, hence the force on the left face is $p(dy\,dz)$. This force is in the positive x direction. Now recall that pressure varies from point to point in the flow. Hence, there is some change in pressure per unit length, symbolized by the derivative dp/dx. Thus, if we move away from the left face by a distance dx along the x axis, the *change* in pressure is $(dp/dx)\,dx$. Consequently, the pressure on the right face is $p + (dp/dx)\,dx$. The area of the right face is also $dy\,dz$, hence the force on the right face is $[p + (dp/dx)\,dx](dy\,dz)$. This force acts in the negative x direction, as shown in Figure 4.4. The net force in the x direction, F, is the sum of the two:

$$F = p\,dy\,dz - \left(p + \frac{dp}{dx}\,dx \right) dy\,dz$$

or
$$F = -\frac{dp}{dx}(dx\,dy\,dz) \tag{4.5}$$

Equation (4.5) gives the force on the fluid element due to pressure. Because of the convenience of choosing the x axis in the flow direction, the pressures on the faces parallel to the streamlines do not affect the motion of the element along the streamline.

The mass of the fluid element is the density ρ multiplied by the volume $dx\,dy\,dz$:

$$m = \rho(\,dx\,dy\,dz\,) \qquad (4.6)$$

Also, the acceleration a of the fluid element is, by definition of acceleration (rate of change of velocity), $a = dV/dt$. Noting that, also by definition, $V = dx/dt$, we can write

$$a = \frac{dV}{dt} = \frac{dV}{dx}\frac{dx}{dt} = \frac{dV}{dx}V \qquad (4.7)$$

Equations (4.5) to (4.7) give the force, mass, and acceleration, respectively, that go into Newton's second law, Eq. (4.4):

$$F = ma$$

$$-\frac{dp}{dx}(\,dx\,dy\,dz\,) = \rho(\,dx\,dy\,dz\,)V\frac{dV}{dx}$$

or
$$\boxed{dp = -\rho V\,dV} \qquad (4.8)$$

Equation (4.8) is *Euler's equation*. Basically, it relates rate of change of momentum to the force; hence it can also be designated as the *momentum equation*. It is important to keep in mind the assumptions utilized in obtaining Eq. (4.8); we neglected friction and gravity. For flow which is frictionless, aerodynamicists sometimes use another term, *inviscid flow*. Equation (4.8) is the momentum equation for inviscid (frictionless) flow. Moreover, the flow field is assumed to be steady, i.e., invariant with respect to time.

Please note that Eq. (4.8) relates pressure and velocity (in reality, it relates a change in pressure, dp, to a change in velocity, dV). Equation (4.8) is a differential equation, and hence it describes the phenomena in an infinitesimally small neighborhood around the given point P in Figure 4.4. Now consider two points, 1 and 2, far removed from each other in the flow but on the same streamline. In order to relate p_1 and V_1 at point 1 to p_2 and V_2 at the other, far-removed point 2, Eq. (4.8) must be integrated between points 1 and 2. This integration is different depending on whether the flow is compressible or incompressible. Euler's equation itself, Eq. (4.8), holds for both cases. For compressible flow, ρ in Eq. (4.8) is a variable; for incompressible flow, ρ is a constant.

First, consider the case of incompressible flow. Let points 1 and 2 be located along a given streamline, such as that shown over an airfoil in Figure 4.5. From

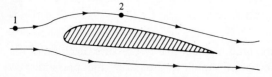

Figure 4.5 Two points at different locations along a streamline.

Eq. (4.8),

$$dp + \rho V \, dV = 0$$

where ρ = const. Integrating between points 1 and 2, we obtain

$$\int_{p_1}^{p_2} dp + \rho \int_{V_1}^{V_2} V \, dV = 0$$

$$(p_2 - p_1) + \rho \left(\frac{V_2^2}{2} - \frac{V_1^2}{2} \right) = 0$$

$$p_2 + \rho \frac{V_2^2}{2} = p_1 + \rho \frac{V_1^2}{2} \qquad (4.9a)$$

$$p + \rho \frac{V^2}{2} = \text{const along streamline} \qquad (4.9b)$$

Either form, Eq. (4.9a) or (4.9b), is called *Bernoulli's equation*. Historically, Bernoulli's equation is one of the most fundamental equations in fluid mechanics. The following important points should be noted:

1. Equations (4.9a) and (4.9b) hold only for inviscid (frictionless), incompressible flow.
2. Equations (4.9a) and (4.9b) relate properties between different points along a streamline.
3. For a compressible flow, Eq. (4.8) must be used, with ρ treated as a variable. Bernoulli's equation *must not* be used for compressible flow.
4. Remember that Eqs. (4.8) and (4.9a) and (4.9b) say that $F = ma$ for a fluid flow. They are essentially Newton's second law applied to fluid dynamics.

Returning to Figure 4.5, if all the streamlines have the same value of p and V far upstream (far to the left in Figure 4.5), then the constant in Bernoulli's equation is the *same for all streamlines*. This would be the case, for example, if the flow far upstream was uniform flow, such as that encountered in flight through the atmosphere and in the test sections of well-designed wind tunnels. In such cases, Eqs. (4.9a) and (4.9b) are not limited to the same streamline. Instead, points 1 and 2 can be anywhere in the flow, even on different streamlines.

For the case of compressible flow also, Euler's equation, Eq. (4.8), can be integrated between points 1 and 2; however, because ρ is a variable, we must in principle have some extra information on how ρ varies with V before the integration can be carried out. This information can be obtained; however, there is an alternate, more convenient route to treating many practical problems in compressible flow that does not explicitly require the use of the momentum equation. Hence, in this case, we will not pursue the integration of Eq. (4.8) further.

4.4 A COMMENT

It is important to make a philosophical distinction between the nature of the equation of state, Eq. (2.3), and the flow equations of continuity, Eq. (4.2), and momentum, such as Eq. (4.9a). The equation of state relates p, T, and ρ to each other at the *same* point; in contrast, the flow equations relate ρ and V (as in the continuity equation) and p and V (as in Bernoulli's equation) at one point in the flow to the same quantities at another point in the flow. There is a basic difference here, and it is well to keep it in mind when setting up the solution of aerodynamic problems.

Example 4.2 Consider an airfoil (the cross section of a wing as shown in Figure 4.5) in a flow of air, where far ahead (upstream) of the airfoil, the pressure, velocity, and density are 2116 lb/ft^2, 100 mi/h, and 0.002377 slug/ft^3, respectively. At a given point A on the airfoil, the pressure is 2070 lb/ft^2. What is the velocity at point A?

SOLUTION First, we must deal in consistent units; $V_1 = 100$ mi/h is *not* in consistent units. However, a convenient relation to remember is that 60 mi/h = 88 ft/s. Hence, $V_1 = 100(88/60)$ = 146.7 ft/s. This velocity is low enough that we can assume incompressible flow. Hence, Bernoulli's equation, Eq. (4.9), is valid.

$$p_1 + \frac{\rho V_1^2}{2} = p_A + \frac{\rho V_A^2}{2}$$

Thus
$$V_A = \left[\frac{2(p_1 - p_A)}{\rho} + V_1^2 \right]^{1/2}$$

$$= \left[\frac{2(2116 - 2070)}{0.002377} + (146.7)^2 \right]^{1/2}$$

$$\boxed{V_A = 245.4 \text{ ft/s}}$$

Example 4.3 Consider the same convergent duct and conditions as in Example 4.1. If the air pressure and temperature at the inlet are $p_1 = 1.2 \times 10^5$ N/m^2 and $T_1 = 330$ K, calculate the pressure at the exit.

SOLUTION First, we must obtain the density. From the equation of state,

$$\rho_1 = \frac{p_1}{RT_1} = \frac{1.2 \times 10^5}{287(330)} = 1.27 \text{ kg/m}^3$$

Still assuming incompressible flow, Eq. (4.9) gives

$$p_1 + \frac{\rho V_1^2}{2} = p_2 + \frac{\rho V_2^2}{2}$$

$$p_2 = p_1 + \tfrac{1}{2}\rho\left(V_1^2 - V_2^2 \right) = 1.2 \times 10^5 + \left(\tfrac{1}{2}\right)(1.27)\left[(10)^2 - (30)^2 \right]$$

$$\boxed{p_2 = 1.195 \times 10^5 \text{ N/m}^2}$$

Note: In accelerating from 10 to 30 m/s, the air pressure decreases only a small amount, less than 0.45%. This is a characteristic of very low velocity airflow.

4.5 ELEMENTARY THERMODYNAMICS

As stated earlier, when the airflow velocity exceeds 100 m/s, the flow can no longer be treated as incompressible. Later on, we shall restate this criterion in terms of the Mach number, which is the ratio of the flow velocity to the speed of sound, and we will show that the flow must be treated as compressible when the Mach number exceeds 0.3. This is the situation with the vast majority of current aerodynamic applications; hence, the study of compressible flow is of extreme importance.

A high-speed flow of gas is also a high-energy flow. The kinetic energy of the fluid elements in a high-speed flow is large and must be taken into account. When high-speed flows are slowed down, the consequent reduction in kinetic energy appears as a substantial increase in temperature. As a result, high-speed flows, compressibility, and vast energy changes are all related. Thus, to study compressible flows, we must first examine some of the fundamentals of energy changes in a gas and the consequent response of pressure and temperature to these energy changes. Such fundamentals are the essence of the science of thermodynamics.

Here, the assumption is made that the reader is not familiar with thermodynamics. Therefore, the purpose of this section is to introduce those ideas and results of thermodynamics which are absolutely necessary for our further analysis of high-speed, compressible flows.

The pillar of thermodynamics is a relationship called the first law, which is an empirical observation of natural phenomena. It can be developed as follows. Consider a fixed mass of gas (for convenience, say a unit mass) contained within a flexible *boundary*, as shown in Figure 4.6. This mass is called the *system*, and everything outside the boundary is the *surroundings*. Now, as in Chap. 2, consider the gas making up the system to be composed of individual molecules moving about with random motion. The energy of this molecular motion, summed over all the molecules in the system, is called the *internal energy* of the system. Let e denote the internal energy per unit mass of gas. The *only* means by which e can be increased (or decreased) are the following:

1. Heat added to (or taken away from) the system. This heat comes from the surroundings and is added to the system across the boundary. Let δq be an incremental amount of heat added per unit mass.

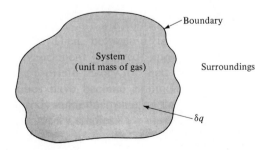

Figure 4.6 System of unit mass.

2. Work done on (or by) the system. This work can be manifested by the boundary of the system being pushed in (work done on the system) or pushed out (work done by the system). Let δw be an incremental amount of work done on the system per unit mass.

Also, let de be the corresponding change in internal energy per unit mass. Then, simply on the basis of common sense, confirmed by laboratory results, we can write

$$\boxed{\delta q + \delta w = de} \tag{4.10}$$

Equation (4.10) is termed the *first law of thermodynamics*. It is an energy equation which states that the change in internal energy is equal to the sum of the heat added to and the work done on the system. (Note in the above that δ and d both represent infinitesimally small quantities; however, d is a "perfect differential" and δ is not.)

Equation (4.10) is very fundamental; however, it is not in a practical form for use in aerodynamics, which speaks in terms of pressures, velocities, etc. To obtain more useful forms of the first law, we must first derive an expression for δw in terms of p and v (specific volume), as follows. Consider the system sketched in Figure 4.7. Let dA be an incremental surface area of the boundary. Assume that work ΔW is being done on the system by dA being pushed in a small distance s, as also shown in Figure 4.7. Since work is defined as force times distance, we have

$$\Delta W = (\text{force})(\text{distance})$$
$$\Delta W = (p\, dA)s \tag{4.11}$$

Now assume that many elemental surface areas of the type shown in Figure 4.7 are distributed over the total surface area A of the boundary. Also, assume that all the elemental surfaces are being simultaneously displaced a small distance s into the system. Then, the total work δw done on the unit mass of gas inside the system is the sum (integral) of each elemental surface over the whole boundary, that is, from Eq. (4.11),

$$\delta w = \int_A (p\, dA)s = \int_A ps\, dA \tag{4.12}$$

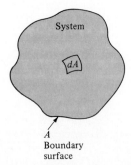

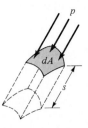

Figure 4.7 Work being done on the system by pressure.

Assume p is constant everywhere in the system (which, in thermodynamic terms, contributes to a state of thermodynamic equilibrium). Then, from Eq. (4.12),

$$\delta w = p \int_A s \, dA \qquad (4.13)$$

The integral $\int_A s \, dA$ has physical meaning. Geometrically, it is the change in volume of the unit mass of gas inside the system, created by the boundary surface being displaced inward. Let dv be the change in volume. Since the boundary is pushing in, the volume decreases (dv is a negative quantity) and work is done on the gas (hence δw is a positive quantity in our development). Thus

$$\int_A s \, dA \equiv -dv \qquad (4.14)$$

Substituting Eq. (4.14) into Eq. (4.13), we obtain

$$\boxed{\delta w = -p \, dv} \qquad (4.15)$$

Equation (4.15) gives the relation for work done strictly in terms of the thermodynamic variables p and v.

When Eq. (4.15) is substituted into Eq. (4.10), the first law becomes

$$\boxed{\delta q = de + p \, dv} \qquad (4.16)$$

Equation (4.16) is an alternate form of the first law of thermodynamics.

It is convenient to define a new quantity called *enthalpy h* as

$$h = e + pv = e + RT \qquad (4.17)$$

where $pv = RT$, assuming a perfect gas. Then, differentiating the definition, Eq. (4.17), we find

$$dh = de + p \, dv + v \, dp \qquad (4.18)$$

Substituting Eq. (4.18) into (4.16), we obtain

$$\delta q = de + p \, dv = (dh - p \, dv - v \, dp) + p \, dv$$

$$\boxed{\delta q = dh - v \, dp} \qquad (4.19)$$

Equation (4.19) is yet another alternate form of the first law.

Before we go further, remember that a substantial part of science and engineering is simply the language. In this section, we are presenting some of the language of thermodynamics essential to our future aerodynamic applications. We continue developing this language.

Figures 4.6 and 4.7 illustrate systems to which heat δq is added and on which work δw is done. At the same time, δq and δw may cause the pressure, temperature, and density of the system to change. The way (or means) by which changes of the thermodynamic variables (p, T, ρ, v) of a system take place is

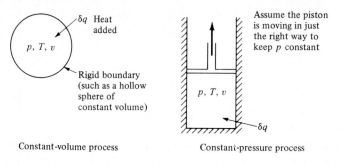

Constant-volume process   Constant-pressure process

Figure 4.8 Illustration of constant-volume and constant-pressure processes.

called a *process*. For example, a *constant-volume process* is illustrated at the left of Figure 4.8. Here, the system is a gas inside a rigid boundary, such as a hollow steel sphere, and therefore the volume of the system always remains constant. If an amount of heat δq is added to this system, p and T will change. Thus, by definition, such changes take place at constant volume; this is a constant-volume process. Another example is given at the right of Figure 4.8. Here, the system is a gas inside a cylinder-piston arrangement. Consider that heat δq is added to the system and at the same time assume the piston is moved in just exactly the right way to maintain a constant pressure inside the system. When δq is added to this system, T and v (hence ρ) will change. By definition, such changes take place at constant pressure; this is a constant-pressure process. There are many different kinds of processes treated in thermodynamics. The above are only two examples.

The last concept to be introduced in this section is that of specific heat. Consider a system to which a small amount of heat δq is added. The addition of δq will cause a small change in temperature dT of the system. By definition, specific heat is the heat added per unit change in temperature of the system. Let c denote specific heat. Thus,

$$c \equiv \frac{\delta q}{dT}$$

However, with this definition, c is multivalued. That is, for a fixed quantity δq, the resulting value of dT can be different, depending on the type of process in which δq is added. In turn, the value of c depends on the type of process. Therefore, we can in principle define more precisely a different specific heat for each type of process. We will be interested in only two types of specific heat, one at constant volume and the other at constant pressure, as follows.

If the heat δq is added at *constant volume* and it causes a change in temperature dT, the *specific heat at constant volume* c_v is defined as

$$c_v \equiv \left(\frac{\delta q}{dT} \right)_{\text{at constant volume}}$$

or
$$\delta q = c_v \, dT \text{ (constant volume)} \tag{4.20}$$

On the other hand, if δq is added at constant pressure and it causes a change in temperature dT (whose value is different from the dT above), the *specific heat at constant pressure* c_p is defined as

$$c_p \equiv \left(\frac{\delta q}{dT} \right)_{\text{at constant pressure}}$$

or
$$\delta q = c_p \, dT \text{ (constant pressure)} \qquad (4.21)$$

The above definitions of c_v and c_p, when combined with the first law, yield useful relations for internal energy e and enthalpy h as follows. First, consider a constant-volume process, where by definition $dv = 0$. Thus, from the alternate form of the first law, Eq. (4.16),

$$\delta q = de + p \, dv = de + 0 = de \qquad (4.22)$$

Substituting the definition of c_v, Eq. (4.20), into Eq. (4.22), we get

$$\boxed{de = c_v \, dT} \qquad (4.23)$$

Assuming that c_v is a constant, which is very reasonable for air at normal conditions, and letting $e = 0$ when $T = 0$, Eq. (4.23) may be integrated to

$$\boxed{e = c_v T} \qquad (4.24)$$

Next, consider a constant-pressure process, where by definition $dp = 0$. From the alternate form of the first law, Eq. (4.19),

$$\delta q = dh - v \, dp = dh - 0 = dh \qquad (4.25)$$

Substituting the definition of c_p, Eq. (4.21), into Eq. (4.25), we find

$$\boxed{dh = c_p \, dT} \qquad (4.26)$$

Again, assuming that c_p is constant and letting $h = 0$ at $T = 0$, we see that Eq. (4.26) yields

$$\boxed{h = c_p T} \qquad (4.27)$$

Equations (4.23) to (4.27) are very important relationships. They have been derived from the first law, into which the definitions of specific heat have been inserted. Look at them! They relate thermodynamic variables *only* (e to T and h to T); work and heat do not appear in these equations. In fact, Eqs. (4.23) to (4.27) are quite general. Even though we used examples of constant volume and constant pressure to obtain them, they hold *in general* as long as the gas is a perfect gas (no intermolecular forces). Hence, for *any* process,

$$de = c_v \, dT$$
$$dh = c_p \, dT$$
$$e = c_v T$$
$$h = c_p T$$

This generalization of Eqs. (4.23) to (4.27) to any process may not seem logical and may be hard to accept; nevertheless, it is valid, as can be shown by good thermodynamic arguments beyond the scope of this book. For the remainder of our discussions, we will make frequent use of these equations to relate internal energy and enthalpy to temperature.

4.6 ISENTROPIC FLOW

We are almost ready to return to our consideration of aerodynamics. However, there is one more concept we must introduce, a concept that bridges both thermodynamics and compressible aerodynamics, namely, that of *isentropic flow*.

First, consider three more definitions:

1. *Adiabatic process*—a process in which no heat is added or taken away, $\delta q = 0$
2. *Reversible process*—a process in which no frictional or other dissipative effects occur
3. *Isentropic process*—a process which is both adiabatic and reversible

Thus, an isentropic process is one in which there is neither heat exchange nor any effect due to friction. (The source of the word "isentropic" comes from another defined thermodynamic variable called *entropy*. The entropy is constant for an isentropic process. A discussion of entropy is not vital to our discussion here; therefore, no further elaboration will be given.)

Isentropic processes are very important in aerodynamics. For example, consider the flow of air over the airfoil shown in Figure 4.5. Imagine a fluid element moving along one of the streamlines. There is *no* heat being added or taken away from this fluid element; heat exchange mechanisms such as heating by a flame, cooling in a refrigerator, or intense radiation absorption are all ruled out by the nature of the physical problem we are considering. Thus, the flow of the fluid element along the streamline is *adiabatic*. At the same time, the shearing stress exerted on the surface of the fluid element due to friction is generally quite small and can be neglected (except very near the surface, as will be discussed later). Thus, the flow is also frictionless. [Recall that this same assumption was used in obtaining the momentum equation, Eq. (4.8).] Hence, the flow of the fluid element is both adiabatic and reversible (frictionless); i.e., the flow is *isentropic*. Other aerodynamic flows can also be treated as isentropic, e.g., the flows through wind-tunnel nozzles and rocket engines.

Note that, even though the flow is adiabatic, the temperature need not be constant. Indeed, the temperature of the fluid element can vary from point to point in an adiabatic, compressible flow. This is because the volume of the fluid element (of fixed mass) changes as it moves through regions of different density along the streamline; when the volume varies, work is done [Eq. (4.15)], hence the internal energy changes [Eq. (4.10)], and hence the temperature changes [Eq. (4.23)]. This argument holds for compressible flows, where the density is variable.

On the other hand, for incompressible flow, where ρ = const, the volume of the fluid element of fixed mass does not change as it moves along a streamline; hence no work is done and no change in temperature occurs. If the flow over the airfoil in Figure 4.5 were incompressible, the entire flow field would be at constant temperature. For this reason, temperature is not an important quantity for frictionless incompressible flow. Moreover, our present discussion of isentropic flows is relevant to *compressible* flows only, as explained below.

An isentropic process is more than just another definition. It provides us with several important relationships between the thermodynamic variables T, p, and ρ at two different points (say points 1 and 2 in Figure 4.5) along a given streamline. These relations are obtained as follows. Since the flow is isentropic (adiabatic and reversible), $\delta q = 0$. Thus, from Eq. (4.16),

$$\delta q = de + p \, dv = 0$$
$$- p \, dv = de \tag{4.28}$$

Substitute Eq. (4.23) into (4.28):

$$- p \, dv = c_v \, dT \tag{4.29}$$

In the same manner, using the fact that $\delta q = 0$ in Eq. (4.19), we also obtain

$$\delta q = dh - v \, dp = 0$$
$$v \, dp = dh \tag{4.30}$$

Substitute Eq. (4.26) into (4.30):

$$v \, dp = c_p \, dT \tag{4.31}$$

Divide Eq. (4.29) by (4.31):

$$\frac{-p \, dv}{v \, dp} = \frac{c_v}{c_p}$$

or

$$\frac{dp}{p} = -\left(\frac{c_p}{c_v}\right)\frac{dv}{v} \tag{4.32}$$

The ratio of specific heats c_p/c_v appears so frequently in compressible flow equations that it is given a symbol all its own, usually γ; $c_p/c_v \equiv \gamma$. For air at normal conditions, which apply to the applications treated in this book, both c_p and c_v are constants, and hence γ = const = 1.4 (for air). $c_p/c_v \equiv \gamma = 1.4$ (for air at normal conditions). Thus, Eq. (4.32) can be written as

$$\frac{dp}{p} = -\gamma\frac{dv}{v} \tag{4.33}$$

Referring to Figure 4.5, integrate Eq. (4.33) between points 1 and 2:

$$\int_{p_1}^{p_2} \frac{dp}{p} = -\gamma \int_{v_1}^{v_2} \frac{dv}{v}$$

$$\ln \frac{p_2}{p_1} = -\gamma \ln \frac{v_2}{v_1}$$

$$\frac{p_2}{p_1} = \left(\frac{v_2}{v_1}\right)^{-\gamma} \tag{4.34}$$

Since $v_1 = 1/\rho_1$ and $v_2 = 1/\rho_2$, Eq. (4.34) becomes

$$\frac{p_2}{p_1} = \left(\frac{\rho_2}{\rho_1}\right)^\gamma \qquad \text{isentropic flow} \tag{4.35}$$

From the equation of state, we have $\rho = p/(RT)$. Thus, Eq. (4.35) yields

$$\frac{p_2}{p_1} = \left(\frac{p_2}{RT_2}\frac{RT_1}{p_1}\right)^\gamma$$

$$\left(\frac{p_2}{p_1}\right)^{1-\gamma} = \left(\frac{T_1}{T_2}\right)^\gamma = \left(\frac{T_2}{T_1}\right)^{-\gamma}$$

or

$$\frac{p_2}{p_1} = \left(\frac{T_2}{T_1}\right)^{\gamma/(\gamma-1)} \qquad \text{isentropic flow} \tag{4.36}$$

Combining Eqs. (4.35) and (4.36), we obtain

$$\frac{p_2}{p_1} = \left(\frac{\rho_2}{\rho_1}\right)^\gamma = \left(\frac{T_2}{T_1}\right)^{\gamma/(\gamma-1)} \qquad \text{isentropic flow} \tag{4.37}$$

The relationships given in Eq. (4.37) are powerful. They provide important information for p, T, and ρ between two different points on a streamline in an isentropic flow. Moreover, if the streamlines all emanate from a uniform flow far upstream (far to the left in Figure 4.5), then Eq. (4.37) holds for any two points in the flow, not necessarily those on the same streamline.

Emphasis is again made that the isentropic flow relations, Eq. (4.37), are relevant to compressible flows only. By contrast, the assumption of incompressible flow (remember, incompressible flow is a myth, anyway) is not consistent with the same physics that went into the development of Eq. (4.37). To analyze incompressible flows, we need only the continuity equation [say, Eq. (4.3)] and the momentum equation [Bernoulli's equation, Eqs. (4.9a) and (4.9b)]. To analyze compressible flows, we need the continuity equation, Eq. (4.2), the momentum equation [Euler's equation, Eq. (4.8)], and another soon-to-be-derived relation called the energy equation. If the compressible flow is isentropic, then Eq. (4.37) can be used to replace either the momentum or the energy equation. Since Eq. (4.37) is a simpler, more useful algebraic relation than Euler's equation, Eq. (4.8), which is a differential equation, we will frequently use Eq. (4.37) in place of Eq. (4.8) for the analysis of compressible flows in this book.

As mentioned above, to complete the development of the fundamental relations for the analysis of compressible flow, we must now consider the energy equation.

Example 4.4 An airplane is flying at standard sea-level conditions. The temperature at a point on the wing is 250 K. What is the pressure at this point?

SOLUTION The air pressure and temperature, p_1 and T_1, far upstream of the wing correspond to standard sea level. Hence, $p_1 = 1.01 \times 10^5$ N/m² and $T_1 = 288.16$ K. Assume the flow is isentropic (hence compressible). Then, the relation between points 1 and 2 is obtained from Eq. (4.37):

$$\frac{p_2}{p_1} = \left(\frac{T_2}{T_1} \right)^{\gamma/(\gamma - 1)}$$

$$p_2 = p_1 \left(\frac{T_2}{T_1} \right)^{\gamma/(\gamma - 1)} = (1.01 \times 10^5) \left(\frac{250}{288.16} \right)^{1.4/(1.4 - 1)}$$

$$\boxed{p_2 = 6.14 \times 10^4 \text{ N/m}^2}$$

4.7 THE ENERGY EQUATION

Recall that our approach to the derivation of the fundamental equations for fluid flow is to state a fundamental principle and then to proceed to cast that principle in terms of the flow variables p, T, ρ, and V. Also recall that compressible flow, high-speed flow, and massive changes in energy also go hand in hand. Therefore, our last fundamental physical principle that we must take into account is as follows.

Physical principle: Energy can be neither created nor destroyed. It can only change in form.

In quantitative form, this principle is nothing more than the first law of thermodynamics, Eq. (4.10). To apply this law to fluid flow, consider again a fluid element moving along a streamline, as shown in Figure 4.4. Let us apply the first law of thermodynamics,

$$\delta q + \delta w = de$$

to this fluid element. Recall that an alternate form of the first law is Eq. (4.19),

$$\delta q = dh - v \, dp$$

Again, we consider an adiabatic flow, where $\delta q = 0$. Hence, from Eq. (4.19),

$$dh - v \, dp = 0 \tag{4.38}$$

Recalling Euler's equation, Eq. (4.8),

$$dp = -\rho V \, dV$$

we can combine Eqs. (4.38) and (4.8) to obtain

$$dh + v\rho V \, dV = 0 \tag{4.39}$$

However, $v = 1/\rho$, hence Eq. (4.39) becomes

$$dh + V \, dV = 0 \tag{4.40}$$

Integrating Eq. (4.40) between two points along the streamline, we obtain

$$\int_{h_1}^{h_2} dh + \int_{V_1}^{V_2} V\,dV = 0$$

$$(h_2 - h_1) + \left(\frac{V_2^2}{2} - \frac{V_1^2}{2}\right) = 0$$

$$h_1 + \frac{V_1^2}{2} = h_2 + \frac{V_2^2}{2}$$

$$h + \frac{V^2}{2} = \text{const}$$

(4.41)

Equation (4.41) is the energy equation for frictionless, adiabatic flow. It can be written in terms of T by using Eq. (4.27), $h = c_p T$. Hence, Eq. (4.41) becomes

$$c_p T_1 + \tfrac{1}{2} V_1^2 = c_p T_2 + \tfrac{1}{2} V_2^2$$

$$c_p T + \tfrac{1}{2} V^2 = \text{const}$$

(4.42)

Equation (4.42) relates the temperature and velocity at two different points along a streamline. Again, if all the streamlines emanate from a uniform flow far upstream, then Eq. (4.42) holds for any two points in the flow, not necessarily on the same streamline. Moreover, Eq. (4.42) is just as powerful and necessary for the analysis of compressible flow as is Eq. (4.37).

Example 4.5 A supersonic wind tunnel is sketched in Figure 4.22. The air temperature and pressure in the reservoir of the wind tunnel are $T_0 = 1000$ K and $p_0 = 10$ atm, respectively. The static temperatures at the throat and exit are $T^* = 833$ K and $T_e = 300$ K, respectively. The mass flow through the nozzle is 0.5 kg/s. For air, $c_p = 1008$ J/(kg)(K). Calculate:
  (a) Velocity at the throat, V^*
  (b) Velocity at the exit, V_e
  (c) Area of the throat, A^*
  (d) Area of the exit, A_e

SOLUTION Since the problem deals with temperatures and velocities, the energy equation seems useful.
  (a) From Eq. (4.42), written between the reservoir and the throat,

$$c_p T_0 + \tfrac{1}{2} V_0^2 = c_p T^* + \tfrac{1}{2} V^{*2}$$

However, in the reservoir, $V_0 \approx 0$. Hence,

$$V^* = \sqrt{2 c_p (T_0 - T^*)}$$

$$= \sqrt{2(1008)(1000 - 833)} = \boxed{580 \text{ m/s}}$$

(b) From Eq. (4.42) written between the reservoir and the exit,

$$c_p T_0 = c_p T_e + \tfrac{1}{2} V_e^2$$

$$V_e = \sqrt{2 c_p (T_0 - T_e)}$$

$$= \sqrt{2(1008)(1000 - 300)} = \boxed{1188 \text{ m/s}}$$

(c) The basic equation dealing with mass flow and area is the continuity equation, Eq. (4.2). Note that the velocities are certainly large enough that the flow is compressible, hence Eq. (4.2), rather than Eq. (4.3), is appropriate.

$$\dot{m} = \rho^* A^* V^*$$

or

$$A^* = \frac{\dot{m}}{\rho^* V^*}$$

In the above, $\dot{m}$ is given and V^* is known from part (a). However, ρ^* must be obtained before we can calculate A^* as desired. To obtain ρ^*, note that, from the equation of state,

$$\rho_0 = \frac{p_0}{RT_0} = \frac{(10)(1.01 \times 10^5)}{(287)(1000)} = 3.52 \text{ kg/m}^3$$

Assuming the nozzle flow is isentropic, which is a very good approximation for the real case, from Eq. (4.37), we get

$$\left(\frac{\rho^*}{\rho_0} \right) = \left(\frac{T^*}{T_0} \right)^{1/(\gamma - 1)}$$

$$\rho^* = \rho_0 \left(\frac{T^*}{T_0} \right)^{1/(\gamma - 1)} = (3.52) \left(\frac{833}{1000} \right)^{1/(1.4 - 1)} = 2.23 \text{ kg/m}^3$$

Thus

$$A^* = \frac{\dot{m}}{\rho^* V^*} = \frac{0.5}{(2.23)(580)} = \boxed{3.87 \times 10^{-4} \text{ m}^2 = 3.87 \text{ cm}^2}$$

(d) Finding A_e is similar to the above solution for A^*

$$\dot{m} = \rho_e A_e V_e$$

where, for isentropic flow,

$$\rho_e = \rho_0 \left(\frac{T_e}{T_0} \right)^{1/(\gamma - 1)} = (3.52) \left(\frac{300}{1000} \right)^{1/(1.4 - 1)} = 0.174 \text{ kg/m}^3$$

Thus

$$A_e = \frac{\dot{m}}{\rho_e V_e} = \frac{0.5}{(0.174)(1188)} = \boxed{24.2 \times 10^{-3} \text{ m}^2 = 2.42 \text{ cm}^2}$$

4.8 SUMMARY OF EQUATIONS

We have just finished applying some very basic physical principles to obtain equations for the analysis of flowing gases. The reader is cautioned not to be confused by the multiplicity of equations; they are useful, indeed necessary, tools to examine and solve various aerodynamic problems of interest. It is important for an engineer or scientist to look at such equations and see not just a mathematical relationship, but primarily a physical relationship. These equations talk! For example, Eq. (4.2) says that mass is conserved; Eq. (4.42) says that energy is conserved for an adiabatic, frictionless flow; etc. Never lose sight of the physical implications and limitations of these equations.

To help set these equations in your mind, here is a compact summary of our results so far:

1. For the steady incompressible flow of a frictionless fluid in a stream tube of varying area, p and V are the meaningful flow variables; ρ and T are constants throughout the flow. To solve for p and V, use

$$A_1 V_1 = A_2 V_2 \qquad \text{continuity}$$

$$p_1 + \tfrac{1}{2}\rho V_1^2 = p_2 + \tfrac{1}{2}\rho V_2^2 \qquad \text{Bernoulli's equation}$$

2. For steady isentropic (adiabatic and frictionless) compressible flow in a stream tube of varying area, p, ρ, T, and V are all variables. They are obtained from

$$\rho_1 A_1 V_1 = \rho_2 A_2 V_2 \qquad \text{continuity}$$

$$\frac{p_1}{p_2} = \left(\frac{\rho_1}{\rho_2}\right)^{\gamma} = \left(\frac{T_1}{T_2}\right)^{\gamma/(\gamma-1)} \qquad \text{isentropic relations}$$

$$c_p T_1 + \tfrac{1}{2} V_1^2 = c_p T_2 + \tfrac{1}{2} V_2^2 \qquad \text{energy}$$

$$p_1 = \rho_1 R T_1$$

$$p_2 = \rho_2 R T_2 \qquad \text{equation of state}$$

Let us now apply these relations to study some basic aerodynamic phenomena and problems.

4.9 THE SPEED OF SOUND

Sound waves travel through the air at a definite speed—the speed of sound. This is obvious from natural observation; a lightning bolt is observed in the distance, and the thunder is heard at some later instant. In many aerodynamic problems, the speed of sound plays a pivotal role. How do we calculate the speed of sound? What does it depend on: pressure, temperature, density, or some combination thereof? Why is it so important? Answers to these questions are discussed in this section.

First let us derive a formula to calculate the speed of sound. Consider a sound wave moving into a stagnant gas, as shown in Figure 4.9. This sound wave is created by some source, say a small firecracker in the corner of a room. The air in the room is motionless and has density ρ, pressure p, and temperature T. If you are standing in the middle of the room, the sound wave sweeps by you at velocity a in m/s, ft/s, etc. The sound wave itself is a thin region of disturbance

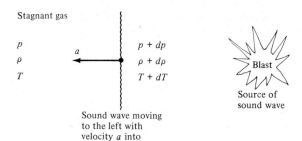

Figure 4.9 Model of a sound wave moving into a stagnant gas.

in the air, across which the pressure, temperature, and density change slightly. (The change in pressure is what activates your eardrum and allows you to hear the sound wave.) Imagine that you now hop on the sound wave and move with it. As you are sitting on the moving wave, look to the left in Figure 4.9, i.e., look in the direction in which the wave is moving. From your vantage point on the wave, the sound wave seems to stand still and the air in front of the wave appears to be coming at you with velocity a; i.e., you see the picture shown in Figure 4.10, where the sound wave is standing still and the air ahead of the wave is moving toward the wave with velocity a. Now, return to Figure 4.9 for a moment. Sitting on top of and riding with the moving wave, look to the right, i.e., look behind the wave. From your vantage point, the air appears to be moving away from you. This appearance is sketched in Figure 4.10, where the wave is standing still. Here, the air behind the motionless wave is moving to the right, away from the wave. However, in passing through the wave, the pressure, temperature, and density of the air are slightly changed by the amounts dp, dT, and $d\rho$. From our previous discussions, you would then expect the air speed q to change slightly, say by an amount da. Thus, the air behind the wave is moving away from the wave with velocity $a + da$, as shown in Figure 4.10. Figures 4.9 and 4.10 are completely analogous pictures; only their perspectives are different. Figure 4.9 is what you see by standing in the middle of the room and watching the wave go by; Figure 4.10 is what you see by riding on top of the wave and watching the air go by. Both pictures are equivalent. However, Figure 4.10 is easier to work with, so we will concentrate on it.

Let us apply our fundamental equations to the gas flow shown in Figure 4.10. Our objective will be to obtain an equation for a, where a is the speed of the

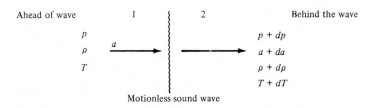

Figure 4.10 Model with the sound wave stationary.

sound wave, the speed of sound. Let points 1 and 2 be ahead of and behind the wave, respectively, as shown in Figure 4.10. Applying the continuity equation, Eq. (4.2), we find

$$\rho_1 A_1 V_1 = \rho_2 A_2 V_2$$

or

$$\rho A_1 a = (\rho + d\rho) A_2 (a + da) \qquad (4.43)$$

Here, A_1 and A_2 are the areas of a stream tube running through the wave. Just looking at the picture shown in Figure 4.10, there is no geometric reason why the stream tube should change area in passing through the wave. Indeed, it does not; the area of the stream tube is constant, hence $A = A_1 = A_2 = $ const. (This is an example of a type of flow called one-dimensional, or constant-area, flow.) Thus, Eq. (4.43) becomes

$$\rho a = (\rho + d\rho)(a + da)$$

or

$$\rho a = \rho a + a\, d\rho + \rho\, da + d\rho\, da \qquad (4.44)$$

The product of two small quantities $d\rho\, da$ is very small in comparison to the other terms in Eq. (4.44) and hence can be ignored. Thus, from Eq. (4.44),

$$a = -\rho \frac{da}{d\rho} \qquad (4.45)$$

Now, apply the momentum equation in the form of Euler's equation, Eq. (4.8):

$$dp = -\rho a\, da$$

or

$$da = -\frac{dp}{\rho a} \qquad (4.46)$$

Substitute Eq. (4.46) into (4.45):

$$a = \frac{\rho}{d\rho} \frac{dp}{\rho a}$$

or

$$a^2 = \frac{dp}{d\rho} \qquad (4.47)$$

On a physical basis, the flow through a sound wave involves no heat addition, and the effect of friction is negligible. Hence, the flow through a sound wave is isentropic. Thus, from Eq. (4.47), the speed of sound is given by

$$\boxed{a = \sqrt{\left(\frac{dp}{d\rho}\right)_{\text{isentropic}}}} \qquad (4.48)$$

Equation (4.48) is fundamental and important. However, it does not give us a straightforward formula for computing a number for a. We must proceed further.

For isentropic flow, Eq. (4.37) gives

$$\frac{p_2}{p_1} = \left(\frac{\rho_2}{\rho_1}\right)^{\gamma}$$

or

$$\frac{p_2}{\rho_2^{\gamma}} = \frac{p_1}{\rho_1^{\gamma}} = \text{const} = c \qquad (4.49)$$

Equation (4.49) says that the ratio p/ρ^γ is the same constant value at every point in an isentropic flow. Thus, we can write everywhere

$$\frac{p}{\rho^\gamma} = c \qquad (4.50)$$

Hence

$$\left(\frac{dp}{d\rho}\right)_{\text{isentropic}} = \frac{d}{d\rho} c\rho^\gamma = c\gamma\rho^{\gamma-1} \qquad (4.51)$$

Substituting for c in Eq. (4.51) the ratio of Eq. (4.50), we obtain

$$\left(\frac{dp}{d\rho}\right)_{\text{isentropic}} = \frac{p}{\rho^\gamma}\gamma\rho^{\gamma-1} = \frac{\gamma p}{\rho} \qquad (4.52)$$

Substitute Eq. (4.52) into (4.48):

$$a = \sqrt{\gamma\frac{p}{\rho}} \qquad (4.53)$$

However, for a perfect gas, p and ρ are related through the equation of state; $p = \rho RT$, hence $p/\rho = RT$. Substituting this result into Eq. (4.53) yields

$$a = \sqrt{\gamma RT} \qquad (4.54)$$

Equations (4.48), (4.53), and (4.54) are important results for the speed of sound; however, Eq. (4.54) is the most useful. It also demonstrates a fundamental result, that *the speed of sound in a perfect gas depends only on the temperature of the gas*. This simple result may appear surprising at first. However, it is to be expected on a physical basis, as follows. The propagation of a sound wave through a gas takes place via molecular collisions. For example, consider again a small firecracker in the corner of the room. When the firecracker is set off, some of its energy is transferred to the neighboring gas molecules in the air, thus increasing their kinetic energy. In turn, these energetic gas molecules are moving randomly about, colliding with some of their neighboring molecules and transferring some of their extra energy to these new molecules. Thus, the energy of a sound wave is transmitted through the air by molecules which collide with each other. Each molecule is moving at a different velocity, but summed over a large number of molecules, a mean or average molecular velocity can be defined. Therefore, looking at the collection of molecules as a whole, we see that the sound energy released by the firecrackers will be transferred through the air at something approximating this mean molecular velocity. Recall from Chap. 2 that temperature is a measure of the mean molecular kinetic energy, hence of the mean molecular velocity; then temperature should also be a measure of the speed of a sound wave transmitted by molecular collisions. Eq. (4.54) proves this to be a fact.

For example, consider air at standard sea-level temperature $T_s = 288.16$ K. From Eq. (4.54), the speed of sound is $a = \sqrt{\gamma RT} = \sqrt{1.4(287)(288.16)} = 340.3$

m/s. From the results of the kinetic theory of gases, the mean molecular velocity can be obtained as $\bar{V} = \sqrt{(8/\pi)RT} = \sqrt{(8/\pi)287(288.16)} = 458.9$ m/s. Thus, the speed of sound is of the same order of magnitude as the mean molecular velocity and is smaller by about 26 percent.

Emphasis is made that the speed of sound is a point property of the flow, in the same vein as T is a point property as described in Chap. 2. It is also a thermodynamic property of the gas, defined by Eqs. (4.48) to (4.54). In general, the value of the speed of sound varies from point to point in the flow.

The speed of sound leads to another, vital definition for high-speed gas flows, namely, the *Mach number*. Consider a point B in a flow field. The flow velocity at B is V, and the speed of sound is a. By definition, the Mach number M at point B is the flow velocity divided by the speed of sound:

$$\boxed{M = \frac{V}{a}} \tag{4.55}$$

We will find that M is one of the most powerful quantities in aerodynamics. We can immediately use it to define three different regimes of aerodynamic flows:

1. If $M < 1$, the flow is *subsonic*.
2. If $M = 1$, the flow is *sonic*.
3. If $M > 1$, the flow is *supersonic*.

Each of these regimes is characterized by its own special phenomena, as will be discussed in subsequent sections. In addition, two other specialized aerodynamic regimes are commonly defined, namely, *transonic* flow, where M generally ranges from slightly less than to slightly greater than 1 (for example, $0.8 \leq M \leq 1.2$), and *hypersonic* flow, where generally $M > 5$. The definitions of subsonic, sonic, and supersonic flows in terms of M as given above are precise; the definitions of transonic and hypersonic flows in terms of M are a bit more imprecise and really refer to sets of specific aerodynamic phenomena, rather than to just the value of M. This distinction will be made more clear in subsequent sections.

Example 4.6 A jet transport is flying at a standard altitude of 30,000 ft with a velocity of 550 mi/h. What is its Mach number?

SOLUTION From the standard atmosphere table, Appendix B, at 30,000 ft, $T_\infty = 411.86°R$. Hence, from Eq. (4.54),

$$a_\infty = \sqrt{\gamma RT} = \sqrt{1.4(1716)(411.86)} = 995 \text{ ft/s}$$

The airplane velocity is $V_\infty = 550$ mi/h; however, in consistent units, remembering that 88 ft/s = 60 mi/h, we find that

$$V_\infty = 550 (88/60) = 807 \text{ ft/s}$$

From Eq. (4.55)

$$M_\infty = \frac{V_\infty}{a_\infty} = \frac{807}{995} = \boxed{0.811}$$

Example 4.7 In the nozzle flow described in Example 4.5, calculate the Mach number of the flow at the throat, M^*, and at the exit, M_e.

SOLUTION From Example 4.5, at the throat, $V^* = 580$ m/s and $T^* = 833$ K. Hence, from Eq. (4.54),

$$a^* = \sqrt{\gamma R T^*} = \sqrt{1.4(287)(833)} = 580 \text{ m/s}$$

From Eq. (4.55)

$$M^* = \frac{V^*}{a^*} = \frac{580}{580} = \boxed{1}$$

Note: The flow is sonic at the throat. We will soon prove that the Mach number at the throat is always sonic in supersonic nozzle flows (except in special, nonequilibrium, high-temperature flows, which are beyond the scope of this book).

Also from Example 4.5, at the exit, $V_e = 1188$ m/s and $T_e = 300$ K. Hence,

$$a_e = \sqrt{\gamma R T_e} = \sqrt{1.4(287)(300)} = 347 \text{ m/s}$$

$$M_e = \frac{V_e}{a_e} = \frac{1188}{347} = \boxed{3.42}$$

4.10 LOW-SPEED SUBSONIC WIND TUNNELS

Throughout the remainder of this book the aerodynamic fundamentals and tools (equations) developed in previous sections will be applied to specific problems of interest. The first of these will be a discussion of low-speed subsonic wind tunnels.

In the first place, what are wind tunnels, any kind of wind tunnels? In the most basic sense, they are ground-based experimental facilities designed to produce flows of air (or sometimes other gases) which simulate natural flows occurring outside the laboratory. For most aerospace engineering applications, wind tunnels are designed to simulate flows encountered in the flight of airplanes, missiles, or space vehicles. Since these flows have ranged from the 27 mi/h speed of the early Wright Flyer to the 25,000 mi/h reentry velocity of the Apollo lunar spacecraft, obviously many different types of wind tunnels, from low subsonic to hypersonic, are necessary for the laboratory simulation of actual flight conditions. However, referring again to Figure 1.27, flow velocities of 300 mi/h or less were the flight regime of interest until about 1940. Hence, during the first four decades of human flight, airplanes were tested and developed in wind tunnels designed to simulate low-speed subsonic flight. Such tunnels are still in use today but are complemented by transonic, supersonic, and hypersonic wind tunnels as well.

The essence of a typical low-speed subsonic wind tunnel is sketched in Figure 4.11. The airflow with pressure p_1 enters the nozzle at a low velocity V_1, where the area is A_1. The nozzle converges to a smaller area A_2 at the test section. Since we are dealing with low-speed flows, where M is generally less than 0.3, the flow will be assumed to be incompressible. Hence, Eq. (4.3) dictates that the flow velocity increases as the air flows through the convergent nozzle. The velocity in the test section is then, from Eq. (4.3),

$$V_2 = \frac{A_1}{A_2} V_1 \tag{4.56}$$

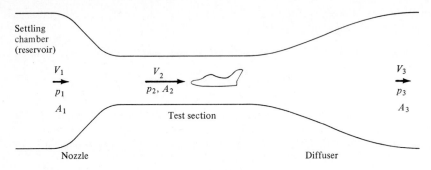

Figure 4.11 Simple schematic of a subsonic wind tunnel.

After flowing over an aerodynamic model (which may be a model of a complete airplane or part of an airplane, such as a wing, tail, or engine nacelle), the air passes into a diverging duct called a diffuser, where the area increases and velocity decreases to A_3 and V_3, respectively. Again, from continuity,

$$V_3 = \frac{A_2}{A_3} V_2$$

The pressure at various locations in the wind tunnel is related to the velocity, through Bernoulli's equation, Eq. (4.9a), for incompressible flow.

$$p_1 + \tfrac{1}{2}\rho V_1^2 = p_2 + \tfrac{1}{2}\rho V_2^2 = p_3 + \tfrac{1}{2}\rho V_3^2 \tag{4.57}$$

From Eq. (4.57), as V increases, p decreases; hence $p_2 < p_1$, that is, the test section pressure is smaller than the reservoir pressure upstream of the nozzle. In many subsonic wind tunnels, all or part of the test section is open, or vented, to the surrounding air in the laboratory. In such cases, the outside air pressure is communicated directly to the flow in the test section, and $p_2 = 1$ atm. Downstream of the test section, in the diverging area diffuser, the pressure increases as velocity decreases. Hence, $p_3 > p_2$. If $A_3 = A_1$, then from Eq. (4.56), $V_3 = V_1$; and from Eq. (4.57), $p_3 = p_1$. (Note: In actual wind tunnels, the aerodynamic drag created by the flow over the model in the test section causes a loss of momentum not included in the derivation of Bernoulli's equation; hence in reality, p_3 is slightly less than p_1, because of such losses.)

In practical operation of this type of wind tunnel, the test section velocity is governed by the pressure difference $p_1 - p_2$ and the area ratio of the nozzle, A_2/A_1, as follows. From Eq. (4.57),

$$V_2^2 = \frac{2}{\rho}(p_1 - p_2) + V_1^2 \tag{4.58}$$

From Eq. (4.56), $V_1 = (A_2/A_1)V_2$. Substituting this into the right-hand side of Eq. (4.58), we obtain

$$V_2^2 = \frac{2}{\rho}(p_1 - p_2) + \left(\frac{A_2}{A_1}\right)^2 V_2^2 \tag{4.59}$$

Solving Eq. (4.59) for V_2 yields

$$V_2 = \sqrt{\frac{2(p_1 - p_2)}{\rho\left[1 - (A_2/A_1)^2\right]}}$$

(4.60)

The area ratio A_2/A_1 is a fixed quantity for a wind tunnel of given design. Hence, the "control knob" of the wind tunnel basically controls $p_1 - p_2$, which allows the wind-tunnel operator to control the value of test section velocity V_2 via Eq. (4.60).

In subsonic wind tunnels, the most convenient method of measuring the pressure difference $p_1 - p_2$, hence of measuring V_2 via Eq. (4.60), is by means of a *manometer*. A basic type of manometer is the U tube shown in Figure 4.12. Here, the left side of the tube is connected to a pressure p_1, the right side of the tube is connected to a pressure p_2, and the difference Δh in the heights of a fluid in both sides of the U tube is a measurement of the pressure difference $p_2 - p_1$. This can easily be demonstrated by considering the force balance on the liquid in the tube at the two cross sections cut by plane B-B, shown in Figure 4.12. Plane B-B is drawn tangent to the top of the column of fluid on the left. If A is the cross-sectional area of the tube, then $p_1 A$ is the force exerted on the left column of fluid. The force on the right column at plane B-B is the sum of the weight of the fluid above plane B-B and the force due to the pressure $p_2 A$. The volume of the fluid in the right column above B-B is $A\Delta h$. The specific weight (weight per unit volume) of the fluid is $w = \rho_l g$, where ρ_l is the density of the fluid and g is the acceleration of gravity. Hence, the total weight of the column of fluid above B-B is the specific weight times the volume, that is, $wA\Delta h$. The total force on the right-hand cross section at plane B-B is then $p_2 A + wA\Delta h$. Since the fluid is stationary in the tube, the forces on the left- and right-hand cross sections must balance, i.e., they are the same. Hence,

$$p_1 A = p_2 A + wA\Delta h$$

or

$$p_1 - p_2 = w\Delta h$$

(4.61)

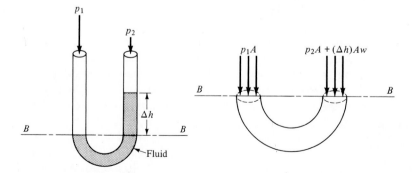

Figure 4.12 Force diagram for a manometer.

If the left-hand side of the U tube manometer were connected to the reservoir in a subsonic tunnel (point 1 in Figure 4.11) and the right-hand side were connected to the test section (point 2), then Δh of the U tube would directly measure the velocity of the airflow in the test section via Eqs. (4.61) and (4.60).

Example 4.8 In a low-speed subsonic wind tunnel, one side of a mercury manometer is connected to the settling chamber (reservoir) and the other side is connected to the test section. The contraction ratio of the nozzle, A_2/A_1, equals 1/15. The reservoir pressure and temperature are $p_1 = 1.1$ atm and $T_1 = 300$ K, respectively. When the tunnel is running, the height difference between the two columns of mercury is 10 cm. The density of liquid mercury is 1.36×10^4 kg/m³. Calculate the airflow velocity in the test section, V_2.

SOLUTION

$$\Delta h = 10 \text{ cm} = 0.1 \text{ m}$$

$$w \text{ (for mercury)} = \rho_l g = (1.36 \times 10^4 \text{ kg/m}^3)(9.8 \text{ m/s}^2)$$

$$w = 1.33 \times 10^5 \text{ N/m}^3$$

From Eq. (4.61)

$$p_1 - p_2 = w\Delta h = (1.33 \times 10^5 \text{ N/m}^3)(0.1 \text{ m}) = 1.33 \times 10^4 \text{ N/m}^2$$

To find the velocity V_2, use Eq. (4.60). However, in Eq. (4.60) we need a value of density ρ. This can be found from the reservoir conditions by using the equation of state. (Remember: 1 atm $= 1.01 \times 10^5$ N/m².)

$$\rho_1 = \frac{p_1}{RT_1} = \frac{1.1(1.01 \times 10^5)}{287(300)} = 1.29 \text{ kg/m}^3$$

Since we are dealing with a low-speed subsonic flow, assume $\rho_1 = \rho = \text{const}$. Hence, from Eq. (4.60)

$$V_2 = \sqrt{\frac{2(p_1 - p_2)}{\rho\left[1 - (A_2/A_1)^2\right]}} = \sqrt{\frac{2(1.33 \times 10^4)}{(1.29)\left[1 - \left(\frac{1}{15}\right)^2\right]}} = \boxed{144 \text{ m/s}}$$

Note: This answer corresponds to approximately a Mach number of 0.4 in the test section, one slightly above the value of 0.3 that bounds incompressible flow. Hence, our assumption of $\rho = \text{const}$ in this example is inaccurate by about 8%.

4.11 MEASUREMENT OF AIRSPEED

In the previous section, we demonstrated that the airflow velocity in the test section of a low-speed wind tunnel (assuming incompressible flow) can be obtained by measuring $p_1 - p_2$. However, the previous analysis implicitly assumes that the flow properties are reasonably constant over any given cross section of the flow in the tunnel (so-called quasi-one-dimensional flow). If the flow is not constant over a given cross section, for example, if the flow velocity in the middle of the test section is higher than near the walls, then V_2 obtained from the preceding section is only a mean value of the test section velocity. For this reason and for many other aerodynamic applications, it is important to obtain a *point* measurement of velocity at a given spatial location in the flow. This measurement can be made by an instrument called a *Pitot-static tube*, as described below.

First, we must add to our inventory of aerodynamic definitions. We have been glibly talking about the pressures at points in flows, such as points 1 and 2 in Figure 4.5. However, these pressures are of a special type, called *static*. Static pressure at a given point is the pressure we would feel if we were moving along with the flow at that point. It is the ramification of gas molecules moving about with random motion and transferring their momentum to or across surfaces, as discussed in Chap. 2. If we look more closely at the molecules in a flowing gas, we see that they have a purely random motion superimposed on a directed motion due to the velocity of the flow. Static pressure is a consequence of just the purely random motion of the molecules. When an engineer or scientist uses the word "pressure," it always means static pressure unless otherwise identified, and we will continue such practice here. In all of our previous discussions so far, the pressures have been static pressures.

A second type of pressure is commonly utilized in aerodynamics, namely, *total* pressure. To define and understand total pressure, consider again a fluid element moving along a streamline, as shown in Figure 4.4. The pressure of the gas in this fluid element is the static pressure. However, now imagine that we grab hold of this fluid element and slow it down to zero velocity. Moreover, imagine that we do this isentropically. Intuitively, the thermodynamic properties p, T, and ρ of the fluid element would change as we bring the element to rest; they would follow the conservation laws we have discussed previously in this chapter. Indeed, as the fluid element is isentropically brought to rest, p, T, and ρ would all increase above their original values when the element was moving freely along the streamline. The values of p, T, and ρ of the fluid element after it has been brought to rest are called *total* values, i.e., total pressure p_0, total temperature T_0, etc. Thus, we are led to the following precise definition:

Total pressure at a given point in a flow is that pressure that would exist if the flow were slowed down isentropically to zero velocity.

There is a perspective to be gained here. Total pressure p_0 is a property of the gas flow at a given point. It is something that is associated with the flow itself. The process of isentropically bringing the fluid element to rest is just an imaginary mental process we use to define the total pressure. It does not mean that we actually have to do it in practice. In other words, considering again the flow sketched in Figure 4.5, there are *two* pressures we can consider at points 1, 2, etc., associated with each point of the flow: a static pressure p and a total pressure p_0, where $p_0 > p$.

For the special case of a gas that is not moving, i.e., the fluid element has no velocity in the first place, then static and total pressures are synonymous: $p_0 = p$. This is the case in common situations such as the stagnant air in the room, gas confined in a cylinder, etc.

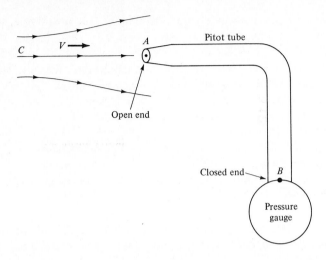

Figure 4.13 Sketch of a Pitot tube.

There is an aerodynamic instrument that actually measures the total pressure at a point in the flow, namely, a *Pitot tube*. A basic sketch of a Pitot tube is shown in Figure 4.13. It consists of a tube placed parallel to the flow and open to the flow at one end (point A). The other end of the tube (point B) is closed. Now imagine that the flow is first started. Gas will pile up inside the tube. After a few moments, there will be no motion inside the tube because the gas has nowhere to go—the gas will stagnate once steady-state conditions have been reached. In fact, the gas will be stagnant everywhere inside the tube, including at point A. As a result, the flow field sees the open end of the Pitot tube (point A) as an obstruction, and a fluid element moving along the streamline, labeled C, has no choice but to stop when it arrives at point A. Since no heat has been exchanged, and friction is negligible, this process will be isentropic, i.e., a fluid element moving along streamline C will be isentropically brought to rest at point A by the very presence of the Pitot tube. Therefore, the pressure at point A is, truly speaking, the total pressure p_0. This pressure will be transmitted throughout the Pitot tube, and if a pressure gauge is placed at point B, it will in actuality measure the *total pressure* of the flow. In this fashion, a Pitot tube is an instrument that measures the total pressure of a flow.

By definition, any point of a flow where $V = 0$ is called a *stagnation point*. In Figure 4.13, point A is a stagnation point.

Consider the arrangement shown in Figure 4.14. Here we have a uniform flow with velocity V_1 moving over a flat surface parallel to the flow. There is a small hole in the surface at point A called a *static pressure orifice*. Since the surface is parallel to the flow, only the random motion of the gas molecules will be felt by the surface itself. In other words, the surface pressure is indeed the static pressure p. This will be the pressure at the orifice at point A. On the other hand, the Pitot tube at point B in Figure 4.14 will feel the total pressure p_0, as discussed above. If the static pressure orifice at point A and the Pitot tube at point B are

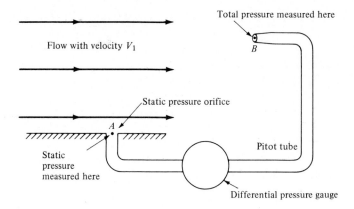

Total pressure measured here

Flow with velocity V_1

B

Static pressure orifice

A

Pitot tube

Static pressure measured here

Differential pressure gauge

Figure 4.14 Schematic of a Pitot-static measurement.

connected across a pressure gauge, as shown in Figure 4.14, the gauge will measure the difference between total and static pressure, $p_0 - p$.

Now we arrive at the main thrust of this section. The pressure difference $p_0 - p$, as measured in Figure 4.14, gives a measure of the flow velocity V_1. A combination of a total pressure measurement and a static pressure measurement allows us to measure the velocity at a given point in a flow. These two measurements can be combined in the same instrument, a *Pitot-static probe*, as illustrated in Figure 4.15. A Pitot-static probe measures p_0 at the nose of the probe and p at a point on the probe surface downstream of the nose. The pressure difference $p_0 - p$ yields the velocity V_1, but the quantitative formulation differs depending on whether the flow is low-speed (incompressible), high-speed subsonic, or supersonic.

A Incompressible Flow

Consider again the sketch shown in Figure 4.14. At point A, the pressure is p and the velocity is V_1. At point B, the pressure is p_0 and the velocity is zero. Applying Bernoulli's equation, Eq. (4.9*a*), at points A and B, we obtain

$$\underset{\substack{\text{Static} \\ \text{pressure}}}{p} + \underset{\substack{\text{Dynamic} \\ \text{pressure}}}{\tfrac{1}{2}\rho V_1^{\,2}} = \underset{\substack{\text{Total} \\ \text{pressure}}}{p_0} \qquad (4.62)$$

Total pressure felt here

Static pressure felt here

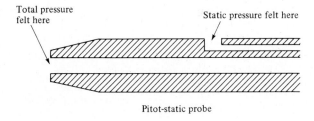

Pitot-static probe

Figure 4.15   Schematic of a Pitot-static probe.

In Eq. (4.62), for *dynamic pressure q* we have the definition

$$q \equiv \tfrac{1}{2}\rho V^2 \qquad (4.63)$$

which is frequently employed in aerodynamics; the grouping $\tfrac{1}{2}\rho V^2$ is termed the dynamic pressure for flows of all types, incompressible to hypersonic. From Eq. (4.62),

$$\boxed{p_0 = p + q} \qquad (4.64)$$

This relation holds for incompressible flow only. The total pressure equals the sum of the static plus the dynamic pressure. Also from Eq. 4.62,

$$\boxed{V_1 = \sqrt{2\frac{p_0 - p}{\rho}}} \qquad (4.65)$$

Equation (4.65) is the desired result; it allows the calculation of flow velocity from a measurement of $p_0 - p$, obtained from a Pitot-static tube. Again, we emphasize that Eq. (4.65) holds only for incompressible flow.

A Pitot tube can be used to measure the flow velocity at various points in the test section of a low-speed wind tunnel, as shown in Figure 4.16. The total pressure at point B is obtained by the Pitot probe, and the static pressure, also at point B, is obtained from a static pressure orifice located at point A on the wall of the closed test section, assuming that the static pressure is constant throughout the test section. This assumption of constant static pressure is fairly good for subsonic wind-tunnel test sections and is commonly made. If the test section is

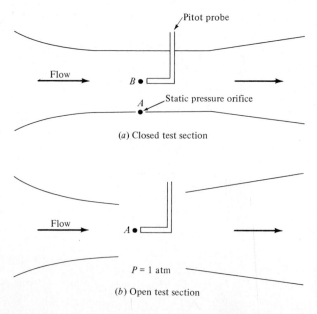

(a) Closed test section

(b) Open test section

Figure 4.16 Pressure measurements in open and closed test sections of subsonic wind tunnels.

open to the room, as also sketched in Figure 4.16, then the static pressure at all points in the test section is $p = 1$ atm. In either case, the velocity at point A is calculated from Eq. (4.65). The density ρ in Eq. (4.65) is a constant (incompressible flow). Its value can be obtained by measurements of p and T somewhere in the tunnel, using the equation of state to calculate $\rho = p/RT$. These measurements are usually made in the reservoir upstream of the nozzle.

Either a Pitot tube or a Pitot-static tube can be used to measure the airspeed of airplanes. Such tubes can be seen extending from airplane wingtips, with the tube oriented in the flight direction, as shown in Figure 4.17. If a Pitot tube is used, then the ambient static pressure in the atmosphere around the airplane is obtained from a static pressure orifice placed strategically on the airplane surface. Its location is placed where the surface pressure is nearly the same as the pressure of the surrounding atmosphere. Such a location is found by experience. It is generally on the fuselage somewhere between the nose and the wing. The values of p_0 obtained from the wingtip Pitot probe and p obtained from the static pressure orifice on the surface allow the calculation of the airplane's speed through the air using Eq. (4.65), *as long as the airplane's velocity is low enough to justify the assumption of incompressible flow*, i.e., for velocities less than 300 ft/s. In actual practice, the measurements of p_0 and p are joined across a differential pressure gauge which is calibrated in terms of airspeed, using Eq. (4.65). This airspeed indicator is a dial in the cockpit, with units of velocity, say miles per hour, on the dial. However, in determining the calibration, i.e., in determining what values of miles per hour go along with given values of $p_0 - p$, the engineer must decide what value of ρ to use in Eq. (4.65). If ρ is the true value, somehow measured in the actual air around the airplane, then Eq. (4.65) gives the *true airspeed* of the airplane:

$$V_{\text{true}} = \sqrt{\frac{2(p_0 - p)}{\rho}} \qquad (4.66)$$

However, the measurement of atmospheric air density directly at the airplane's location is difficult. Therefore, for practical reasons, the airspeed indicators on low-speed airplanes are calibrated by using the standard sea-level value of ρ_s in

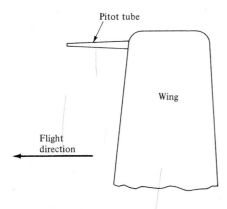

Pitot tube

Wing

Flight direction

Figure 4.17 Sketch of wing-mounted Pitot probe.

Eq. (4.65). This gives a value of velocity called the *equivalent airspeed*:

$$V_e = \sqrt{\frac{2(p_0 - p)}{\rho_s}} \tag{4.67}$$

The equivalent airspeed V_e differs slightly from V_{true}, the difference being the factor $(\rho/\rho_s)^{1/2}$. At altitudes near sea level, this difference is small.

> **Example 4.9** The altimeter on a low-speed Cessna 150 private aircraft reads 5000 ft. By an independent measurement, the outside air temperature is 505°R If a Pitot tube mounted on the wingtip measures a pressure of 1818 lb/ft², what is the true velocity of the airplane? What is the equivalent airspeed?
>
> SOLUTION An altimeter measures the pressure altitude (see discussion in Chap. 3). From the standard atmosphere table, Appendix B, at 5000 ft, $p = 1761$ lb/ft². Also, the Pitot tube measures total pressure; hence
>
> $$p_0 - p = 1818 - 1761 = 57 \text{ lb/ft}^2$$
>
> The true airspeed can be obtained from Eq. (4.66); however, we need ρ, which is obtained from the equation of state. For the outside, ambient air,
>
> $$\rho = \frac{p}{RT} = \frac{1761}{1716(505)} = 2.03 \times 10^{-3} \text{ slug/ft}^3$$
>
> From Eq. (4.66),
>
> $$V_{true} = \sqrt{\frac{2(p_0 - p)}{\rho}} = \sqrt{\frac{2(57)}{2.03 \times 10^{-3}}} = \boxed{237 \text{ ft/s}}$$
>
> Note: Since 88 ft/s = 60 mi/h, $V_{true} = 237(60/88) = 162$ mi/h.
>   The equivalent airspeed (that which would be read on the airspeed indicator in the cockpit) is obtained from Eq. (4.67), where $\rho_s = 0.002377$ slug/ft³ (the standard sea-level value). Hence, from Eq. (4.67),
>
> $$V_e = \sqrt{\frac{2(p_0 - p)}{\rho_s}} = \sqrt{\frac{2(57)}{2.377 \times 10^{-3}}} = \boxed{219 \text{ ft/s}}$$
>
> Note that there exists a 7.6 percent difference between V_{true} and V_e.

B Subsonic Compressible Flow

The results of the previous section are valid for airflows where $M < 0.3$, that is, where the flow can be reasonably assumed to be incompressible. This is the flight regime of, for example, small, piston engine private aircraft. For higher-speed flows, but where the Mach number is still less than 1 (high-speed subsonic flows), other equations must be used. This is the flight regime of commercial jet transports such as the Boeing 747 and the McDonnel-Douglas DC-10 and of many military aircraft. For these cases, compressibility must be taken into account, as follows.

Consider the definition of enthalpy $h = e + pv$. Since $h = c_p T$ and $e = c_v T$, then $c_p T = c_v T + RT$, or

$$c_p - c_v = R \tag{4.68}$$

Divide Eq. (4.68) by c_p:

$$1 - \frac{1}{c_p/c_v} = \frac{R}{c_p}$$

$$1 - \frac{1}{\gamma} = \frac{\gamma - 1}{\gamma} = \frac{R}{c_p}$$

or

$$\boxed{c_p = \frac{\gamma R}{\gamma - 1}}$$

(4.69)

Equation (4.69) holds for a perfect gas with constant specific heats. It is a necessary thermodynamic relation for use in the energy equation, as follows.

Consider again a Pitot tube in a flow, as shown in Figures 4.13 and 4.14. Assume the flow velocity V_1 is high enough that compressibility must be taken into account. As usual, the flow is isentropically compressed to zero velocity at the stagnation point on the nose of the probe. The values of the stagnation, or total, pressure and temperature at this point are p_0 and T_0, respectively. From the energy equation, Eq. (4.42), written between a point in the freestream flow where the temperature and velocity are T_1 and V_1, respectively, and the stagnation point, where the velocity is zero and the temperature is T_0,

$$c_p T_1 + \tfrac{1}{2} V_1^2 = c_p T_0$$

or

$$\frac{T_0}{T_1} = 1 + \frac{V_1^2}{2 c_p T_1}$$

(4.70)

Substitute Eq. (4.69) for c_p in Eq. (4.70):

$$\frac{T_0}{T_1} = 1 + \frac{V_1^2}{2[\gamma R/(\gamma - 1)] T_1} = 1 + \frac{\gamma - 1}{2} \frac{V_1^2}{\gamma R T_1}$$

(4.71)

However, from Eq. (4.54) for the speed of sound,

$$a_1^2 = \gamma R T_1$$

Thus, Eq. (4.71) becomes

$$\frac{T_0}{T_1} = 1 + \frac{\gamma - 1}{2} \frac{V_1^2}{a_1^2}$$

(4.72)

Since the Mach number $M_1 = V_1/a_1$, Eq. (4.72) becomes

$$\boxed{\frac{T_0}{T_1} = 1 + \frac{\gamma - 1}{2} M_1^2}$$

(4.73)

Since the gas is *isentropically* compressed at the nose of the Pitot probe in Figures 4.13 and 4.14, Eq. (4.37) holds between the freestream and the stagnation point.

That is, $p_0/p_1 = (\rho_0/\rho_1)^\gamma = (T_0/T_1)^{\gamma/(\gamma-1)}$. Therefore, from Eq. (4.73), we obtain

$$\frac{p_0}{p_1} = \left(1 + \frac{\gamma - 1}{2}M_1^2\right)^{\gamma/(\gamma-1)} \tag{4.74}$$

$$\frac{\rho_0}{\rho_1} = \left(1 + \frac{\gamma - 1}{2}M_1^2\right)^{1/(\gamma-1)} \tag{4.75}$$

Equations (4.73) to (4.75) are fundamental and important relations for compressible, isentropic flow. They apply to many other practical problems in addition to the Pitot tube. It should be noted that Eq. (4.73) holds for adiabatic flow, whereas Eqs. (4.74) and (4.75) contain the additional assumption of friction-less (hence isentropic) flow. Also, from a slightly different perspective, Eqs. (4.73) to (4.75) determine the total temperature, density, and pressure, T_0, ρ_0, and p_0, at any point in the flow where the static temperature, density, and pressure are T_1, ρ_1, and p_1 and where the Mach number is M_1. In other words, reflecting the earlier discussion of the definition of total conditions, Eqs. (4.73) to (4.75) give the values of p_0, T_0, and ρ_0 that are associated with a point in the flow where the pressure, temperature, density, and Mach number are p_1, T_1, ρ_1, and M_1. These equations also demonstrate the powerful influence of Mach number in aerodynamic flow calculations. It is very important to note that the ratios T_0/T_1, p_0/p_1, and ρ_0/ρ_1 are functions of M_1 only (assuming γ is known; $\gamma = 1.4$ for normal air).

Returning to our objective of measuring airspeed, and solving Eq. (4.74) for M_1, we obtain

$$M_1^2 = \frac{2}{\gamma - 1}\left[\left(\frac{p_0}{p_1}\right)^{(\gamma-1)/\gamma} - 1\right] \tag{4.76}$$

Hence, for subsonic compressible flow, the ratio of total to static pressure, p_0/p_1, is a direct measure of Mach number. Thus, individual measurements of p_0 and p_1 in conjunction with Eq. (4.76) can be used to calibrate an instrument in the cockpit of an airplane called a *Mach meter*, where the dial reads directly in terms of the flight Mach number of the airplane.

To obtain the actual flight velocity, recall that $M_1 = V_1/a_1$, hence Eq. (4.76) becomes

$$V_1^2 = \frac{2a_1^2}{\gamma - 1}\left[\left(\frac{p_0}{p_1}\right)^{(\gamma-1)/\gamma} - 1\right] \tag{4.77a}$$

Equation (4.77) can be rearranged algebraically as

$$V_1^2 = \frac{2a_1^2}{\gamma - 1}\left[\left(\frac{p_0 - p_1}{p_1} + 1\right)^{(\gamma-1)/\gamma} - 1\right] \tag{4.77b}$$

Equations (4.77a) and (4.77b) give the *true* airspeed of the airplane. However, they require a knowledge of a_1, hence T_1. The static temperature in the air surrounding the airplane is difficult to measure. Hence, all high-speed (but subsonic) airspeed indicators are calibrated from Eq. (4.77b), assuming that a_1 is equal to the standard sea-level value $a_s = 340.3$ m/s $= 1116$ ft/s. Moreover, the airspeed indicator is designed to sense the actual pressure *difference* $p_0 - p_1$, in Eq. (4.77b), not the pressure *ratio* p_0/p_1, as appears in Eq. (4.77a). Hence, the form of Eq. (4.77b) is used to define a calibrated airspeed as follows:

$$V^2_{cal} = \frac{2a_s^2}{\gamma - 1}\left[\left(\frac{p_0 - p_1}{p_s} + 1\right)^{(\gamma-1)/\gamma} - 1\right] \tag{4.78}$$

where a_s and p_s are the standard sea-level values of the speed of sound and static pressure, respectively.

Again, emphasis is made that Eqs. (4.76) to (4.78) must be used to measure airspeed when $M_1 > 0.3$, that is, when the flow is compressible. Equations based on Bernoulli's equation, such as Eqs. (4.66) and (4.67), *are not valid* when $M_1 > 0.3$.

Example 4.10 A high-speed subsonic McDonnel-Douglas DC-10 airliner is flying at a pressure altitude of 10 km. A Pitot tube on the wingtip measures a pressure of 4.24×10^4 N/m². Calculate the Mach number at which the airplane is flying. If the ambient air temperature is 230 K, calculate the true airspeed and the calibrated airspeed.

SOLUTION From the standard atmosphere table, Appendix A, at an altitude of 10,000 m, $p = 2.65 \times 10^4$ N/m². Hence, from Eq. (4.76),

$$M_1^2 = \frac{2}{\gamma - 1}\left[\left(\frac{p_0}{p_1}\right)^{(\gamma-1)/\gamma} - 1\right] = \frac{2}{1.4-1}\left[\left(\frac{4.24 \times 10^4}{2.65 \times 10^4}\right)^{0.286} - 1\right]$$

$$M_1^2 = 0.719$$

Thus $\boxed{M_1 = 0.848}$

It is given that $T_1 = 230$ K, hence

$$a_1 = \sqrt{\gamma R T_1} = \sqrt{1.4(287)(230)} = 304.0 \text{ m/s}$$

From Eq. (4.77)

$$V_1^2 = \frac{2a_1^2}{\gamma - 1}\left[\left(\frac{p_0}{p_1}\right)^{(\gamma-1)/\gamma} - 1\right] = \frac{2(304.0)^2}{1.4-1}\left[\left(\frac{4.24}{2.65}\right)^{0.286} - 1\right]$$

$\boxed{V_1 = 258 \text{ m/s}}$ true airspeed

Note: As a check, from the definition of Mach number,

$$V_1 = M_1 a_1 = 0.848(304.0) = 258 \text{ m/s}$$

The calibrated airspeed can be obtained from Eq. (4.78).

$$V^2_{\text{cal}} = \frac{2a_s^2}{\gamma - 1}\left[\left(\frac{p_0 - p_1}{p_s} + 1\right)^{(\gamma - 1)/\gamma} - 1\right]$$

$$= \frac{2(340.3)^2}{1.4 - 1}\left[\left(\frac{4.24 \times 10^4 - 2.65 \times 10^4}{1.01 \times 10^5} + 1\right)^{0.286} - 1\right]$$

$$\boxed{V_{\text{cal}} = 157 \text{ m/s}}$$

Note that the difference between true and calibrated airspeeds is 39 percent. Note: Just out of curiosity, let us calculate V_1 the *wrong* way, i.e., let us apply Eq. (4.66), which was obtained from Bernoulli's equation for incompressible flow. Equation (4.66) does *not* apply to the high-speed case of this problem, but let us see what result we get anyway.

$$\rho = \frac{p_1}{RT_1} = \frac{2.65 \times 10^4}{287(230)} = 0.4 \text{ kg/m}^3$$

From Eq. (4.66)

$$V_{\text{true}} = \sqrt{\frac{2(p_0 - p)}{\rho}} = \sqrt{\frac{2(4.24 - 2.65) \times 10^4}{0.4}} = 282 \text{ m/s} \qquad \text{(incorrect answer)}$$

Comparing with $V_1 = 258$ m/s obtained above, an error of 9.3 percent is introduced in the calculation of true airspeed by using the *incorrect* assumption of incompressible flow. This error grows very rapidly as the Mach number approaches unity, as discussed in a subsequent section.

C Supersonic Flow

Airspeed measurements in supersonic flow, that is, $M > 1$, are qualitatively different from those for subsonic flow. In supersonic flow, a *shock wave* will form ahead of the Pitot tube, as shown in Figure 4.18. Shock waves are very thin regions of the flow (for example, 10^{-4} cm), across which some very severe changes in the flow properties take place. Specifically, as a fluid element flows through a shock wave:

1. The Mach number *decreases*.
2. The static pressure *increases*.
3. The static temperature *increases*.

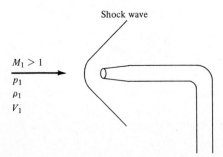

Figure 4.18 Pitot tube in supersonic flow.

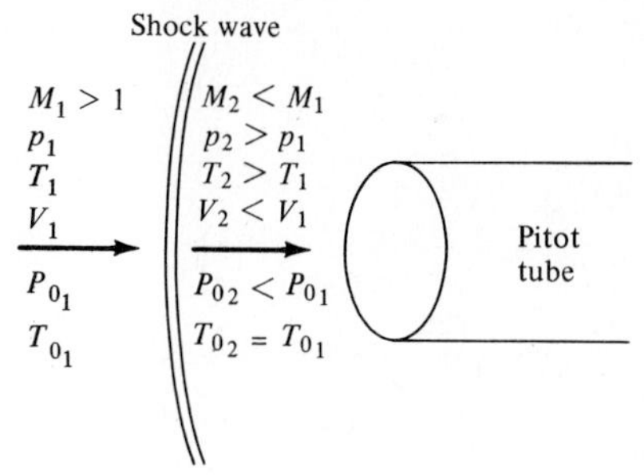

$M_1 > 1$
p_1
T_1
V_1

P_{0_1}

T_{0_1}

Shock wave

$M_2 < M_1$
$p_2 > p_1$
$T_2 > T_1$
$V_2 < V_1$

$P_{0_2} < P_{0_1}$

$T_{0_2} = T_{0_1}$

Pitot tube

Figure 4.19 Changes across a shock wave in front of a Pitot tube in supersonic flow.

4. The flow velocity *decreases*.
5. The total pressure p_0 *decreases*.
6. The total temperature T_0 stays the same for a perfect gas.

These changes across a shock wave are shown in Figure 4.19.

How and why does a shock wave form in supersonic flow? There are various answers with various degrees of sophistication. However, the essence is as follows. Refer to Figure 4.13, which shows a Pitot tube in subsonic flow. The gas molecules that collide with the probe set up a disturbance in the flow. This disturbance is communicated to other regions of the flow, away from the probe, by means of weak pressure waves (essentially sound waves) propagating at the local speed of sound. If the flow velocity V_1 is less than the speed of sound, as in Figure 4.13, then the pressure disturbances (which are traveling at the speed of sound) will work their way upstream and eventually be felt in all regions of the flow. On the other hand, refer to Figure 4.18, which shows a Pitot tube in supersonic flow. Here, V_1 is greater than the speed of sound. Thus, pressure disturbances which are created at the probe surface and which propagate away at the speed of sound *cannot* work their way upstream. Instead, these disturbances coalesce at a finite distance from the probe and form a natural phenomenon called a shock wave, as shown in Figures 4.18 and 4.19. The flow upstream of the shock wave (to the left of the shock) does not feel the pressure disturbance; i.e., the presence of the Pitot tube is not communicated to the flow upstream of the shock. The presence of the Pitot tube is felt only in the regions of flow behind the shock wave. Thus, the shock wave is a thin boundary in a supersonic flow, across which major changes in flow properties take place and which divides the region of undisturbed flow upstream from the region of disturbed flow downstream.

Whenever a solid body is placed in a supersonic stream, shock waves will occur. An example is shown in Figure 4.20, which shows photographs of the supersonic flow over several aerodynamic shapes. The shock waves, which are generally not visible to the naked eye, are made visible in Figure 4.20 by means of a specially designed optical system called a *schlieren system*. (An example where shock waves are sometimes visible to the naked eye is on the wing of a high speed subsonic transport such as a Boeing 707. As we will discuss shortly, there are regions of local supersonic flow on the upper surface of the wing, and these

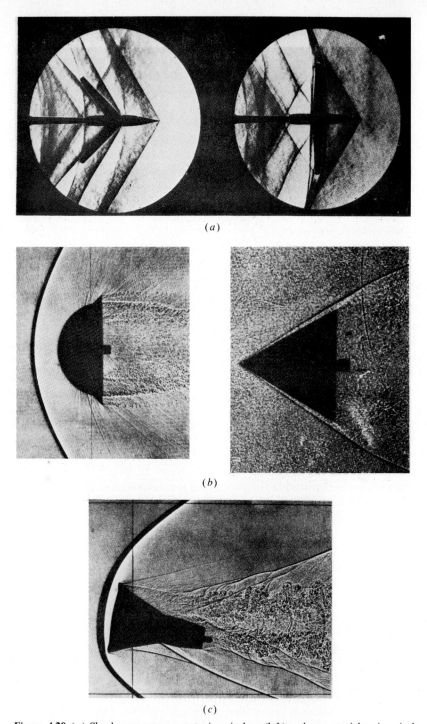

Figure 4.20 (*a*) Shock waves on a swept-wing airplane (left) and on a straight-wing airplane (right). Schlieren pictures taken in a supersonic wind tunnel at NASA Ames Research Center. (*b*) Shock waves on a blunt body (left) and sharp-nosed body (right). (*c*) Shock waves on a model of the Gemini manned space capsule. *(Courtesy of NASA Ames Research Center.)* Parts *b* and *c* are shadowgraphs of the flow.

supersonic regions are generally accompanied by weak shock waves. If the sun is almost directly overhead and if you look out the window along the span of the wing, you can sometimes see these waves dancing back and forth on the wing surface.)

Consider again the measurement of airspeed in a supersonic flow. The measurement is complicated by the presence of the shock wave in Figure 4.18 because the flow through a shock wave is *nonisentropic*. Within the thin structure of a shock wave itself, very large friction and thermal conduction effects are taking place. Hence, neither adiabatic nor frictionless conditions hold, and hence the flow is *not* isentropic. As a result, Eq. (4.74) and hence Eqs. (4.76) and (4.77a) do not hold across the shock wave. A major consequence is that the total pressure p_0 is smaller behind the shock than in front of it. In turn, the total pressure measured at the nose of the Pitot probe in supersonic flow will *not* be the same value as associated with the freestream, i.e., as associated with M_1. Consequently, a separate shock wave theory must be applied to relate the Pitot tube measurement to the value of M_1. This theory is beyond the scope of our presentation, but the resulting formula is given below for the sake of completeness:

$$\frac{p_{0_2}}{p_1} = \left(\frac{(\gamma + 1)^2 M_1^2}{4\gamma M_1^2 - 2(\gamma - 1)} \right)^{\gamma/(\gamma-1)} \frac{1 - \gamma + 2\gamma M_1^2}{\gamma + 1} \tag{4.79}$$

This equation is called the *Rayleigh Pitot tube formula*. It relates the Pitot tube measurement of total pressure behind the shock, p_{0_2}, and a measurement of freestream static pressure (again obtained by a static pressure orifice somewhere on the surface of the airplane) to the freestream supersonic Mach number M_1. In this fashion, measurements of p_{0_2} and p_1, along with Eq. (4.79), allow the calibration of a Mach meter for supersonic flight.

Example 4.11 An experimental rocket-powered aircraft is flying with a velocity of 3000 mi/h at an altitude where the ambient pressure and temperature are 151 lb/ft^2 and 390°R, respectively. A Pitot tube is mounted in the nose of the aircraft. What is the pressure measured by the Pitot tube?

SOLUTION First, we ask the question, is the flow supersonic or subsonic, i.e., what is M_1? From Eq. (4.54)

$$a_1 = \sqrt{\gamma R T_1} = \sqrt{1.4(1716)(390)} = 968.0 \text{ ft/s}$$
$$V_1 = 3000(88/60) = 4400 \text{ ft/s}$$
$$M_1 = \frac{V_1}{a_1} = \frac{4400}{968.0} = 4.55$$

Hence, $M_1 > 1$; the flow is supersonic. There is a shock wave in front of the Pitot tube; therefore, Eq. (4.74) developed for isentropic flow does *not* hold. Instead, Eq. (4.79) must be used.

$$\frac{p_{0_2}}{p_1} = \left(\frac{(\gamma + 1)^2 M_1^2}{4\gamma M_1^2 - 2(\gamma - 1)} \right)^{\gamma/(\gamma-1)} \frac{1 - \gamma + 2\gamma M_1^2}{\gamma + 1}$$

$$= \left(\frac{(2.4)^2 (4.54)^2}{4(1.4)(4.54)^2 - 2(0.4)} \right)^{3.5} \frac{1 - 1.4 + 2(1.4)(4.54)^2}{2.4} = 27$$

Thus $\qquad p_{0_2} = 27 p_1 = 27(151) = \boxed{4077 \text{ lb/ft}^2}$

Note: Again, out of curiosity, let us calculate the *wrong* answer. If we did *not* take into account the shock wave in front of the Pitot tube at supersonic speeds, then Eq. (4.74) would give

$$\frac{p_0}{p_1} = \left(1 + \frac{\gamma-1}{2} M_1^2\right)^{\gamma/(\gamma-1)} = \left(1 + \frac{0.4}{2} (4.54)^2\right)^{3.5} = 304.2$$

Thus $\quad p_0 = 304.2\, p_1 = 304.2(151) = 45931 \text{ lb/ft}^2 \quad$ incorrect answer

Note that the incorrect answer is off by a factor of more than 10!

D Summary

As a summary on the measurement of airspeed, note that different results apply to different regimes of flight: low-speed (incompressible), high-speed subsonic, and supersonic. These differences are fundamental and serve as excellent examples of the application of the different laws of aerodynamics developed in previous sections. Moreover, many of the formulas developed in this section apply to other practical problems, as discussed below.

4.12 SUPERSONIC WIND TUNNELS AND ROCKET ENGINES

For more than a century, projectiles such as bullets and artillery shells have been fired at supersonic velocities. However, the main aerodynamic interest in supersonic flows occurred after World War II with the advent of jet aircraft and rocket-propelled guided missiles. As a result, almost every aerodynamic laboratory has an inventory of supersonic and hypersonic wind tunnels to simulate modern high-speed flight. In addition to their practical importance, supersonic wind tunnels are an excellent example of the application of the fundamental laws of aerodynamics. The flow through rocket engine nozzles is another example of the same laws. In fact, the basic aerodynamics of supersonic wind tunnels and rocket engines are essentially the same, as discussed below.

First, consider isentropic flow in a stream tube, as sketched in Figure 4.1. From the continuity equation, Eq. (4.2),

$$\rho A V = \text{const}$$

or $\qquad \ln \rho + \ln A + \ln V = \ln(\text{const})$

Differentiating, we obtain

$$\frac{d\rho}{\rho} + \frac{dA}{A} + \frac{dV}{V} = 0 \tag{4.80}$$

Recalling the momentum equation, Eq. (4.8) (Euler's), we obtain

$$dp = -\rho V \, dV$$

Hence $\qquad\qquad \rho = -\frac{dp}{V \, dV} \tag{4.81}$

Substitute Eq. (4.81) into (4.80):

$$-\frac{d\rho V \, dV}{dp} + \frac{dA}{A} + \frac{dV}{V} = 0 \tag{4.82}$$

Since the flow is isentropic,

$$\frac{d\rho}{dp} = \frac{1}{dp/d\rho} \equiv \frac{1}{(dp/d\rho)_{\text{isentropic}}} \equiv \frac{1}{a^2}$$

Thus, Eq. (4.82) becomes

$$-\frac{V\,dV}{a^2} + \frac{dA}{A} + \frac{dV}{V} = 0$$

Rearranging, we get

$$\frac{dA}{A} = \frac{V\,dV}{a^2} - \frac{dV}{V} = \left(\frac{V^2}{a^2} - 1\right)\frac{dV}{V}$$

or

$$\boxed{\frac{dA}{A} = (M^2 - 1)\frac{dV}{V}} \tag{4.83}$$

Equation (4.83) is called the *area-velocity relation*, and it contains a wealth of information about the flow in the stream tube shown in Figure 4.1. First, note the mathematical convention that an increasing velocity and an increasing area correspond to positive values of dV and dA, respectively, whereas a decreasing velocity and a decreasing area correspond to negative values of dV and dA. This is the normal convention for differentials from differential calculus. With this in mind, Eq. (4.83) yields the following physical phenomena:

1. If the flow is subsonic ($M < 1$), for the velocity to increase (dV positive), the area must decrease (dA negative); i.e., when the flow is subsonic, the area must converge for the velocity to increase. This is sketched in Figure 4.21a. This same result was observed in Sec. 4.2 for incompressible flow. Of course, incompressible flow is, in a sense, a singular case of subsonic flow, where $M \to 0$.
2. If the flow is supersonic ($M > 1$), for the velocity to increase (dV positive), the area must also increase (dA positive); i.e., when the flow is supersonic, the area must diverge for the velocity to increase. This is sketched in Figure 4.21b.
3. If the flow is sonic ($M = 1$), then Eq. (4.83) yields for the velocity

$$\frac{dV}{V} = \frac{1}{M^2 - 1}\frac{dA}{A} = \frac{1}{0}\frac{dA}{A} \tag{4.84}$$

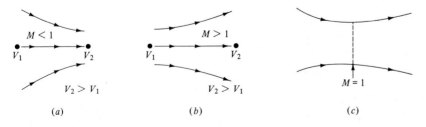

(a) (b) (c)

Figure 4.21 Results from the area-velocity relation.

which at first glance says that dV/V is infinitely large. However, on a physical basis, the velocity, and hence the change in velocity dV, must at all times be finite. This is only common sense. Thus, looking at Eq. (4.84), the only way for dV/V to be finite is to have $dA/A = 0$, so

$$\frac{dV}{V} = \frac{1}{0}\frac{dA}{A} = \frac{0}{0} = \text{finite number}$$

i.e., in the language of differential calculus, dV/V is an indeterminate form of zero over zero and hence can have a finite value. In turn, if $dA/A = 0$, the stream tube has a *minimum* area at $M = 1$. This minimum area is called a *throat* and is sketched in Figure 4.21c.

Therefore, in order to expand a gas to supersonic speeds, starting with a stagnant gas in a reservoir, the above discussion says that a duct of a sufficiently converging-diverging shape must be used. This is sketched in Figure 4.22, where typical shapes for supersonic wind-tunnel nozzles and rocket engine nozzles are shown. In both cases, the flow starts out with a very low velocity, $V \approx 0$, in the reservoir, expands to high subsonic speeds in the convergent section, reaches Mach 1 at the throat, and then goes supersonic in the divergent section downstream of the throat. In a supersonic wind tunnel, smooth, uniform flow at the nozzle exit is usually desired, and therefore a long, gradually converging and diverging nozzle is employed, as shown at the top of Figure 4.22. For rocket engines, the flow quality at the exit is not quite as important, but the weight of the nozzle is a major concern. For the weight to be minimized, the engine's length is minimized, which gives rise to a rapidly diverging, bell-like shape for the supersonic section, as shown at the bottom of Figure 4.22. A photograph of a typical rocket engine is shown in Figure 4.23.

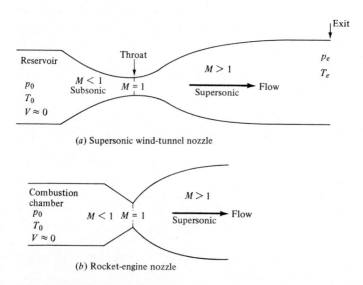

(a) Supersonic wind-tunnel nozzle

(b) Rocket-engine nozzle

Figure 4.22 Supersonic nozzle shapes.

The real flow through nozzles such as those sketched in Figure 4.22 is closely approximated by isentropic flow because little or no heat is added or taken away through the nozzle walls and a vast core of the flow is virtually frictionless. Therefore, Eqs. (4.73) to (4.75) apply to nozzle flows. Here, the total pressure and temperature p_0 and T_0 remain constant throughout the flow, and Eqs. (4.73) to (4.75) can be interpreted as relating conditions at any point in the flow to the stagnation conditions in the reservoir. For example, consider Figure 4.22, which illustrates the reservoir conditions p_0 and T_0 where $V \approx 0$. Consider any cross section downstream of the reservoir. The static temperature, density, and pressure at this section are T_1, ρ_1, and p_1, respectively. If the Mach number M_1 is known at this point, T_1, ρ_1, and p_1 can be found from Eqs. (4.73) to (4.75) as:

$$T_1 = T_0 \left[1 + \tfrac{1}{2}(\gamma - 1) M_1^2 \right]^{-1} \tag{4.85}$$

$$\rho_1 = \rho_0 \left[1 + \tfrac{1}{2}(\gamma - 1) M_1^2 \right]^{-1/(\gamma-1)} \tag{4.86}$$

$$p_1 = p_0 \left[1 + \tfrac{1}{2}(\gamma - 1) M_1^2 \right]^{-\gamma/(\gamma-1)} \tag{4.87}$$

Again, Eqs. (4.85) to (4.87) demonstrate the power of the Mach number in making aerodynamic calculations. The variation of Mach number itself through the nozzle is strictly a function of the ratio of the cross-sectional area to the throat area, A/A_t. This relation can be developed from the aerodynamic fundamentals already discussed; the resulting form is

$$\left(\frac{A}{A_t} \right)^2 = \frac{1}{M^2} \left[\frac{2}{\gamma + 1} \left(1 + \frac{\gamma - 1}{2} M^2 \right) \right]^{(\gamma+1)(\gamma-1)} \tag{4.88}$$

Therefore, the analysis of isentropic flow through a nozzle is relatively straightforward. The procedure is summarized in Figure 4.24. Consider that the nozzle shape, hence A/A_t, is given as shown in Figure 4.24a. Then, from Eq.

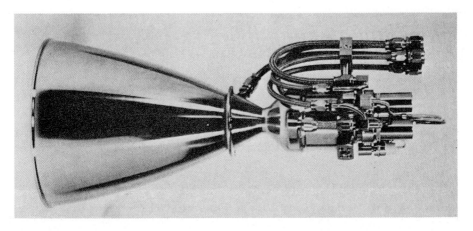

Figure 4.23 A typical rocket engine. Shown is a small rocket designed by Messerschmitt-Bolkow-Blohm for European satellite launching.

(4.88), the Mach number can be obtained (implicitly). Its variation is sketched in Figure 4.24b. Since M is now known through the nozzle, then Eqs. (4.85) to (4.87) give the variations of T, ρ, and p, which are sketched in Figure 4.24c to e. The directions of these variations are important and should be noted. From Figure 4.24, the Mach number continuously increases through the nozzle, going from near zero in the reservoir to $M = 1$ at the throat and to supersonic values downstream of the throat. In turn, p, T, and ρ begin with their stagnation values in the reservoir and continuously decrease to low values at the nozzle exit. Hence, a supersonic nozzle flow is an expansion process, where pressure decreases through the nozzle. In fact, it is this pressure decrease that provides the mechanical force for pushing the flow through the nozzle. If the nozzle shown in Figure 4.24a is simply set out by itself in a laboratory, obviously nothing will happen; the air will not start to rush through the nozzle by its own accord. Instead, to establish the flow sketched in Figure 4.24, we must provide a high-pressure source at the inlet, and/or a low-pressure source at the exit, with the pressure ratio just the right value, as prescribed by Eq. (4.87) and as sketched in Figure 4.24c.

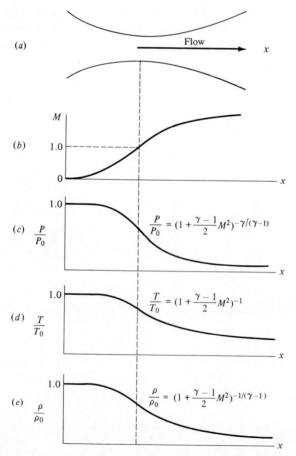

Figure 4.24 Variation of Mach number, pressure, temperature, and density through a supersonic nozzle.

Example 4.12 You are given the job of designing a supersonic wind tunnel that has a Mach 2 flow at standard sea-level conditions in the test section. What reservoir pressure and temperature and what area ratio A_e/A_t are required to obtain these conditions?

SOLUTION The static pressure $p_e = 1$ atm $= 1.01 \times 10^5$ N/m^2, and the static temperature $T_e = 288.16$ K, from conditions at standard sea level. These are the desired conditions at the exit of the nozzle (the entrance to the test section). The necessary reservoir conditions are obtained from Eqs. (4.85) and (4.87):

$$\frac{T_0}{T_e} = 1 + \frac{\gamma - 1}{2} M_e^2 = 1 + \frac{1.4 - 1}{2} (2)^2 = 1.8$$

Thus,

$$T_0 = 1.8 T_e = 1.8(288.16) = \boxed{518.7 \text{ K}}$$

$$\frac{p_0}{p_e} = \left[1 + \frac{\gamma - 1}{2} M_e^2 \right]^{\gamma/(\gamma - 1)} = (1.8)^{3.5} = 7.82$$

Thus,

$$p_0 = 7.82 p_e = 7.82(1.01 \times 10^5) = \boxed{7.9 \times 10^5 \text{ N/m}^2}$$

The area ratio is obtained from Eq. (4.88):

$$\left(\frac{A_e}{A_t} \right)^2 = \frac{1}{M^2} \left[\frac{2}{\gamma + 1} \left(1 + \frac{\gamma - 1}{2} M^2 \right) \right]^{(\gamma + 1)/(\gamma - 1)}$$

$$= \frac{1}{2^2} \left[\frac{2}{2.4} \left(1 + \frac{0.4}{2} 2^2 \right) \right]^{(2.4)/(0.4)} = 2.85$$

Hence

$$\boxed{\frac{A_e}{A_t} = 1.69}$$

4.13 DISCUSSION ON COMPRESSIBILITY

We have been stating all along that flows where $M < 0.3$ can be treated at essentially incompressible and, conversely, flows where $M \geq 0.3$ should be treated as compressible. We are now in a position to prove this.

Consider a gas at rest $(V = 0)$ with density ρ_0. Now accelerate this gas isentropically to some velocity V and Mach number M. Obviously, the thermodynamic properties of the gas will change, including the density. In fact, the change in density will be given by Eq. (4.75):

$$\frac{\rho_0}{\rho} = \left(1 + \frac{\gamma - 1}{2} M^2 \right)^{1/(\gamma - 1)}$$

For $\gamma = 1.4$, this variation of ρ/ρ_0 is given in Figure 4.25. Note that, for $M < 0.3$, the density change in the flow is less than 5%; that is, the density is essentially constant for $M < 0.3$, and for all practical purposes the flow is incompressible. Therefore, we have just demonstrated the validity of the statement

For $M < 0.3$, the flow can be treated as incompressible.

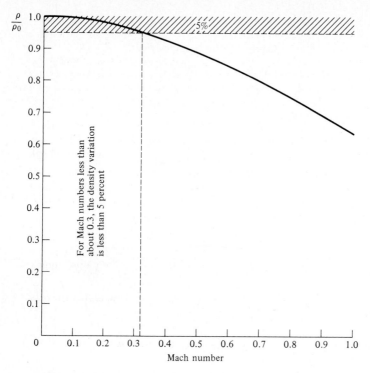

Figure 4.25 Density variation with Mach number for $\gamma = 1.4$, showing region where the density change is less than 5 percent.

4.14 INTRODUCTION TO VISCOUS FLOW

To this point, we have dealt exclusively with frictionless flows, and we have adequately treated a number of practical problems with this assumption. However, there are numerous other problems in which the effect of friction is dominant, indeed, in which the complete nature of the problem is governed by the presence of friction between the airflow and a solid surface. A classic example is sketched in Figure 4.26, which shows the low-speed flow over a sphere. At the left is sketched the flow field for frictionless flow. The streamlines are symmetrical

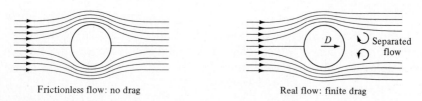

Figure 4.26 Comparison between ideal frictionless flow and real flow with the effects of friction.

and amazingly, there is no aerodynamic force on the sphere. The pressure distribution over the forward surface exactly balances that over the rearward surface, and hence there is no drag (no force in the flow direction). However, this purely theoretical result is contrary to common sense; in real life there is a drag force on the sphere tending to retard the motion of the sphere. The failure of the theory to predict drag was bothersome to early nineteenth century aerodynamicists and was even given a name, *d'Alembert's paradox*. The problem is caused by not including friction in the theory. The real flow over a sphere is sketched on the right in Figure 4.26. The flow separates on the rearward surface of the sphere, setting up a complicated flow in the wake and causing the pressure on the rearward surface to be less than on the forward surface. Hence, there is a drag force exerted on the sphere, as shown by D in Figure 4.26. The difference between the two flows in Figure 4.26 is simply friction, but what a difference!

Consider the flow of a gas over a solid surface, such as the airfoil sketched in Figure 4.27. According to our previous considerations of frictionless flows, we have considered the flow velocity at the surface as being a finite value, such as V_2 shown in Figure 4.27; i.e., due to the lack of friction, the streamline right at the surface slips over the surface. In fact, we stated that if the flow is incompressible, V_2 can be calculated from Bernoulli's equation:

$$p_1 + \tfrac{1}{2}\rho V_1{}^2 = p_2 + \tfrac{1}{2}\rho V_2{}^2$$

However, in real life, *the flow at the surface adheres to the surface* because of friction between the gas and the solid material; i.e., right at the surface, the flow velocity is zero, and there is a thin region of retarded flow in the vicinity of the surface, as sketched in Figure 4.28. This region of viscous flow which has been retarded due to friction at the surface is called a *boundary layer*. The inner edge of the boundary layer is the solid surface itself, such as point a in Figure 4.28, where $V = 0$. The outer edge of the boundary layer is given by point b, where the flow velocity is essentially the value given by V_2 in Figure 4.27. That is, point b in Figure 4.28 is essentially equivalent to point 2 in Figure 4.27. In this fashion, the flow properties at the outer edge of the boundary layer in Figure 4.28 can be calculated from a frictionless flow analysis, as pictured in Figure 4.27. This leads to an important conceptual point in theoretical aerodynamics: a flow field can be split into two regions, one region where friction is important, namely, in the boundary layer near the surface, and another region of frictionless flow (sometimes called potential flow) outside the boundary layer. This concept was first

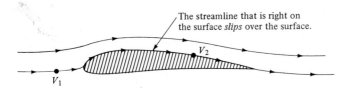

Figure 4.27 Frictionless flow.

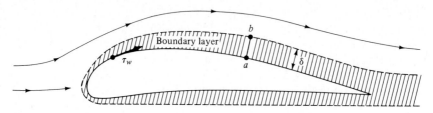

Figure 4.28 Flow in real life, with friction. The thickness of the boundary layer is greatly overemphasized for clarity.

introduced by Ludwig Prandtl in 1904, and it revolutionized modern theoretical aerodynamics.

It can be shown experimentally and theoretically that the pressure through the boundary layer in a direction perpendicular to the surface is constant. That is, letting p_a and p_b be the static pressures at points a and b, respectively, in Figure 4.28, then $p_a = p_b$. This is an important phenomenon. This is why a surface pressure distribution calculated from frictionless flow (Figure 4.27) many times gives accurate results for the real-life surface pressures; it is because the frictionless calculations give the correct pressures at the outer edge of the boundary layer (point b), and these pressures are impressed without change through the boundary layer right down to the surface (point a). The above statements are reasonable for slender aerodynamic shapes such as the airfoil in Figure 4.28; they do not hold for regions of separated flow over blunt bodies, as previously sketched in Figure 4.26. Such separated flows will be discussed in a subsequent section.

Refer again to Figure 4.28. The *boundary layer thickness* δ grows as the flow moves over the body; i.e., more and more of the flow is affected by friction as the distance along the surface increases. In addition, the presence of friction creates a *shear stress* at the surface, τ_w. This shear stress has dimensions of force/area and acts in a direction tangential to the surface. Both δ and τ_w are important quantities, and a large part of boundary layer theory is devoted to their calculation. As we will see, τ_w gives rise to a drag force called *skin friction drag*, hence attesting to its importance. In subsequent sections, equations for the calculation of δ and τ_w will be given.

Looking more closely at the boundary layer, a *velocity profile* through the boundary layer is sketched in Figure 4.29. The velocity starts out at zero at the surface and increases continuously to its value of V_2 at the outer edge. Let us set up coordinate axes x and y such that x is parallel to the surface and y is normal to the surface, as shown in Figure 4.29. By definition, a *velocity profile* gives the variation of velocity in the boundary layer as a function of y. In general, the velocity profiles at different x stations are different.

The slope of the velocity profile at the wall is of particular importance, because it governs the wall shear stress. Let $(dV/dy)_{y=0}$ be defined as the velocity gradient at the wall. Then the shear stress at the wall is given by

$$\tau_w = \mu \left(\frac{dV}{dy} \right)_{y=0} \tag{4.89}$$

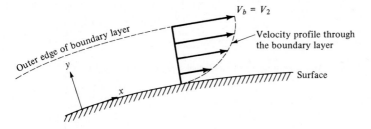

Figure 4.29 Velocity profile through a boundary layer.

where μ is called the *absolute viscosity coefficient* (or simply the *viscosity*) of the gas. The viscosity coefficient has dimensions of mass/(length)(time), as can be verified from Eq. (4.89) combined with Newton's second law. It is a physical property of the fluid; μ is different for different gases and liquids. Also, μ varies with T. For liquids, μ decreases as T increases (we all know that oil gets "thinner" when the temperature is increased). But for gases, μ increases as T increases (air gets "thicker" when temperature is increased). For air at standard sea level temperature,

$$\mu = 1.7894 \times 10^{-5} \text{ kg/(m)(s)} = 3.7373 \times 10^{-7} \text{ slug/(ft)(s)}$$

The variation of μ with temperature for air is given in Figure 4.30.

In this section, we are simply introducing the fundamental concepts of boundary layer flows; such concepts are essential to the practical calculation of

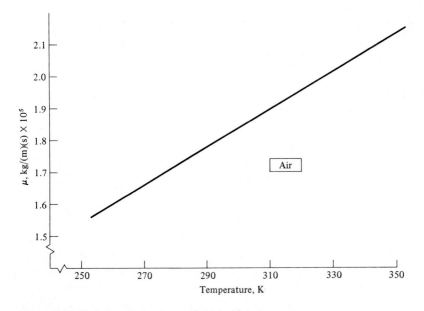

Figure 4.30 Variation of viscosity coefficient with temperature.

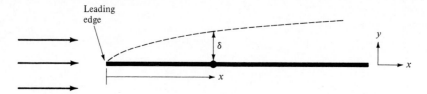

Figure 4.31 Growth of the boundary layer thickness.

aerodynamic drag, as we will soon appreciate. In this spirit, we introduce another important dimensionless "number," a number of importance and impact on aerodynamics equal to those of the Mach number discussed earlier—the *Reynolds number*. Consider the development of a boundary layer on a surface, such as the flat plate sketched in Figure 4.31. Let x be measured from the leading edge, i.e., the front tip of the plate. Let V_∞ be the flow velocity far upstream of the plate. (The subscript ∞ is commonly used to denote conditions far upstream of an aerodynamic body, so-called *freestream conditions*.) The *Reynolds number* Re_x is defined as

$$Re_x = \frac{\rho_\infty V_\infty x}{\mu_\infty} \tag{4.90}$$

Note that Re_x is dimensionless and that it varies linearly with x. For this reason, Re_x is sometimes called a "local" Reynolds number, because it is based on the local coordinate x.

To this point in our discussion on aerodynamics, we have always considered flow streamlines to be smooth and regular curves in space. However, in a viscous flow, and particularly in boundary layers, life is not quite so simple. There are two basic types of viscous flow:

1. *Laminar flow*, where the streamlines are smooth and regular and a fluid element moves smoothly along a streamline
2. *Turbulent flow*, where the streamlines break up and a fluid element moves in a random, irregular, and tortuous fashion

If you observe the smoke rising from a lit cigarette, as sketched in Figure 4.32, you see first a region of smooth flow—laminar flow—and then a transition to irregular, mixed-up flow—turbulent flow. The differences between laminar and turbulent flow are dramatic, and they have major impact on aerodynamics. For example, consider the velocity profiles through a boundary layer, as sketched in Figure 4.33. The profiles are different, depending on whether the flow is laminar or turbulent. The turbulent profile is "fatter," or fuller, than the laminar profile. For the turbulent profile, from the outer edge to a point near the surface, the velocity remains reasonably close to the freestream velocity; it then rapidly

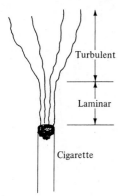

Turbulent

Laminar

Cigarette

Figure 4.32 Smoke pattern from a cigarette.

decreases to zero at the surface. In contrast, the laminar velocity profile gradually decreases to zero from the outer edge to the surface. Now, consider the velocity gradient at the wall, $(dV/dy)_{y=0}$, which is the reciprocal of the slope of the curves shown in Figure 4.33 evaluated at $y = 0$. From Figure 4.33, it is clear that

$$\left[\left(\frac{dV}{dy}\right)_{y=0} \text{ for laminar flow}\right] < \left[\left(\frac{dV}{dy}\right)_{y=0} \text{ for turbulent flow}\right]$$

Recalling Eq. (4.89) for τ_w leads us to the fundamental and highly important fact

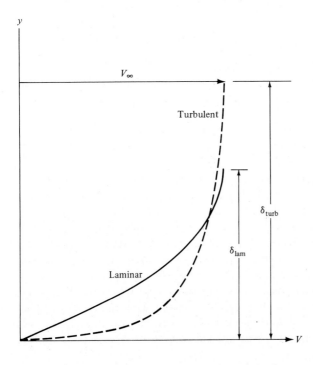

Figure 4.33 Velocity profiles for laminar and turbulent boundary layers. Note that the turbulent boundary layer thickness is larger than the laminar boundary layer thickness.

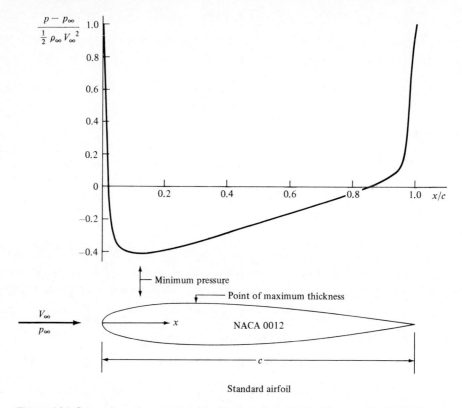

Figure 4.34 Comparison of conventional and laminar flow airfoils. The pressure distributions shown are the theoretical results obtained by the NACA and are for 0° angle of attack. The airfoil shapes are drawn to scale.

that *laminar shear stress is less than turbulent shear stress.*

$$\tau_{w \text{ laminar}} < \tau_{w \text{ turbulent}}$$

This obviously implies that the skin friction exerted on an airplane wing or body will depend on whether the boundary layer on the surface is laminar or turbulent, with laminar flow yielding the smaller skin friction drag.

It appears to be almost universal in nature that systems with the maximum amount of *disorder* are favored. For aerodynamics, this means that the vast majority of practical viscous flows are turbulent. The boundary layers on most practical airplanes, missiles, ship hulls, etc., are turbulent, with the exception of small regions near the leading edge, as we will soon see. Consequently, the skin friction on these surfaces is the higher, turbulent value. For the aerodynamicist, who is usually striving to reduce drag, this is unfortunate. However, the skin friction on slender shapes, such as wing cross sections (airfoils), can be reduced

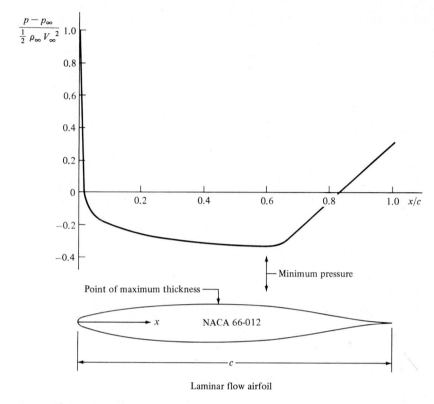

Laminar flow airfoil

Figure 4.34 (continued)

by designing the shape in such a manner to encourage laminar flow. Figure 4.34 indicates how this can be achieved. Here, two airfoils are shown; the standard airfoil on the left has a maximum thickness near the leading edge, whereas the laminar flow airfoil has its maximum thickness near the middle of the airfoil. The pressure distributions on the top surface of the airfoils are sketched above the airfoils in Figure 4.34. Note that for the standard airfoil, the minimum pressure occurs near the leading edge, and there is a long stretch of increasing pressure from this point to the trailing edge. Turbulent boundary layers are encouraged by such increasing pressure distributions. Hence, the standard airfoil is generally bathed in long regions of turbulent flow, with the attendant high skin friction drag. On the other hand, note that for the laminar flow airfoil, the minimum pressure occurs near the trailing edge, and there is a long stretch of decreasing pressure from the leading edge to the point of minimum pressure. Laminar boundary layers are encouraged by such decreasing pressure distributions. Hence, the laminar flow airfoil is generally bathed in long regions of laminar flow, thus benefiting from the reduced skin friction drag.

The North American P-51 Mustang, designed at the outset of World War II, was the first production aircraft to employ a laminar flow airfoil. However, laminar flow is a sensitive phenomenon; it readily gets unstable and tries to change into turbulent flow. For example, the slightest roughness of the airfoil surface caused by such real-life effects as protruding rivets, imperfections in machining, and bug spots can cause a premature transition to turbulent flow in advance of the design condition. Therefore, most laminar flow airfoils used on production aircraft do not yield the extensive regions of laminar flow that are obtained in controlled laboratory tests using airfoil models with highly polished, smooth surfaces. From this point of view, the early laminar flow airfoils were not successful. However, they were successful from an entirely different point of view; namely, they were found to have excellent high-speed properties, postponing to a higher flight Mach number the large drag rise due to shock waves and flow separation encountered near Mach 1. (Such high-speed effects are discussed in Secs. 5.9–5.11.) As a result, the early laminar flow airfoils were extensively used on jet-propelled airplanes during the 1950s and 1960s and are still employed today on some modern high-speed aircraft.

Given a laminar or turbulent flow over a surface, how do we actually calculate the skin friction drag? The answer is given in the following two sections.

4.15 RESULTS FOR A LAMINAR BOUNDARY LAYER

Consider again the boundary layer flow over a flat plate as sketched in Figure 4.31. Assume that the flow is laminar. The two physical quantities of interest are the boundary layer thickness δ and shear stress τ_w at location x. Formulas for these quantities can be obtained from laminar boundary layer theory, which is beyond the scope of this book. However, the results, which have been verified by experiment, are as follows. The laminar boundary layer thickness is

$$\boxed{\delta = \frac{5.2x}{\sqrt{Re_x}} \text{ laminar}} \tag{4.91}$$

where $Re_x = \rho_\infty V_\infty x/\mu_\infty$ as defined in Eq. (4.90). It is remarkable that a phenomenon as complex as the development of a boundary layer, which depends at least on density, velocity, viscosity, and length of the surface, should be described by a formula as simple as Eq. (4.91). In this vein, Eq. (4.91) demonstrates the powerful influence of Reynolds number Re_x in aerodynamic calculations.

Note from Eq. (4.91) that laminar boundary layer thickness varies inversely as the square root of the Reynolds number. Also, since $Re_x = \rho_\infty V_\infty x/\mu_\infty$, then from Eq. (4.91) $\delta \propto x^{1/2}$; that is, the laminar boundary layer grows *parabolically*.

The local shear stress τ_w is also a function of x, as sketched in Figure 4.35. Rather than dealing with τ_w directly, aerodynamicists find it more convenient to

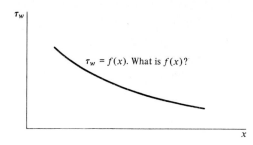

$\tau_w = f(x)$. What is $f(x)$?

Figure 4.35 Variation of shear stress with distance along the surface.

define a local skin friction *coefficient* c_{f_x} as

$$c_{f_x} \equiv \frac{\tau_w}{\frac{1}{2}\rho_\infty V_\infty^{\,2}} \equiv \frac{\tau_w}{q_\infty} \qquad (4.92)$$

The skin friction coefficient is dimensionless and is defined as the local shear stress divided by the dynamic pressure at the outer edge of the boundary. From laminar boundary layer theory,

$$c_{f_x} = \frac{0.664}{\sqrt{\mathrm{Re}_x}} \quad \text{laminar} \qquad (4.93)$$

where, as usual, $\mathrm{Re}_x = \rho_\infty V_\infty x/\mu_\infty$. Equation (4.93) demonstrates the convenience of defining a dimensionless skin friction coefficient. On one hand, the dimensional shear stress τ_w (as sketched in Figure 4.35) depends on several quantities such as ρ_∞, V_∞, and Re_x; on the other hand, from Eq. (4.93), c_{f_x} is a function of Re_x *only*. This convenience obtained from using dimensionless coefficients and numbers reverberates throughout aerodynamics. Relations between dimensionless quantities such as those given in Eq. (4.93) can be substantiated by *dimensional analysis*, a formal procedure to be discussed later.

Combining Eqs. (4.92) and (4.93), we can obtain values of τ_w from

$$\tau_w = f(x) = \frac{0.664 q_\infty}{\sqrt{\mathrm{Re}_x}} \qquad (4.94)$$

Note from Eqs. (4.93) and (4.94) that both c_{f_x} and τ_w for laminar boundary layers vary as $x^{-1/2}$; that is, c_{f_x} and τ_w decrease along the surface in the flow direction, as sketched in Figure 4.35. The shear stress near the leading edge of a flat plate is greater than near the trailing edge.

The variation of local shear stress τ_w along the surface allows us to calculate the total skin friction drag due to the airflow over an aerodynamic shape. Recall from Sec. 2.2 that the net aerodynamic force on any body is fundamentally due to the pressure and shear stress distributions on the surface. In many cases, it is this total aerodynamic force that is of primary interest. For example, if you mount a flat plate parallel to the airstream in a wind tunnel and measure the force exerted on the plate, by means of a balance of some sort, you are not measuring the local

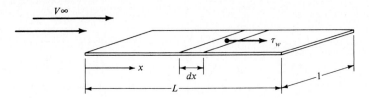

Figure 4.36 Total drag is the integral of the local shear stress over the surface.

shear stress τ_w; rather, you are measuring the total drag due to skin friction being exerted over all the surface. This *total skin friction drag* can be obtained as follows.

Consider a flat plate of length L and unit width oriented parallel to the flow, as shown in perspective in Figure 4.36. Consider also an infinitesimally small surface element of the plate of length dx and width unity, as shown in Figure 4.36. The local shear stress on this element is τ_x, a function of x. Hence, the force on this element due to skin friction is $\tau_w\,dx\,(1) = \tau_w\,dx$. The total skin friction drag is the sum of the forces on all the infinitesimal elements from the leading to the trailing edge; i.e., the total skin friction drag D_f is obtained by integrating τ_x along the surface, as

$$D_f = \int_0^L \tau_w\, dx \tag{4.95}$$

Combining Eqs. (4.94) and (4.95) yields

$$D_f = 0.664 q_\infty \int_0^L \frac{dx}{\sqrt{\mathrm{Re}_x}} = \frac{0.664 q_\infty}{\sqrt{\rho_\infty V_\infty/\mu_\infty}} \int_0^L \frac{dx}{\sqrt{x}}$$

$$D_f = \frac{1.328 q_\infty L}{\sqrt{\rho_\infty V_\infty L/\mu}} \tag{4.96}$$

Let us define a *total skin friction drag coefficient* C_f as

$$C_f \equiv \frac{D_f}{q_\infty S} \tag{4.97}$$

where S is the total area of the plate, $S = L(1)$. Thus, from Eqs. (4.96) and (4.97),

$$C_f = \frac{D_f}{q_\infty L(1)} = \frac{1.328 q_\infty L}{q_\infty L (\rho_\infty V_\infty L/\mu_\infty)^{1/2}}$$

or $$C_f = \frac{1.328}{\sqrt{\mathrm{Re}_L}} \quad \text{laminar} \tag{4.98}$$

where the Reynolds number is now based on the total length L, that is, $\mathrm{Re}_L \equiv \rho_\infty V_\infty L/\mu_\infty$.

Do not confuse Eq. (4.98) with Eq. (4.93); they are different quantities. The local skin friction coefficient c_{f_x} in Eq. (4.93) is based on the local Reynolds number $\mathrm{Re}_x = \rho_\infty V_\infty x/\mu_\infty$ and is a function of x. On the other hand, the total skin friction coefficient C_f is based on Reynolds number for the plate length L, $\mathrm{Re}_L = \rho_\infty V_\infty L/\mu_\infty$.

Emphasis is made that Eqs. (4.91), (4.93), and (4.98) apply to laminar boundary layers only; for turbulent flow the expressions are different. Also, these equations are exact only for low-speed (incompressible) flow. However, they have been shown to be reasonably accurate for high-speed subsonic flows as well. For supersonic and hypersonic flows, where the velocity gradients within the boundary layer are so extreme and where the presence of frictional dissipation creates very large temperatures within the boundary layer, the form of these equations can still be used for engineering approximations, but ρ and μ must be evaluated at some reference conditions germane to the flow inside the boundary layer. Such matters are beyond the scope of this book.

Example 4.13 Consider the flow of air over a small flat plate which is 5 cm long in the flow direction and 1 m wide. The freestream conditions correspond to standard sea level, and the flow velocity is 120 m/s. Assuming laminar flow, calculate:
(a) The boundary layer thickness at the downstream edge (the trailing edge)
(b) The drag force on the plate

SOLUTION
(a) At the trailing edge of the plate, where $x = 5$ cm $= 0.05$ m, the Reynolds number is, from Eq. (4.90),

$$\mathrm{Re}_x = \frac{\rho_\infty V_\infty x}{\mu_\infty} = \frac{(1.225 \text{ kg/m}^3)(120 \text{ m/s})(0.05 \text{ m})}{1.789 \times 10^{-5} \text{ kg/(m)(s)}}$$

$$= 4.11 \times 10^5$$

From Eq. (4.91),

$$\delta = \frac{5.2x}{\mathrm{Re}_x^{1/2}} = \frac{5.2(0.05)}{(4.11 \times 10^5)^{1/2}} = \boxed{4.06 \times 10^{-4} \text{ m}}$$

Note how thin the boundary layer is—only 0.0406 cm at the trailing edge.
(b) To obtain the skin friction drag, Eq. (4.98) gives, with $L = 0.05$ m,

$$C_f = \frac{1.328}{\mathrm{Re}_L^{1/2}} = \frac{1.328}{(4.11 \times 10^5)^{1/2}} = 2.07 \times 10^{-3}$$

The drag can be obtained from the definition of the skin friction drag coefficient, Eq. (4.97), once q_∞ and S are known.

$$q_\infty = \tfrac{1}{2}\rho_\infty V_\infty^2 = \tfrac{1}{2}(1.225)(120)^2 = 8820 \text{ N/m}^2$$

$$S = 0.05(1) = 0.05 \text{ m}^2$$

Thus, from Eq. (4.97), the drag *on one surface* of the plate (say the top surface) is

$$(\text{top}) \, D_f = q_\infty S C_f = 8820(0.05)(2.07 \times 10^{-3}) = 0.913 \text{ N}$$

Since both the top and bottom surfaces are exposed to the flow, the total friction drag will be double the above result:

$$\text{Total } D_f = 2(0.913) = \boxed{1.826 \text{ N}}$$

4.16 RESULTS FOR A TURBULENT BOUNDARY LAYER

Under the same flow conditions, a turbulent boundary layer will be *thicker* than a laminar boundary layer. This comparison is sketched in Figure 4.37. Unlike

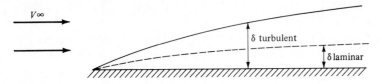

Figure 4.37 Turbulent boundary layers are thicker than laminar boundary layers.

laminar flows, no exact theoretical results can be presented for turbulent boundary layers. The study of turbulence is a major effort in fluid dynamics today; so far, turbulence is still an unsolved theoretical problem and is likely to remain so for an indefinite time. In fact, turbulence is one of the major unsolved problems in theoretical physics. As a result, our knowledge of δ and τ_w for turbulent boundary layers must rely on experimental results. Such results yield the following approximate formulas for turbulent flow.

$$\delta = \frac{0.37x}{Re_x^{0.2}} \quad \text{turbulent} \tag{4.99}$$

Note from Eq. (4.99) that a turbulent boundary grows approximately as $x^{4/5}$. This is in contrast to the slower $x^{1/2}$ variation for a laminar boundary layer. As a result, turbulent boundary layers grow faster and are thicker than laminar boundary layers.

The total skin friction coefficient is given approximately as

$$C_f = \frac{0.074}{Re_L^{0.2}} \quad \text{turbulent} \tag{4.100}$$

Note that for turbulent flow, C_f varies as $x^{-1/5}$; this is in contrast to the $x^{-1/2}$ variation for laminar flow. Hence C_f is larger for turbulent flow, which precisely confirms our reasoning at the end of Sec. 4.14, where we noted that τ_w(laminar) $< \tau_w$(turbulent). Also note that C_f in Eq. (4.100) is once again a function of Re_L. Values of C_f for both laminar and turbulent flows are commonly plotted in the form shown in Figure 4.38. Note the magnitude of the numbers involved in Figure 4.38. The values of Re_L for actual flight situations may vary from 10^5 to 10^8 or higher; the values of C_f are generally much less than unity, on the order of 10^{-2}.

Example 4.14 Consider the same flow over the same flat plate as in Example 4.13; however, assume that the boundary layer is now completely turbulent. Calculate the boundary layer thickness at the trailing edge and the drag force on the plate.

SOLUTION From Example 4.13, $Re_x = 4.11 \times 10^5$. From Eq. (4.99), for turbulent flow,

$$\delta = \frac{0.37x}{Re_x^{0.2}} = \frac{0.37(0.05)}{(4.11 \times 10^5)^{0.2}} = \boxed{1.39 \times 10^{-3} \text{ m}}$$

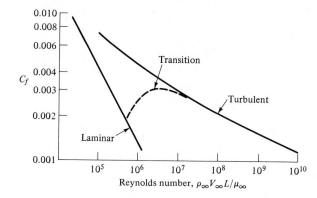

Figure 4.38 Variation of skin friction coefficient with Reynolds number for low-speed flow. Comparison of laminar and turbulent flow.

Note: Compare this result with the laminar flow result from Example 4.13.

$$\frac{\delta_{\text{turb}}}{\delta_{\text{lam}}} = \frac{1.39 \times 10^{-3}}{4.06 \times 10^{-4}} = 3.42$$

Note that the turbulent boundary layer at the trailing edge is 3.42 times thicker than the laminar boundary layer—quite a sizable amount! From Eq. (4.100)

$$C_f = \frac{0.074}{\text{Re}_L^{0.2}} = \frac{0.074}{(4.11 \times 10^5)^{0.2}} = 0.00558$$

On the top surface,

$$D_f = q_\infty S C_f = 8820(0.05)(0.00558) = 2.46 \text{ N}$$

Considering both top and bottom surface, we have

$$\text{Total } D_f = 2(2.46) = \boxed{4.92 \text{ N}}$$

Note that the turbulent drag is 2.7 times larger than the laminar drag.

4.17 TRANSITION

In Sec. 4.15 we discussed the flow over a flat plate as if it were all laminar. Similarly, in Sec. 4.16 we assumed all turbulent flow. In reality, the flow *always* starts out from the leading edge as laminar. Then, at some point downstream of the leading edge, the laminar boundary layer becomes unstable and small "bursts" of turbulent flow begin to grow in the flow. Finally, over a certain region called the *transition region*, the boundary layer becomes completely turbulent. For purposes of analysis, we usually draw the picture shown in Figure 4.39, where a laminar boundary starts out from the leading edge of a flat plate and grows parabolically downstream. Then, at the *transition point*, it becomes a turbulent boundary layer growing at a faster rate, on the order of $x^{4/5}$ downstream. The value of x where transition is said to take place is the *critical value* x_{cr}. In turn, x_{cr} allows the definition of a *critical Reynolds number* for transition as

$$\text{Re}_{x_{\text{cr}}} = \frac{\rho_\infty V_\infty x_{\text{cr}}}{\mu_\infty} \tag{4.101}$$

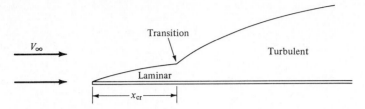

Figure 4.39 Transition from laminar to turbulent flow. The boundary layer thickness is exaggerated for clarity.

Volumes of literature have been written on the phenomenon of transition from laminar to turbulent flow. Obviously, because τ_w is different for the two flows, knowledge of where on the surface transition takes place is vital to an accurate prediction of skin friction drag. The location of the transition point (in reality, a finite region) depends on many quantities, such as the Reynolds number, Mach number, heat transfer to or from the surface, turbulence in the freestream, surface roughness, pressure gradient, etc. A comprehensive discussion of transition is beyond the scope of this book. However, if the critical Reynolds number is given to you (usually from experiments for a given type of flow), then the location of transition x_{cr} can be obtained directly from the definition, Eq. (4.101).

For example, assume that you have an airfoil of given surface roughness in a flow at a freestream velocity of 150 m/s and you wish to predict how far from the leading edge transition will take place. After searching through the literature for low-speed flows over such surfaces, you may find that the critical Reynolds number determined by experience is approximately $\text{Re}_{x_{cr}} = 5 \times 10^5$. Applying this "experience" to your problem, using Eq. (4.101) and assuming the thermodynamic conditions of the airflow correspond to standard sea level, you find

$$x_{cr} = \frac{\mu_\infty \text{Re}_{x_{cr}}}{\rho_\infty V_\infty} = \frac{(1.789 \times 10^{-5} \text{ kg/ms})(5 \times 10^5)}{(1.225 \text{ kg/m}^3)(150 \text{ m/s})} = 0.047 \text{ m}$$

Note that the region of laminar flow in this example is small—only 4.7 cm between the leading edge and the transition point. If now you double the freestream velocity to 300 m/s, the transition point is still governed by the critical Reynolds number $\text{Re}_{x_{cr}} = 5 \times 10^5$. Thus,

$$x_{cr} = \frac{(1.789 \times 10^{-5})(5 \times 10^5)}{1.225(300)} = 0.0235 \text{ m}$$

Hence, when the velocity is doubled, the transition point moves forward half the distance to the leading edge.

In summary, once you know the critical Reynolds number, you can find x_{cr} from Eq. (4.101). However, an accurate value of $\text{Re}_{x_{cr}}$ applicable to your problem must come from somewhere—experiment, free flight, or some semiempirical theory—and this may be difficult to obtain. This situation provides a little insight into why basic studies of transition and turbulence are needed to advance our understanding of such flows and to allow us to apply more valid reasoning to the prediction of transition in practical problems.

Example 4.15 The wingspan of the Wright Flyer I biplane is 40 ft 4 in, and the planform area of each wing is 255 ft^2 (see Figures 1.1 and 1.2). Assume the wing shape is rectangular (obviously not quite the case, but not bad), as shown in Figure 4.40. If the Flyer is moving with a velocity of 30 mi/h at standard sea-level conditions, calculate the skin friction drag on the wings. Assume the transition Reynolds number is 6.5×10^5. The areas of laminar and turbulent flow are illustrated by areas A and B, respectively, in Figure 4.40.

SOLUTION The general procedure is:

(a) Calculate D_f for the combined area $A + B$ assuming the flow is completely turbulent.

(b) Obtain the turbulent D_f for area B only by calculating the turbulent D_f for area A and subtracting this from the result of part (a).

(c) Calculate the laminar D_f for area A.

(d) Add results from parts (b) and (c) to obtain total drag on the complete surface $A + B$.

First, obtain some useful numbers in consistent units: $b = 40$ ft 4 in $= 40.33$ ft. Let $S =$ planform area $= A + B = 255$ ft^2. Hence $c = S/b = 255/40.33 = 6.32$ ft. At standard sea level, $\rho_\infty = 0.002377$ slug/ft^3 and $\mu_\infty = 3.7373 \times 10^{-7}$ slug/(ft)(s). Also $V_\infty = 30$ mi/h $= 30(88/60) = 44$ ft/s.

Thus

$$\text{Re}_c = \frac{\rho_\infty V_\infty c}{\mu_\infty} = \frac{0.002377(44)(6.32)}{3.7373 \times 10^{-7}}$$

$$= 1.769 \times 10^6$$

This is the Reynolds number at the trailing edge. To find x_{cr},

$$\text{Re}_{x_{cr}} = \frac{\rho_\infty V_\infty x_{cr}}{\mu_\infty}$$

$$x_{cr} = \frac{\text{Re}_{x_{cr}} \mu_\infty}{\rho_\infty V_\infty}$$

$$= \frac{(6.5 \times 10^5)(3.7373 \times 10^{-7})}{0.002377(44)} = 2.32 \text{ ft}$$

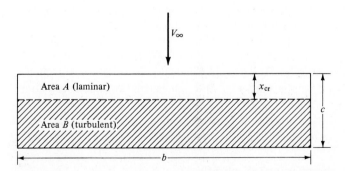

Figure 4.40 Planform view of surface experiencing transition from laminar to turbulent flow.

We are now ready to calculate drag. Assume that the wings of the Wright Flyer I are thin enough that the flat plate formulas apply.

(a) To calculate turbulent drag over the complete surface $S = A + B$, use Eq. (4.100):

$$C_f = \frac{0.074}{\mathrm{Re}_L^{0.2}} = \frac{0.074}{(1.769 \times 10^6)^{0.2}} = 0.00417$$

$$q_\infty = \tfrac{1}{2}\rho_\infty V_\infty^2 = \tfrac{1}{2}(0.002377)(44)^2 = 2.30 \ \mathrm{lb/ft^2}$$

$$(D_f)_s = q_\infty S C_f = 2.30(255)(0.00417) = 2.446 \ \mathrm{lb}$$

(b) For area A only, assuming turbulent flow,

$$C_f = \frac{0.074}{\left(\mathrm{Re}_{x_{cr}}\right)^{0.2}} = \frac{0.074}{(6.5 \times 10^5)^{0.2}} = 0.00509$$

$$(D_f)_A = q_\infty A C_f = 2.30(2.32 \times 40.33)(0.00509) = 1.095 \ \mathrm{lb}$$

Hence, the turbulent drag on area B only is

$$(D_f)_B = (D_f)_s - (D_f)_A = 2.446 - 1.095 = 1.351 \ \mathrm{lb}$$

(c) Considering the drag on area A, which is in reality a laminar drag, we obtain from Eq. (4.98)

$$C_f = \frac{1.328}{\left(\mathrm{Re}_{x_{cr}}\right)^{0.5}} = \frac{1.328}{(6.5 \times 10^5)^{0.5}} = 0.00165$$

$$(D_f)_A = q_\infty A C_f = 2.30(2.32 \times 40.33)(0.00165) = 0.354 \ \mathrm{lb}$$

(d) The total drag D_f on the surface is

$$D_f = (\text{laminar drag on } A) + (\text{turbulent drag on } B)$$

$$= 0.354 \ \mathrm{lb} + 1.351 \ \mathrm{lb} = 1.705 \ \mathrm{lb}$$

This is the drag on one surface. Each wing has a top and bottom surface, and there are two wings. Hence, the total skin friction drag on the complete biplane wing configuration is

$$D_f = 4(1.705) = \boxed{6.820 \ \mathrm{lb}}$$

4.18 FLOW SEPARATION

We have seen that the presence of friction in the flow causes a shear stress at the surface of a body, which in turn contributes to the aerodynamic drag of the body: skin friction drag. However, friction also causes another phenomenon called *flow separation*, which in turn creates another source of aerodynamic drag called *pressure drag due to separation*. The real flow field about a sphere sketched in Figure 4.26 is dominated by the separated flow on the rearward surface. Consequently, the pressure on the rearward surface is less than the pressure on the forward surface, and this imbalance of pressure forces causes a drag, hence the term "pressure drag due to separation." In comparison, the skin friction drag on the sphere is very small.

Another example of where flow separation is important is the flow over an airfoil. Consider an airfoil at a low angle of attack (low angle of incidence) to the

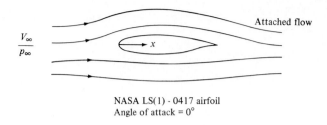

NASA LS(1) - 0417 airfoil
Angle of attack = 0°

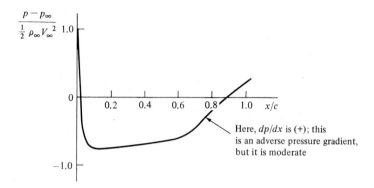

Figure 4.41 Pressure distribution over the top surface for attached flow over an airfoil. Theoretical data for a modern NASA low-speed airfoil, from NASA Conference Publication 2046, *Advanced Technology Airfoil Research*, Vol. II, March 1978, p. 11. *(After McGhee, Beasley, and Whitcomb.)*

flow, as sketched in Figure 4.41. The streamlines move smoothly over the airfoil. The pressure distribution over the top surface is also shown in Figure 4.41. Note that the pressure at the leading edge is high; the leading edge is a stagnation region, and the pressure is essentially stagnation pressure. This is the highest pressure anywhere on the airfoil. As the flow expands around the top surface of the airfoil, the surface pressure decreases dramatically, dipping to a minimum pressure, which is below the freestream static pressure p_∞. Then, as the flow moves farther downstream, the pressure gradually increases, reaching a value slightly above freestream pressure at the trailing edge. This region of increasing pressure is called a region of *adverse* pressure gradient, defined as a region where dp/dx is positive. This region is so identified in Figure 4.41. The adverse pressure gradient is moderate, that is, dp/dx is small, and for all practical purposes the flow remains attached to the airfoil surface, as sketched in Figure 4.41. The drag on this airfoil is therefore mainly skin friction drag D_f.

Now consider the same airfoil at a very high angle of attack, as shown in Figure 4.42. First, assume that we had some magic fluid that would remain attached to the surface—purely an artificial situation. If this were the case, then the pressure distribution on the top surface would follow the dashed line in Figure 4.42. The pressure would drop precipitously downstream of the leading

Separated flow

NASA LS(1) - 0417 airfoil
Angle of attack = 18.4°

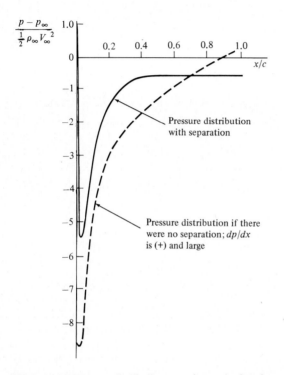

Pressure distribution
with separation

Pressure distribution if there
were no separation; dp/dx
is (+) and large

Figure 4.42 Pressure distribution over the top surface for separated flow over an airfoil. Theoretical data for a modern NASA low-speed airfoil, from NASA Conference Publication 2045, Part 1, *Advanced Technology Airfoil Research*, vol. I, March 1978, p. 380. *(After Zumwalt and Nack.)*

edge to a value far below the freestream static pressure p_∞. Farther downstream the pressure would rapidly recover to a value above p_∞. However, in this recovery, the adverse pressure gradient would no longer be moderate, as was the case in Figure 4.41. Instead, in Figure 4.42, the adverse pressure gradient would be severe, that is, dp/dx would be large. In such cases, the *real* flow field tends to separate from the surface. Therefore, in Figure 4.42, the real flow field is sketched with a large region of separated flow over the top surface of the airfoil. In this real separated flow, the *actual* surface pressure distribution is given by the *solid* curve.

In comparison to the dashed curve, note that the actual pressure distribution does not dip to as low a pressure minimum and that the pressure near the trailing edge does not recover to a value above p_∞. This has two major consequences, as can be seen from Figure 4.43. Here, the airfoil at a large angle of attack (thus, with flow separation) is shown with the real surface pressure distribution symbolized by the solid arrows. Pressure always acts normal to a surface. Hence the arrows are all perpendicular to the local surface. The length of the arrows denotes the magnitude of the pressure. A solid curve is drawn through the base of the arrows to form an "envelope" to make the pressure distribution easier to visualize. On the other hand, if the flow were *not* separated, i.e., if the flow were attached, then the pressure distribution would be that shown by the dashed arrows (and the dashed envelope). The solid and dashed arrows in Figure 4.43 qualitatively correspond to the solid and dashed pressure distribution curves, respectively, in Figure 4.42.

The solid and dashed arrows in Figure 4.43 should be looked at carefully. They explain the two major consequences brought about by the separated flow over the airfoil. The first consequence is a loss of lift. The aerodynamic lift (the vertical force shown in Figure 4.43) is derived from the net component of a pressure distribution in the vertical direction. High lift is obtained when the pressure on the bottom surface is large and the pressure on the top surface is small. Separation does not affect the bottom surface pressure distribution. How-

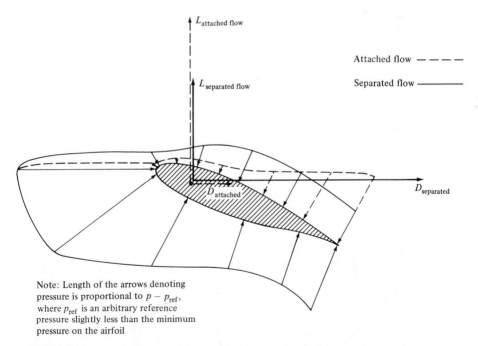

Note: Length of the arrows denoting pressure is proportional to $p - p_{ref}$, where p_{ref} is an arbitrary reference pressure slightly less than the minimum pressure on the airfoil

Figure 4.43 Qualitative comparison of pressure distribution, lift, and drag for attached and separated flow. Note that for separated flow, the lift decreases and the drag increases.

ever, comparing the solid and dashed arrows on the top surface just downstream of the leading edge, we find the solid arrows indicate a higher pressure, hence lower lift, when the flow is separated. In addition, the pressure drag is derived from the net component of the pressure distribution in the horizontal direction. This is shown as the horizontal force D in Figure 4.43. If the flow were *not* separated, the horizontal force due to pressure (the dashed arrows) would exactly cancel in the left and right directions, and there would be no drag due to pressure (again we have d'Alembert's paradox). However, in the real separated flow, the horizontal components of pressure do not cancel, and for the conditions in Figure 4.43, there is a net pressure force in the drag direction.

Therefore, two major consequences of the flow separating over an airfoil are:

1. A drastic loss of lift (stalling)
2. A major increase in drag, caused by pressure drag due to separation

When the wing of an airplane is pitched to a high angle of attack, the wing can stall, i.e., there can be a sudden loss of lift. Our discussion above gives the physical reasons for this stalling phenomenon. Additional ramifications of stalling will be discussed in Chap. 5.

Before ending this discussion of separated flow, let us ask the question; why does a flow separate from a surface? The answer is combined in the concept of an adverse pressure gradient (dp/dx is positive) and the velocity profile through the boundary layer, as shown in Figure 4.33. If dp/dx is positive, then the fluid elements moving along a streamline have to work their way "uphill" against an increasing pressure. Consequently, the fluid elements will slow down under the influence of an adverse pressure gradient. For the fluid elements moving outside the boundary layer, where the velocity (and hence kinetic energy) is high, this is not much of a problem. The fluid element keeps moving downstream. However, consider a fluid element deep inside the boundary layer. Looking at Figure 4.33, we see its velocity is small. It has been retarded by friction forces. The fluid element still encounters the same adverse pressure gradient, but its velocity is too low to negotiate the increasing pressure. As a result, the element comes to a stop somewhere downstream and then reverses its direction. Such reversed flow causes the flow field in general to separate from the surface, as shown in Figure 4.42. This is physically how separated flow develops.

Reflecting once again on Figure 4.33, note that turbulent boundary layers have fuller velocity profiles. At a given distance from the surface (a given value of y), the velocity of a fluid element in a turbulent boundary is higher than in a laminar boundary layer. Hence, in turbulent boundary layers, there is more flow kinetic energy nearer the surface. Hence, the flow is less inclined to separate. This leads to a very fundamental fact. Laminar boundary layers separate more easily than turbulent boundary layers. *Hence, to help prevent flow field separation, we want a turbulent boundary layer.*

4.19 SUMMARY OF VISCOUS EFFECTS ON DRAG

We have seen that the presence of friction in a flow produces two sources of drag:

1. Skin friction drag D_f due to shear stress at the wall
2. Pressure drag due to flow separation, D_p, sometimes identified as *form* drag

The *total* drag which is caused by viscous effects is then

$$\underset{\substack{\text{Total drag} \\ \text{due to viscous} \\ \text{effects}}}{D} = \underset{\substack{\text{Drag due} \\ \text{to skin} \\ \text{friction}}}{D_f} + \underset{\substack{\text{Drag due to} \\ \text{separation} \\ \text{(pressure drag)}}}{D_p} \qquad (4.102)$$

Equation (4.102) contains one of the classic compromises of aerodynamics. In previous sections, we have pointed out that skin friction drag is reduced by maintaining a laminar boundary layer over a surface. However, we have also pointed out at the end of Sec. 4.18 that turbulent boundary layers inhibit flow separation; hence pressure drag due to separation is reduced by establishing a turbulent boundary layer on the surface. Therefore, in Eq. (4.102) we have the following compromise:

$$D = \underset{\substack{\text{Less for laminar,} \\ \text{more for turbulent}}}{D_f} + \underset{\substack{\text{More for laminar,} \\ \text{less for turbulent}}}{D_p}$$

Consequently, as discussed at the end of Sec. 4.14, it cannot be said in general that either laminar or turbulent flow is preferable over the other. Any preference depends on the specific application. For example, for a blunt body such as the sphere in Figure 4.26, the drag is mainly pressure drag due to separation; hence, turbulent boundary layers reduce the drag on spheres and are therefore preferable. (We will discuss this again in Chap. 5). On the other hand, for a slender body such as a sharp, slender cone or a thin airfoil at small angles of attack to the flow, the drag is mainly skin friction drag; hence, laminar boundary layers are preferable in this case. For in-between cases, the ingenuity of the designer along with practical experience helps to determine what compromises are best.

As a final note to this section, the total drag D given by Eq. (4.102) is called *profile drag* because both skin friction and pressure drag due to separation are ramifications of the shape and size of the body, i.e., the "profile" of the body. The profile drag D is the total drag on an aerodynamic shape due to viscous effects. However, it is not in general the total aerodynamic drag on the body. There is one more source of drag, induced drag, which will be discussed in the next chapter.

4.20 HISTORICAL NOTE: BERNOULLI AND EULER

Equation (4.9) is one of the oldest and most powerful equations in fluid dynamics. It is credited to Daniel Bernoulli, who lived during the eighteenth century; little did Bernoulli know that his concept would find widespread application in the aeronautics of the twentieth century. Who was Bernoulli, and how did Bernoulli's equation come about? Let us briefly look into these questions; the answers will lead us to a rather unexpected conclusion.

Daniel Bernoulli (1700–1782) was born in Groningen, Netherlands, on January 29, 1700. He was a member of a remarkable family. His father, Johann Bernoulli, was a noted mathematician who made contributions to differential and integral calculus and who later became a doctor of medicine. Jakob Bernoulli, who was Johann's brother (Daniel's uncle), was an even more accomplished mathematician; he made major contributions to the calculus; he coined the term "integral." Sons of both Jakob and Johann, including Daniel, went on to become noted mathematicians and physicists. The entire family was Swiss and made their home in Basel, Switzerland, where they held various professorships at the University of Basel. Daniel Bernoulli was born away from Basel only because his father spent 10 years as professor of mathematics in the Netherlands. With this type of pedigree, Daniel could hardly avoid making contributions to mathematics and science himself.

And indeed he did make contributions. For example, he had insight into the kinetic theory of gases; he theorized that a gas was a collection of individual particles moving about in an agitated fashion and correctly associated the increased temperature of a gas with increased energy of the particles. These ideas, originally published in 1738, were to lead a century later to a mature understanding of the nature of gases and heat and helped to lay the foundation for the elegant kinetic theory of gases.

Daniel's thoughts on the kinetic motion of gases were published in his book *Hydrodynamica* (1738). However, this book was to etch his name more deeply in association with fluid mechanics than kinetic theory. The book was started in 1729, when Daniel was a professor of mathematics at Leningrad (then St. Petersburg) in Russia. By this time he was already well recognized; he had won 10 prizes offered by the Royal Academy of Sciences in Paris for his solution of various mathematical problems. In his *Hydrodynamica* (which was published completely in Latin), Bernoulli ranged over such topics as jet propulsion, manometers, and flow in pipes. He also attempted to obtain a relationship between pressure and velocity, but his derivation was obscure. In fact, even though Bernoulli's equation, Eq. (4.9), is usually ascribed to Daniel via his *Hydrodynamica*, the precise equation is not to be found in the book! The picture is further complicated by his father, Johann, who published a book in 1743 entitled *Hydraulica*. It is clear from this latter book that the father understood Bernoulli's theorem better than the son; Daniel thought of pressure strictly in terms of the height of a manometer column, whereas Johann had the more fundamental understanding that pressure was a force acting on the fluid. However, neither of the Bernoullis understood that pressure is a point property. That was to be left to Leonhard Euler.

Leonhard Euler (1707–1783) was also a Swiss mathematician. He was born at Basel, Switzerland, on April 15, 1707, seven years after the birth of Daniel

Bernoulli. Euler went on to become one of the mathematical giants of history, but his contributions to fluid dynamics are of interest here. Euler was a close friend of the Bernoullis; indeed, he was a student of Johann Bernoulli at the University of Basel. Later, Euler followed Daniel to St. Petersburg, where he became a professor of mathematics. It was here that Euler was influenced by the work of the Bernoullis in hydrodynamics, but more influenced by Johann than by Daniel. Euler originated the concept of pressure acting at a point in a gas. This quickly led to his differential equation for a fluid accelerated by gradients in pressure, the same equation we have derived as Eq. (4.8). In turn, Euler integrated the differential equation to obtain, for the first time in history, Bernoulli's equation, just as we have obtained Eq. (4.9). Hence we see that Bernoulli's equation, Eq. (4.9), is really a historical misnomer. Credit for Bernoulli's equation is legitimately shared by Euler.

4.21 HISTORICAL NOTE: THE PITOT TUBE

The use of a Pitot tube to measure airspeed is described in Sec. 4.11; indeed, the Pitot tube is today so commonly used in aerodynamic laboratories and on aircraft that it is almost taken for granted. However, this simple little device has had a rather interesting and somewhat obscure history, as follows.

The Pitot tube is named after its inventor, Henri Pitot (1695–1771). Born in Aramon, France, in 1695, Pitot began his career as an astronomer and mathematician. He was accomplished enough to be elected to the Royal Academy of Sciences, Paris, in 1724. About this time, Pitot became interested in hydraulics and, in particular, in the flow of water in rivers and canals. However, he was not satisfied with the existing technique of measuring the flow velocity, which was to observe the speed of a floating object on the surface of the water. So he devised an instrument consisting of two tubes; one was simply a straight tube open at one end which was inserted vertically into the water (to measure static pressure), and the other was a tube with one end bent at right angles, with the open end facing directly into the flow (to measure total pressure). In 1732, between two piers of a bridge over the Seine River in Paris, he used this instrument to measure the flow velocity of the river. This invention and first use of the Pitot tube was announced by Pitot to the Academy on November 12, 1732. In his presentation, he also presented some data of major importance on the variation of water flow velocity with depth. Contemporary theory, based on experience of some Italian engineers, held that the flow velocity at a given depth was proportional to the mass above it; hence the velocity was thought to increase with depth. Pitot reported the stunning (and correct) results, measured with his instrument, that in reality the flow velocity decreased as the depth increased. Hence, the Pitot tube was introduced with style.

Interestingly enough, Pitot's invention soon fell into disfavor with the engineering community. A number of investigators attempted to use just the Pitot tube itself, without a local static pressure measurement. Others, using the device under uncontrolled conditions, produced spurious results. Various shapes and forms other than a simple tube were sometimes used for the mouth of the instrument. Moreover, there was no agreed-upon rational theory of the Pitot tube. Note that Pitot developed his instrument in 1732, six years *before* Daniel Bernoulli's *Hydrodynamica* and well before Euler had developed the Bernoullis' concepts into Eq. (4.9), as discussed in Sec. 4.20. Hence, Pitot used intuition, not theory, to establish that the pressure difference measured by his instrument was an indication of the square of the local flow velocity. Of course, as described in Sec. 4.11, we now clearly understand that a Pitot-static device measures the difference between total and static pressure and that for incompressible flow, this difference is related to the velocity squared through Bernoulli's equation; i.e., from Eq. (4.62),

$$p_0 - p = \tfrac{1}{2}\rho V^2$$

However, for more than 150 years after Pitot's introduction of the instrument, various engineers attempted to interpret readings in terms of

$$p_0 - p = \tfrac{1}{2}K\rho V^2$$

where K was an empirical constant, generally much different than unity. Controversy was still raging as late as 1913, when John Airey, a professor of mechanical engineering from the University of Michigan, finally performed a series of well-controlled experiments in a water tow tank, using Pitot probes of six different shapes. These shapes are shown in Figure 4.44, which is taken from Airey's paper in the April 17, 1913, issue of the *Engineering News* entitled "Notes on the Pitot Tube." In this paper, Airey states that all his measurements indicate that $K = 1.0$ within 1 percent accuracy, independent of the shape of the tube. Moreover, he presents a rational theory based on Bernoulli's equation. Further comments on these results are made in a paper entitled "Origin and Theory of the Pitot Tube" by A. E. Guy, the chief engineer of a centrifugal pump company in Pittsburgh, in a later, June 5, 1913, issue of the *Engineering News*. This paper also helped to establish the Pitot tube on firmer technical grounds.

It is interesting to note that neither of these papers in 1913 mentioned what was to become the most prevalent use of the Pitot tube, namely, the measurement of airspeed for airplanes and wind tunnels. The first practical airspeed indicator, a Venturi tube, was used on an aircraft by the French Captain A. Eteve in January 1911, more than seven years after the first powered flight. Later in 1911, British engineers at the Royal Aircraft Establishment (RAE) at Farnborough employed a Pitot tube on an airplane for the first time. This was to eventually evolve into the primary instrument for flight speed measurement.

There was still controversy over Pitot tubes, as well as the need for reliable airspeed measurements, in 1915, when the brand-new National Advisory Committee for Aeronautics (NACA) stated in its First Annual Report that "an important problem to aviation in general is the devising of accurate, reliable and durable air speed meters.... The Bureau of Standards is now engaged in

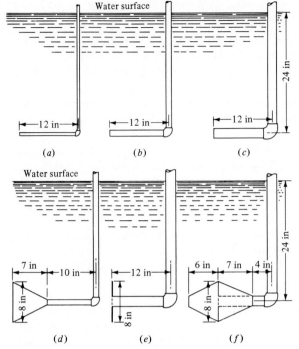

Figure 4.44 Six forms of Pitot tubes tested by John Airey. *(From En-gineering News, vol. 69, no. 16, p. 783, April 1913.)*

investigation of such meters, and attention is invited to the report of Professor Herschel and Dr. Buckingham of the bureau on Pitot tubes." The aforementioned report was NACA report no. 2, part 1, "The Pitot Tube and other Anemometers for Aeroplanes," by W. H. Herschel, and part 2, "The Theory of the Pitot and Venturi Tubes," by E. Buckingham. Part 2 is of particular interest. In clear terms, it gives a version of the theory we have developed in Sec. 4.11 for the Pitot tube, and moreover it develops for the first time the theory for *compressible* subsonic flow—quite unusual for 1915! Buckingham shows that, to obtain 0.5 percent accuracy with the incompressible relations, V_∞ should not exceed 148 mi/h = 66.1 m/s. However, he goes on to state that "since the accuracy of better than 1.0 percent can hardly be demanded of an airplane speedometer, it is evident that for all ordinary speeds of flight, no correction for compressibility is needed... ." This was certainly an appropriate comment for the "ordinary" airplanes of that day; indeed, it was accurate for most aircraft until the 1930s.

In retrospect, we see that the Pitot tube was invented almost 250 years ago but that its use was controversial and obscure until the second decade of powered flight. Then, between 1911 and 1915, one of those "explosions" in technical advancement occurred. Pitot tubes found a major home on airplanes, and the appropriate theory for their correct use was finally established. Since then, Pitot tubes have become commonplace. Indeed, the Pitot tube is usually the first aerodynamic instrument introduced to students of aerospace engineering in their laboratory studies.

4.22 HISTORICAL NOTE: THE FIRST WIND TUNNELS

Aerospace engineering in general, and aerodynamics in particular, is an empirically based discipline. Discovery and development by experimental means have been its lifeblood, extending all the way back to George Cayley (see Chap. 1). In turn, the workhorse for such experiments has been predominantly the wind tunnel, so much so that today most aerospace industrial, government, and university laboratories have a complete spectrum of wind tunnels ranging from low subsonic to hypersonic speeds.

It is interesting to reach back briefly into history and look at the evolution of wind tunnels. Amazingly enough, this history goes back more than 400 years, because the cardinal principle of wind-tunnel testing was stated by Leonardo da Vinci near the beginning of the sixteenth century as follows:

> For since the action of the medium upon the body is the same whether the body moves in a quiescent medium, or whether the particles of the medium impinge with the same velocity upon the quiescent body; let us consider the body as if it were quiescent and see with what force it would be impelled by the moving medium.

This is almost self-evident today, that the lift and drag of an aerodynamic body are the same whether it moves through the stagnant air at 100 mi/h or whether the air moves over the stationary body at 100 mi/h. This concept is the very foundation of wind-tunnel testing.

The first actual wind tunnel in history was designed and built over 100 years ago by Francis Wenham at Greenwich, England, in 1871. We have met Mr. Wenham once before, in Sec. 1.4, where his activity in the Aeronautical Society of Great Britain was noted. Wenham's tunnel was nothing more than a 10-ft-long wooden box with a square cross section, 18 in on a side. A steam-driven fan at the front end blew air through the duct. There was no contour, hence no aerodynamic control or enhancement of the flow. Plane aerodynamic surfaces were placed in the airstream at the end of the box, where Wenham measured the lift and drag on weighing beams linked to the model.

Thirteen years later, Horatio F. Phillips, also an Englishman, built the second known wind tunnel in history. Again, the flow duct was a box, but Phillips used steam ejectors (high-speed steam nozzles) downstream of the test section to suck air through the tunnel. Phillips went on to conduct some pioneering airfoil testing in his tunnel, which will be mentioned again in Chap. 5.

Other wind tunnels were built before the turning point in aviation in 1903. For example, the first wind tunnel in Russia was due to Nikolai Joukowski at the University of Moscow in 1891 (it had a 2-in diameter). A larger, 7 in × 10 in tunnel was built in Austria in 1893 by Ludwig Mach, son of the famed scientist and philosopher Ernst Mach, after whom the Mach number is named. The first tunnel in the United States was built at the Massachusetts Institute of Technology in 1896 by Alfred J. Wells, who used the machine to measure the drag on a flat plate as a check on the whirling arm measurements of Langley (see Sec. 1.8).

Another tunnel in the United States was built by Dr. A. Heb Zahm at the Catholic University of America in 1901. In light of these activities, it is obvious that at the turn of the twentieth century aerodynamic testing in wind tunnels was poised and ready to burst forth with the same energy that accompanied the development of the airplane itself.

It is fitting that the same two people responsible for getting the airplane off the ground should also have been responsible for the first concentrated series of wind-tunnel tests. As noted in Sec. 1.9, the Wright brothers in late 1901 concluded that a large part of the existing aerodynamic data was erroneous. This led to their construction of a 6-ft-long 16-in-square wind tunnel powered by a two-bladed fan connected to a gasoline engine. An original photograph of the Wrights' wind tunnel in their Dayton bicycle shop is shown in Figure 4.45. They designed and built their own balance to measure the ratios of lift to drag. Using this apparatus, Wilbur and Orville undertook a major program of aeronautical research between September 1901 and August 1902. During this time, they tested over 200 different airfoil shapes manufactured out of wax. The results from these tests constitute the first major impact of wind-tunnel testing on the development of a successful airplane. As we quoted in Sec. 1.9, Orville said about their results: "Our tables of air pressure which we made in our wind tunnel would enable us to

Figure 4.45 An original photograph of the Wright brothers' wind tunnel in their bicycle shop in Dayton, Ohio, 1901 to 1902.

calculate in advance the performance of a machine." What a fantastic development! This was a turning point in the history of wind-tunnel testing, and it had as much impact on that discipline as the December 17, 1903, flight had on the airplane.

The rapid growth in aviation after 1903 was paced by the rapid growth of wind tunnels, both in numbers and in technology. For example, tunnels were built at the National Physical Laboratory in London in 1903, in Rome in 1903, in Moscow in 1905, in Göttingen, Germany (by the famous Dr. Ludwig Prandtl, originator of the boundary layer concept in fluid dynamics) in 1908, in Paris in 1909 (including two built by Gustave Eiffel, of tower fame), and again at the National Physical Laboratory in 1910 and 1912.

All of these tunnels, quite naturally, were low-speed facilities, but they were pioneering for their time. Then, in 1915, with the creation of the NACA (see Sec. 2.6), the foundation was laid for some major spurts in wind-tunnel design. The first NACA wind tunnel became operational at the Langley Memorial Aeronautical Laboratory at Hampton, Virginia, in 1920. It had a 5-ft-diameter test section which accommodated models up to 3.5 ft wide. Then in 1923, in order to simulate the higher Reynolds numbers associated with flight, the NACA built the first variable-density wind tunnel, a facility that could be pressurized to 20 atm in the flow and therefore obtain a 20-fold increase in density, hence Re, in the test section. During the 1930s and 1940s, subsonic wind tunnels grew larger and larger. In 1931, an NACA wind tunnel with a 30 ft × 60 ft oval test section went into operation at Langley with a 129-mi/h maximum flow velocity. This was the first million-dollar tunnel in history. Later, in 1944, a 40 ft × 80 ft tunnel with a flow velocity of 265 mi/h was initiated at Ames Aeronautical Laboratory at Moffett Field, California. This is still the largest wind tunnel in the world today. Figure 4.46 shows the magnitude of such tunnels: whole airplanes can be mounted in the test section!

The tunnels mentioned above were low-speed, essentially incompressible flow tunnels. They were the cornerstone of aeronautical testing until the 1930s and remain an important part of the aerodynamic scene today. However, airplane speeds were progressively increasing, and new wind tunnels with higher velocity capability were needed. Indeed, the first requirement for high-speed subsonic tunnels was established by propellers—in the 1920s and 1930s the propeller diameters and rotational speeds were both increasing so as to encounter compressibility problems at the tips. This problem led the NACA to build a 12-in-diameter high-speed tunnel at Langley in 1927. It could produce a test section flow of 765 mi/h. In 1936, to keep up with increasing airplane speeds, Langley built a large 8-ft high-speed wind tunnel providing 500 mi/h. This was increased to 760 mi/h in 1945. An important facility was built at Ames in 1941, a 16-ft tunnel with an airspeed of 680 mi/h. A photograph of the Ames 16-ft tunnel is shown in Figure 4.47 just to provide a feeling for the massive size involved with such a facility.

In the early 1940s, the advent of the V-2 rocket as well as the jet engine put supersonic flight in the minds of aeronautical engineers. Suddenly, the require-

Figure 4.46 A subsonic wind tunnel large enough to test a full-size airplane. The NASA Langley Research Center 30 ft × 60 ft tunnel.

Figure 4.47 The Ames 16-ft high-speed subsonic wind tunnel, illustrating the massive size that goes along with such a wind tunnel complex. *(Courtesy NASA Ames Research Center.)*

ment for supersonic tunnels became a major factor. However, supersonic flows in the laboratory and in practice date farther back than this. The first supersonic nozzle was developed by Laval about 1880 for use with steam turbines. This is why convergent-divergent nozzles are frequently called Laval nozzles. In 1905, Prandtl built a small Mach 1.5 tunnel at Göttingen to be used to study steam turbine flows and (of all things) the moving of sawdust around sawmills.

The first practical supersonic wind tunnel for aerodynamic testing was developed by Dr. A. Busemann at Braunschweig, Germany, in the mid-1930s. Using the "method of characteristics" technique, which he had developed in 1929, Busemann designed the first smooth supersonic nozzle contour which produced shock-free isentropic flow. He had a diffuser with a second throat downstream to decelerate the flow and to obtain efficient operation of the tunnel. A photograph of Busemann's tunnel is shown in Figure 4.48. All supersonic tunnels today look essentially the same.

Working from Busemann's example, the Germans built two major supersonic tunnels at their research complex at Peenemünde during World War II. These were used for research and development of the V-2 rocket. After the war, these tunnels were moved almost in total to the U.S. Naval Ordnance Laboratory (later, one was moved to the University of Maryland), where they are still in use today. However, the first supersonic tunnel built in the United States was designed by Theodore von Karman and his colleagues at the California Institute of Technology in 1944 and was built and operated at the Army Ballistics Research Laboratory at Aberdeen, Maryland, under contract with Cal Tech. Then, the

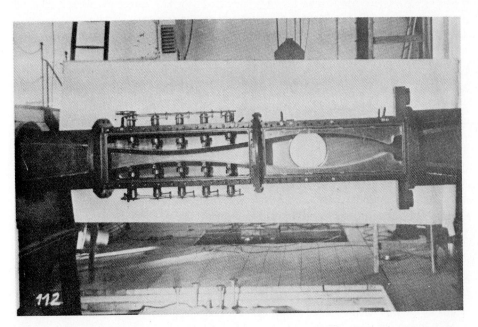

Figure 4.48 The first practical supersonic wind tunnel, built by A. Busemann in the mid-1930s. *(Courtesy of A. Busemann.)*

1950s saw a virtual harvest of supersonic wind tunnels, one of the largest being the 16 ft × 16 ft continuously operated supersonic tunnel of the Air Force at the Arnold Engineering Development Center (AEDC) in Tennessee.

About this time, the development of the intercontinental ballistic missile (ICBM) was on the horizon, soon to be followed by the space program of the 1960s. Flight vehicles were soon to encounter velocities as high as 36,000 ft/s in the atmosphere, hypersonic velocities. In turn, hypersonic wind tunnels ($M > 5$) were suddenly in demand. The first hypersonic wind tunnel was operated by the NACA at Langley in 1947. It had an 11-in-square test section capable of Mach 7. Three years later, another hypersonic tunnel went into operation at the Naval Ordnance Laboratory. These tunnels are distinctly different from their supersonic relatives in that, to obtain hypersonic speeds, the flow has to be expanded so far that the temperature decreases to the point of liquifying the air. To prevent this, all hypersonic tunnels, both old and new, have to have the reservoir gas heated to temperatures far above room temperature before its expansion through the nozzle. Heat transfer is a problem to high-speed flight vehicles, and such heating problems feed right down to the ground-testing facilities for such vehicles.

In summary, modern wind-tunnel facilities slash across the whole spectrum of flight velocities, from low subsonic to hypersonic speeds. These facilities are part of the everyday life of aerospace engineering; hopefully, this brief historical sketch has provided some insight into their tradition and development.

4.23 HISTORICAL NOTE: OSBORNE REYNOLDS AND HIS NUMBER

In Secs. 4.14 to 4.17, we observed that the Reynolds number, defined in Eq. (4.90) as $\text{Re} = \rho_\infty V_\infty x / \mu_\infty$, was the governing parameter for viscous flow. Boundary layer thickness, skin friction drag, transition to turbulent flow, and many other characteristics of viscous flow depend explicitly on the Reynolds number. Indeed, we can readily show that the Reynolds number itself has physical meaning—it is proportional to the ratio of inertia forces to viscous forces in a fluid flow. Clearly, the Reynolds number is an extremely important dimensionless parameter in fluid dynamics. Where did the Reynolds number come from? When was it first introduced, and under what circumstances? The Reynolds number is named after a man—Osborne Reynolds. Who was Reynolds? The purpose of this section is to address these questions.

First, let us look at Osborne Reynolds, the man. He was born on October 23, 1842, in Belfast, Ireland. He was raised in an intellectual family atmosphere; his father had been a fellow of Queens College, Cambridge, a principal of Belfast Collegiate School, headmaster of Dedham Grammar School in Essex, and finally rector at Debach-with-Boulge in Suffolk. Indeed, Anglican clerics were a tradition in the Reynolds family; in addition to his father, his grandfather and great-grandfather had also been rectors at Debach. Against this background, Osborne Reynolds's early education was carried out by his father at Dedham. In his teens, he already showed an intense interest in the study of mechanics, for which he had a natural aptitude. At the age of 19, he served a short apprenticeship in mechanical engineering before attending Cambridge University a year later. Reynolds was a highly successful student at Cambridge, graduating with the highest honors in mathematics. In 1867, he was elected a fellow of Queens College, Cambridge (an honor earlier bestowed upon his father). He went on to serve one year as a practicing civil engineer in the office of John Lawson in London. However, in 1868, Owens College in Manchester (later to become the University of Manchester) established its chair of engineering—the second of its kind in any English university (the first was the chair of civil engineering established at the University College, London, in 1865). Reynolds applied for this chair, writing in his application:

From my earliest recollection I have had an irresistible liking for mechanics and the physical laws on which mechanics as a science are based. In my boyhood I had the advantage of the constant guidance of my father, also a lover of mechanics and a man of no mean attainment in mathematics and their application to physics.

Despite his youth and relative lack of experience, Reynolds was appointed to the chair at Manchester. For the next 37 years he would serve as a professor at Manchester until his retirement in 1905.

During those 37 years, Reynolds distinguished himself as one of history's leading practitioners of classical mechanics. For his first years at Manchester, he worked on problems involving electricity, magnetism, and the electromagnetic properties of solar and cometary phenomena. After 1873, he focused on fluid mechanics—the area in which he made his lasting contributions. For example, he (1) developed Reynolds's analogy in 1874, a relation between heat transfer and frictional shear stress in a fluid, (2) measured the average specific heat of water between freezing and boiling, which ranks among the classic determinations of physical constants, (3) studied water currents and waves in estuaries, (4) developed turbines and pumps, and (5) studied the propagation of sound waves in fluids. However, his most important work, and the one which gave birth to the concept of the Reynolds number, was reported in 1883 in a paper entitled "An Experimental Investigation of the Circumstances which Determine whether the Motion of Water in Parallel Channels Shall Be Direct or Sinuous, and of the Law of Resistance in Parallel Channels." Published in the *Proceedings of the Royal Society*, this paper was the first to demonstrate the transition from laminar to turbulent flow and relate this transition to a critical value of a dimensionless parameter—later to become known as the Reynolds number. Reynolds studied this phenomenon in the water flow through pipes. His experimental apparatus is illustrated in Figure 4.49, taken from his original 1883 paper. (Note that before the day of modern photographic techniques, some technical papers contained

Figure 4.49 Osborne Reynolds's apparatus for his famous pipe-flow experiments. This figure is from his original paper, referenced in the text.

rather elegant hand sketches of experimental apparatus, of which Figure 4.49 is an example.) Reynolds filled a large reservoir with water, which fed into a glass pipe through a larger bell-mouth entrance. As the water flowed through the pipe, Reynolds introduced dye into the middle of the stream, at the entrance of the bell mouth. What happened to this thin filament of dye as it flowed through the pipe is illustrated in Figure 4.50, also from Reynolds original paper. The flow is from right to left. If the flow velocity was small, the thin dye filament would travel downstream in a smooth, neat, orderly fashion, with a clear demarcation between the dye and the rest of the water, as illustrated in Figure 4.50*a*. However, if the flow velocity was increased beyond a certain value, the dye filament would suddenly become unstable and would fill the entire pipe with color, as shown in Figure 4.50*b*. Reynolds clearly pointed out that the smooth dye filament in Figure 4.50*a* corresponded to laminar flow in the pipe, whereas the agitated and totally diffused dye filament in Figure 4.50*b* was due to turbulent flow in the pipe. Furthermore, Reynolds studied the details of this turbulent flow by visually observing the pipe flow illuminated by a momentary electric spark, much as we would use a strobe light today. He saw that the turbulent flow consisted of a large number of distinct eddies, as sketched in Figure 4.50*c*. The transition from laminar to turbulent flow occurred when the parameter defined by $\rho VD/\mu$ exceeded a certain critical value, where ρ was the density of the water, V was the mean flow velocity, μ was the viscosity coefficient, and D was the diameter of the pipe. This dimensionless parameter, first introduced by Reynolds, later became known as the Reynolds number. Reynolds measured the critical value of this number, above which turbulent flow occurred, as 2300. This original work of

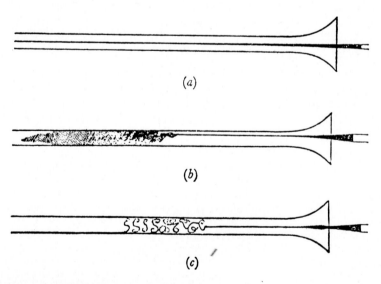

Figure 4.50 Development of turbulent flow in pipes, as observed and sketched by Reynolds. This figure is from his original paper, referenced in the text.

Reynolds initiated the study of transition from laminar to turbulent flow as a new field of research in fluid dynamics—a field which is still today one of the most important and insufficiently understood areas of aerodynamics.

Reynolds was a scholarly man with high standards. Engineering education was new to English universities at that time, and Reynolds had definite ideas about its proper form. He felt that all engineering students, no matter what their specialty, should have a common background based on mathematics, physics, and in particular the fundamentals of classical mechanics. At Manchester, he organized a systematic engineering curriculum covering the basics of civil and mechanical engineering. Ironically, despite his intense interest in education, as a lecturer in the classroom Reynolds left something to be desired. His lectures were hard to follow, and his topics frequently wandered with little or no connection. He was known to have new ideas during the course of a lecture and to spend the remainder of the lecture working out these ideas on the board, seemingly oblivious to the students in the classroom. That is, he did not "spoon-feed" his students, and many of the poorer students did not pass his courses. In contrast, the best students enjoyed his lectures and found them stimulating. Many of Reynolds's successful students went on to become distinguished engineers and scientists, the most notable being Sir J. J. Thomson, later the Cavendish Professor of Physics at Cambridge; Thomson is famous for first demonstrating the existence of the electron in 1897, for which he received the Nobel prize in 1906.

In regard to Reynolds's interesting research approach, his student, colleague, and friend, Professor A. H. Gibson, had this to say in his biography of Reynolds, written for the British Council in 1946:

> Reynolds' approach to a problem was essentially individualistic. He never began by reading what others thought about the matter, but first thought this out for himself. The novelty of his approach to some problems made some of his papers difficult to follow, especially those written during his later years. His more descriptive physical papers, however, make fascinating reading, and when addressing a popular audience, his talks were models of clear exposition.

At the turn of the century, Reynolds's health began to fail, and he subsequently had to retire in 1905. The last years of his life were ones of considerably diminished physical and mental capabilities, a particularly sad state for such a brilliant and successful scholar. He died at Somerset, England, in 1912. Sir Horace Lamb, one of history's most famous fluid dynamicists and a long-time colleague of Reynolds, wrote after Reynolds's death:

> The character of Reynolds was, like his writings, strongly individual. He was conscious of the value of his work, but was content to leave it to the mature judgement of the scientific world. For advertisement he had no taste, and undue pretensions on the part of others only elicited a tolerant smile. To his pupils he was most generous in the opportunities for valuable work which he put in their way, and in the share of co-operation. Somewhat reserved in serious or personal matters and occasionally combative and tenacious in debate, he was in the ordinary relations of life the most kindly and genial of companions. He had a keen sense of humor and delighted in startling paradoxes, which he would maintain, half seriously and half playfully, with astonishing ingenuity and resource. The illness which at length compelled his retirement was felt as a grievous calamity by his pupils, his colleagues and other friends throughout the country.

The purpose of this section has been to relate the historical beginnings of the Reynolds number in fluid mechanics. From now on, when you use the Reynolds number, hopefully you will view it not only as a powerful dimensionless parameter governing viscous flow, but also as a testimonial to its originator—one of the famous fluid dynamicists of the nineteenth century.

4.24 HISTORICAL NOTE: PRANDTL AND THE DEVELOPMENT OF THE BOUNDARY LAYER CONCEPT

The modern science of aerodynamics has its roots as far back as Isaac Newton, who devoted the entire second book of his *Principia* (1687) to fluid dynamics—especially to the formulation of "laws of resistance" (drag). He noted that drag is a function of fluid density, velocity, and shape of the body in motion. However, Newton was unable to formulate the correct equation for drag. Indeed, he derived a formula which gave the drag on an inclined object as proportional to the sine squared of the angle of attack. Later, Newton's sine-squared law was used to demonstrate the "impossibility of heavier-than-air flight" and served to hinder the intellectual advancement of flight in the nineteenth century. Ironically, the physical assumptions used by Newton in deriving his sine-squared law approximately reflect the conditions of hypersonic flight, and the "newtonian law" has been used since 1950 in the design of high Mach number vehicles. However, Newton correctly reasoned the mechanism of shear stress in a fluid. In section 9 of book 2 of the *Principia*, Newton states the following hypothesis: "The resistance arising from want of lubricity in the parts of a fluid is...proportional to the velocity with which the parts of the fluid are separated from each other." This is the first statement in history of the friction law for laminar flow; it is embodied in Eq. (4.89), which describes a so-called "newtonian fluid."

Further attempts to understand fluid dynamic drag were made by the French mathematician Jean le Rond d'Alembert, who is noted for developing the calculus of partial differences (leading to the mathematics of partial differential equations). In 1768, d'Alembert applied the equations of motion for an incompressible, inviscid (frictionless) flow about a two-dimensional body in a moving fluid and found that no drag is obtained. He wrote: "I do not see then, I admit, how one can explain the resistance of fluids by the theory in a satisfactory manner. It seems to me on the contrary that this theory, dealt with and studied with profound attention gives, at least in most cases, resistance absolutely zero: a singular paradox which I leave to geometricians to explain." That this theoretical result of zero drag is truly a paradox was clearly recognized by d'Alembert, who also conducted experimental research on drag and who was among the first to discover that drag was proportional to the square of the velocity, as derived in Sec. 5.3 and given in Eq. (5.18).

D'Alembert's paradox arose due to the neglect of friction in the classical theory. It was not until a century later that the effect of friction was properly incorporated in the classical equations of motion by the work of M. Navier (1785–1836) and Sir George Stokes (1819–1903). The so-called Navier-Stokes equations stand today as the classical formulation of fluid dynamics. However, in general they are nonlinear equations and are extremely difficult to solve; indeed, only with the numerical power of modern high-speed digital computers are "exact" solutions of the Navier-Stokes equations finally being obtained for general flow fields. Also in the nineteenth century, the first experiments on transition from laminar to turbulent flow were carried out by Osborne Reynolds (1842–1912), as related in Sec. 4.23. In his classic paper of 1883 entitled "An Experimental Investigation of the Circumstances which Determine whether the Motion of Water in Parallel Channels Shall Be Direct or Sinuous, and of the Law of Resistance in Parallel Channels," Reynolds observed a filament of colored dye in a pipe flow and noted that transition from laminar to turbulent flow always corresponded to approximately the same value of a dimensionless number, $\rho VD/\mu$, where D was the diameter of the pipe. This was the origin of the Reynolds number, defined in Sec. 4.14, and discussed at length in Sec. 4.23.

Therefore, at the beginning of the twentieth century, when the Wright brothers were deeply involved in the development of the first successful airplane, the development of theoretical fluid dynamics still had not led to practical results for aerodynamic drag. It was this environment into which Ludwig Prandtl was born on February 4, 1875, at Freising, in Bavaria, Germany. Prandtl was a genius who had the talent of cutting through a maze of complex physical phenomena to extract the most salient points and put them in simple mathematical form. Educated as a physicist, Prandtl was appointed in 1904 as professor of applied mechanics at Göttingen University in Germany, a post he occupied until his death in 1953.

In the period 1902–1904, Prandtl made one of the most important contributions to fluid dynamics. Thinking about the viscous flow over a body, he reasoned that the flow velocity right at the surface was zero and that if the Reynolds number was high enough, the influence of friction was limited to a thin layer (Prandtl first called it a transition layer) near the surface. Therefore, the analysis of the flow field could be divided into two distinct regions—one close to the surface, which included friction, and the other farther away, in which friction could be neglected. In one of the most important fluid dynamics papers in history, entitled "Uber Flussigkeitsbewegung bei sehr kleiner Reibung," Prandtl reported his thoughts to the Third International Mathematical Congress at Heidelberg in 1904. In this paper, Prandtl observed:

A very satisfactory explanation of the physical process in the boundary layer (Grenzschicht) between a fluid and a solid body could be obtained by the hypothesis of an adhesion of the fluid to the walls, that is, by the hypothesis of a zero relative velocity between fluid and wall. If the viscosity is very small and the fluid path along the wall not too long, the fluid velocity ought to resume its normal value at a very short distance from the wall. In the thin transition layer however, the sharp changes of velocity, even with small coefficient of friction, produce marked results.

In the same paper, Prandtl's theory is applied to the prediction of flow separation:

In given cases, in certain points fully determined by external conditions, the fluid flow ought to separate from the wall. That is, there ought to be a layer of fluid which, having been set in rotation by the friction on the wall, insinuates itself into the free fluid, transforming completely the motion of the latter.....

Prandtl's boundary layer hypothesis allows the Navier-Stokes equations to be reduced to a simpler form; by 1908, Prandtl and one of his students, H. Blasius, had solved these simpler boundary layer equations for laminar flow over a flat plate, yielding the equations for boundary layer thickness and skin friction drag given by Eqs. (4.91) and (4.93). Finally, after centuries of effort, the first rational resistance laws describing fluid dynamic drag due to friction had been obtained.

Prandtl's work was a stroke of genius, and it revolutionized theoretical aerodynamics. However, possibly due to the language barrier, it only slowly diffused through the worldwide technical community. Serious work on boundary layer theory did not emerge in England and the United States until the 1920s. By that time, Prandtl and his students at Göttingen had applied it to various aerodynamic shapes and were including the effects of turbulence.

Prandtl has been called the "father of aerodynamics," and rightly so. His contributions extend far beyond boundary layer theory; e.g., he pioneered the development of wing lift and drag theory, as will be seen in the next chapter. Moreover, he was interested in more fields than just fluid dynamics—he made several important contributions to structural mechanics as well.

As a note on Prandtl's personal life, he had the singleness of purpose which seems to drive many giants of humanity. However, his almost complete preoccupation with his work led to a somewhat naive outlook on life. Theodore von Karman, one of Prandtl's most illustrious students, relates that Prandtl would rather find fancy in the examination of children's toys than participate in social gatherings. When Prandtl was near 40, he suddenly decided that it was time to get married, and he wrote to a friend for the hand of one of his two daughters—Prandtl did not care which one! During the 1930s and early 1940s, Prandtl had mixed emotions about the political problems of the day. He continued his research work at Göttingen under Hitler's Nazi regime but became continually confused about the course of events. Von Karman writes about Prandtl in his autobiography:

I saw Prandtl once again for the last time right after the Nazi surrender. He was a sad figure. The roof of his house in Göttingen, he mourned, had been destroyed by an American bomb. He couldn't understand why this had been done to him! He was also deeply shaken by the collapse of Germany. He lived only a few years after that, and though he did engage in some research work in meteorology, he died, I believe, a broken man, still puzzled by the ways of mankind.

Prandtl died in Göttingen on August 15, 1953. Of any fluid dynamicist or aerodynamicist in history, Prandtl came closest to deserving a Nobel prize. Why he never received one is an unanswered question. However, as long as there are flight vehicles, and as long as students study the discipline of fluid dynamics, the name of Ludwig Prandtl will be enshrined for posterity.

4.25 CHAPTER SUMMARY

A few of the important concepts from this chapter are summarized as follows:

1. The basic equations of aerodynamics, in the form derived here, are:

Continuity: $$\rho_1 A_1 V_1 = \rho_2 A_2 V_2 \tag{4.2}$$

Momentum: $$dp = -\rho V\, dV \tag{4.8}$$

Energy: $$c_p T_1 + \tfrac{1}{2} V_1^2 = c_p T_2 + \tfrac{1}{2} V_2^2 \tag{4.42}$$

These equations hold for a compressible flow. For an incompressible flow, we have:

Continuity: $$A_1 V_1 = A_2 V_2 \tag{4.3}$$

Momentum: $$p_1 + \rho \frac{V_1^2}{2} = p_2 + \rho \frac{V_2^2}{2} \tag{4.9a}$$

Equation (4.9a) is called Bernoulli's equation.

2. The change in pressure, density, and temperature between two points in an isentropic process is given by

$$\frac{p_2}{p_1} = \left(\frac{\rho_2}{\rho_1}\right)^{\gamma} = \left(\frac{T_2}{T_1}\right)^{\gamma/(\gamma-1)}$$

3. The speed of sound is given by

$$a = \sqrt{\left(\frac{dp}{d\rho}\right)_{\text{isentropic}}} \tag{4.48}$$

For a perfect gas, this becomes

$$a = \sqrt{\gamma R T} \tag{4.54}$$

4. The speed of a gas flow can be measured by a Pitot tube, which senses the total pressure p_0. For incompressible flow,

$$V_1 = \sqrt{\frac{2(p_0 - p_1)}{\rho}} \tag{4.66}$$

For subsonic compressible flow,

$$V_1^2 = \frac{2a_1^2}{\gamma - 1}\left[\left(\frac{p_0}{p_1}\right)^{(\gamma-1)/\gamma} - 1\right]$$ (4.77a)

For supersonic flow, a shock wave exists in front of the Pitot tube, and Eq. (4.79) must be used in lieu of Eq. (4.77a) in order to find the Mach number of the flow.

5. The area-velocity relation for isentropic flow is

$$\frac{dA}{A} = (M^2 - 1)\frac{dV}{V}$$ (4.83)

From this relation, we observe that (1) for a subsonic flow, the velocity increases in a convergent duct and decreases in a divergent duct, (2) for a supersonic flow, the velocity increases in a divergent duct and decreases in a convergent duct, and (3) the flow is sonic only at the minimum area.

6. The isentropic flow of a gas is governed by

$$\frac{p_0}{p_1} = \left(1 + \frac{\gamma - 1}{2}M_1^2\right)^{\gamma/(\gamma-1)}$$ (4.74)

$$\frac{T_0}{T_1} = 1 + \frac{\gamma - 1}{2}M_1^2$$ (4.73)

$$\frac{\rho_0}{\rho_1} = \left(1 + \frac{\gamma - 1}{2}M_1^2\right)^{1/(\gamma-1)}$$ (4.75)

Here T_0, p_0, and ρ_0 are the total temperature, pressure, and density, respectively. For an isentropic flow, $p_0 = $ const throughout the flow. Similarly, $\rho_0 = $ const and $T_0 = $ const throughout the flow.

7. Viscous effects create a boundary layer along a solid surface in a flow. In this boundary layer, the flow moves slowly and the velocity goes to zero right at the surface. The shear stress at the wall is given by

$$\tau_w = \mu\left(\frac{dV}{dy}\right)_{y=0}$$ (4.89)

The shear stress is larger for a turbulent boundary layer than for a laminar boundary layer.

8. For a laminar boundary layer, on a flat plate,

$$\delta = \frac{5.2x}{\sqrt{Re_x}}$$ (4.91)

and

$$C_f = \frac{1.328}{\sqrt{Re_L}}$$ (4.98)

where δ is the boundary layer thickness, C_f is the total skin friction drag coefficient, and Re is the Reynolds number;

$$\text{Re}_x = \frac{\rho_\infty V_\infty x}{\mu_\infty} \qquad \text{(local Reynolds number)}$$

$$\text{Re}_L = \frac{\rho_\infty V_\infty L}{\mu_\infty} \qquad \text{(plate Reynolds number)}$$

Here, x is the running length along the plate, and L is the total length of the plate.

9. For a turbulent boundary layer on a flat plate,

$$\delta = \frac{0.37x}{\text{Re}_x^{0.2}} \qquad (4.99)$$

$$C_f = \frac{0.074}{\text{Re}_L^{0.2}} \qquad (4.100)$$

10. Any real flow along a surface first starts out as laminar but then changes into a turbulent flow. The point where this transition effectively occurs (in reality, transition occurs over a finite length) is designated x_{cr}. In turn, the critical Reynolds number for transition is defined as

$$\text{Re}_{x_{cr}} = \frac{\rho_\infty V_\infty x_{cr}}{\mu_\infty} \qquad (4.101)$$

11. Whenever a boundary layer encounters an adverse pressure gradient (a region of increasing pressure in the flow direction), it can readily separate from the surface. On an airfoil or wing, such flow separation decreases the lift and increases the drag.

BIBLIOGRAPHY

Airey, J., "Notes on the Pitot Tube," *Engineering News*, vol. 69, no. 16, April 17, 1913, pp. 782–783.

Goin, K. L., "The History, Evolution, and Use of Wind Tunnels," *AIAA Student Journal*, February 1971, pp. 3–13.

Guy, A. E., "Origin and Theory of the Pitot Tube," *Engineering News*, vol. 69, no. 23, June 5, 1913, pp. 1172–1175.

Kuethe, A. M., and Schetzer, J. D., *Foundations of Aerodynamics*, Wiley, New York, 1959.

Pope, A. *Aerodynamics of Supersonic Flight*, Pitman, New York, 1958.

von Karman, T., *Aerodynamics*, McGraw-Hill, New York, 1963.

PROBLEMS

4.1 Consider the incompressible flow of water through a divergent duct. The inlet velocity and area are 5 ft/s and 10 ft^2, respectively. If the exit area is four times the inlet area, calculate the water flow velocity at the exit.

4.2 In the above problem, calculate the pressure difference between the exit and the inlet. The density of water is 62.4 lb_m/ft^3.

4.3 Consider an airplane flying with a velocity of 60 m/s at a standard altitude of 3 km. At a point on the wing, the airflow velocity is 70 m/s. Calculate the pressure at this point. Assume incompressible flow.

4.4 An instrument used to measure the airspeed on many early low-speed airplanes, principally during the 1919–1930 time period, was the venturi tube. This simple device is a convergent-divergent duct. (The front section's cross-sectional area A decreases in the flow direction, and the back section's cross-sectional area increases in the flow direction. Somewhere in between the inlet and exit of the duct, there is a minimum area, called the throat.) Let A_1 and A_2 denote the inlet and throat areas, respectively. Let p_1 and p_2 be the pressures at the inlet and throat, respectively. The venturi tube is mounted at a specific location on the airplane (generally on the wing or near the front of the fuselage), where the inlet velocity V_1 is essentially the same as the freestream velocity, i.e., the velocity of the airplane through the air. With a knowledge of the area ratio A_2/A_1 (a fixed design feature) and a measurement of the pressure difference $p_1 - p_2$, the airplane's velocity can be determined. For example, assume $A_2/A_1 = 1/4$, and $p_1 - p_2 = 80$ lb/ft^2. If the airplane is flying at standard sea level, what is its velocity?

4.5 Imagine that you have designed a low-speed airplane with a maximum velocity at sea level of 90 m/s. For your airspeed instrument, you plan to use a venturi tube with a 1.3 : 1 area ratio. Inside the cockpit is an airspeed indicator—a dial that is connected to a pressure gauge sensing the venturi tube pressure difference $p_1 - p_2$ and properly calibrated in terms of velocity. What is the maximum pressure difference you would expect the gauge to experience?

4.6 A supersonic nozzle is also a convergent-divergent duct, which is fed by a large reservoir at the inlet to the nozzle. In the reservoir of the nozzle, the pressure and temperature are 10 atm and 300 K, respectively. At the nozzle exit, the pressure is 1 atm. Calculate the temperature and density of the flow at the exit. Assume the flow is isentropic and, of course, compressible.

4.7 Derive an expression for the exit velocity of a supersonic nozzle in terms of the pressure ratio between the reservoir and exit, p_0/p_e, and the reservoir temperature T_0.

4.8 Consider an airplane flying at a standard altitude of 5 km with a velocity of 270 m/s. At a point on the wing of the airplane, the velocity is 330 m/s. Calculate the pressure at this point.

4.9 The mass flow of air through a supersonic nozzle is 1.5 lb_m/s. The exit velocity is 1500 ft/s, and the reservoir temperature and pressure are 1000°R and 7 atm, respectively. Calculate the area of the nozzle exit. For air, $c_p = 6000$ ft · lb/(slug) (°R).

4.10 Calculate the Mach number at the exit of the nozzle in Prob. 4.9.

4.11 A Boeing 747 is cruising at a velocity of 250 m/s at a standard altitude of 13 km. What is its Mach number?

4.12 A high-speed missile is traveling at Mach 3 at standard sea level. What is its velocity in miles per hour?

4.13 Consider a low-speed subsonic wind tunnel with a nozzle contraction ratio of 1/20. One side of a mercury manometer is connected to the settling chamber, and the other side to the test section. The pressure and temperature in the test section are 1 atm and 300 K, respectively. What is the height difference between the two columns of mercury when the test section velocity is 80 m/s?

4.14 A Pitot tube is mounted in the test section of a low-speed subsonic wind tunnel. The flow in the test section has a velocity, static pressure, and temperature of 150 mi/h, 1 atm, and 70°F, respectively. Calculate the pressure measured by the Pitot tube.

4.15 The altimeter on a low-speed Piper Aztec reads 8000 ft. A Pitot tube mounted on the wingtip measures a pressure of 1650 lb/ft^2. If the outside air temperature is 500°R, what is the true velocity of the airplane? What is the equivalent airspeed?

4.16 The altimeter on a low-speed airplane reads 2 km. The airspeed indicator reads 50 m/s. If the outside air temperature is 280 K, what is the true velocity of the airplane?

4.17 A Pitot tube is mounted in the test section of a high-speed subsonic wind tunnel. The pressure and temperature of the airflow are 1 atm and 270 K, respectively. If the flow velocity is 250 m/s, what is the pressure measured by the Pitot tube?

4.18 A high-speed subsonic Boeing 707 airliner if flying at a pressure altitude of 12 km. A Pitot tube on the vertical tail measures a pressure of 2.96×10^4 N/m². At what Mach number is the airplane flying?

4.19 A high-speed subsonic airplane is flying at Mach 0.65. A Pitot tube on the wingtip measures a pressure of 2339 lb/ft². What is the altitude reading on the altimeter.

4.20 A high-performance F-16 fighter is flying at Mach 0.96 at sea level. What is the air temperature at the stagnation point at the leading edge of the wing?

4.21 An airplane is flying at a pressure altitude of 10 km with a velocity of 596 m/s. The outside air temperature is 220 K. What is the pressure measured by a Pitot tube mounted on the nose of the airplane?

4.22 Consider the flow of air through a supersonic nozzle. The reservoir pressure and temperature are 5 atm and 500 K, respectively. If the Mach number at the nozzle exit is 3, calculate the exit pressure, temperature, and density.

4.23 Consider a supersonic nozzle across which the pressure ratio is $p_e/p_0 = 0.2$. Calculate the ratio of exit area to throat area.

4.24 The wing of the Fairchild Republic A-10A twin-jet close-support airplane is approximately a rectangular shape with a wingspan (the length perpendicular to the flow direction) of 17.5 m and a chord (the length parallel to the flow direction) of 3 m. The airplane is flying at standard sea level with a velocity of 200 m/s. If the flow is considered to be completely laminar, calculate the boundary layer thickness at the trailing edge and the total skin friction drag. Assume the wing is approximated by a flat plate.

4.25 In Prob. 4.24, assume the flow is completely turbulent. Calculate the boundary layer thickness at the trailing edge and the total skin friction drag. Compare these turbulent results with the above laminar results.

4.26 If the critical Reynolds number for transition is 10^6, calculate the skin friction drag for the wing in Prob. 4.24.

CHAPTER FIVE

AIRFOILS, WINGS, AND OTHER AERODYNAMIC SHAPES

There can be no doubt that the inclined plane is the true principle of aerial navigation by mechanical means.

Sir George Cayley, 1843

5.1 INTRODUCTION

It is remarkable that the modern airplane as we know it today, with its fixed wing and vertical and horizontal tail surfaces, was first conceived by George Cayley in 1799, more than 175 years ago. He inscribed his first concept on a silver disc (presumably for permanence) shown in Figure 1.5. It is also remarkable that Cayley recognized that a curved surface (as shown on the silver disc) creates more lift than a flat surface. Cayley's fixed-wing concept was a true revolution in the development of heavier-than-air flight machines. Prior to his time, aviation enthusiasts had been doing their best to imitate mechanically the natural flight of birds, which led to a series of human-powered flapping-wing designs (ornithopters), which never had any real possibility of working. In fact, even Leonardo da Vinci devoted a considerable effort to the design of many types of ornithopters in the late fifteenth century, of course to no avail. In such ornithopter designs, the flapping of the wings was supposed to provide simultaneously both lift (to sustain the machine in the air) and propulsion (to push it along in flight). Cayley is responsible for directing people's minds away from imitating bird flight and for separating the two principles of lift and propulsion. He proposed and demonstrated that lift can be obtained from a fixed, straight wing inclined to the airstream, while propulsion can be provided by some independent mechanism such as paddles or airscrews. For this concept and for his many other thoughts and inventions in aeronautics, Sir George Cayley is truly the parent of modern aviation. A more detailed discussion of Cayley's contributions is given in Chap. 1. However, emphasis is made that much of the technology discussed in the present chapter had its origins at the beginning of the nineteenth century—technology that came to fruition on December 17, 1903, near Kitty Hawk, North Carolina.

The following sections develop some of the terminology and basic aerodynamic fundamentals of airfoils and wings. These concepts form the heart of airplane flight and they represent a major excursion into aeronautical engineering.

5.2 AIRFOIL NOMENCLATURE

Consider the wing of an airplane, as sketched in Figure 5.1. The cross-sectional shape obtained by the intersection of the wing with the perpendicular plane shown in Figure 5.1 is called an *airfoil*. Such an airfoil is sketched in Figure 5.2, which defines some basic terminology. The major design feature of an airfoil is the *mean camber line*, which is the locus of points halfway between the upper and lower surfaces as measured perpendicular to the mean camber line itself. The most forward and rearward points of the mean camber line are the *leading* and *trailing edges*, respectively. The straight line connecting the leading and trailing edges is the *chord line* of the airfoil, and the precise distance from the leading to the trailing edge measured along the chord line is simply designated the *chord* of the airfoil, given by the symbol c. The *camber* is the maximum distance between the mean camber line and the chord line, measured perpendicular to the chord line. The camber, the shape of the mean camber line, and to a lesser extent, the thickness distribution of the airfoil essentially control the lift and moment characteristics of the airfoil.

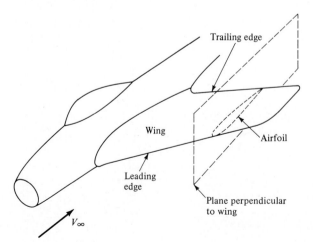

Figure 5.1 Sketch of a wing and airfoil.

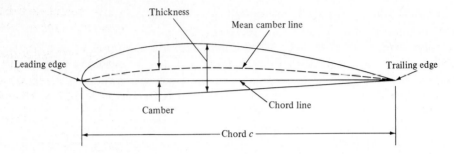

Figure 5.2 Airfoil nomenclature. The shape shown here is a NACA 4415 airfoil.

More definitions are illustrated in Figure 5.3, which shows an airfoil inclined to a stream of air. The freestream velocity V_∞ is the velocity of the air far upstream of the airfoil. The *direction* of V_∞ is defined as the *relative wind*. The angle between the relative wind and the chord line is the *angle of attack* α of the airfoil. As described in Chaps. 2 and 4, there is an aerodynamic force created by the pressure and shear stress distributions over the wing surface. This resultant force is shown by the vector R in Figure 5.3. In turn, the aerodynamic force R can be resolved into two forces, parallel and perpendicular to the relative wind. The *drag D* is always defined as the component of the aerodynamic force *parallel to the relative wind*. The *lift L* is always defined as the component of the aerodynamic force *perpendicular to the relative wind*.

In addition to lift and drag, the surface pressure and shear stress distributions also create a *moment M* which tends to *rotate* the wing. To see more clearly how this moment is created, consider the surface pressure distribution over an airfoil

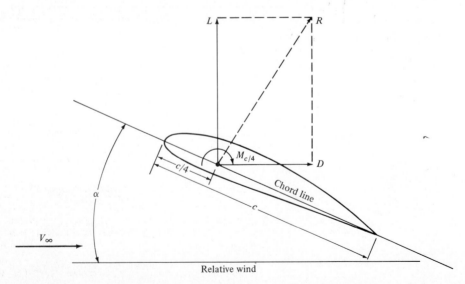

Figure 5.3 Sketch showing the definitions of lift, drag, moments, angle of attack, and relative wind.

as sketched in Figure 5.4 (we will ignore the shear stress for this discussion). Consider just the pressure on the top surface of the airfoil. This pressure gives rise to a net force F_1 in the general downward direction. Moreover, F_1 acts through a given point on the chord line, point 1, which can be found by integrating the pressure times distance over the surface (analogous to finding the centroid or center of pressure from integral calculus). Now consider just the pressure on the bottom surface of the airfoil. This pressure gives rise to a net force F_2 in the general upward direction, acting through point 2. The total aerodynamic force on the airfoil is the *summation* of F_1 and F_2, and lift is obtained when $F_2 > F_1$. However, note from Figure 5.4 that F_1 and F_2 will create a moment which will tend to rotate the airfoil. Moreover, the value of this aerodynamically induced moment depends on the point about which we chose to take moments. For example, if we take moments about the leading edge, the aerodynamic moment is designated M_{LE}. It is more common in the case of subsonic airfoils to take moments about a point on the chord at a distance $c/4$ from the leading edge, the *quarter-chord point*, as illustrated in Figure 5.3. This moment about the quarter chord is designated $M_{c/4}$. In general, $M_{LE} \neq M_{c/4}$. Intuition will tell you that lift, drag, and moments on a wing will change as the angle of attack α changes. In fact, the variations of these aerodynamic quantities with α represent some of the most important information that an airplane designer needs to know. We will address this matter in the following sections. However, it should be pointed out that, although M_{LE} and $M_{c/4}$ are both functions of α, there exists a certain point on the airfoil about which moments essentially *do not* vary with α. This point is defined as the *aerodynamic center*, and the moment about the aerodynamic center is designated M_{ac}. By definition,

$$M_{ac} = \text{const}$$

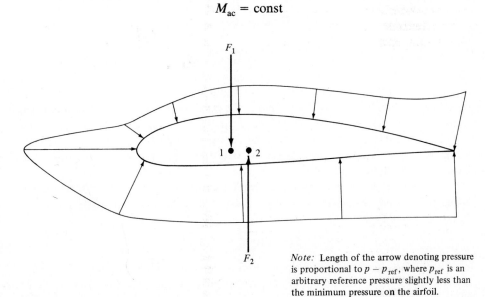

Note: Length of the arrow denoting pressure is proportional to $p - p_{ref}$, where p_{ref} is an arbitrary reference pressure slightly less than the minimum pressure on the airfoil.

Figure 5.4 The physical origin of moments on an airfoil.

independent of angle of attack. The location of the aerodynamic center for real aerodynamic shapes can be found from experiment. For low-speed subsonic airfoils, the aerodynamic center is generally very close to the quarter-chord point.

5.3 LIFT, DRAG, AND MOMENT COEFFICIENTS

Again appealing to intuition, we note that it makes sense that for an airplane in flight, the actual magnitudes of L, D, and M depend not only on α, but on velocity and altitude as well. In fact, we can expect that the variations of L, D, and M depend at least on:

1. Freestream velocity V_∞.
2. Freestream density ρ_∞, that is, on altitude.
3. Size of the aerodynamic surface. For airplanes, we will use the *wing area S* to indicate size.
4. Angle of attack α.
5. Shape of the airfoil.
6. Viscosity coefficient μ_∞ (because the aerodynamic forces are generated in part from skin friction distributions).
7. Compressibility of the airflow. In Chap. 4 we demonstrated that compressibility effects are governed by the value of the freestream Mach number, $M_\infty = V_\infty/a_\infty$. Since V_∞ is already listed above, we can therefore designate a_∞ as our index for compressibility.

Hence, we can write that, for a given shape airfoil at a given angle of attack,

$$L = f(V_\infty, \rho_\infty, S, \mu_\infty, a_\infty) \tag{5.1}$$

and D and M are similar functions.

In principle, for a given airfoil at a given angle of attack, we could find the variation of L by performing a myriad of wind-tunnel experiments wherein V_∞, ρ_∞, S, μ_∞, and a_∞ are individually varied and then try to make sense out of the resulting huge collection of data. This is the hard way. Instead, we could ask the question, are there *groupings* of the quantities V_∞, ρ_∞, S, μ_∞, a_∞, and L such that Eq. (5.1) can be written in terms of fewer parameters? The answer is yes. In the process of developing this answer, we will gain some insight into the beauty of nature as applied to aerodynamics.

The technique we will apply is a simple example of a more general theoretical approach called *dimensional analysis*. Let us assume that Eq. (5.1) is of the functional form

$$L = ZV_\infty^a \rho_\infty^b S^d a_\infty^e \mu_\infty^f \tag{5.2}$$

where Z, a, b, d, e, and f are dimensionless constants. However, no matter what the values of these constants may be, it is a physical fact that the dimensions of the left- and right-hand sides of Eq. (5.2) must match; i.e., if L is a force (say in newtons), then the net result of all the exponents and multiplication on the right-hand side must also produce a result with dimensions of a force. This constraint will ultimately give us information on the values of a, b, etc. If we designate the basic dimensions of mass, length, and time by m, l, and t, respectively, then the dimensions of various physical quantities are as given below.

Physical Quantity	Dimensions
L	ml/t^2 (from Newton's second law)
V_∞	l/t
ρ_∞	m/l^3
S	l^2
a_∞	l/t
μ_∞	m/lt

Thus, equating the *dimensions* of the left- and right-hand sides of Eq. (5.2), we obtain

$$\frac{ml}{t^2} = \left(\frac{l}{t}\right)^a \left(\frac{m}{l^3}\right)^b (l^2)^d \left(\frac{l}{t}\right)^e \left(\frac{m}{lt}\right)^f \qquad (5.3)$$

Consider mass m. The exponent of m on the left-hand side is 1. Thus, the exponents of m on the right must add up to 1. Hence

$$1 = b + f \qquad (5.4)$$

Similarly, for time t we have

$$-2 = -a - e - f \qquad (5.5)$$

and for length l,

$$1 = a - 3b + 2d + e - f \qquad (5.6)$$

Solving Eqs. (5.4) to (5.6) for a, b, and d in terms of e and f yields

$$b = 1 - f \qquad (5.7)$$

$$a = 2 - e - f \qquad (5.8)$$

$$d = 1 - \frac{f}{2} \qquad (5.9)$$

Substituting Eqs. (5.7) to (5.9) into (5.2) gives

$$L = Z(V_\infty)^{2-e-f} \rho_\infty^{1-f} S^{1-f/2} a_\infty^{e} \mu_\infty^{f} \qquad (5.10)$$

Rearranging Eq. (5.10), we find

$$L = Z\rho_\infty V_\infty^2 S \left(\frac{a_\infty}{V_\infty}\right)^e \left(\frac{\mu_\infty}{\rho_\infty V_\infty S^{1/2}}\right)^f \qquad (5.11)$$

Note that $a_\infty/V_\infty = 1/M_\infty$, where M_∞ is the freestream Mach number. Also note that the dimensions of S are l^2; hence the dimension of $S^{1/2}$ is l, purely a length. Let us choose this length to be the chord c by convention. Hence, $\mu_\infty/\rho_\infty V_\infty S^{1/2}$ can be replaced in our consideration by the equivalent quantity

$$\mu_\infty/\rho_\infty V_\infty c$$

However, $\mu_\infty/\rho_\infty V_\infty c \equiv 1/\text{Re}$, where Re is based on the chord length c. Hence, Eq. (5.11) becomes

$$L = Z\rho_\infty V_\infty^2 S\left(\frac{1}{M_\infty}\right)^e \left(\frac{1}{\text{Re}}\right)^f \tag{5.12}$$

We now *define* a new quantity called the *lift coefficient* c_l as

$$c_l/2 \equiv Z\left(\frac{1}{M_\infty}\right)^e \left(\frac{1}{\text{Re}}\right)^f \tag{5.13}$$

Then, Eq. (5.12) becomes

$$L = \tfrac{1}{2}\rho_\infty V_\infty^2 S c_l \tag{5.14}$$

Recalling from Chap. 4 that the dynamic pressure is $q_\infty \equiv \tfrac{1}{2}\rho_\infty V_\infty^2$, we transform Eq. (5.14) into

$$
\begin{array}{ccccccc}
L & = & q_\infty & \times & S & \times & c_l \\
\uparrow & & \uparrow & & \uparrow & & \uparrow \\
\text{Lift} & & \text{Dynamic} & & \text{Wing} & & \text{Lift} \\
& & \text{pressure} & & \text{area} & & \text{coefficient}
\end{array}
\tag{5.15}
$$

Look what has happened! Equation (5.1), written from intuition, but not very useful, has cascaded to the simple, direct form of Eq. (5.15), which contains a tremendous amount of information. In fact, Eq. (5.15) is one of the most important relations in applied aerodynamics. It says that the lift is directly proportional to the dynamic pressure (hence to the square of the velocity). It is also directly proportional to the wing area S and to the lift coefficient c_l. In fact, Eq. (5.15) can be turned around and used as a *definition* for the lift coefficient:

$$c_l \equiv \frac{L}{q_\infty S} \tag{5.16}$$

That is, the lift coefficient is always defined as the aerodynamic lift divided by the dynamic pressure and some reference area (for wings, the convenient reference area S, as we have been using).

The lift coefficient is a function of M_∞ and Re as reflected in Eq. (5.13). Moreover, since M_∞ and Re are dimensionless and since Z was assumed initially as a dimensionless constant, from Eq. (5.13) c_l is dimensionless. This is also consistent with Eqs. (5.15) and (5.16). Also, recall that the above derivation was carried out for an airfoil of given shape and at a given angle of attack α. If α were to vary, then c_l would also vary. Hence, for a given airfoil,

$$c_l = f(\alpha, M_\infty, \text{Re}) \tag{5.17}$$

This relation is important. Fix in your mind that lift coefficient is a function of angle of attack, Mach number, and Reynolds number.

Performing a similar dimensional analysis on drag and moments, beginning with relations analogous to Eq. (5.1), we find that

$$D = q_\infty S c_d \qquad (5.18)$$

where c_d is a dimensionless *drag coefficient*, and

$$M = q_\infty S c c_m \qquad (5.19)$$

where c_m is a dimensionless *moment coefficient*. Note that Eq. (5.19) differs slightly from Eqs. (5.15) and (5.18) by the inclusion of the chord length c. This is because L and D have dimensions of a force, whereas M has dimensions of a force-length product.

The importance of Eqs. (5.15) to (5.19) cannot be emphasized too much. They are fundamental to all of applied aerodynamics. They are readily obtained from dimensional analysis, which essentially takes us from loosely defined functional relationships [such as Eq. (5.1)] to well-defined relations between dimensionless quantities [Eqs. (5.15) to (5.19)]. In summary, for an airfoil of given shape, the dimensionless lift, drag, and moment coefficients have been defined as

$$c_l = \frac{L}{q_\infty S} \qquad c_d = \frac{D}{q_\infty S} \qquad c_m = \frac{M}{q_\infty S c} \qquad (5.20)$$

where

$$c_l = f_1(\alpha, M_\infty, \mathrm{Re}) \qquad c_d = f_2(\alpha, M_\infty, \mathrm{Re}) \qquad c_m = f_3(\alpha, M_\infty, \mathrm{Re})$$

$$(5.21)$$

Reflecting for an instant, we find there may appear to be a conflict in our aerodynamic philosophy. On the one hand, Chaps. 2 and 4 emphasized that lift, drag, and moments on an aerodynamic shape stem from the detailed pressure and shear stress distributions on the surface and that measurements and/or calculations of these distributions, especially for complex configurations, are not trivial undertakings. On the other hand, Eqs. (5.20) indicate that lift, drag, and moments can be quickly obtained from simple formulas. The bridge between these two outlooks is, of course, the lift, drag, and moment coefficients. All the physical complexity of the flow field around an aerodynamic body is implicitly buried in c_l, c_d, and c_m. Before the simple Eqs. (5.20) can be used to calculate lift, drag, and moments for an airfoil, wing, or body, the appropriate aerodynamic coefficients must be known. From this point of view, the simplicity of Eqs. (5.20) is a bit deceptive. These equations simply shift the forces of aerodynamic rigor from the forces and moments themselves to the appropriate coefficients instead. So we are now led to the questions, how do we obtain values of c_l, c_d, and c_m for given configurations, and how do they vary with α, M_∞, and Re? The answers are introduced in the following sections.

5.4 AIRFOIL DATA

A goal of theoretical aerodynamics is to predict values of c_l, c_d, and c_m from the basic equations and concepts of physical science, some of which were discussed in previous chapters. However, simplifying assumptions are usually necessary to make the mathematics tractable. Therefore, when theoretical results are obtained, they are generally not "exact." The use of high-speed digital computers to solve the governing flow equations is now bringing us much closer to the accurate calculation of aerodynamic characteristics; however, there are still limitations imposed by the numerical methods themselves, and the storage and speed capacity of current computers is still not sufficient to solve many complex aerodynamic flows. As a result, the practical aerodynamicist has to rely upon direct *experimental* measurements of c_l, c_d, and c_m for specific bodies of interest.

A large bulk of experimental airfoil data was compiled over the years by the National Advisory Committee for Aeronautics (NACA), which was absorbed in the creation of the National Aeronautics and Space Administration (NASA) in 1958. Lift, drag, and moment coefficients were systematically measured for many airfoil shapes in low-speed subsonic wind tunnels. These measurements were carried out on straight, constant-chord wings which completely spanned the tunnel test section from one side wall to the other. In this fashion, the flow essentially "saw" a wing with no wingtips, and the experimental airfoil data were thus obtained for "infinite wings." (The distinction between infinite and finite wings will be made in subsequent sections.) Some results of these airfoil measurements are given in Appendix D. The first page of Appendix D gives data for c_l and $c_{m,c/4}$ versus angle of attack for the NACA 1408 airfoil. The second page gives c_d and $c_{m,\mathrm{ac}}$ versus c_l for the same airfoil. Since c_l is known as a function of α from the first page, then the data from both pages can be cross-plotted to obtain the variation of c_d and $c_{m,\mathrm{ac}}$ versus α. The remaining pages of Appendix D give the same type of data for different standard NACA airfoil shapes.

Let us examine the variation of c_l with α more closely. This variation is sketched in Figure 5.5. The experimental data indicate that c_l varies *linearly* with α over a large range of angle of attack. Thin-airfoil theory, which is the subject of more advanced books on aerodynamics, also predicts the same type of linear variation. The slope of the linear portion of the lift curve is designated as $a_0 \equiv dc_l/d\alpha \equiv$ lift slope. Note that in Figure 5.5, when $\alpha = 0$, there is still a positive value of c_l, that is, there is still some lift even when the airfoil is at zero angle of attack to the flow. This is due to the positive camber of the airfoil. All airfoils with such camber have to be pitched to some negative angle of attack before zero lift is obtained. The value of α when lift is zero is defined as the *zero lift angle of attack* $\alpha_{L=0}$ and is illustrated in Figure 5.5. This effect is further demonstrated in Figure 5.6, where the lift curve for a cambered airfoil is

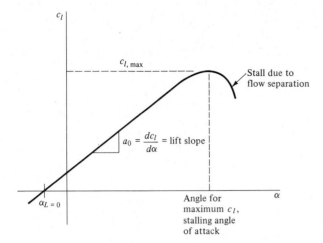

Figure 5.5 Sketch of a typical lift curve.

compared with that for a symmetric (no camber) airfoil. Note that the lift curve for a symmetric airfoil goes through the origin. Refer again to Figure 5.5, at the other extreme: for large values of α, the linearity of the lift curve breaks down. As α is increased beyond a certain value, c_l peaks at some maximum value, $c_{l,\max}$, and then drops precipitously as α is further increased. In this situation, where the lift is rapidly decreasing at high α, the airfoil is *stalled*.

The phenomenon of airfoil stall is of critical importance in airplane design. It is caused by flow separation on the upper surface of the airfoil. This is illustrated in Figure 5.7, which again shows the variation of c_l versus α for an airfoil. At point 1 on the linear portion of the lift curve, the flow field over the airfoil is attached to the surface, as pictured in Figure 5.7. However, as discussed in Chap. 4, the effect of friction is to slow the airflow near the surface; in the presence of

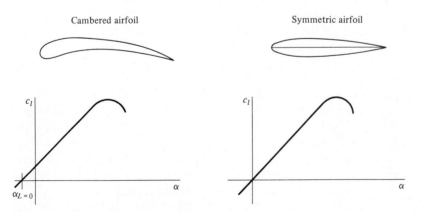

Figure 5.6 Comparison of lift curves for cambered and symmetric airfoils.

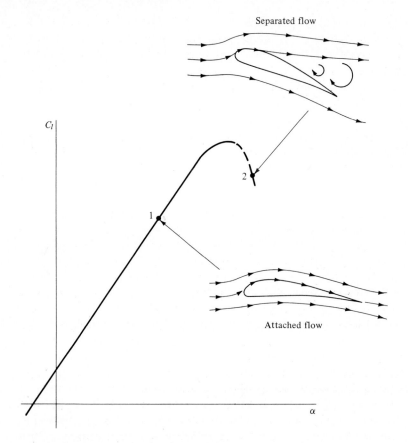

Figure 5.7 Flow mechanism associated with stalling.

an adverse pressure gradient, there will be a tendency for the boundary layer to separate from the surface. As the angle of attack is increased, the adverse pressure gradient on the top surface of the airfoil will become stronger, and at some value of α—the stalling angle of attack—the flow becomes separated from the top surface. When separation occurs, the lift decreases drastically and the drag increases suddenly. This is the picture associated with point 2 in Figure 5.7. (It would be well for the reader to review at this stage the discussion on flow separation and its effect on pressure distribution, lift, and drag in Sec. 4.18.)

The nature of the flow field over the wing of an airplane that is below, just beyond, and way beyond the stall is shown in Figures 5.8a, b, and c, respectively. These figures are photographs of a wind-tunnel model with a wingspan of 6 ft. The entire model has been painted with a mixture of mineral oil and a fluorescent powder, which glows under ultraviolet light. After the wind tunnel is turned on, the fluorescent oil indicates the streamline pattern on the surface of the model. In Figure 5.8a, the angle of attack is below the stall; the flow is fully attached, as evidenced by the fact that the high surface shear stress has scrubbed most of the oil from the surface. In Figure 5.8b, the angle of attack is slightly beyond the stall. A large, mushroom-shaped, separated flow pattern has developed over the wing, with attendant highly three-dimensional, low-energy recirculating flow. In Figure 5.8c, the angle of attack is far beyond the stall. The flow over almost the entire wing has separated. These photographs are striking examples of different types of flow that can occur over an airplane wing at different angles of attack, and they graphically show the extent of the flow field separation that can occur.

The lift curves sketched in Figures 5.5 to 5.7 illustrate the type of variation observed experimentally in the data of Appendix D. Returning to Appendix D, note that the lift curves are all virtually linear up to the stall. Singling out a given airfoil, say the NACA 2412 airfoil, also note that c_l versus α is given for three different values of the Reynolds number from 3.1×10^6 to 8.9×10^6. The lift curves for all three values of Re fall on top of each other in the linear region, i.e., Re has little influence on c_l when the flow is attached. However, flow separation is a viscous effect, and as discussed in Chap. 4, Re is a governing parameter for viscous flow. Therefore, it is not surprising that the experimental data for $c_{l,\max}$ in the stalling region are affected by Re, as can be seen by the slightly different variations of c_l at high α for different values of Re. In fact, these lift curves at different Re answer part of the question posed in Eq. (5.17): the data represent $c_l = f(\mathrm{Re})$. Again, Re exerts little or no effect on c_l except in the stalling region.

On the same page as c_l versus α, the variation of $c_{m,c/4}$ versus α is also given. It has only a slight variation with α and is almost completely unaffected by Re. Also note that the values of $c_{m,c/4}$ are slightly negative. By convention, a positive moment is in a clockwise direction; it pitches the airfoil towards larger angles of attack, as shown in Figure 5.3. Therefore, for the NACA 2412 airfoil, with $c_{m,c/4}$ negative, the moments are counterclockwise, and the airfoil tends to pitch downward. This is characteristic of all airfoils with positive camber.

On the page following c_l and $c_{m,c/4}$, the variation of c_d and $c_{m,\mathrm{ac}}$ is given versus c_l. Because c_l varies linearly with α, the reader can visualize these curves of c_d and $c_{m,\mathrm{ac}}$ as being plotted versus α as well; the shapes will be the same. Note that the drag curves have a "bucket" type of shape, with minimum drag occurring at small values of c_l (hence there are small angles of attack). As α goes to large negative or positive values, c_d increases. Also note that c_d is strongly affected by Re, there being a distinct drag curve for each Re. This is to be expected because the drag for a slender aerodynamic shape is mainly skin friction drag, and from Chap. 4 we have seen that Re strongly governs skin friction. With regard to $c_{m,\mathrm{ac}}$, the definition of the aerodynamic center is clearly evident; $c_{m,\mathrm{ac}}$ is constant with respect to α. Also, it is insensitive to Re and has a small negative value.

Refer to Eqs. (5.21); the airfoil data in Appendix D experimentally provide the variation of c_l, c_d, and c_m with α and Re. The effect of M_∞ on the airfoil coefficients will be discussed later. However, emphasis is made that the data in Appendix D were measured in low-speed subsonic wind tunnels. Hence, the flow was essentially incompressible. Thus, c_l, $c_{m,c/4}$, c_d, and $c_{m,\mathrm{ac}}$ given in Appendix D are incompressible flow values. It is important to keep this in mind during our subsequent discussions.

Figure 5.8 Surface oil-flow patterns on a wind-tunnel model of a Grumman American Yankee taken by Dr. Allen Winkelmann in the Glenn L. Martin Wind Tunnel at the University of Maryland. The mixture is mineral oil and a fluorescent powder, and the photographs were taken under ultraviolet light. (a) Below the stall. The wing is at $\alpha = 4°$, where the flow is attached. (b) Very near the stall. The wing is at $\alpha = 11°$, where the highly three-dimensional separated flow is developing in a mushroom cell pattern. (c) Far above the stall. The wing is at $\alpha = 24°$, where the flow over almost the entire wing has separated.

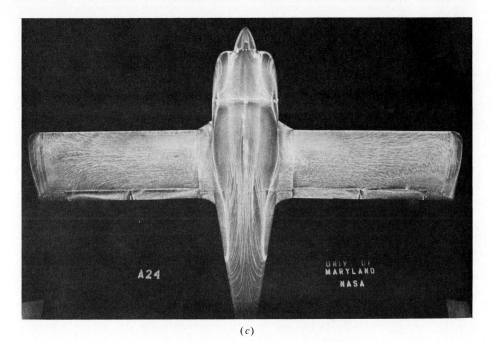

(c)

Example 5.1 A model wing of constant chord length is placed in a low-speed subsonic wind tunnel, spanning the test section. The wing has a NACA 2412 airfoil and a chord length of 1.3 m. The flow in the test section is at a velocity of 50 m/s at standard sea-level conditions. If the wing is at a 4° angle of attack, calculate (a) c_l, c_d, and $c_{m,c/4}$ and (b) the lift, drag, and moments about the quarter chord, per unit span.

SOLUTION

(a) From Appendix D, for a NACA 2412 airfoil at a 4° angle of attack,

$$\boxed{\begin{aligned} c_l &= 0.63 \\ c_{m,c/4} &= -0.035 \end{aligned}}$$

To obtain c_d, we must first check the value of the Reynolds number:

$$\mathrm{Re} = \frac{\rho_\infty V_\infty c}{\mu_\infty} = \frac{(1.225 \text{ kg/m}^3)(50 \text{ m/s})(1.3 \text{ m})}{1.789 \times 10^{-5} \text{ kg/(m)(s)}}$$

$$= 4.45 \times 10^6$$

For this value of Re and for $c_l = 0.63$, from Appendix D,

$$\boxed{c_d = 0.007}$$

(b) Since the chord is 1.3 m and we want the aerodynamic forces and moments *per unit span* (a unit length along the wing, perpendicular to the flow), then $S = c(1) = 1.3(1) = 1.3 \text{ m}^2$. Also

$$q_\infty = \tfrac{1}{2}\rho_\infty V_\infty{}^2 = \tfrac{1}{2}(1.225)(50)^2 = 1531 \text{ N/m}^2$$

From Eq. (5.20),

$$L = q_\infty S c_l = 1531(1.3)(0.63) = \boxed{1254 \text{ N}}$$

Since 1 N = 0.2248 lb, then also

$$L = (1254 \text{ N})(0.2248 \text{ lb/N}) = 281.9 \text{ lb}$$

$$D = q_\infty S c_d = 1531(1.3)(0.007) = \boxed{13.9 \text{ N}}$$

$$= 13.9(0.2248) = 3.13 \text{ lb}$$

Note: The ratio of lift to drag, which is an important aerodynamic quantity, is

$$\frac{L}{D} = \frac{c_l}{c_d} = \frac{1254}{13.9} = 90.2$$

$$M_{c/4} = q_\infty S c_{m,c/4} c = 1531(1.3)(-0.035)(1.3)$$

$$\boxed{M_{c/4} = -90.6 \text{ (N)(m)}}$$

Example 5.2 The same wing in the same flow as in Example 5.1 is pitched to an angle of attack such that the lift per unit span is 700 N (157 lb).
(a) What is the angle of attack?
(b) To what angle of attack must the wing be pitched to obtain zero lift?

SOLUTION
(a) From the previous example,

$$q_\infty = 1531 \text{ N/m}^2$$

$$S = 1.3 \text{ m}^2$$

Thus

$$c_l = \frac{L}{q_\infty S} = \frac{700}{1531(1.3)} = 0.352$$

From Appendix D for the NACA 2412 airfoil, the angle of attack corresponding to $c_l = 0.352$ is

$$\boxed{\alpha = 1°}$$

(b) Also from Appendix D, for zero lift, that is, $c_l = 0$,

$$\boxed{\alpha_{L=0} = -2.2°}$$

5.5 INFINITE VERSUS FINITE WINGS

As stated in Sec. 5.4, the airfoil data in Appendix D were measured in low-speed subsonic wind tunnels where the model wing spanned the test section from one side wall to the other. In this fashion, the flow sees essentially a wing with no wingtips, i.e., the wing could in principle be stretching from plus infinity to minus infinity in the spanwise direction. Such an *infinite wing* is sketched in Figure 5.9, where the wing stretches to $\pm \infty$ in the z direction. The flow about this wing varies only in the x and y directions; for this reason the flow is called *two-dimen-*

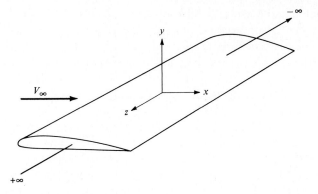

Figure 5.9 Infinite (two-dimensional) wing.

sional. Thus, the airfoil data in Appendix D apply only to such infinite (or two-dimensional) wings. This is an important point to keep in mind.

On the other hand, all real airplane wings are obviously finite, as sketched in Figure 5.10. Here, the top view (planform view) of a finite wing is shown, where the distance between the two wingtips is defined as the *wingspan b*. The area of the wing in this planform view is designated, as before, by S. This leads to an important definition which pervades all aerodynamic wing considerations, namely, the aspect ratio AR.

$$\boxed{\text{Aspect ratio} \equiv AR \equiv \frac{b^2}{S}} \tag{5.22}$$

The importance of AR will come to light in subsequent sections.

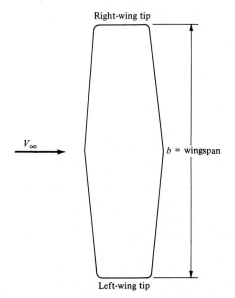

Figure 5.10 Finite wing; plan view (top).

The flow field about a finite wing is three-dimensional and is therefore inherently different from the two-dimensional flow about an infinite wing. As a result, the lift, drag, and moment coefficients for a finite wing with a given airfoil shape at a given α are different from the lift, drag, and moment coefficients for an infinite wing with the same airfoil shape at the same α. For this reason, the aerodynamic coefficients for a finite wing are designated by capital letters, C_L, C_D, C_M; this is in contrast to those for an infinite wing, which we have been designating as c_l, c_d, and c_m. Note that the data in Appendix D are for infinite (two-dimensional) wings, i.e., the data are for c_l, c_d, and c_m. In a subsequent section, we will show how to obtain the finite wing aerodynamic coefficients from the infinite wing data in Appendix D. Our purpose in this section is simply to underscore that there is a difference.

5.6 PRESSURE COEFFICIENT

We continue with our parade of aerodynamic definitions. Consider the pressure distribution over the top surface of an airfoil. Instead of plotting the actual pressure (say in units of N/m^2), let us define a new dimensionless quantity called the *pressure coefficient* C_p as

$$C_p \equiv \frac{p - p_\infty}{q_\infty} \equiv \frac{p - p_\infty}{\frac{1}{2}\rho_\infty V_\infty^2} \tag{5.23}$$

The pressure distribution is sketched in terms of C_p in Figure 5.11. This figure is worth close attention, because pressure distributions found in the aerodynamic literature are usually given in terms of the dimensionless pressure coefficient. Note from Figure 5.11 that C_p at the leading edge is positive because $p > p_\infty$. However, as the flow expands around the top surface of the airfoil, p decreases rapidly, and

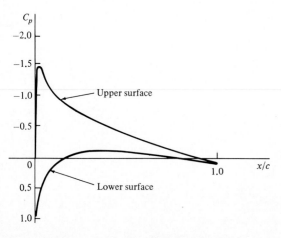

Figure 5.11 Distribution of pressure coefficient over the top and bottom surfaces of a NACA 0012 airfoil at 3.93° angle of attack. $M_\infty = 0.345$, Re $=3.245 \times 10^6$. Experimental data from Ohio State University, in NACA Conference Publication 2045, part I, *Advanced Technology Airfoil Research*, vol. I, p. 1590. *(After Freuler and Gregorek.)*

C_p goes negative in those regions where $p < p_\infty$. By convention, plots of C_p for airfoils are usually shown with negative values above the abscissa, as shown in Figure 5.11.

The pressure coefficient is an important quantity; for example, the distribution of C_p over the airfoil surface leads directly to the value of c_l, as will be discussed in a subsequent section. Moreover, considerations of C_p lead directly to the calculation of the effect of Mach number M_∞ on the lift coefficient. To set the stage for this calculation, consider C_p at a given point on an airfoil surface. The airfoil is a given shape at a fixed angle of attack. The value of C_p can be measured by testing the airfoil in a wind tunnel. Assume that, at first, V_∞ in the tunnel test section is low, say, $M_\infty < 0.3$, such that the flow is essentially incompressible. The measured value of C_p at the point on the airfoil will therefore be a low-speed value. Let us designate the low-speed (incompressible) value of C_p by $C_{p,0}$. If V_∞ is increased but M_∞ is still less than 0.3, C_p will not change, that is, C_p is essentially constant with velocity at low speeds. However, if we now increase V_∞ such that $M_\infty > 0.3$, then compressibility becomes a factor, and the effect of compressibility is to increase the absolute magnitude of C_p as M_∞ increases. This variation of C_p with M_∞ is shown in Figure 5.12. Note that, at $M_\infty \approx 0$, $C_p = C_{p,0}$. As M_∞ increases to $M_\infty \approx 0.3$, C_p is essentially constant. However, as M_∞ is increased beyond 0.3, C_p increases dramatically. (That is, the absolute magnitude increases; if $C_{p,0}$ is negative, then C_p will become an increasingly more negative number as M_∞ increases, whereas if $C_{p,0}$ is positive, then C_p will become an increasingly more positive number as M_∞ increases.) The variation of C_p with

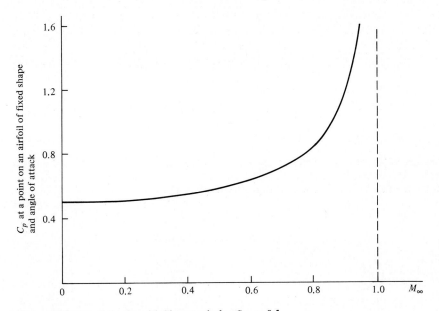

Figure 5.12 Plot of the Prandtl-Glauert rule for $C_{p,0} = 0.5$.

M_∞ for high subsonic Mach numbers was a major focus of aerodynamic research after World War II. An approximate theoretical analysis yields the result

$$\boxed{C_p = \frac{C_{p,0}}{\sqrt{1 - M_\infty{}^2}}} \tag{5.24}$$

Equation (5.24) is called the *Prandtl-Glauert rule*. It is reasonably accurate for $0.3 < M_\infty < 0.7$. For $M_\infty > 0.7$, its accuracy rapidly diminishes; indeed, Eq. (5.24) predicts that C_p becomes infinite as M_∞ goes to unity—an impossible physical situation. (It is well to note that nature abhors infinities as well as discontinuities that are sometimes predicted by mathematical, but approximate, theories in physical science.) There are more accurate, but more complicated, formulas than Eq. (5.24) for near-sonic Mach numbers. However, Eq. (5.24) will be sufficient for our purposes.

Formulas such as Eq. (5.24), which attempt to predict the effect of M_∞ on C_p for subsonic speeds, are called *compressibility corrections*, i.e., they modify (correct) the low-speed pressure coefficient $C_{p,0}$ to take into account the effects of compressibility which are so important at high subsonic Mach numbers.

Example 5.3 The pressure at a point on the wing of an airplane is 7.58×10^4 N/m². The airplane is flying with a velocity of 70 m/s at conditions associated with a standard altitude of 2000 m. Calculate the pressure coefficient at this point on the wing.

SOLUTION For a standard altitude of 2000 m,

$$p_\infty = 7.95 \times 10^4 \text{ N/m}^2$$

$$\rho_\infty = 1.0066 \text{ kg/m}^3$$

Thus, $q_\infty = \frac{1}{2}\rho_\infty V_\infty{}^2 = \frac{1}{2}(1.0066)(70)^2 = 2466$ N/m². From Eq. (5.23),

$$C_p = \frac{p - p_\infty}{q_\infty} = \frac{(7.58 - 7.95) \times 10^4}{2466}$$

$$\boxed{C_p = -1.5}$$

Example 5.4 Consider an airfoil mounted in a low-speed subsonic wind tunnel. The flow velocity in the test section is 100 ft/s, and the conditions are standard sea level. If the pressure at a point on the airfoil is 2102 lb/ft², what is the pressure coefficient?

SOLUTION

$$q_\infty = \frac{1}{2}\rho_\infty V_\infty{}^2 = \frac{1}{2}\left(0.002377 \text{ slug/ft}^3\right)(100 \text{ ft/s})^2$$

$$= 11.89 \text{ lb/ft}^2$$

From Eq. (5.23),

$$C_p = \frac{p - p_\infty}{q_\infty} = \frac{2102 - 2116}{11.89} = \boxed{-1.18}$$

Example 5.5 In Example 5.4, if the flow velocity is increased such that the freestream Mach number is 0.6, what is the pressure coefficient at the same point on the airfoil?

SOLUTION First of all, what is the Mach number of the flow in Example 5.4? At standard sea level,

$$T_s = 518.69 \, R$$

Hence,

$$a_\infty = \gamma RT_\infty = \sqrt{1.4(1716)(518.69)} = 1116 \, \text{ft/s}$$

Thus, in Example 5.4, $M_\infty = V_\infty/a_\infty = 100/1116 = 0.09$—a very low value. Hence, the flow in Example 5.4 is essentially incompressible, and the pressure coefficient is a low-speed value, that is, $C_{p,0} = -1.18$. Thus, if the flow Mach number is increased to 0.6, from the Prandtl-Glauert rule, Eq. (5.24),

$$C_p = \frac{C_{p,0}}{(1 - M_\infty^2)^{1/2}} = -\frac{1.18}{(1 - 0.6^2)^{1/2}}$$

$$\boxed{C_p = -1.48}$$

5.7 OBTAINING LIFT COEFFICIENT FROM C_P

If you are given the distribution of pressure coefficient over the top and bottom surfaces of an airfoil, you can calculate c_l in a straightforward manner. Consider a segment of an infinite wing, as shown in Figure 5.13. Assume the segment has unit span and chord c. Let the x direction correspond to the direction of V_∞, that is, the relative wind, and let c be aligned parallel to V_∞ for the time being; i.e., let $\alpha = 0$. Also, let s be the distance measured along the surface from the leading edge. Also, the angle between the normal to the airfoil surface and a line

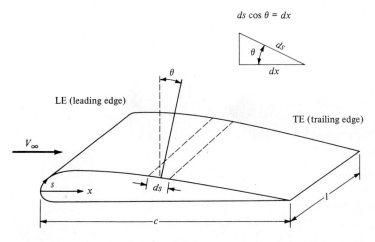

Figure 5.13 Sketch showing how the pressure distribution can be integrated to obtain lift per unit span.

perpendicular to the freestream (but in the plane of the airfoil section) is given by θ. Also consider an infinitesimal strip of the surface area with length ds and unit width, as shown by the dotted lines in Figure 5.13. The aerodynamic force due to pressure on this strip is $p\,ds\,(1)$, which acts normal to the surface. Its component in the lift direction (perpendicular to the relative wind) is $(p\cos\theta)\,ds$. Adding a subscript u to designate the pressure on the upper surface of the airfoil, as well as a negative sign to indicate the force is directed downward (we use the convention that a positive force is directed upward), the contribution to lift of the pressure on the infinitesimal strip is $-p_u\cos\theta\,ds$. If the contributions from all the strips on the upper surface are added from the leading to the trailing edges, we obtain, by letting ds approach zero, the integral

$$-\int_{LE}^{TE}p_u\cos\theta\,ds$$

This is the force in the lift direction due to the pressure distribution on the upper surface. A similar term is obtained for the lower surface of the airfoil, and hence the total lift acting on an airfoil of unit span is given by

$$L = \int_{LE}^{TE}p_l\cos\theta\,ds - \int_{LE}^{TE}p_u\cos\theta\,ds \tag{5.25}$$

where p_l denotes the pressure on the lower surface. From the small triangle in the corner of Figure 5.13, we see the geometric relationship $ds\cos\theta = dx$. Thus, Eq. (5.25) becomes

$$L = \int_0^c p_l\,dx - \int_0^c p_u\,dx \tag{5.26}$$

Adding and subtracting p_∞, we find Eq. (5.26) becomes

$$L = \int_0^c (p_l - p_\infty)\,dx - \int_0^c (p_u - p_\infty)\,dx \tag{5.27}$$

From the definition of lift coefficient, Eq. (5.16),

$$c_l \equiv \frac{L}{q_\infty S} = \frac{L}{q_\infty c(1)} = \frac{L}{q_\infty c} \tag{5.28}$$

Combining Eqs. (5.27) and (5.28) yields

$$c_l = \frac{1}{c}\int_0^c \frac{(p_l - p_\infty)}{q_\infty}\,dx - \frac{1}{c}\int_0^c \frac{p_u - p_\infty}{q_\infty}\,dx \tag{5.29}$$

Note that

$$\frac{p_l - p_\infty}{q_\infty} \equiv C_{p,l} \equiv \text{pressure coefficient on lower surface}$$

$$\frac{p_u - p_\infty}{q_\infty} \equiv C_{p,u} \equiv \text{pressure coefficient on upper surface}$$

Hence, Eq. (5.29) becomes

$$\boxed{c_l = \frac{1}{c}\int_0^c (C_{p,l} - C_{p,u})\,dx} \tag{5.30}$$

Equation (5.30) is a useful relationship; it demonstrates that the lift coefficient can be obtained by integrating C_p over the airfoil surface. If you have plots of pressure coefficient data on the upper and lower surface vs. distance x, then the lift coefficient is equal to the net area between the curves, divided by the chord length.

Recall that the derivation of Eq. (5.30) assumed that c, x, and V_∞ were all in the same direction. Hence Eq. (5.30) is valid only for $\alpha = 0$, and it is a good approximation for a small α. For a more detailed and accurate derivation which is valid for any angle of attack, see Chap. 1 of Anderson, *Fundamentals of Aerodynamics* (McGraw-Hill, 1984).

5.8 COMPRESSIBILITY CORRECTION FOR LIFT COEFFICIENT

The pressure coefficients in Eq. (5.30) can be replaced by the compressibility correction given in Eq. (5.24), as follows

$$c_l = \frac{1}{c} \int_0^c \frac{(C_{p,l} - C_{p,u})_0}{\sqrt{1 - M_\infty^2}} \, dx = \frac{1}{\sqrt{1 - M_\infty^2}} \frac{1}{c} \int_0^c (C_{p,l} - C_{p,u})_0 \, dx \quad (5.31)$$

where again the subscript 0 denotes low-speed incompressible flow values. However, referring to the form of Eq. (5.30), we see that

$$\frac{1}{c} \int_0^c (C_{p,l} - C_{p,u})_0 \, dx \equiv c_{l,0}$$

where $c_{l,0}$ is the low-speed value of lift coefficient. Thus, Eq. (5.31) becomes

$$c_l = \frac{c_{l,0}}{\sqrt{1 - M_\infty^2}} \quad (5.32)$$

Equation (5.32) gives the compressibility correction for lift coefficient. It is subject to the same approximations and accuracy restrictions as the Prandtl-Glauert rule, Eq. (5.24). Also note that the airfoil data in Appendix D were obtained at low speeds, hence the values of lift coefficient obtained from Appendix D are $c_{l,0}$.

Finally, in reference to Eq. (5.17), we now have a reasonable answer to how c_l varies with Mach number. For subsonic speeds, except near Mach 1, the lift coefficient varies inversely as $(1 - M_\infty^2)^{1/2}$.

Example 5.6 Consider a NACA 4412 airfoil at an angle of attack of 4°. If the freestream Mach number is 0.7, what is the lift coefficient?

SOLUTION From Appendix D, for $\alpha = 4°$, $c_l = 0.83$. However, the data in Appendix D were obtained at low speeds, hence the lift coefficient value obtained above, namely 0.83, is really $c_{l,0}$:

$$c_{l,0} = 0.83$$

For high Mach numbers, this must be corrected according to Eq. (5.32):

$$c_l = \frac{c_{l,0}}{\left(1 - M_\infty^2\right)^{1/2}} = \frac{0.83}{\left(1 - .7^2\right)^{1/2}}$$

$$\boxed{c_l = 1.16 \text{ at } M_\infty = 0.7}$$

5.9 CRITICAL MACH NUMBER AND CRITICAL PRESSURE COEFFICIENT

Consider the flow of air over an airfoil. We know that, as the gas expands around the top surface near the leading edge, the velocity and hence the Mach number will increase rapidly. Indeed, there are regions on the airfoil surface where the local Mach number is greater than M_∞. Imagine that we put a given airfoil in a wind tunnel where $M_\infty = 0.3$ and that we observe the peak local Mach number on the top surface of the airfoil to be 0.435. This is sketched in Figure 5.14a. Imagine that we now increase M_∞ to 0.5; the peak local Mach number will correspondingly increase to 0.772, as shown in Figure 5.14b. If we further increase M_∞ to a value of 0.61, we observe that the peak local Mach number is 1.0, locally sonic flow on the surface of the airfoil. This is sketched in Figure 5.14c. Note that the flow over an airfoil can locally be sonic (or higher), even though the freestream Mach number is subsonic. By definition, that freestream Mach number at which sonic flow is first obtained somewhere on the airfoil surface is called the *critical Mach number* of the airfoil. In the above example, the critical Mach number M_{cr} for the airfoil is 0.61. As we will see later, M_{cr} is an

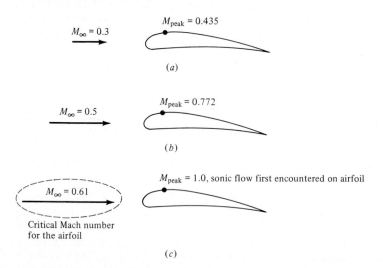

Figure 5.14 Illustration of critical Mach number.

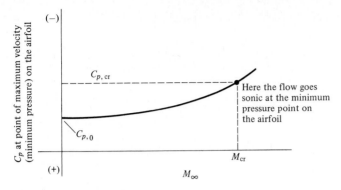

Figure 5.15 Illustration of critical pressure coefficient.

important quantity, because at some freestream Mach number above M_{cr} the airfoil will experience a dramatic increase in drag.

Returning to Figure 5.14, the point on the airfoil where the local M is a peak value is also the point of minimum surface pressure. From the definition of pressure coefficient, Eq. (5.23), C_p will correspondingly have its most negative value at this point. Moreover, according to the Prandtl-Glauert rule, Eq. (5.24), as M_∞ is increased from 0.3 to 0.61, the value of C_p at this point will become increasingly negative. This is sketched in Figure 5.15. The specific value of C_p that corresponds to sonic flow is defined as the *critical pressure coefficient* $C_{p,cr}$. In Figures 5.14a and 5.14b C_p at the minimum pressure point on the airfoil is less negative than $C_{p,cr}$; however, in Figure 5.14c, $C_p = C_{p,cr}$ (by definition).

Consider now three different airfoils ranging from thin to thick, as shown in Figure 5.16. Concentrate first on the thin airfoil. Because of the thin, streamlined profile, the flow over the thin airfoil is only slightly perturbed from its freestream values. The expansion over the top surface is mild, the velocity increases only slightly, the pressure decreases only a relative small amount, and hence the magnitude of C_p at the minimum pressure point is small. Thus, the variation of C_p with M_∞ is shown as the bottom curve in Figure 5.16. For the thin airfoil, $C_{p,0}$ is small in magnitude, and the rate of increase of C_p as M_∞ increases is also relatively small. In fact, because the flow expansion over the thin airfoil surface is mild, M_∞ can be increased to a large subsonic value before sonic flow is encountered on the airfoil surface. The point corresponding to sonic flow conditions on the thin airfoil is labeled point *a* in Figure 5.16. The values of C_p and M_∞ at point *a* are $C_{p,cr}$ and M_{cr}, respectively, for the thin airfoil, by definition. Now consider the airfoil of medium thickness. The flow expansion over the leading edge for this medium airfoil will be stronger, the velocity will increase to larger values, the pressure will decrease to lower values, and the absolute magnitude of C_p is larger. Thus, the pressure coefficient curve for the medium thickness airfoil will lie above that for a thin airfoil, as demonstrated in Figure 5.16. Moreover, because the flow expansion is stronger, sonic conditions will be obtained sooner (at a lower M_∞). Sonic conditions for the medium airfoil are

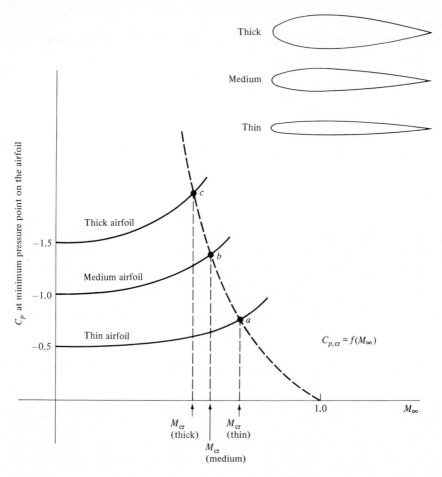

Figure 5.16 Critical pressure coefficient and critical Mach numbers for airfoils of different thicknesses.

labeled as point b in Figure 5.16. Note that point b is to the left of point a, that is, the critical Mach number for the medium-thickness airfoil is less than M_{cr} for the thin airfoil. The same logic holds for the pressure coefficient curve for the thick airfoil, where $C_{p,cr}$ and M_{cr} are given by point c. Emphasis is made that the thinner airfoils have higher values of M_{cr}. As we will see, this is desirable, and hence all airfoils on modern, high-speed airplanes are thin.

The pressure coefficient curves in Figure 5.16 are shown as solid curves. On these curves, only points a, b, and c are critical pressure coefficients, by definition. However, these critical points by themselves form a locus represented by the dotted curve in Figure 5.16; i.e., the critical pressure coefficients themselves are given by a curve of $C_{p,cr} = f(M_\infty)$, as labeled in Figure 5.16. Let us proceed to derive this function. It is an important result, and it also represents an interesting application of our aerodynamic relationships developed in Chap. 4.

First, consider the definition of C_p from Eq. (5.23):

$$C_p = \frac{p - p_\infty}{q_\infty} = \frac{p_\infty}{q_\infty}\left(\frac{p}{p_\infty} - 1\right) \qquad (5.33)$$

From the definition of dynamic pressure,

$$q_\infty \equiv \tfrac{1}{2}\rho_\infty V_\infty^{\,2} = \frac{1}{2}\frac{\rho_\infty}{\gamma p_\infty}(\gamma p_\infty)V_\infty^{\,2} = \frac{1}{2}\frac{V_\infty^{\,2}}{\gamma p_\infty/\rho_\infty}(\gamma p_\infty)$$

However, from Eq. (4.53), $a_\infty^{\,2} = \gamma p_\infty/\rho_\infty$. Thus

$$q_\infty = \frac{1}{2}\frac{V_\infty^{\,2}}{a_\infty^{\,2}}\gamma p_\infty = \frac{\gamma}{2}p_\infty M_\infty^{\,2} \qquad (5.34)$$

We will return to Eq. (5.34) in a moment. Now, recall Eq. (4.74) for isentropic flow,

$$\frac{p_0}{p} = \left(1 + \frac{\gamma - 1}{2}M^2\right)^{\gamma/(\gamma-1)}$$

This relates the total pressure p_0 at a point in the flow to the static pressure p and local Mach number M at the same point. Also, from the same relation,

$$\frac{p_0}{p_\infty} = \left(1 + \frac{\gamma - 1}{2}M_\infty^{\,2}\right)^{\gamma/(\gamma-1)}$$

This relates the total pressure p_0 in the freestream to the freestream static pressure p_∞ and Mach number M_∞. For an isentropic flow, which is a close approximation to the actual, real-life, subsonic flow over an airfoil, the total pressure remains constant throughout. (We refer to more advanced books in aerodynamics for a proof of this fact.) Thus, if the two previous equations are divided, p_0 will cancel, yielding

$$\frac{p}{p_\infty} = \left(\frac{1 + \tfrac{1}{2}(\gamma - 1)M_\infty^{\,2}}{1 + \tfrac{1}{2}(\gamma - 1)M^2}\right)^{\gamma/(\gamma-1)} \qquad (5.35)$$

Substitute Eqs. (5.34) and (5.35) into (5.33):

$$C_p = \frac{p_\infty}{q_\infty}\left(\frac{p}{p_\infty} - 1\right) = \frac{p_\infty}{\tfrac{1}{2}\gamma p_\infty M_\infty^{\,2}}\left[\left(\frac{1 + \tfrac{1}{2}(\gamma - 1)M_\infty^{\,2}}{1 + \tfrac{1}{2}(\gamma - 1)M^2}\right)^{\gamma/(\gamma-1)} - 1\right]$$

$$\boxed{C_p = \frac{2}{\gamma M_\infty^{\,2}}\left[\left(\frac{1 + \tfrac{1}{2}(\gamma - 1)M_\infty^{\,2}}{1 + \tfrac{1}{2}(\gamma - 1)M^2}\right)^{\gamma/(\gamma-1)} - 1\right]} \qquad (5.36)$$

For a given freestream Mach number M_∞, Eq. (5.36) relates the local value of C_p to the local M at any given point in the flow field, hence at any given point on the

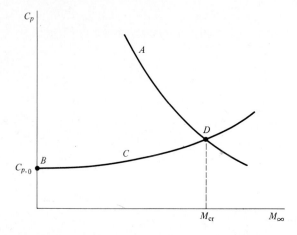

Figure 5.17 Determination of critical Mach number.

airfoil surface. Let us pick that particular point on the surface where $M = 1$. Then, by definition, $C_p = C_{p,\text{cr}}$. Putting $M = 1$ into Eq. (5.36), we obtain

$$C_{p,\text{cr}} = \frac{2}{\gamma M_\infty^2}\left[\left(\frac{2 + (\gamma - 1)M_\infty^2}{\gamma + 1}\right)^{\gamma/(\gamma-1)} - 1\right] \qquad (5.37)$$

Equation (5.37) gives the desired relation $C_{p,\text{cr}} = f(M_\infty)$. When numbers are fed into Eq. (5.37), the dotted curve in Figure 5.16 results. Note that, as M_∞ increases, $C_{p,\text{cr}}$ decreases.

Example 5.7 Given a specific airfoil, how can you estimate its critical Mach number?

SOLUTION There are several steps to this process, as follows.
  (a) Obtain a plot of $C_{p,\text{cr}}$ versus M from Eq. (5.37). This is illustrated by curve A in Figure 5.17. This curve is a fixed "universal" curve, which you can keep for all such problems.
  (b) From measurement or theory for low-speed flow, obtain the minimum pressure coefficient on the top surface of the airfoil. This is $C_{p,0}$ shown as point B in Figure 5.17.
  (c) Using Eq. (5.24), plot the variation of this minimum pressure coefficient versus M_∞. This is illustrated by curve C.
  (d) When curve C intersects curve A, then the minimum pressure coefficient on the top surface of the airfoil is equal to the critical pressure coefficient, and the corresponding M_∞ is the critical Mach number. Hence, point D is the solution for M_{cr}.

5.10 DRAG-DIVERGENCE MACH NUMBER

We now turn our attention to the airfoil drag coefficient c_d. Figure 5.18 sketches the variation of c_d with M_∞. At low Mach numbers, less than M_{cr}, c_d is virtually constant and is equal to its low-speed value given in Appendix D. The flow field

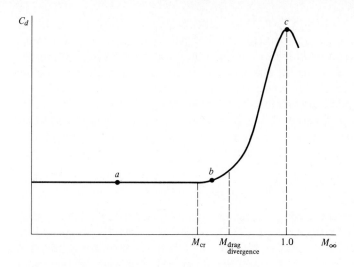

Figure 5.18 Variation of drag coefficient with Mach number.

about the airfoil for this condition (say, point a in Figure 5.18) is noted in Figure 5.19a, where $M < 1$ everywhere in the flow. If M_∞ is increased slightly above M_{cr}, a "bubble" of supersonic flow will occur surrounding the minimum pressure point, as shown in Figure 5.19b. Correspondingly, c_d will still remain reasonably low, as indicated by point b in Figure 5.18. However, if M_∞ is still further increased, a very sudden and dramatic rise in the drag coefficient will be observed,

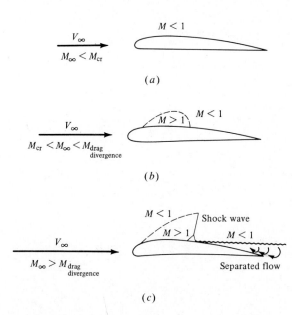

Figure 5.19 Physical mechanism of drag divergence. (a) Flow field associated with point a in Figure 5.18. (b) Flow field associated with point b in Figure 5.18. (c) Flow field associated with point c in Figure 5.18.

as noted by point c in Figure 5.18. Here, shock waves suddenly appear in the flow, as sketched in Figure 5.19c. The shock waves themselves are dissipative phenomena, which result in an increase in drag on the airfoil. But in addition, the sharp pressure increase across the shock waves creates a strong adverse pressure gradient, hence causing the flow to separate from the surface. As discussed in Sec. 4.18, such flow separation can create substantial increases in drag. Thus, the sharp increase in c_d shown in Figure 5.18 is a combined effect of shock waves and flow separation. The *freestream* Mach number at which c_d begins to increase rapidly is defined at the *drag-divergence* Mach number and is noted in Figure 5.18. Note that

$$M_{cr} < M_{drag\ divergence} < 1.0$$

The shock pattern sketched in Figure 5.19c is characteristic of a flight regime called transonic. When $0.8 \le M_{\infty} \le 1.2$, the flow is generally designated as transonic flow and it is characterized by some very complex effects only hinted at in Figure 5.19c. The analysis of transonic flows has been one of the major challenges in modern aerodynamics. Only in recent years, since about 1970, have computer solutions for transonic flows over airfoils come into practical use; these numerical solutions are still in a state of development and improvement. Transonic flow has been a hard nut to crack.

The airfoils on modern subsonic jet aircraft, such as the McDonnell-Douglas DC-10 wide-body transport, are relatively thin profiles designed to increase the drag-divergence Mach number. In fact, NASA (mainly under the direction of Richard Whitcomb at the Langley Research Center) has developed a new *supercritical* airfoil, designed to place the drag-divergence Mach number ex-

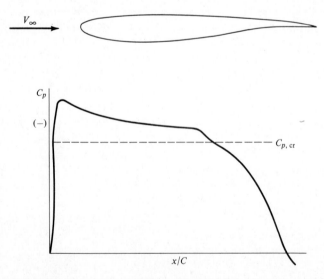

Figure 5.20 Shape of a typical supercritical airfoil and its pressure coefficient distribution over the top surface.

tremely close to 1.0. Supercritical airfoils are of the general shape shown in Figure 5.20. Here, the maximum thickness is designed close to the trailing edge. The flow over the airfoil is largely supersonic, and the airfoil shape is designed to discourage the formation of shock waves.

These considerations of M_{cr} and drag-divergence Mach number are important to subsonic transport design. The cruising speeds of such airplanes can be increased by incorporating airfoils with high values of $M_{drag\ divergence}$. Alternatively, in the spirit of energy conservation, supercritical airfoils should allow less drag, hence less fuel consumption, at the same cruising speed. In either case, the advantages are obvious, and this helps to explain why there has been a modern resurgence of interest in high-speed airfoil design.

As a final note in this section, the next time you have an opportunity to fly in a jet airliner and the sun is directly overhead (around noon and early afternoon), look out along the span of the wing. Due to the refraction of light waves through shock waves, you can sometimes see with the naked eye the transonic shock waves dancing about on the wing surface.

5.11 WAVE DRAG (AT SUPERSONIC SPEEDS)

To this point, we have discussed airfoil properties at subsonic speeds. When M_∞ is supersonic, a major new physical phenomenon is introduced: shock waves. We have previously alluded to shock waves in Sec. 4.11 C in conjunction with the Pitot-tube measurement of supersonic airspeeds. With respect to airfoils (as well as all other aerodynamic bodies) shock waves in supersonic flow create a new source of drag called *wave drag*. In this section, we will simply highlight some of the ideas involving shock waves and the consequent wave drag; a detailed study of shock wave phenomena is left to more advanced texts in aerodynamics.

To obtain a feel for how a shock is produced, imagine that we have a small source of sound waves, a tiny "beeper" (something like a tuning fork). At time $t = 0$, assume the beeper is at point P in Figure 5.21. At this point, let the beeper emit a sound wave, which will propagate in all directions at the speed of sound a. Also, let the beeper move with velocity V, where V is less than the speed of sound. At time t, the sound wave will have moved outward by a distance at, as shown in Figure 5.21. At the same time t, the beeper will have moved a distance Vt, to point Q. Since $V < a$, the beeper will always stay inside the sound wave. If the beeper is constantly emitting sound waves as it moves along, these waves will constantly move outward, ahead of the beeper. As long as $V < a$, the beeper will always be inside the envelope formed by the sound waves.

On the other hand, assume the beeper is moving at supersonic speed, that is, $V > a$. At time $t = 0$, assume the beeper is at point R in Figure 5.22. At this point, let the beeper emit a sound wave, which, as before, will propagate in all directions at the speed of sound a. At time t the sound wave will have moved outward by a distance at, as shown in Figure 5.22. At the same time t, the beeper will have moved a distance Vt, to point S. However, since $V > a$, the beeper will

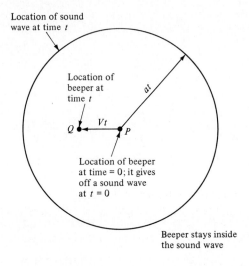

Location of sound
wave at time t

Location of
beeper at
time t

at

Vt

Q P

Location of beeper
at time = 0; it gives
off a sound wave
at $t = 0$

Beeper stays inside
the sound wave

Figure 5.21 Beeper moving less than the speed
of sound.

now be outside the sound wave. If the beeper is constantly emitting sound waves
as it moves along, these waves will now pile up inside an envelope formed by a
line from point S tangent to the circle formed by the first sound wave, centered at
point R. This tangent line, the line where the pressure disturbances are piling up,
is called a *Mach wave*. The vertex of the wave is fixed to the moving beeper at
point S. In supersonic flight, the air ahead of the beeper in Figure 5.22 has no
warning of the approach of the beeper. Only the air behind the Mach wave has
felt the presence of the beeper, and this presence is communicated by pressure

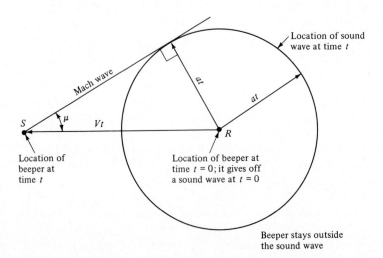

Location of sound
wave at time t

Mach wave

at

at

S μ Vt R

Location of
beeper at
time t

Location of beeper at
time $t = 0$; it gives off
a sound wave at $t = 0$

Beeper stays outside
the sound wave

Figure 5.22 The origin of Mach waves and shock waves. Beeper is moving faster than the speed of
sound.

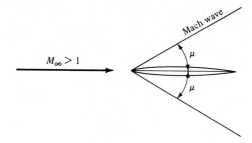

Figure 5.23 Mach waves on a needlelike body.

(sound) waves confined inside the conical region bounded by the Mach wave. In contrast, in subsonic flight, the air ahead of the beeper in Figure 5.21 is forewarned about the oncoming beeper by the sound waves. In this case, there is no piling up of pressure waves; there is no Mach wave.

Hence, we can begin to feel that the coalescing, or piling up, of pressure waves in supersonic flight can create sharply defined waves of some sort. In Figure 5.22, the Mach wave that is formed makes an angle μ with the direction of movement of the beeper. This angle, defined as the *Mach angle*, is easily obtained from the geometry of Figure 5.22, as follows:

$$\sin \mu = \frac{at}{Vt} = \frac{a}{V} = \frac{1}{M}$$

Hence
$$\boxed{\text{Mach angle} \equiv \mu \equiv \arcsin \frac{1}{M}}$$
(5.38)

In real life, a very thin object (such as a thin needle) moving at $M_\infty > 1$ creates a very weak disturbance in the flow, limited to a Mach wave. This is sketched in Figure 5.23. On the other hand, an object with some reasonable thickness, such as the wedge shown in Figure 5.24, moving at supersonic speeds will create a strong disturbance called a *shock wave*. The shock wave will be inclined at an oblique angle β, where $\beta > \mu$, as shown in Figure 5.24. As the flow

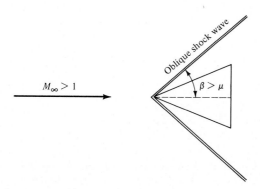

Figure 5.24 Oblique shock waves on a wedge-type body.

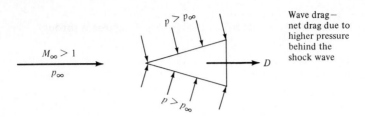

Figure 5.25 Pressure distribution on a wedge at supersonic speeds; origin of wave drag.

moves across the oblique shock wave, the pressure, temperature, and density increase, and the velocity and Mach number decrease.

Consider now the pressure on the surface of the wedge, as sketched in Figure 5.25. Since p increases across the oblique shock wave, then at the wedge surface, $p > p_\infty$. Since the pressure acts normal to the surface and the surface itself is inclined to the relative wind, there will be a net drag produced on the wedge, as seen by simple inspection of Figure 5.25. This drag is called *wave drag*, because it is inherently due to the pressure increase across the shock wave.

In order to minimize the strength of the shock wave, all supersonic airfoil profiles are thin, with relatively sharp leading edges. (The leading edge of the Lockheed F-104 supersonic fighter is almost razor thin.) Let us approximate a thin supersonic airfoil by the flat plate illustrated in Figure 5.26. The flat plate is

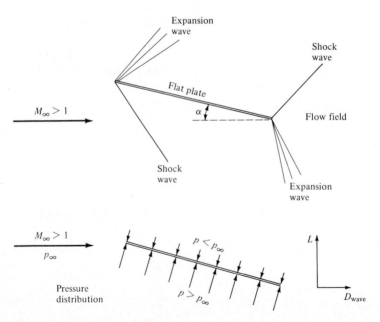

Figure 5.26 Flow field and pressure distribution for a flat plate at angle of attack in supersonic flow. There is a net lift and drag due to the pressure distribution set up by the shock and expansion waves.

inclined at a small angle of attack α to the supersonic freestream. On the top surface of the plate, the flow field is turned away from the freestream through an *expansion wave* at the leading edge; an expansion wave is a fan-shaped region through which the pressure decreases. At the trailing edge on the top side, the flow is turned back into the freestream direction through an oblique shock wave. On the bottom surface of the plate, the flow is turned into the freestream, causing an oblique shock wave with an increase in pressure. At the trailing edge, the flow is turned back to the freestream direction through an expansion wave. (Details and theory for expansion waves, as well as shock waves, are beyond the scope of this book—you will have to simply accept on faith the flow field sketched in Figure 5.26 until your study of aerodynamics becomes more advanced.) The expansion and shock waves at the leading edge result in a surface pressure distribution where the pressure on the top surface is less than p_∞, whereas the pressure on the bottom surface is greater than p_∞. The net effect is an aerodynamic force normal to the plate. The components of this force perpendicular and parallel to the relative wind are the lift and supersonic wave drag, respectively. Approximate relations for the lift and drag coefficients are

$$c_l = \frac{4\alpha}{\left(M_\infty^2 - 1\right)^{1/2}} \tag{5.39}$$

and

$$c_{d,w} = \frac{4\alpha^2}{\left(M_\infty^2 - 1\right)^{1/2}} \tag{5.40}$$

A subscript w has been added to the drag coefficient to emphasize that it is the wave drag coefficient. Equations (5.39) and (5.40) are approximate expressions, useful for thin airfoils at small to moderate angles of attack in supersonic flow. Note that as M_∞ increases, both c_l and c_d decrease. This is not to say that the lift and drag forces themselves decrease with M_∞. Quite the contrary, for any flight regime, as the flight velocity increases, L and D usually increase because the dynamic pressure, $q_\infty = \frac{1}{2}\rho V_\infty^2$, increases. In the supersonic regime, L and D increase with velocity, even though c_l and $c_{d,w}$ decrease with M_∞ according to Eqs. (5.39) and (5.40).

5.12 SUMMARY OF AIRFOIL DRAG

Amplifying on Eq. (4.102), we can write the total drag of an airfoil as the sum of three contributions:

$$D = D_f + D_p + D_w$$

where D = total drag on the airfoil
D_f = skin friction drag
D_p = pressure drag due to flow separation
D_w = wave drag (present only at transonic and supersonic speeds; zero for subsonic speeds below the drag-divergence Mach number)

In terms of the drag coefficients, we can write:

$$c_d = c_{d,f} + c_{d,p} + c_{d,w}$$

where c_d, $c_{d,f}$, $c_{d,p}$, and $c_{d,w}$ are the total drag, skin friction drag, pressure drag, and wave drag coefficients, respectively. The sum $c_{d,f} + c_{d,p}$ is called the *profile drag coefficient*; this is the quantity that is given by the data in Appendix D. The profile drag coefficient is relatively constant with M_∞ at subsonic speeds.

The variation of c_d with M_∞ from incompressible to supersonic speeds is sketched in Figure 5.27. It is important to note the qualitative variation of this curve. For M_∞ ranging from zero to drag divergence, c_d is relatively constant; it consists entirely of profile drag. For M_∞ from drag divergence to slightly above 1, the value of c_d skyrockets; indeed, the peak value of c_d around $M_\infty = 1$ can be an order of magnitude larger than the profile drag itself. This large increase in c_d is due to wave drag associated with the presence of shock waves. For supersonic Mach numbers, c_d decreases approximately as $(M_\infty^2 - 1)^{-1/2}$.

The large increase in the drag coefficient near Mach 1 gave rise to the term "sound barrier" in the 1940s. There was a camp of professionals who at that time felt that the sound barrier could not be pierced, that we could not fly faster than the speed of sound. Certainly, a glance at Eq. (5.24) for the pressure coefficient in subsonic flow, as well as Eq. (5.40) for wave drag in supersonic flow, would hint that the drag coefficient might become infinitely large as M_∞ approaches 1 from either the subsonic or supersonic sides. However, such reasoning is an example of a common pitfall in science and engineering, namely, the application of equations outside their ranges of validity. Neither Eq. (5.24) nor (5.40) is valid in the transonic range near $M_\infty = 1$. Moreover, remember that nature abhors infinities. In real life, c_d does not become infinitely large. To get past the sound barrier, all that is needed (in principle) is an engine with enough thrust to overcome the high (but finite) drag.

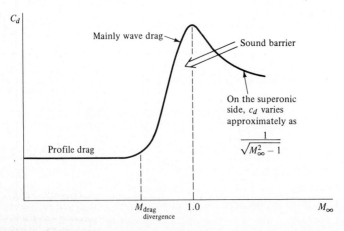

Figure 5.27 Variation of drag coefficient with Mach number for subsonic and supersonic speeds.

5.13 FINITE WINGS

We now return to the discussion initiated in Sec. 5.5. Our considerations so far have dealt mainly with airfoils, where the aerodynamic properties are directly applicable to infinite wings. However, all real wings are finite, and for practical reasons we must translate our knowledge about airfoils to the case where the wing has wingtips. This is the purpose of the next two sections.

The fundamental difference between flows over finite wings as opposed to infinite wings can be seen as follows. Consider the front view of a finite wing as sketched in Figure 5.28a. If the wing has lift, then obviously the average pressure over the bottom surface is greater than that over the top surface. Consequently, there is some tendency for the air to "leak," or flow, around the wingtips from the high- to the low-pressure sides, as shown in Figure 5.28a. This flow establishes a circulatory motion which trails downstream of the wing. The trailing circular motion is called a *vortex*. There is a major trailing vortex from each wingtip, as sketched in Figure 5.28b and as shown in the photograph in Figure 5.29.

These wingtip vortices downstream of the wing induce a small downward component of air velocity in the neighborhood of the wing itself. This can be seen intuitively from Figure 5.28b; the two wingtip vortices tend to drag the surrounding air around with them, and this secondary movement induces a small velocity component in the downward direction at the wing. This downward component is called *downwash*, given by the symbol w.

An effect of downwash can be seen in Figure 5.30. As usual, V_∞ designates the relative wind. However, in the immediate vicinity of the wing, V_∞ and w add

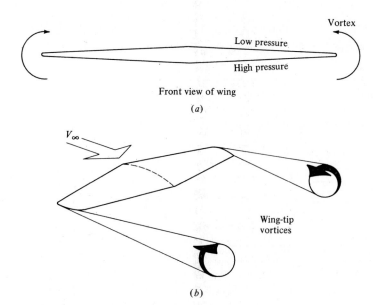

Figure 5.28 Origin of wingtip vortices on a finite wing.

Figure 5.29 Wingtip vortices made visible by smoke ejected at the wing tips of a Boeing 727 test airplane. *(NASA.)*

vectorally to produce a "local" relative wind which is canted downward from the original direction of V_∞. This has several consequences:

1. The angle of attack of the airfoil sections of the wing is effectively reduced in comparison to the angle of attack of the wing referenced to V_∞.
2. There is an increase in the drag. This increase is called *induced drag*, which has at least three physical interpretations. First, the wingtip vortices simply

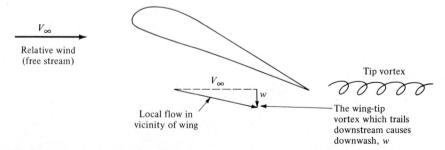

Figure 5.30 The origin of downwash.

alter the flow field about the wing in such a fashion as to change the surface pressure distributions in the direction of increased drag. An alternate explanation is that, because the local relative wind is canted downward (see Figure 5.30), the lift vector itself is "tilted back," hence it contributes a certain component of force parallel to V_∞, that is, a drag force. A third physical explanation of the source of induced drag is that the wingtip vortices contain a certain amount of rotational kinetic energy. This energy has to come from somewhere; indeed, it is supplied by the aircraft propulsion system, where extra power has to be added to overcome the extra increment in drag due to induced drag. All three of these outlooks on the physical mechanism of induced drag are synonymous.

5.14 CALCULATION OF INDUCED DRAG

Consider a section of a finite wing as shown in Figure 5.31. The angle of attack defined between the mean chord of the wing and the direction of V_∞ (the relative wind) is called the *geometric angle of attack* α. However, in the vicinity of the wing the local flow is (on the average) deflected downward by the angle α_i because of downwash. This angle α_i, defined as the *induced angle of attack*, is the difference between the local flow direction and the freestream direction. Hence, contrary to the naked eye, which sees the wing at an angle of attack α, the airfoil section itself is seeing an *effective angle of attack* which is smaller than α. Letting α_{eff} denote the effective angle of attack, we see from Figure 5.31 that $\alpha_{\text{eff}} = \alpha - \alpha_i$.

Let us now adopt the point of view that, because the local flow direction in the vicinity of the wing is inclined downward with respect to the freestream, the lift vector remains perpendicular to the local relative wind and is therefore tilted back through the angle α_i. This is shown in Figure 5.31. However, still considering drag to be parallel to the freestream, we see that the tilted-lift vector

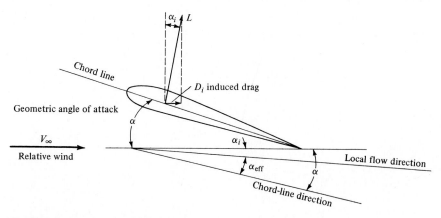

Figure 5.31 The origin of induced drag.

contributes a certain component of drag. This drag is the *induced drag* D_i. From Figure 5.31,

$$D_i = L \sin \alpha_i$$

Values of α_i are generally small, hence $\sin \alpha_i \approx \alpha_i$. Thus

$$D_i = L\alpha_i \qquad (5.41)$$

Note that in Eq. (5.41) α_i must be in radians. Hence, D_i can be calculated from Eq. (5.41) once α_i is obtained.

The calculation of α_i is beyond the scope of this book. However, it can be shown that the value of α_i for a given section of a finite wing depends on the distribution of downwash along the span of the wing. In turn, the downwash distribution is governed by the distribution of lift over the span of the wing. To see this more clearly, consider Figure 5.32, which shows the front view of a finite wing. The lift per unit span may vary as a function of distance along the wing because:

1. The chord may vary in length along the wing.
2. The wing may be twisted such that each airfoil section of the wing is at a different geometric angle of attack.
3. The shape of the airfoil section may change along the span.

Shown in Figure 5.32 is the case of an elliptical lift distribution (the lift per unit span varies elliptically along the span), which in turn produces a uniform downwash distribution. For this case, incompressible flow theory predicts that

$$\alpha_i = \frac{C_L}{\pi \text{AR}} \qquad (5.42)$$

where C_L is the lift coefficient of the finite wing and $\text{AR} = b^2/S$ is the aspect ratio defined in Eq. (5.22). Substituting Eq. (5.42) into (5.41) yields

$$D_i = L\alpha_i = L\frac{C_L}{\pi \text{AR}} \qquad (5.43)$$

However, $L = q_\infty S C_L$; hence from Eq. (5.43),

$$D_i = q_\infty S \frac{C_L{}^2}{\pi \text{AR}}$$

or

$$\frac{D_i}{q_\infty S} = \frac{C_L{}^2}{\pi \text{AR}} \qquad (5.44)$$

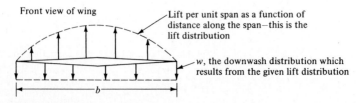

Front view of wing

Lift per unit span as a function of distance along the span—this is the lift distribution

w, the downwash distribution which results from the given lift distribution

b

Figure 5.32 Lift distribution and downwash distribution.

Defining the *induced drag coefficient* as $C_{D,i} = D_i/q_\infty S$, we can write Eq. (5.44) as

$$C_{D,i} = \frac{C_L^2}{\pi\mathrm{AR}} \tag{5.45}$$

This result holds for an elliptical lift distribution, as sketched in Figure 5.32. For a wing with the same airfoil shape across the span and with no twist, an elliptical lift distribution is characteristic of an elliptical wing planform. (The famous British Spitfire of World War II was one of the few aircraft in history designed with an elliptical wing planform. Wings with straight leading and trailing edges are more economical to manufacture.)

For all wings in general, a *span efficiency factor e* can be defined such that

$$C_{D,i} = \frac{C_L^2}{\pi e\mathrm{AR}} \tag{5.46}$$

For elliptical planforms, $e = 1$; for all other planforms, $e < 1$. Thus, $C_{D,i}$ and hence induced drag is a *minimum for an elliptical planform*. For typical subsonic aircraft, e ranges from 0.85 to 0.95. Equation (5.46) is an important relation. It demonstrates that induced drag varies as the square of the lift coefficient; at high lift, such as near $C_{L,\mathrm{max}}$, the induced drag can be a substantial portion of the total drag. Equation (5.46) also demonstrates that as AR is increased, induced drag is decreased. Hence, subsonic airplanes designed to minimize induced drag have high–aspect ratio wings (such as the long, narrow wings of the Lockheed U-2 high-altitude reconnaissance aircraft).

It is clear from Eq. (5.46) that induced drag is intimately related to lift. In fact, another expression for induced drag is *drag due to lift*. In a fundamental sense, that power provided by the engines of the airplane to overcome induced drag is the power required to sustain a heavier-than-air vehicle in the air, the power necessary to produce lift equal to the weight of the airplane in flight.

In light of Eq. (5.46), we can now write the total drag coefficient for a finite wing at subsonic speeds as

$$C_D = c_d + \frac{C_L^2}{\pi e\mathrm{AR}} \tag{5.47}$$

| Total | Profile | Induced |
| drag | drag | drag |

Keep in mind that profile drag is composed of two parts; drag due to skin friction, $c_{d,f}$, and pressure drag due to separation, $c_{d,p}$; that is, $c_d = c_{d,f} + c_{d,p}$. Also keep in mind that c_d can be obtained from the data in Appendix D. The quadratic variation of C_D with C_L given in Eq. (5.47), when plotted on a graph, leads to a curve as shown in Figure 5.33. Such a plot of C_D versus C_L is called a *drag polar*. Much of the basic aerodynamics of an airplane is reflected in the drag polar, and such curves are essential to the design of airplanes. You should become familiar with the concept of drag polar. Note that the drag data in Appendix D are given in terms of drag polars for infinite wings, that is, c_d is plotted versus c_l. However, induced drag is not included in Appendix D because $C_{D,i}$ for an infinite wing (infinite aspect ratio) is zero.

Example 5.8 Consider the Northrop F-5 fighter airplane, which has a wing area of 170 ft². The wing is generating 18,000 lb of lift. For a flight velocity of 250 mi/h at standard sea level, calculate the lift coefficient.

SOLUTION The velocity in consistent units is

$$V_\infty = 250\frac{88}{60} = 366.7 \text{ ft/s}$$

$$q_\infty = \tfrac{1}{2}\rho_\infty V_\infty^2 = \tfrac{1}{2}(0.002377)(250)^2 = 74.28 \text{ lb/ft}^2$$

Hence
$$C_L = \frac{L}{q_\infty S} = \frac{18000}{(74.28)(170)} = \boxed{1.425}$$

Example 5.9 The wingspan of the Northrop F-5 is 25.25 ft. Calculate the induced drag coefficient and the induced drag itself for the conditions of Example 5.8. Assume $e = 0.8$.

SOLUTION The aspect ratio is AR $= b^2/S = (25.25)^2/170 = 3.75$. Since $C_L = 1.425$, from Example 5.8, then from Eq. (5.46),

$$C_{D,i} = \frac{C_L^2}{\pi e \text{AR}} = \frac{(1.425)^2}{\pi(0.8)(3.75)} = \boxed{0.215}$$

From Example 5.8, $q_\infty = 74.28$ lb/ft². Hence

$$D_i = q_\infty S C_{D,i} = (74.28)(170)(0.215) = \boxed{2715 \text{ lb}}$$

Example 5.10 Consider a flying wing (such as the Northrop YB-49 of the early 1950s) with a wing area of 206 m², an aspect ratio of 10, a span effectiveness factor of 0.95, and a NACA 4412 airfoil. The weight of the airplane is 7.5×10^5 N. If the density altitude is 3 km and the flight velocity is 100 m/s, calculate the total drag on the aircraft.

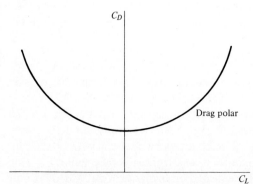

Figure 5.33 Sketch of a drag polar, i.e., a plot of drag coefficient vs. lift coefficient.

SOLUTION First, obtain the lift coefficient. At a density altitude of 3 km = 3000 m, $\rho_\infty = 0.909$ kg/m^3 (from Appendix A).

$$q_\infty = \tfrac{1}{2}\rho_\infty V_\infty^2 = \tfrac{1}{2}(0.909)(100)^2 = 4545 \text{ N/m}^2$$

$$L = W = 7.5 \times 10^5 \text{ N}$$

$$C_L = \frac{L}{q_\infty S} = \frac{7.5 \times 10^5}{4545(206)} = 0.8$$

Note: This is a rather high lift coefficient, but the velocity is low—near the landing speed—hence the airplane is pitched to a rather high angle of attack to generate enough lift to keep the airplane flying.

Next, obtain the induced drag coefficient:

$$C_{D,i} = \frac{C_L^2}{\pi e \text{AR}} = \frac{0.8^2}{\pi(0.95)(10)} = 0.021$$

The profile drag coefficient must be estimated from the aerodynamic data in Appendix D. Assume that c_d is given by the highest Reynolds number data shown for the NACA 4412 airfoil in Appendix D, and furthermore, assume that it is in the "drag bucket." Hence, from Appendix D,

$$c_d \approx 0.006$$

Thus, from Eq. (5.47), the total drag coefficient is

$$C_D = c_d + C_{D,i}$$

$$= 0.006 + 0.021 = 0.027$$

Note that the induced drag is about 3.5 times larger than profile drag for this case, thus underscoring the importance of induced drag.

Therefore, the total drag is

$$D = q_\infty S C_D = 4545(206)(0.027) = \boxed{2.53 \times 10^4 \text{ N}}$$

5.15 CHANGE IN THE LIFT SLOPE

The aerodynamic properties of a finite wing differ in two major respects from the data of Appendix D, which apply to infinite wings. The first difference has already been discussed, namely, the addition of induced drag for a finite wing. The second difference is that the lift curve for a finite wing has a smaller slope than the corresponding lift curve for an infinite wing with the same airfoil cross section. This change in the lift slope can be examined as follows. Recall that because of the presence of downwash, which is induced by the trailing wingtip vortices, the flow in the local vicinity of the wing is canted downward with respect to the freestream relative wind. As a result, the angle of attack which the airfoil section effectively sees, called the effective angle of attack α_{eff}, is less than the geometric angle of attack α. This situation is sketched in Figure 5.34. The difference between α and α_{eff} is the *induced angle of attack* α_i, first introduced in Sec. 5.14, where $\alpha_i = \alpha - \alpha_{\text{eff}}$. Moreover, for an elliptical lift distribution, Eq. (5.42) gives values for the induced angle of attack $\alpha_i = C_L/\pi \text{AR}$. Extending Eq.

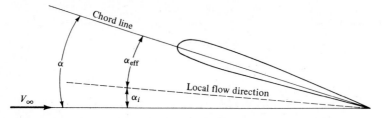

Figure 5.34 Relation between the geometric, effective, and induced angles of attack.

(5.42) to wings of any general planform, a new span effectiveness factor e_1 can be defined such that

$$\alpha_i = \frac{C_L}{\pi e_1 \text{AR}} \tag{5.48}$$

where e_1 and e [defined for induced drag in Eq. (5.46)] are theoretically different but are in practice approximately the same value for a given wing. Note that Eq. (5.48) gives α_i in radians. For α_i in degrees,

$$\alpha_i = \frac{57.3 C_L}{\pi e_1 \text{AR}} \tag{5.48'}$$

Emphasis is made that the flow over a finite wing at an angle of attack α is essentially the same as the flow over an infinite wing at an angle of attack α_{eff}. Keeping this in mind, assume that we plot the lift coefficient for the finite wing C_L versus the effective angle of attack, $\alpha_{\text{eff}} = \alpha - \alpha_i$, as shown in Figure 5.35a. Because we are using α_{eff}, the lift curve should correspond to that for an infinite wing; hence the lift curve slope in Figure 5.35a is a_0, obtained from Appendix D for the given airfoil. However, in real life our naked eyes cannot see α_{eff}; instead, what we actually observe is a finite wing at the geometric angle of attack α (the actual angle between the freestream relative wind and the mean chord line). Hence, for a finite wing, it makes much more sense to plot C_L versus α, as shown in Figure 5.35b, rather than versus α_{eff}, as shown in Figure 5.35a. For example, C_L versus α would be the result most directly obtained from testing a finite wing in a wind tunnel, because α (and not α_{eff}) can be measured directly. Hence, the lift curve slope for a finite wing is defined as $a \equiv dC_L/d\alpha$, where $a \neq a_0$. Noting that $\alpha > \alpha_{\text{eff}}$ from Figure 5.34, the abscissa of Figure 5.35b is stretched out more than the abscissa of Figure 5.34a; hence the lift curve of Figure 5.35b is less inclined, that is, $a < a_0$. *The effect of a finite wing is to reduce the lift curve slope.* However, when the lift is zero, $C_L = 0$, and from Eq. (5.48), $\alpha_i = 0$. Thus, at zero lift, $\alpha = \alpha_{\text{eff}}$. In terms of Figures 5.35a and 5.35b, this means that the angle of attack for zero lift, $\alpha_{L=0}$, is the same for the finite and infinite wings. Thus, for finite wings, $\alpha_{L=0}$ can be obtained directly from Appendix D.

Question: if we know a_0 (say, from Appendix D), how do we find a for a finite wing with a given aspect ratio? The answer can be obtained by examining

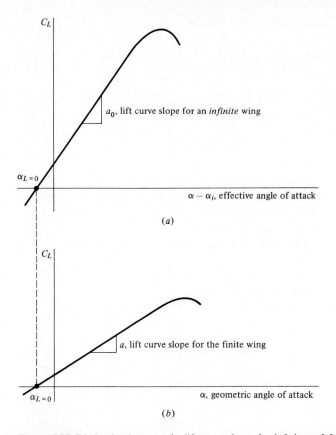

Figure 5.35 Distinction between the lift curve slopes for infinite and finite wings.

Figure 5.35. From Figure 5.35a,

$$\frac{dC_L}{d(\alpha - \alpha_i)} = a_0$$

Integrating, we find

$$C_L = a_0(\alpha - \alpha_i) + \text{const} \tag{5.49}$$

Substituting Eq. (5.48′) into Eq. (5.49), we obtain

$$C_L = a_0\left(\alpha - \frac{57.3C_L}{\pi e_1 \text{AR}}\right) + \text{const} \tag{5.50}$$

Solving Eq. (5.50) for C_L yields

$$C_L = \frac{a_0\alpha}{1 + 57.3a_0/\pi e_1 \text{AR}} + \frac{\text{const}}{1 + 57.3a_0/\pi e_1 \text{AR}} \tag{5.51}$$

Differentiating Eq. (5.51) with respect to α, we get

$$\frac{dC_L}{d\alpha} = \frac{a_0}{1 + 57.3a_0/\pi e_1 \text{AR}} \tag{5.52}$$

However, from Figure 5.35b, by definition, $dC_L/d\alpha = a$. Hence, from Eq. (5.52)

$$\boxed{a = \frac{a_0}{1 + 57.3a_0/\pi e_1 \text{AR}}} \tag{5.53}$$

Equation (5.53) gives the desired lift slope for a finite wing of given aspect ratio AR when we know the corresponding slope a_0 for an infinite wing. Remember: a_0 is obtained from airfoil data such as in Appendix D. Also note that Eq. (5.53) verifies our previous qualitative statement that $a < a_0$.

In summary, a finite wing introduces two major changes to the airfoil data in Appendix D:

1. Induced drag must be added to the finite wing:

$$\underset{\substack{\text{Total} \\ \text{drag}}}{C_D} = \underset{\substack{\text{Profile} \\ \text{drag}}}{c_d} + \underset{\substack{\text{Induced} \\ \text{drag}}}{\frac{C_L{}^2}{\pi e \text{AR}}}$$

2. The slope of the lift curve for a finite wing is less than that for an infinite wing; $a < a_0$.

Example 5.11 Consider a wing with an aspect ratio of 10 and a NACA 23012 airfoil section. Assume Re $\approx 5 \times 10^6$. The span efficiency factor is $e = e_1 = 0.95$. If the wing is at a $4°$ angle of attack, calculate C_L and C_D.

SOLUTION Since we are dealing with a finite wing but have airfoil data (Appendix D) for infinite wings only, the first job is to obtain the slope of this lift curve for the finite wing, modifying the data from Appendix D.

The infinite wing lift slope can be obtained from any two points on the linear curve. For the NACA 23012 airfoil, for example (from Appendix D),

$$c_l = 1.2 \text{ at } \alpha_{\text{eff}} = 10°$$

$$c_l = 0.14 \text{ at } \alpha_{\text{eff}} = 0°$$

Hence

$$a_0 = \frac{dc_l}{d\alpha} = \frac{1.2 - 0.14}{10 - 0} = \frac{1.06}{10} = 0.106 \text{ per degree}$$

Also from Appendix D,

$$\alpha_{L=0} = -1.5 \quad \text{and} \quad c_d \approx 0.006$$

The lift slope for the finite wing can now be obtained from Eq. (5.53).

$$a = \frac{a_0}{1 + 57.3a_0/\pi e_1 \text{AR}} = \frac{0.106}{1 + 57.3(0.106)/\pi(0.95)(10)} = 0.088 \text{ per degree}$$

At $\alpha = 4°$,

$$C_L = a(\alpha - \alpha_{L=0})$$
$$= 0.088[4° - (-1.5)]$$
$$= 0.088(5.5)$$

$$\boxed{C_L = 0.484}$$

The total drag coefficient is given by Eq. (5.47):

$$C_D = c_d + \frac{C_L{}^2}{\pi e \text{AR}} = 0.006 + \frac{0.484^2}{\pi(0.95)(10)}$$

$$= 0.006 + 0.0078 = \boxed{0.0138}$$

5.16 SWEPT WINGS

Almost all modern high-speed aircraft have swept-back wings, such as shown in Figure 5.36b. Why? We are now in a position to answer this question.

We will first consider subsonic flight. Consider the planview of a straight wing, as sketched in Figure 5.36a. Assume this wing has an airfoil cross section with a critical Mach number $M_{cr} = 0.7$. (Remember from Sec. 5.10 that for M_∞ slightly above M_{cr}, there is a large increase in drag. Hence, it is desirable to increase M_{cr} as much as possible in high-speed subsonic airplane design.) Now assume that we sweep the wing back through an angle of, say, 30°, as shown in Figure 5.36b. The airfoil, which still has a value of $M_{cr} = 0.7$, now "sees" essentially only the component of the flow normal to the leading edge of the wing; i.e., the aerodynamic properties of the local section of the swept wing are governed mainly by the flow normal to the leading edge. Hence, if M_∞ is the freestream Mach number, the airfoil in Figure 5.36b is seeing effectively a smaller Mach number, $M_\infty \cos 30°$. As a result, the actual freestream Mach number can be increased *above* 0.7 before critical phenomena on the airfoil are encountered. In fact, we could expect that the critical Mach number for the *swept wing itself* would be as high as $0.7/\cos 30° = 0.808$, as shown in Figure 5.36b. This means that the large increase in drag (as sketched in Figure 5.18) would be delayed to M_∞ much larger than M_{cr} for the airfoil—in terms of Figure 5.36, something much larger than 0.7, and maybe even as high as 0.808. Therefore, we see the main function of a swept wing. *By sweeping the wings of subsonic aircraft, drag divergence is delayed to higher Mach numbers.*

In real life, the flow over the swept wing sketched in Figure 5.36b is a fairly complex three-dimensional flow, and to say that the airfoil sees only the component normal to the leading edge is a sweeping simplification. However, it leads to a good rule of thumb. If Ω is the sweep angle, as shown in Figure 5.36b, the actual critical Mach number for the swept wing is bracketed by:

$$M_{cr} \text{ for airfoil} < \frac{\text{actual } M_{cr}}{\text{for swept wing}} < \frac{M_{cr} \text{ for airfoil}}{\cos \Omega}$$

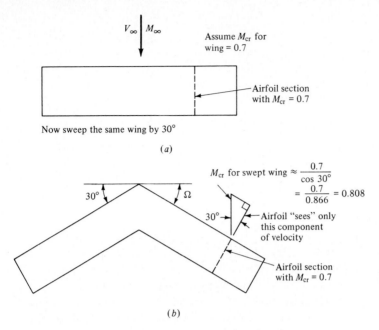

Figure 5.36 Effect of a swept wing on critical Mach number.

For supersonic flight, swept wings are also advantageous, but not quite from the same point of view as described above for subsonic flow. Consider the two swept wings sketched in Figure 5.37. For a given $M_\infty > 1$, there is a Mach cone with vertex angle μ, equal to the Mach angle [recall Eq. (5.38)]. If the leading edge of a swept wing is *outside* the Mach cone, as shown in Figure 5.37a, the component of the Mach number normal to the leading edge is supersonic. As a result, a fairly strong oblique shock wave will be created by the wing itself, with an attendant large wave drag. On the other hand, if the leading edge of the swept wing is *inside* the Mach cone, as shown in Figure 5.37b, the component of the Mach number normal to the leading edge is subsonic. As a result, the wave drag produced by the wing is less. Therefore, the advantage of sweeping the wings for supersonic flight is in general to obtain a decrease in wave drag, and if the wing is swept inside the Mach cone, a considerable decrease can be obtained.

As a final note, we might observe that not all supersonic aircraft have swept wings. In supersonic flight, wave drag can also be reduced by using stubby, very *low* aspect ratio, straight wings with thin, sharp airfoils (such as on the Lockheed F-104). In supersonic flight, wave drag is generally much larger, hence much more important, than induced drag. Thus, low–aspect ratio wings for supersonic aircraft reduce wave drag much more than they increase induced drag. This results in a net decrease in total drag. Thus, low–aspect ratio wings are efficient for supersonic flight, whereas high–aspect ratio wings are efficient for subsonic flight.

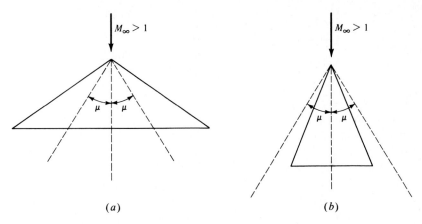

Figure 5.37 Swept wings for supersonic flow. (a) Wing swept outside the Mach cone. (b) Wing swept inside the Mach cone.

5.17 FLAPS—A MECHANISM FOR HIGH LIFT

An airplane normally encounters its lowest flight velocities at takeoff or landing, two periods that are most critical for aircraft safety. The slowest speed at which an airplane can fly in straight and level flight is defined as the *stalling speed* V_{stall}. Hence, the calculation of V_{stall}, as well as aerodynamic methods of making V_{stall} as small as possible, are of vital importance.

The stalling velocity is readily obtained in terms of the maximum lift coefficient, as follows. From the definition of C_L,

$$L = q_\infty S C_L = \tfrac{1}{2}\rho_\infty V_\infty^2 S C_L$$

Thus
$$V_\infty = \sqrt{\frac{2L}{\rho_\infty S C_L}} \tag{5.54}$$

In steady, level flight, the lift is just sufficient to support the weight W of the aircraft, that is, $L = W$. Thus

$$V_\infty = \sqrt{\frac{2W}{\rho_\infty S C_L}} \tag{5.55}$$

Examining Eq. (5.55), for an airplane of given weight and size at a given altitude, we find the only recourse to minimize V_∞ is to maximize C_L. Hence, stalling speed corresponds to the angle of attack that produces $C_{L,\text{max}}$:

$$V_{\text{stall}} = \sqrt{\frac{2W}{\rho_\infty S C_{L,\text{max}}}} \tag{5.56}$$

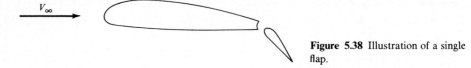

Figure 5.38 Illustration of a single flap.

In order to decrease V_{stall}, $C_{L,\max}$ must be increased. However, for a wing with a given airfoil shape, $C_{L,\max}$ is fixed by nature; i.e., the lift properties of an airfoil, including maximum lift, depend on the physics of the flow over the airfoil. In order to assist nature, the lifting properties of a given airfoil can be greatly enhanced by the use of "artificial" high-lift devices. The most common of these devices is the simple flap at the trailing edge of the wing, as sketched in Figure 5.38. When the flap is rotated downward, the camber of the airfoil is effectively

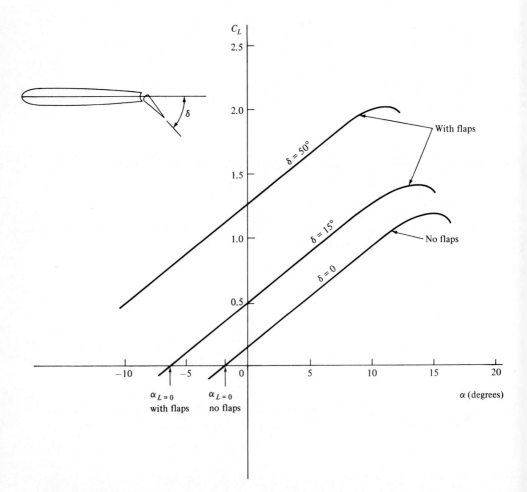

Figure 5.39 Illustration of the effect of flaps on the lift curve. The numbers shown are typical of a modern medium-range jet transport.

increased, with a consequent increase in $C_{L,\text{max}}$ for the wing. At the same time, the zero-lift angle of attack is shifted to a more negative value. These trends are illustrated in Figure 5.39, which compares the variation of C_L with α for a wing with and without flaps. Also, for some of the airfoils given in Appendix D, lift curves are shown with the effect of flap deflection included.

The increase in $C_{L,\text{max}}$ due to flaps can be dramatic. If the flap is designed not only to rotate downward, but also to translate rearward so as to increase the effective wing area, $C_{L,\text{max}}$ can be increased by approximately a factor of 2. If additional high-lift devices are used, such as slats at the leading edge, slots in the surface, or mechanical means of boundary layer control, then $C_{L,\text{max}}$ can sometimes be increased by a factor of 3 or more. For an interesting and more detailed discussion of various high-lift devices, the reader is referred to the books by McCormick and Shevell (see the Bibliography at the end of this chapter).

5.18 THE AERODYNAMICS OF CYLINDERS AND SPHERES

Consider the low-speed subsonic flow over a sphere or an infinite cylinder with its axis normal to the flow. If the flow were inviscid (frictionless), the theoretical flow pattern would look qualitatively as sketched in Figure 5.40a. The streamlines would form a symmetrical pattern; hence the pressure distributions over the front and rear surfaces would also be symmetrical, as sketched in Figure 5.40b. This symmetry creates a momentous phenomenon, namely, that there is *no pressure drag on the sphere* if the flow is frictionless. This can be seen by simple inspection

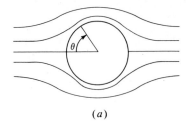

(a)

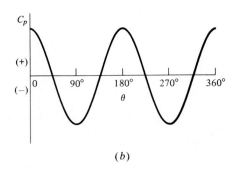

(b)

Figure 5.40 Ideal frictionless flow over a sphere. (a) Flow field. (b) Pressure coefficient distribution.

of Figure 5.40b: the pressure distribution on the front face $(-90° \leq \theta \leq 90°)$ creates a force in the drag direction, but the pressure distribution on the rear face $(90° \leq \theta \leq 270°)$, which is identical to that on the front face, creates an equal and opposite force. Thus, we obtain the curious theoretical result that there is no drag on the body, quite contrary to everyday experience. This conflict between theory and experiment was well known at the end of the nineteenth century and is called *d'Alembert's paradox*.

The actual flow over a sphere or cylinder is sketched in Figure 4.26; as discussed in Sec. 4.18, the presence of friction leads to separated flows in regions of adverse pressure gradients. Examining the theoretical inviscid pressure distribution shown in Figure 5.40b, we find on the rear surface $(90° \leq \theta \leq 270°)$ the pressure increases in the flow direction; i.e., an adverse pressure gradient exists. Thus, it is entirely reasonable that the real-life flow over a sphere or cylinder would be separated from the rear surface. This is indeed the case, as first shown in Figure 4.26 and as sketched again in Figure 5.41a. The real pressure distribution that corresponds to this separated flow is shown as the solid curve in Figure 5.41b. Note that the average pressure is much higher on the front face $(-90° < \theta < 90°)$ than on the rear face $(90° < \theta < 270°)$. As a result, there is a net drag force exerted on the body. Hence, nature and experience are again reconciled, and d'Alembert's paradox is removed by a proper account of the presence of friction.

It is emphasized that the flow over a sphere or cylinder, and therefore the drag, is dominated by flow separation on the rear face. This leads to an interesting variation of C_D with the Reynolds number. Let the Reynolds number be defined in terms of the sphere diameter D: $\mathrm{Re} = \rho_\infty V_\infty D / \mu_\infty$. If a sphere is

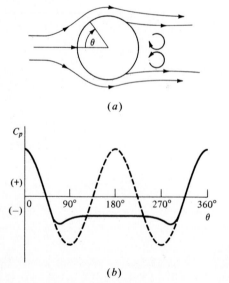

(a)

(b)

Figure 5.41 Real separated flow over a sphere; separation is due to friction. (a) Flow field. (b) Pressure-coefficient distribution.

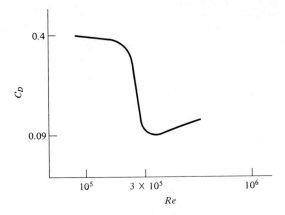

Figure 5.42 Variation of drag coefficient with Reynolds number for a sphere in low-speed flow.

mounted in a low-speed subsonic wind tunnel and the freestream velocity is varied such that Re increases from 10^5 to 10^6, a curious, almost discontinuous drop in C_D is observed at about Re $= 3 \times 10^5$. This behavior is sketched in Figure 5.42. What causes this precipitous decrease in drag? The answer lies in the different effects of laminar and turbulent boundary layers on flow separation. At the end of Sec. 4.18 it was noted that laminar boundary layers separate much more readily than turbulent boundary layers. In the flow over a sphere at Re $< 3 \times 10^5$, the boundary layer is laminar. Hence, the flow is totally separated from the rear face, and the wake behind the body is large, as sketched in Figure 5.43*a*. In turn, the value of C_D is large, as noted at the left of Figure 5.42 for Re $< 3 \times 10^5$. On the other hand, as Re is increased above 3×10^5, transition takes place on the front face, the boundary layer becomes turbulent, and the separation point moves rearward. (Turbulent boundary layers remain attached for longer distances in the face of adverse pressure gradients.) In this case, the wake behind the body is much smaller, as sketched in Figure 5.43*b*. In turn, the pressure drag is less and C_D decreases, as noted at the right of Figure 5.42.

Therefore, to decrease the drag on a sphere or cylinder, a turbulent boundary layer must be obtained on the front surface. This can be made to occur naturally by increasing Re until transition occurs on the front face. It can also be forced artificially at lower values of Re by using a rough surface to encourage early transition or by wrapping wire or other protuberances around the surface to create turbulence. (The use of such artificial devices is sometimes called "tripping the boundary layer.")

It is interesting to note that the dimples on the surface of a golf ball are designed to promote turbulence and hence reduce the drag on the ball in flight. Indeed, some very recent research has shown that polygonal dimples result in less drag than the conventional circular dimples on golf balls; but a dimple of any shape leads to less drag than a smooth surface (table tennis balls have more drag than golf balls).

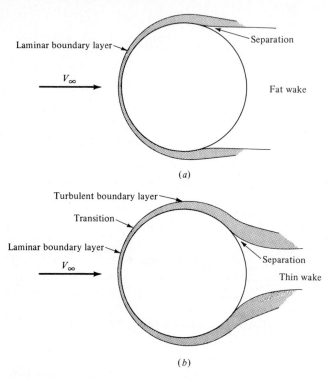

Figure 5.43 Laminar and turbulent flow over a sphere.

5.19 HOW LIFT IS PRODUCED—SOME ALTERNATE EXPLANATIONS

As an epilogue to our discussions of airfoils and wings, it is fitting to examine once again the source of aerodynamic lift. As emphasized in the previous chapters, the fundamental source of lift is the pressure distribution over the wing surface. This is sketched in Figure 5.44, which illustrates the low pressure on the top surface, the high pressure on the bottom surface, and the resulting net lift.

An alternate explanation is sometimes given: The wing deflects the airflow such that the mean velocity vector behind the wing is canted slightly downward, as sketched in Figure 5.45. Hence, the wing imparts a downward component of momentum to the air; i.e., the wing exerts a force on the air, pushing the flow downward. From Newton's third law, the equal and opposite reaction produces a lift. However, this explanation really involves the *effect* of lift, and not the cause. In reality, the air pressure on the surface (see Figure 5.44) is pushing on the surface, hence creating lift. As a result of the equal-and-opposite principle, the airfoil surface pushes on the air, imparting a force on the airflow which deflects the velocity downward. Hence, the net rate of change of downward momentum created in the airflow because of the presence of the wing can be thought of as an

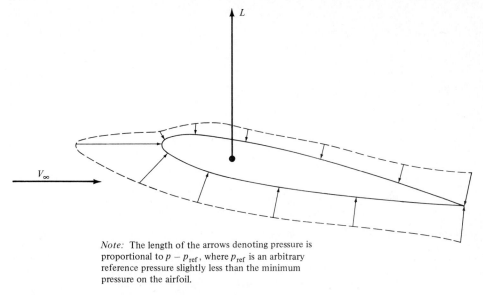

Note: The length of the arrows denoting pressure is proportional to $p - p_{ref}$, where p_{ref} is an arbitrary reference pressure slightly less than the minimum pressure on the airfoil.

Figure 5.44 Pressure distribution is the source of lift.

effect due to the surface pressure distribution; the pressure distribution by itself is the fundamental cause of lift.

A third argument, called the *circulation theory of lift*, is sometimes given for the source of lift. However, this turns out to be not so much an *explanation* of lift per se, but rather more of a mathematical formulation for the calculation of lift for an airfoil of given shape. Moreover, it is mainly applicable to incompressible flow. The circulation theory of lift is elegant and well-developed; it is also beyond the scope of this book. However, some of its flavor is given as follows.

Consider the flow over a given airfoil, as shown in Figure 5.46. Imagine a closed curve C drawn around the airfoil. At a point on this curve, the flow

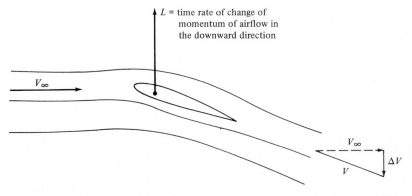

Figure 5.45 Relationship of lift to the time rate of change of momentum of the airflow.

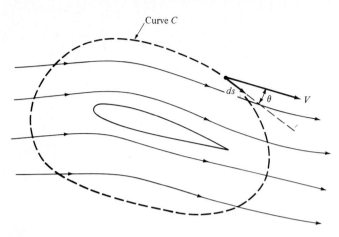

Figure 5.46 Diagram for the circulation theory of lift.

velocity is V and the angle between V and a tangent to the curve is θ. Let ds be an incremental distance along C. A quantity called the *circulation* Γ is defined as

$$\Gamma \equiv \oint_C V \cos \theta \, ds \tag{5.57}$$

that is, Γ is the line integral of the component of flow velocity along the closed curve C. After a value of Γ is obtained, the lift *per unit span* can be calculated from

$$\boxed{L = \rho_\infty V_\infty \Gamma_\infty} \tag{5.58}$$

Equation (5.57) is the *Kutta-Joukowsky theorem*; it is a pivotal relation in the circulation theory of lift. The object of the theory is to (somehow) calculate Γ for a given V_∞ and airfoil shape. Then Eq. (5.58) yields the lift. A major thrust of ideal incompressible flow theory, many times called *potential flow theory*, is to calculate Γ. Such matters are discussed in more-advanced aerodynamic texts (see, for example, Anderson, *Fundamentals of Aerodynamics*, McGraw-Hill, 1984).

The circulation theory of lift is compatible with the true physical nature of the flow over an airfoil, as any successful mathematical theory must be. In the simplest sense, we can visualize the true flow over an airfoil, shown at the right of Figure 5.47, as the superposition of a uniform flow and a circulatory flow, shown at the left of Figure 5.47. The circulatory flow is clockwise, which when added to the uniform flow, yields a higher velocity above the airfoil and a lower velocity below the airfoil. From Bernoulli's equation, this implies a lower pressure on the top surface of the airfoil and a higher pressure on the bottom surface, hence generating lift in the upward direction. The strength of the circulatory contribution, defined by Eq. (5.57), is just the precise value such that when it is added to the uniform flow contribution, the actual flow over the airfoil leaves the trailing

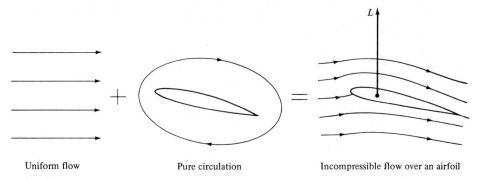

Uniform flow Pure circulation Incompressible flow over an airfoil

Figure 5.47 Addition of two elementary flows to synthesize a more complex flow. If one or more of the elementary flows has circulation, then the synthesized flow also has the same circulation. The lift is directly proportional to the circulation.

edge smoothly, as sketched at the right of Figure 5.47. This is called the *Kutta condition* and is one of the major facets of the circulation theory of lift.

Again, keep in mind that the actual mechanism that nature has of communicating a lift to the airfoil is the pressure distribution over the surface of the airfoil, as sketched in Figure 5.44. In turn, this pressure distribution ultimately causes a time rate of change of momentum of the airflow, as shown in Figure 5.45, a principle which can be used as an alternate way of visualizing the generation of lift. Finally, even the circulation theory of lift stems from the pressure distribution over the surface of the airfoil, because the derivation of the Kutta-Joukowsky theorem, Eq. (5.58), involves the surface pressure distribution. Again, for more details, consult Anderson, *Fundamentals of Aerodynamics* (McGraw-Hill, 1984).

5.20 HISTORICAL NOTE: AIRFOILS AND WINGS

We know that George Cayley introduced the concept of a fixed-wing aircraft in 1799; this has been discussed at length in Secs. 1.3 and 5.1. Moreover, Cayley appreciated the fact that lift is produced by a region of low pressure on the top surface of the wing and high pressure on the bottom surface and that a cambered shape produces more lift than a flat surface. Indeed, Figure 1.5 shows that Cayley was thinking of a curved surface for a wing, although the curvature was due to the wind billowing against a loosely fitting fabric surface. However, neither Cayley nor any of his immediate followers performed work even closely resembling airfoil research or development.

It was not until 1884 that the first serious airfoil developments were made. In this year Horatio F. Phillips, an Englishman, was granted a patent for a series of double-surface, cambered airfoils. Figure 5.48 shows Phillips's patent drawings for his airfoil section. Phillips was an important figure in late nineteenth century aeronautical engineering; we have met him before, in Sec. 4.22, in conjunction

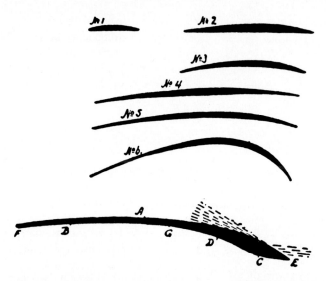

Figure 5.48 Double-surface airfoil sections by Phillips. The six upper shapes were patented by Phillips in 1884; the lower airfoil was patented in 1891.

with his ejector-driven wind tunnel. In fact, the airfoil shapes in Figure 5.48 were the result of numerous wind-tunnel experiments in which Phillips examined curved wings of "every conceivable form and combination of forms." Phillips widely published his results, which had a major impact on the aeronautics community. Continuing with his work, Phillips patented more airfoil shapes in 1891. Then, moving into airplane design in 1893, he built and tested a large multiplane model, consisting of a large number of wings, each with a 19-ft span and a chord of only $1\frac{1}{2}$ in, which looked like a venetian blind! The airplane was powered by a steam engine with a 6.5-ft propeller. The vehicle ran on a circular track and actually lifted a few feet off the ground, momentarily. After this demonstration, Phillips gave up until 1907, when he made the first tentative hop flight in England in a similar, but gasoline-powered, machine, staying airborne for about 500 ft. This was his last contribution to aeronautics. However, his pioneering work during the 1880s and 1890s clearly designates Phillips as the grandparent of the modern airfoil.

After Phillips, the work on airfoils shifted to a search for the most efficient shapes. Indeed, work is still being done today on this very problem, although much progress has been made. This progress covers several historical periods, as described below.

A The Wright Brothers

As noted in Secs. 1.8 and 4.22, Wilbur and Orville Wright, after their early experience with gliders, concluded in 1901 that many of the existing "air pressure" data on airfoil sections were inadequate and frequently incorrect. In order to rectify these deficiencies, they constructed their own wind tunnel (see Figure 4.45), in which they tested several hundred different airfoil shapes between September 1901 and August 1902. From their experimental results, the Wright brothers chose an airfoil with a 1/20 maximum camber-to-chord ratio for their successful Wright Flyer I in 1903. These wind-tunnel tests by the Wright brothers constituted a major advance in airfoil technology at the turn of the century.

B British and United States Airfoils (1910 to 1920)

In the early days of powered flight, airfoil design was basically customized and personalized; very little concerted effort was made to find a standardized, efficient section. However, some early work was performed by the British Government at the National Physical Laboratory (NPL), leading to a series of Royal Aircraft Factory (RAF) airfoils used on World War I airplanes. Figure 5.49 illustrates the shape of the RAF 6 airfoil. In the United States, most aircraft until 1915 used either an RAF section or a shape designed by the Frenchman Alexandre Gustave Eiffel. This tenuous status of airfoils led the NACA, in its First Annual Report in 1915, to emphasize the need for "the evolution of more efficient wing sections of practical form, embodying suitable dimensions for an economical structure, with moderate travel of the center of pressure and still affording a large angle of attack combined with efficient action." To this day, more than 70 years later, NASA is still pursuing such work.

The first NACA work on airfoils was reported in NACA report no. 18, "Aerofoils and Aerofoil Structural Combinations," by Lt. Col. Edgar S. Gorrell and Major H. S. Martin, prepared at the Massachusetts Institute of Technology (M.I.T.) in 1917. Gorrell and Martin summarized the contemporary airfoil status as follows:

> Mathematical theory has not, as yet, been applied to the discontinuous motion past a cambered surface. For this reason, we are able to design aerofoils only by consideration of those forms which have been successful, by applying general rules learned by experience, and by then testing the aerofoils in a reliable wind tunnel.

In NACA report No. 18, Gorrell and Martin disclosed a series of tests on the largest single group of airfoils to that date, except for the work done at NPL and by Eiffel. They introduced the U.S.A. airfoil series and reported wind-tunnel data

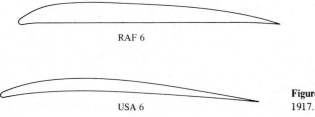

RAF 6

USA 6

Figure 5.49 Typical airfoils in 1917.

for the U.S.A. 1 through U.S.A. 6 sections. Figure 5.49 illustrates the shape of the U.S.A. 6 airfoil. The airfoil models were made of brass and were finite wings with a span of 18 in and a chord of 3 in, that is, AR = 6. Lift and drag coefficients were measured at a velocity of 30 mi/h in the M.I.T. wind tunnel. These airfoils represented the first systematic series originated and studied by the NACA.

C 1920 to 1930

Based on their wind-tunnel observations in 1917, Gorrell and Martin stated that slight variations in airfoil design make large differences in the aerodynamic performance. This is the underlying problem of airfoil research, and it led in the 1920s to a proliferation of airfoil shapes. In fact, as late as 1929, F. A. Louden, in his NACA report no. 331, entitled "Collection of Wind Tunnel Data on Commonly Used Wing Sections," stated that "the wing sections most commonly used in this country are the Clark Y, Clark Y-15, Gottingen G-387, G-398, G-436, N.A.C.A. M-12, Navy N-9, N-10, N-22, R.A.F.-15, Sloane, U.S.A.-27, U.S.A.-35A, U.S.A.-35B." However, help was on its way. As noted in Sec. 4.22, the NACA built a variable-density wind tunnel at Langley Aeronautical Laboratory in 1923, a wind tunnel that was to become a workhorse in future airfoil research, as emphasized below.

D The Early NACA Four-Digit Airfoils

In a classic work in 1933, order and logic were finally brought to airfoil design in the United States. This was reported in NACA report no. 460, "The Characteristics of 78 Related Airfoil Sections from Tests in the Variable-Density Wind Tunnel," by Eastman N. Jacobs, Kenneth E. Ward, and Robert M. Pinkerton. Their philosophy on airfoil design was as follows:

> Airfoil profiles may be considered as made up of certain profile-thickness forms disposed about certain mean lines. The major shape variables then become two, the thickness form and the mean-line form. The thickness form is of particular importance from a structural standpoint. On the other hand, the form of the mean line determines almost independently some of the most important aerodynamic properties of the airfoil section, e.g., the angle of zero lift and the pitching-moment characteristics. The related airfoil profiles for this investigation were derived by changing systematically these shape variables.

They then proceeded to define and study for the first time in history the famous NACA four-digit airfoil series, some of which are given in Appendix D of this book. For example, the NACA 2412 is defined as a shape that has a maximum camber of 2 percent of the chord (the first digit); the maximum camber occurs at a position of 0.4 chord from the leading edge (the second digit), and the maximum thickness is 12 percent (the last two digits). Jacobs and his colleagues tested these airfoils in the NACA variable-density tunnel using a 5 in × 30 in finite wing (again, an aspect ratio of 6). In NACA report no. 460, they gave curves of C_L, C_D, and L/D for the finite wing. Moreover, using the same formulas developed in Sec. 5.15, they corrected their data to give results for the infinite wing case, also. After this work was published, the standard NACA four-digit airfoils were widely used. Indeed, even today the NACA 2412 is used on several light aircraft.

E Later NACA Airfoils

In the late 1930s, the NACA developed a new camber line family to increase maximum lift, with the 230 camber line being the most popular. Combining with the standard NACA thickness distribution, this gave rise to the NACA five-digit airfoil series, such as the 23012, some of which are still flying today, for example, on the Cessna Citation and the Beech King Air. This work was followed by families of high-speed airfoils and laminar flow airfoils in the 1940s.

To reinforce its airfoil development, in 1939 the NACA constructed a new low-turbulence two-dimensional wind tunnel at Langley exclusively for airfoil testing. This tunnel has a rectangular test section 3 ft wide and $7\frac{1}{2}$ ft high and can be pressurized up to 10 atm for high–Reynolds number testing. Most importantly, this tunnel allows airfoil models to span the test section completely, thus directly providing infinite wing data. This is in contrast to the earlier tests described above, which used a finite wing of AR = 6 and then corrected the data to correspond to infinite wing conditions. Such corrections are always compromised by tip effects. (For example, what is the *precise* span efficiency factor for a given wing?) With the new two-dimensional tunnel, vast numbers of tests were performed in the early 1940s on both old and new airfoil shapes over a Reynolds number range from 3 to 9 million and at Mach numbers less than 0.17 (incompressible flow). The airfoil models generally had a 2-ft chord and completely spanned the 3-ft width of the test section. It is interesting to note that the lift and drag are not obtained on a force balance. Rather, the lift is calculated by integrating the measured pressure distribution, and the drag is calculated from Pitot pressure measurements made in the wake downstream of the trailing edge. However, the pitching moments are measured directly on a balance. A vast amount of airfoil data obtained in this fashion from the two-dimensional tunnel at Langley was compiled and published in a book entitled *Theory of Wing Sections Including a Summary of Airfoil Data*, by Abbott and von Doenhoff, in 1949 (see Bibliography at end of this chapter). It is important to note that all the airfoil data in Appendix D are obtained from this reference, i.e., all the data in Appendix D are direct measurements for an infinite wing at essentially incompressible flow conditions.

F Modern Airfoil Work

Priorities for supersonic and hypersonic aerodynamics put a stop to the NACA airfoil development in 1950. Over the next 15 years specialized equipment for airfoil testing was dismantled. Virtually no systematic airfoil research was done in the United States during this period.

However, in 1965 Richard T. Whitcomb made a breakthrough with the NASA supercritical airfoil. This revolutionary development, which allowed the design of wings with high critical Mach numbers (see Sec. 5.10), served to reactivate the interest in airfoils within NASA. Since that time, a healthy program in modern airfoil development has been reestablished. The low-turbulence, pressurized, two-dimensional wind tunnel at Langley is back in operation. Moreover, a new dimension has been added to airfoil research: the high-speed digital computer. In fact, computer programs for calculating the flow field around airfoils at subsonic speeds are so reliable that they shoulder some of the routine testing duties heretofore exclusively carried by wind tunnels. The same cannot yet be said about transonic cases, but current research is focusing on this problem. A recent and interesting survey of modern airfoil activity within NASA is given by Pierpont in *Astronautics and Aeronautics* (see Bibliography).

Of special note is the modern low-speed airfoil series, designated LS(1), developed by NASA for use by general aviation on light airplanes. The shape of a typical LS(1) airfoil is contrasted with a "conventional" airfoil in Figure 5.50. Its lifting characteristics, illustrated in Figure 5.51, are clearly superior and should allow smaller wing areas, hence less drag, for airplanes of the type shown in Figure 5.50.

In summary, airfoil development over the past 100 years has moved from an ad hoc individual process to a very systematic and logical engineering process. It is alive and well today, with the promise of major advancements in the future using both wind tunnels and computers.

G Finite Wings

Some historical comments on the finite wing are in order. Francis Wenham (see Chap. 1), in his classic paper entitled *Aerial Locomotion* given to the Aeronautical

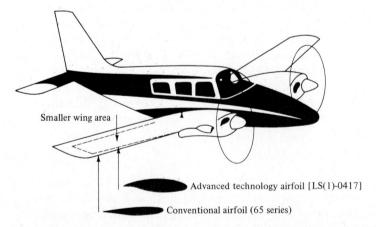

Figure 5.50 Shape comparison between the modern LS(1)-0417 and a conventional airfoil. The higher lift obtained with the LS(1)-0417 allows a smaller wing area and hence lower drag. *(NASA.)*

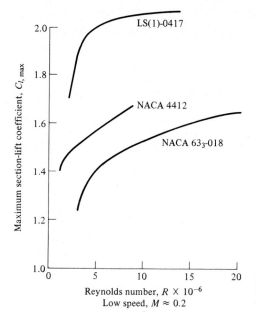

Figure 5.51 Comparison of maximum lift coefficients between the LS(1)-0417 and conventional airfoils. *(NASA.)*

Society of Great Britain on June 27, 1866, theorized (correctly) that most of the lift of a wing occurs from the portion near the leading edge and hence a long, narrow wing would be most efficient. In this fashion, he was the first person in history to appreciate the value of high–aspect ratio wings for subsonic flight. Moreover, he suggested stacking a number of long, thin wings above each other to generate the required lift, and hence he became an advocate of the multiplane concept. In turn, he built two full-size gliders in 1858, both with five wings each, and successfully demonstrated the validity of his ideas.

However, the true aerodynamic theory and understanding of finite wings did not come until 1907, when the Englishman Frederick W. Lanchester published his book *Aerodynamics*. In it, he outlined the circulation theory of lift, developed independently about the same time by Kutta in Germany and Joukowsky in Russia. More importantly, Lanchester discussed for the first time the effect of wingtip vortices on finite wing aerodynamics. Unfortunately, Lanchester was not a clear writer; his ideas were extremely difficult to understand, and hence they did not find application in the aeronautical community.

In 1908, Lanchester visited Göttingen, Germany, and fully discussed his wing theory with Ludwig Prandtl and his student, Theodore von Karman. Prandtl spoke no English, Lanchester spoke no German, and in light of Lanchester's unclear ways of explaining his ideas, there appeared to be little chance of understanding between the two parties. However, in 1914, Prandtl set forth a simple, clear, and correct theory for calculating the effect of tip vortices on the aerodynamic characteristics of finite wings. It is virtually impossible to assess how much Prandtl was influenced by Lanchester, but to Prandtl must go the credit of first establishing a practical finite wing theory, a theory which is fundamental to our discussion of finite wings in Secs. 5.13 to 5.15. Indeed, Prandtl's first published words on the subject were:

> The lift generated by the airplane is, on account of the principle of action and reaction, necessarily connected with a descending current in all its details. It appears that the descending current is formed by a pair of vortices, the vortex filaments of which start from the airplane wingtips. The distance of the two vortices is equal to the span of the airplane, their strength is equal to the circulation of the current around the airplane, and the current in the vicinity of the airplane is fully given by the superposition of the uniform current with that of a vortex consisting of three rectilinear sections.

Prandtl's pioneering work on finite wing theory, along with his ingenious concept of the boundary layer, has earned him the title "parent of aerodynamics." In the four years following 1914, he went on to show that an elliptical lift distribution results in the minimum induced drag. Indeed, the first coining of the terms "induced drag" and "profile drag" was made in 1918 by Max Munk in a note entitled "Contribution to the Aerodynamics of the Lifting Organs of the Airplane." Munk was a colleague of Prandtl's, and the note was one of several classified German wartime reports on airplane aerodynamics.

5.21 HISTORICAL NOTE: ERNST MACH AND HIS NUMBER

Airplanes that fly at Mach 2 are commonplace today. Indeed, high-performance military aircraft such as the Lockheed SR-71 Blackbird can exceed Mach 3. As a result, the term "Mach number" has become part of our general language—the average person in the street understands that Mach 2 means twice the speed of sound. On a more technical basis, the dimensional analysis described in Sec. 5.3 demonstrated that aerodynamic lift, drag, and moments depend on two important dimensionless products—the Reynolds number and the Mach number. Indeed, in a more general treatment of fluid dynamics, the Reynolds number and Mach number can be shown as the major governing parameters for any realistic flow field; they are among a series of governing dimensionless parameters called *similarity parameters*. We have already examined the historical source of the Reynolds number in Sec. 4.23; let us do the same for the Mach number in the present section.

The Mach number is named after Ernst Mach, a famous nineteenth century physicist and philosopher. Mach was an illustrious figure with widely varying interests. He was the first person in history to observe supersonic flow and to understand its basic nature. Let us take a quick look at this man and his contributions to supersonic aerodynamics.

Ernst Mach was born at Turas, Moravia, in Austria, on February 18, 1838. Mach's father and mother were both extremely private and introspective intellectuals. His father was a student of philosophy and classical literature; his mother was a poet and musician. The family was voluntarily isolated on a farm, where

Mach's father pursued an interest of raising silkworms—indeed pioneering the beginning of silkworm culture in Europe. At an early age, Mach was not a particularly successful student. Later, Mach described himself as a "weak pitiful child who developed very slowly." Through extensive tutoring by his father at home, Mach learned Latin, Greek, history, algebra, and geometry. After marginal performances in grade school and high school (due not to any lack of intellectual ability, but rather to a lack of interest in the material usually taught by rote), Mach entered the University of Vienna. There, he blossomed, spurred by interest in mathematics, physics, philosophy, and history. In 1860, he received a Ph.D. in physics, writing a thesis entitled "On Electrical Discharge and Induction." By 1864, he was a professor of physics at the University of Graz. (The variety and depth of his intellectual interests at this time are attested by the fact that he turned down the position of a chair in *surgery* at the University of Salzburg to go to Graz.) In 1867, Mach became a professor of experimental physics at the University of Prague—a position he would occupy for the next 28 years.

In today's modern technological world, where engineers and scientists are virtually forced, out of necessity, to peak their knowledge in very narrow areas of extreme specialization, it is interesting to reflect on the personality of Mach, who was the supreme generalist. Here is only a partial list of Mach's contributions, as demonstrated in his writings: physical optics, history of science, mechanics, philosophy, origins of relativity theory, supersonic flow, physiology, thermodynamics, sugar cycle in grapes, physics of music, and classical literature. He even wrote on world affairs. (One of Mach's papers commented on the "absurdity committed by the statesman who regards the individual as existing solely for the sake of the state"; for this, Mach was severely criticized at that time by Lenin.) We can only sit back with awe and envy for Mach, who—in the words of the American philosopher William James—knew "everything about everything."

Mach's contributions to supersonic aerodynamics were highlighted in a paper entitled "Photographische Fixierung der durch Projektile in der Luft eingeleiten Vorgange," given to the Academy of Sciences in Vienna in 1887. Here, for the first time in history, Mach showed a photograph of a shock wave in front of a bullet moving at supersonic speeds. This historic photograph, taken from Mach's original paper, is shown in Figure 5.52. Also visible are weaker waves at the rear of the projectile and the structure of the turbulent wake downstream of the base region. The two vertical lines are trip wires designed to time the photographic light source (or spark) with the passing of the projectile. Mach was a precise and careful experimentalist; the quality of the picture shown in Figure 5.52, along with the very fact that he was able to make the shock waves visible in the first place (he used an innovative optical technique called the shadowgram), attest to his exceptional experimental abilities. Note that Mach was able to carry out such experiments involving split-second timing without the benefit of electronics—indeed, the vacuum tube had not yet been invented.

Mach was the first to understand the basic characteristics of supersonic flow. He was the first to point out the importance of the flow velocity V relative to the speed of sound a and to note the discontinuous and marked changes in a flow

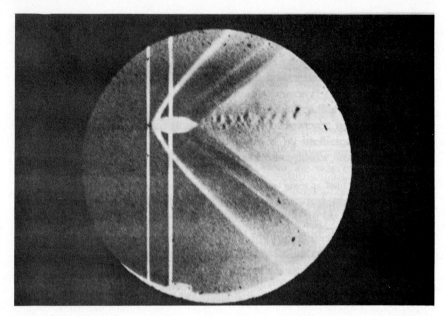

Figure 5.52 Photograph of a bullet in supersonic flight, published by Ernst Mach in 1887.

field as the ratio V/a changes from below 1 to above 1. He did not, however, call this ratio the Mach number. The term "Mach number" was not coined until 1929, when the well-known Swiss engineer Jacob Ackeret introduced this terminology, in honor of Mach, at a lecture at the Eidgenossiche Technische Hochschule in Zurich. Hence, the term "Mach number" is of fairly recent usage, not introduced into the English literature until the early 1930s.

Mach was an active thinker, lecturer, and writer up to the time of his death on February 19, 1916, near Munich, one day after his seventy-eighth birthday. His contributions to human thought were many, and his general philosophy on epistemology—a study of knowledge itself—is still discussed in college classes in philosophy today. Aeronautical engineers know him as the originator of supersonic aerodynamics; the rest of the world knows him as a man who originated the following philosophy, as paraphrased by Richard von Mises, himself a well-known mathematician and aerodynamicist of the early twentieth century:

> Mach does not start out to analyze statements, systems of sentences, or theories, but rather the world of phenomena itself. His elements are not the simplest sentences, and hence the building stones of theories, but rather—at least according to his way of speaking—the simplest facts, phenomena, and events of which the world in which we live and which we know is composed. The world open to our observation and experience consists of "colors, sounds, warmths, pressures, spaces, times, etc." and their components in greater and smaller complexes. All we make statements or assertions about, or formulate questions and answers to, are the relations in which these elements stand to each other. That is Mach's point of view.*

*From Richard von Mises, *Positivism, A Study in Human Understanding*, Braziller, New York, 1956.

Figure 5.53 Ernst Mach (1838–1916).

We end this section with a photograph of Mach, taken about 1910, shown in Figure 5.53. It is a picture of a thoughtful, sensitive man; no wonder that his philosophy of life emphasized observation through the senses, as discussed by von Mises above. In honor of his memory, an entire research institute, the Ernst Mach Institute in West Germany, is named. This institute deals with research in experimental gas dynamics, ballistics, high-speed photography, and cinematography. Indeed, for a much more extensive review of the technical accomplishments of Mach, see the recent paper authored by a member of the Ernst Mach Institute, H. Reichenbach, entitled "Contributions of Ernst Mach to Fluid Mechanics," in *Annual Reviews of Fluid Mechanics*, vol. 15, 1983, pp. 1–28 (published by Annual Reviews, Inc., Palo Alto, California).

5.22 HISTORICAL NOTE: THE FIRST MANNED SUPERSONIC FLIGHT

On October 14, 1947, a human being flew faster than the speed of sound for the first time in history. Imagine the magnitude of this accomplishment—just 60 years after Ernst Mach observed shock waves on supersonic projectiles (see Sec.

5.21) and a scant 44 years after the Wright brothers achieved their first successful powered flight (see Secs. 1.1 and 1.8). It is almost certain that Mach was not thinking at all about heavier-than-air manned flight of any kind, which in his day was still considered to be virtually impossible and the essence of foolish dreams. It is also almost certain that the Wright brothers had not the remotest idea that their fledgling 30-mi/h flight on December 17, 1903, would ultimately lead to a manned supersonic flight in Orville's lifetime (although Wilbur died in 1912, Orville lived an active life until his death in 1948). Compared to the total spectrum of manned flight reaching all the way back to the ideas of Leonardo da Vinci in the fifteenth century (see Sec. 1.2), this rapid advancement into the realm of supersonic flight is truly phenomenal. How did this advancement occur? What are the circumstances surrounding the first supersonic flight? Why was it so important? The purpose of this section is to address these questions.

Supersonic flight did not happen by chance; rather, it was an inevitable result of the progressive advancement of aeronautical technology over the years. On one hand we have the evolution of high-speed aerodynamic theory, starting with the pioneering work of Mach, as described in Sec. 5.21. This was followed by the development of supersonic nozzles by two European engineers, Carl G. P. de Laval in Sweden and A. B. Stodola in Switzerland. In 1887 de Laval used a convergent-divergent supersonic nozzle to produce a high-velocity flow of steam to drive a turbine. In 1903 Stodola was the first person in history to definitely prove (by means of a series of laboratory experiments) that such convergent-divergent nozzles did indeed produce supersonic flow. From 1905 to 1908, Prandtl in Germany took pictures of Mach waves inside supersonic nozzles and developed the first rational theory for oblique shock waves and expansion waves. After World War I, Prandtl studied compressibility effects in high-speed subsonic flow. This work, in conjunction with independent studies by the English aerodynamicist Herman Glauert, led to the publishing of the Prandtl-Glauert rule in the late 1920s (see Sec. 5.6 for a discussion of the Prandtl-Glauert rule and its use as a compressibility correction). These milestones, among others, established a core of aerodynamic theory for high-speed flow—a core that was well-established at least 20 years before the first supersonic flight. (For more historical details concerning the evolution of aerodynamic theory pertaining to supersonic flight, see Anderson, *Modern Compressible Flow: With Historical Perspective*, McGraw-Hill, 1982.)

On the other hand, we also have the evolution of hardware necessary for supersonic flight. The development of high-speed wind tunnels, starting with the small 12-inch-diameter high-speed subsonic tunnel at the NACA Langley Memorial Aeronautical Laboratory in 1927 and continuing with the first practical supersonic wind tunnels developed by Adolf Busemann in Germany in the early 1930s, is described in Sec. 4.22. The exciting developments leading to the first successful rocket engines in the late 1930s are discussed in Sec. 9.12. The concurrent invention and development of the jet engine, which would ultimately provide the thrust necessary for everyday supersonic flight, are related in Sec. 9.11. Hence, on the basis of the theory and hardware which existed at that time, the advent of manned supersonic flight in 1947 was a natural progression in the advancement of aeronautics.

However, in 1947 there was one missing link in both the theory and the hardware—the transonic regime, near Mach 1. The governing equations for transonic flow are highly nonlinear and hence are difficult to solve. No practical solution of these equations existed in 1947. This theoretical gap was compounded by a similar gap in wind tunnels. The sensitivity of a flow near Mach 1 makes the design of a proper transonic tunnel difficult. In 1947, no reliable transonic wind tunnel data were available. This gap of knowledge was of great concern to the aeronautical engineers who designed the first supersonic airplane, and it was the single most important reason for the excitement, apprehension, uncertainty, and outright bravery that surrounded the first supersonic flight.

The unanswered questions about transonic flow did nothing to dispel the myth of the "sound barrier" that arose in the 1930s and 1940s. As discussed in Sec. 5.12, the very rapid increase in drag coefficient beyond the drag-divergence Mach number led some people to believe that humans would never fly faster than sound. Grist was lent to their arguments when, on September 27, 1946, Geoffrey deHavilland, son of the famous British airplane designer, took the D.H. 108 Swallow up for an attack on the world's speed record. The Swallow was an experimental jet-propelled aircraft, with swept wings and no horizontal tail. Attempting to exceed 615 mi/h on its first high-speed, low-altitude run, the Swallow encountered major compressibility problems and broke up in the air. DeHavilland was killed instantly. The sound barrier had taken its toll. It was against this background that the first supersonic flight was attempted in 1947.

During the late 1930s, and all through World War II, some visionaries clearly saw the need for an experimental airplane designed to probe the mysteries of supersonic flight. Finally, in 1944, their efforts prevailed; the Army Air Force, in conjunction with the NACA, awarded a contract to the Bell Aircraft Corporation for the design, construction, and preliminary testing of a manned supersonic airplane. Designated the XS-1 (Experimental Sonic-1), this design had a fuselage shaped like a 50-caliber bullet, mated to a pair of very thin (thickness-to-chord ratio of 0.06), low–aspect ratio, straight wings, as shown in Figure 5.54. The aircraft was powered by a four-chamber liquid-propellant rocket engine mounted in the tail. This engine, made by Reaction Motors and designated the XLR11, produced 6000 lb of thrust by burning a mixture of liquid oxygen and diluted alcohol.

The Bell XS-1 was designed to be carried aloft by a parent airplane, such as the giant Boeing B-29, and then launched at altitude; in this fashion, the extra weight of fuel that would have been necessary for takeoff and climb to altitude was saved, allowing the designers to concentrate on one performance aspect—speed. Three XS-1s were ultimately built, the first one being completed just after Christmas 1945. There followed a year and a half of gliding and then powered tests, wherein the XS-1 was cautiously nudged toward the speed of sound.

Figure 5.54 The Bell XS-1, the first supersonic airplane, 1947. *(National Air and Space Museum.)*

Muroc Dry Lake is a large expanse of flat, hard lake bed in the Mojave Desert in California. Site of a U.S. Army high-speed flight test center during World War II, Muroc was later to become known as Edwards Air Force Base, now the site of the Air Force Test Pilots School and the home of all experimental high-speed flight testing for both the Air Force and NASA. On Tuesday, October 14, 1947, the Bell XS-1, nestled under the fuselage of a B-29, was waiting on the flight line at Muroc. After intensive preparations by a swarm of technicians, the B-29, with its cargo took off at 10 A.M. On board the XS-1 was Captain Charles E. (Chuck) Yeager. That morning, Yeager was in excruciating pain from two broken ribs fractured during a horseback-riding accident earlier that weekend; however, he told virtually no one. At 10:26 A.M. at an altitude of 20,000 ft, the Bell XS-1, with Yeager as its pilot, was dropped from the B-29. What happened next is one of the major milestones in aviation history. Let us see how Yeager himself recalled events, as stated in his written flight report:

Date: 14 October 1947
Pilot: Capt. Charles E. Yeager
Time: 14 Minutes
         9th Powered Flight

1. After normal pilot entry and the subsequent climb, the XS-1 was dropped from the B-29 at 20,000′ and at 250 MPH IAS. This was slower than desired.

2. Immediately after drop, all four cylinders were turned on in rapid sequence, their operation stabilizing at the chamber and line pressures reported in the last flight. The ensuing climb was made at .85–.88 Mach, and, as usual it was necessary to change the stabilizer setting to 2 degrees nose down from its pre-drop setting of 1 degree nose down. Two cylinders were turned off between 35,000′ and 40,000′, but speed had increased to .92 Mach as the airplane was leveled off at 42,000′. Incidentally, during the slight push-over at this altitude, the lox line pressure dropped perhaps 40 psi and the resultant rich mixture caused the chamber pressures to decrease slightly. The effect was only momentary, occurring at .6 G's, and all pressures returned to normal at 1 G.

3. In anticipation of the decrease in elevator effectiveness at all speeds above .93 Mach, longitudinal control by means of the stabilizer was tried during the climb at .83, .88, and .92 Mach. The stabilizer was moved in increments of $\frac{1}{4}-\frac{1}{3}$ degree and proved to be very effective; also, no change in effectiveness was noticed at the different speeds.

4. At 42,000' in approximately level flight, a third cylinder was turned on. Acceleration was rapid and speed increased to .98 Mach. The needle of the machmeter fluctuated at this reading momentarily, then passed off the scale. Assuming that the off-scale reading remained linear, it is estimated that 1.05 Mach was attained at this time. Approximately 30% of fuel and lox remained when this speed was reached and the motor was turned off.

5. While the usual lift buffet and instability characteristics were encountered in the .88–.90 Mach range and elevator effectiveness was very greatly decreased at .94 Mach, stability about all three axes was good as speed increased and elevator effectiveness was regained above .97 Mach. As speed decreased after turning off the motor, the various phenomena occurred in reverse sequence at the usual speeds, and in addition, a slight longitudinal porpoising was noticed from .98–.96 Mach which was controllable by elevators alone. Incidentally, the stabilizer setting was not changed from its 2 degrees nose down position after trial at .92 Mach.

6. After jettisoning the remaining fuel and lox at 1 G stall was performed at 45,000'. The flight was concluded by the subsequent glide and a normal landing on the lakebed.

CHARLES E. YEAGER
Capt. Air Corps

In reality, the Bell SX-1 had reached $M_\infty = 1.06$, as determined from official NACA tracking data. The duration of its supersonic flight was 20.5 s, almost twice as long as the Wright brothers' entire first flight just 44 years earlier. On that day, Chuck Yeager became the first person to fly faster than the speed of sound. It is a fitting testimonial to the aeronautical engineers at that time that the flight was smooth and without unexpected consequences. An aircraft had finally been properly designed to probe the "sound barrier," which it penetrated with relative ease. Less than a month later, Yeager reached Mach 1.35 in the same airplane. The sound barrier had not only been penetrated, it had been virtually destroyed as the myth it really was.

As a final note, the whole story of the human and engineering challenges that revolved about the quest for and eventual achievement of supersonic flight is fascinating, and it is a living testimonial to the glory of aeronautical engineering. The story is brilliantly spelled out by Dr. Richard Hallion, earlier a curator at the Air and Space Museum of the Smithsonian Institution and now chief historian at Edwards Air Force Base, in his book *Supersonic Flight* (see Bibliography at the end of this chapter). The reader should make every effort to study Hallion's story of the events leading to and following Yeager's flight in 1947.

5.23 HISTORICAL NOTE:
THE X-15—FIRST MANNED HYPERSONIC AIRPLANE
AND STEPPING-STONE TO THE SPACE SHUTTLE

Faster and higher—for all practical purposes, this has been the driving potential behind the development of aviation since the Wrights' first successful flight in 1903. (See Sec. 1.11 and Figures 1.27 and 1.28.) This credo was never more true than during the 15 years following Chuck Yeager's first supersonic flight in the Bell XS-1, described in Sec. 5.22. Once the "sound barrier" was broken, it was left far behind in the dust. The next goal became manned *hypersonic* flight—Mach 5 and beyond.

To accomplish this goal, the NACA initiated a series of preliminary studies in the early 1950s for an aircraft to fly beyond Mach 5, the definition of the hypersonic flight regime. This definition is essentially a rule of thumb; unlike the severe and radical flow field changes which take place when an aircraft flies through Mach 1, nothing dramatic happens when Mach 5 is exceeded. Rather, the hypersonic regime is simply a very high Mach number regime, where shock waves are particularly strong and the gas temperatures behind these shock waves are high. For example, consider Eq. (4.73), which gives the total temperature T_0, that is, the temperature of a gas which was initially at a Mach number M_1 and which has been adiabatically slowed to zero velocity. This is essentially the temperature at the stagnation point on a body. If $M_1 = 7$, Eq. (4.73) shows that (for $\gamma = 1.4$), $T_0/T_1 = 10.8$. If the flight altitude is, say, 100,000 ft where $T_1 = 419°R$, then $T_0 = 4525°R = 4065°F$—far above the melting point of stainless steel. Therefore, as flight velocities increase far above the speed of sound, they gradually approach a "thermal barrier," i.e., those velocities beyond which the skin temperatures become too high and structural failure can occur. As in the case of the sound barrier, the thermal barrier is only a figure of speech—it is not an inherent limitation on flight speed. With proper design to overcome the high rates of aerodynamic heating, vehicles have today flown at Mach numbers as high as 36 (the Apollo lunar return capsule, for example). (For more details on high-speed reentry aerodynamic heating, see Sec. 8.11.)

Nevertheless, in the early 1950s, manned hypersonic flight was a goal to be achieved—an untried and questionable regime characterized by high temperatures and strong shock waves. The basic NACA studies fed into an industrywide design competition for a hypersonic airplane. In 1955, North American Aircraft Corporation was awarded a joint NACA–Air Force–Navy contract to design and construct three prototypes of a manned hypersonic research airplane capable of Mach 7 and a maximum altitude of 264,000 ft. This airplane was designated the X-15 and is shown in Figure 5.55. The first two aircraft were powered by Reaction Motors LR11 rocket engines, with 8000 lb of thrust (essentially the same as the engine used for the Bell XS-1). Along with the third prototype, the two aircraft were later reengined with a more powerful rocket motor, the Reaction Motors XLR99, capable of 57,000 lb of thrust. The basic internal structure of the airplane was made from titanium and stainless steel, but the airplane skin was Inconel X—a nickel alloy steel capable of withstanding temperatures up to 1200°F. (Although the theoretical stagnation temperature at Mach 7 is 4065°F, as discussed above, the actual skin temperature is cooler because of heat sink and heat dissipation effects.) The wings had a low aspect ratio of 2.5 and a thickness-to-chord ratio of 0.05—both intended to reduce supersonic wave drag.

The first X-15 was rolled out of the North American factory at Los Angeles on October 15, 1958. Then Vice President Richard M. Nixon was the guest of honor at the rollout ceremonies. The X-15 had become a political, as well as a technical, accomplishment, because the United States was attempting to heal its wounded pride after the Russians had launched the first successful unmanned satellite, *Sputnik I*, just a year earlier (see Sec. 8.15). The next day, the X-15 was transported by truck to the nearby Edwards Air Force Base (the site at Muroc which saw the first supersonic flights of the Bell XS-1).

As in the case of the XS-1, the X-15 was designed to be carried aloft by a parent airplane, this time a Boeing B-52 jet bomber. The first free flight, without power, was made by Scott Crossfield on June 8, 1959. This was soon followed by the first powered flight on September 17, 1959, when the X-15 reached Mach 2.1 in a shallow climb to 52,341 ft. Powered with the smaller LR11 rocket engines, the X-15 set a speed record of Mach 3.31 on August 4, 1960, and an altitude record of 136,500 ft just eight days later. However, these records were just transitory. After November 1960, the X-15 received the more powerful XLR99 engine. Indeed, the first flight with this rocket was made on November 15, 1960; on this flight, with power adjusted to its *lowest* level and with the air brakes fully extended, the X-15 still hit 2000 mi/h. Finally, on June 27, 1962, hypersonic flight was fully achieved when Joseph Walker, chief test pilot for NASA, flew the X-15 at 4159 mi/h. Less than a month later, Major Robert White took the X-15

Figure 5.55 The North American X-15, the first manned hypersonic airplane. *(North American / Rockwell Corporation.)*

to an altitude record of 314,750 ft. This began an illustrious series of hypersonic flight tests, peaked by a flight at Mach 6.72 on October 3, 1967, with Air Force Major Pete Knight at the controls.

Experimental aircraft are just that—vehicles designed for specific experimental purposes, which, after they are achieved, lead to the end of the program. This happened to the X-15 when on October 24, 1968, the last flight was carried out—the 199th of the entire program. A 200th flight was planned, partly for reasons of nostalgia; however, technical problems delayed this planned flight until December 20, when the X-15 was ready to go, attached to its B-52 parent plane as usual. However, of all things, a highly unusual snow squall suddenly hit Edwards, and the flight was cancelled. The X-15 never flew again. In 1969, the first X-15 was given to the National Air and Space Museum of the Smithsonian, where it now hangs with distinction in the Milestones of Flight Gallery, along with the Bell XS-1.

The X-15 opened the world of manned hypersonic flight. The next hypersonic airplane was the space shuttle. The vast bulk of aerodynamic and flight dynamic data generated during the X-15 program carried over to the space shuttle design. The pilots' experience with low-speed flights in a high-speed aircraft with low lift-to-drag ratio set the stage for flight preparations with the space shuttle. In these respects, the X-15 was clearly the major stepping-stone to the space shuttle of the 1980s.

5.24 CHAPTER SUMMARY

Some of the aspects of this chapter are highlighted below.

1. For an airfoil, the lift, drag, and moment coefficients are defined as

$$c_l = \frac{L}{q_\infty S} \qquad c_d = \frac{D}{q_\infty S} \qquad c_m = \frac{M}{q_\infty S c}$$

where L, D, and M are the lift, drag, and moments per unit span and $S = c(1)$.

For a finite wing, the lift, drag, and moment coefficients are defined as

$$C_L = \frac{L}{q_\infty S} \qquad C_D = \frac{D}{q_\infty S} \qquad C_M = \frac{M}{q_\infty S c}$$

where L, D, and M are the lift, drag, and moments for the complete wing, and S is the wing planform area.

For a given shape, these coefficients are a function of angle of attack, Mach number, and Reynolds number.

2. The pressure coefficient is defined as

$$C_p = \frac{p - p_\infty}{\frac{1}{2}\rho_\infty V_\infty^2} \tag{5.23}$$

3. The Prandtl-Glauert rule is a compressibility correction for subsonic flow:

$$C_p = \frac{C_{p,0}}{\sqrt{1 - M_\infty^2}} \qquad (5.24)$$

where $C_{p,0}$ and C_p are the incompressible and compressible pressure coefficients, respectively. The same rule holds for the lift and moment coefficients, e.g.,

$$c_l = \frac{c_{l,0}}{\sqrt{1 - M_\infty^2}} \qquad (5.32)$$

4. The critical Mach number is that freestream Mach number at which sonic flow is first achieved at some point on a body. The drag-divergence Mach number is that freestream Mach number at which the drag coefficient begins to rapidly increase due to the occurrence of transonic shock waves. For a given body, the drag-divergence Mach number is slightly higher than the critical Mach number.
5. The Mach angle is defined as

$$\mu = \arcsin \frac{1}{M} \qquad (5.38)$$

6. The total drag coefficient for a finite wing is equal to

$$C_D = c_d + \frac{C_L^2}{\pi e \text{AR}} \qquad (5.47)$$

where c_d is the profile drag coefficient and $C_L^2/(\pi e \text{AR})$ is the induced-drag coefficient.
7. The lift slope for a finite wing, a, is given by

$$a = \frac{a_0}{1 + 57.3 a_0 / \pi e_1 \text{AR}} \qquad (5.53)$$

where a_0 is the lift slope for the corresponding infinite wing.

BIBLIOGRAPHY

Abbott, I. H., and von Doenhoff, A. E., *Theory of Wing Sections*, McGraw-Hill, New York, 1949 (also Dover, New York, 1959).

Anderson, John D., Jr., *Fundamentals of Aerodynamics*, McGraw-Hill, New York, 1984.

Dommasch, D. O., Sherbey, S. S., and Connolly, T. F., *Airplane Aerodynamics*, 4th ed., Pitman, New York, 1968.

Hallion, R., *Supersonic Flight (The Story of the Bell X-1 and Douglas D-558)*, Macmillan, New York, 1972.

McCormick, B. W., *Aerodynamics, Aeronautics and Flight Mechanics*, Wiley, New York, 1979.

Pierpont, P. K., "Bringing Wings of Change," *Astronautics and Aeronautics*, vol. 13, no. 10, October 1975, pp. 20–27.

Shapiro, A. H., *Shape and Flow: The Fluid Dynamics of Drag*, Anchor, Garden City, New York, 1961.
Shevell, R. S., *Fundamentals of Flight*, Prentice-Hall, Englewood Cliffs, NJ, 1983.
von Karman, T. (with Lee Edson), *The Wind and Beyond*, Little, Brown, Boston, 1967 (an autobiography).

PROBLEMS

5.1 By the method of dimensional analysis, derive the expression $M = q_\infty S c c_m$ for the aerodynamic moment on an airfoil where c is the chord and c_m is the moment coefficient.

5.2 Consider an infinite wing with a NACA 1412 airfoil section and a chord length of 3 ft. The wing is at an angle of attack of 5° in an airflow velocity of 100 ft/s at standard sea-level conditions. Calculate the lift, drag, and moment about the quarter chord per unit span.

5.3 Consider a rectangular wing mounted in a low-speed subsonic wind tunnel. The wing model completely spans the test section so that the flow "sees" essentially an infinite wing. If the wing has a NACA 23012 airfoil section and a chord of 0.3 m, calculate the lift, drag, and moment about the quarter chord per unit span when the airflow pressure, temperature, and velocity are 1 atm, 303 K, and 42 m/s, respectively. The angle of attack is 8°.

5.4 The wing model in Problem 5.3 is pitched to a new angle of attack, where the lift on the entire wing is measured as 200 N by the wind-tunnel force balance. If the wingspan is 2 m, what is the angle of attack?

5.5 Consider a rectangular wing with a NACA 0009 airfoil section spanning the test section of a wind tunnel. The test section airflow conditions are standard sea level with a velocity of 120 mi/h. The wing is at an angle of attack of 4°, and the wind-tunnel force balance measures a lift of 29.5 lb. What is the area of the wing?

5.6 The ratio of lift to drag, L/D, for a wing or airfoil is an important aerodynamic parameter. Indeed, it is a direct measure of the aerodynamic efficiency of the wing. If a wing is pitched through a range of angle of attack, the L/D first increases, then goes through a maximum, and then decreases. Consider an infinite wing with a NACA 2412 airfoil. Estimate the maximum value of L/D. Assume the Reynolds number is 9×10^6.

5.7 Consider an airfoil in a freestream with a velocity of 50 m/s at standard sea-level conditions. At a point on the airfoil, the pressure is 9.5×10^4 N/m^2. What is the pressure coefficient at this point?

5.8 Consider a low-speed airplane flying at a velocity of 55 m/s. If the velocity at a point on the fuselage is 62 m/s, what is the pressure coefficient at this point?

5.9 Consider a wing mounted in the test section of a subsonic wind tunnel. The velocity of the airflow is 160 ft/s. If the velocity at a point on the wing is 195 ft/s, what is the pressure coefficient at this point?

5.10 Consider the same wing in the same wind tunnel as in Problem 5.9. If the test section air temperature is 510°R, and if the flow velocity is increased to 700 ft/s, what is the pressure coefficient at the same point?

5.11 Consider a wing in a high-speed wind tunnel. At a point on the wing, the velocity is 850 ft/s. If the test section flow is at a velocity of 780 ft/s, with a pressure and temperature of 1 atm and 505°R, respectively, calculate the pressure coefficient at the point.

5.12 If the test section flow velocity in Problem 5.11 is reduced to 100 ft/s, what will the pressure coefficient become at the same point on the wing?

5.13 Consider a NACA 1412 airfoil at an angle of attack of 4°. If the freestream Mach number is 0.8, calculate the lift coefficient.

5.14 A NACA 4415 airfoil is mounted in a high-speed subsonic wind tunnel. The lift coefficient is measured as 0.85. If the test section Mach number is 0.7, at what angle of attack is the airfoil?

5.15 Consider an airfoil at a given angle of attack, say α_1. At low speeds, the minimum pressure coefficient on the top surface of the airfoil is -0.90. What is the critical Mach number of the airfoil?

5.16 Consider the airfoil in Problem 5.15 at a smaller angle of attack, say α_2. At low speeds, the minimum pressure coefficient is -0.65 at this lower angle of attack. What is the critical Mach number of the airfoil?

5.17 Consider a uniform flow with a Mach number of 2. What angle does a Mach wave make with respect to the flow direction?

5.18 Consider a supersonic missile flying at Mach 2.5 at an altitude of 10 km. Assume the angle of the shock wave from the nose is approximated by the Mach angle (i.e., a very weak shock). How far behind the nose of the vehicle will the shock wave impinge upon the ground? (Ignore the fact that the speed of sound, hence the Mach angle, changes with altitude.)

5.19 The wing area of the Lockheed F-104 straight-wing supersonic fighter is approximately 210 ft². If the airplane weighs 16,000 lb and is flying in level flight at Mach 2.2 at a standard altitude of 36,000 ft, estimate the wave drag on the wings.

5.20 The Cessna Cardinal, a single-engine light plane, has a wing with an area of 16.2 m² and an aspect ratio of 7.31. Assume the span efficiency factor is 0.62. If the airplane is flying at standard sea-level conditions with a velocity of 251 km/h, what is the induced drag when the total weight is 9800 N?

5.21 For the Cessna Cardinal in Problem 5.20, calculate the induced drag when the velocity is 85.5 km/h (stalling speed at sea level with flaps down).

5.22 Consider a finite wing with an area and aspect ratio of 21.5 m² and 5, respectively (this is comparable to the wing on a Gates Learjet, a twin-jet executive transport). Assume the wing has a NACA 65-210 airfoil, a span efficiency factor of 0.9, and a profile drag coefficient of 0.004. If the wing is at a 6° angle of attack, calculate C_L and C_D.

5.23 During the 1920s and early 1930s, the NACA obtained wind-tunnel data on different airfoils by testing finite wings with an aspect ratio of 6. These data were then "corrected" to obtain infinite wing airfoil characteristics. Consider such a finite wing with an area and aspect ratio of 1.5 ft² and 6, respectively, mounted in a wind tunnel where the test section flow velocity is 100 ft/s at standard sea-level conditions. When the wing is pitched to $\alpha = -2°$, no lift is measured. When the wing is pitched to $\alpha = 10°$, a lift of 17.9 lb is measured. Calculate the lift slope for the airfoil (the infinite wing) if the span effectiveness factor is 0.95.

5.24 A finite wing of area 1.5 ft² and aspect ratio of 6 is tested in a subsonic wind tunnel at a velocity of 130 ft/s at standard sea-level conditions. At an angle of attack of $-1°$, the measured lift and drag are 0 and 0.181 lb, respectively. At an angle of attack of 2°, the lift and drag are measured as 5.0 lb and 0.23 lb, respectively. Calculate the span efficiency factor and the infinite wing lift slope.

5.25 Consider a light, single-engine airplane such as the Piper Super Cub. If the maximum gross weight of the airplane is 7780 N, the wing area is 16.6 m², and the maximum lift coefficient is 2.1 with flaps down, calculate the stalling speed at sea level.

SIX

ELEMENTS OF
AIRPLANE PERFORMANCE

First Europe, and then the globe, will be linked by flight, and nations so knit together that they will grow to be next-door neighbors. This conquest of the air will prove, ultimately, to be man's greatest and most glorious triumph. What railways have done for nations, airways will do for the world.

Claude Grahame-White
British Aviator, 1914

6.1 INTRODUCTION

Henson's aerial steam carriage of the midnineteenth century (see Figure 1.11) was pictured by contemporary artists as flying to all corners of the world. Of course, questions about *how* it would fly to such distant locations were not considered by the designers. As with most early aeronautical engineers of that time, their main concern was simply to lift or otherwise propel the airplane from the ground; what happened once the vehicle was airborne was viewed with secondary importance. However, with the success of the Wright brothers in 1903, and with the subsequent rapid development of aviation during the pre–World War I era, the airborne performance of the airplane suddenly became of primary importance. Some obvious questions were, and still are, asked about a given design. What is the maximum speed of the airplane? How fast can it climb to a given altitude? How far can it fly on a given tank of fuel? How long can it stay in the air? Answers to these and similar questions constitute the study of *airplane performance*, which is the subject of this chapter.

In previous chapters, the physical phenomena producing lift, drag, and moments of an airplane were introduced. Emphasis was made that the aerodynamic forces and moments exerted on a body moving through a fluid stem from two sources:

1. The pressure distribution
2. The shear stress distribution

both acting over the body surface. The physical laws governing such phenomena were examined, with various applications to aerodynamic flows.

In the present chapter, we begin a new phase of study. The airplane will be considered a rigid body on which is exerted four natural forces: lift, drag, propulsive thrust, and weight. Concern will be focused on the movement of the airplane as it responds to these forces. Such considerations form the core of *flight dynamics*, an important discipline of aerospace engineering. Studies of airplane performance (this chapter) and stability and control (Chap. 7) both fall under the heading of flight dynamics.

In these studies, we will no longer be concerned with aerodynamic details; rather, we will generally assume that the aerodynamicists have done their work and provided us with the pertinent aerodynamic data for a given airplane. These data are usually packaged in the form of a *drag polar* for the complete airplane, given as

$$C_D = C_{D,e} + \frac{C_L{}^2}{\pi e \text{AR}} \tag{6.1a}$$

Equation (6.1a) is an extension of Eq. (5.47) to include the whole airplane. Here, C_D is the drag coefficient for the complete airplane; C_L is the total lift coefficient, including the small contributions from the horizontal tail and fuselage; and $C_{D,e}$ is defined as the parasite drag coefficient, which contains not only the profile drag of the wing [c_d in Eq. (5.47)] but also the friction and pressure drag of the tail surfaces, fuselage, engine nacelles, landing gear, and any other component of the airplane which is exposed to the airflow. At transonic and supersonic speeds, $C_{D,e}$ also contains wave drag. Because of changes in the flow field around the airplane —especially changes in the amount of separated flow over parts of the airplane—as the angle of attack is varied, $C_{D,e}$ will change with angle of attack; that is, $C_{D,e}$ is itself a function of lift coefficient. A reasonable approximation for this function is

$$C_{D,e} = C_{D,0} + rC_L{}^2$$

where r is an empirically determined constant. Hence, Eq. (6.1a) can be written as

$$C_D = C_{D,0} + \left(r + \frac{1}{\pi e \text{AR}}\right)C_L{}^2 \tag{6.1b}$$

In Eqs. (6.1a) and (6.1b), e is the familiar span efficiency factor, which takes into account the nonelliptical lift distribution on wings of general shape (see Sect. 5.14). Let us now *redefine e* such that it also includes the effect of the variation of parasite drag with lift; i.e., let us write Eq. (6.1b) in the form

$$C_D = C_{D,0} + \frac{C_L{}^2}{\pi e \text{AR}} \tag{6.1c}$$

where $C_{D,0}$ is the parasite drag coefficient at *zero lift* and the term $C_L^2/\pi e \text{AR}$ includes both induced drag and the contribution to parasite drag due to lift. In Eq. (6.1c), our redefined e, which now includes the effect of r from Eq. (6.1b), is called the *Oswald efficiency factor* (named after W. Bailey Oswald, who first established this terminology in NACA report no. 408 in 1933). In this chapter, the basic aerodynamic properties of the airplane will be described by Eq. (6.1c), and we will consider both $C_{D,0}$ and e as known aerodynamic quantities, obtained from the aerodynamicist. We will continue to designate $C_L^2/\pi e \text{AR}$ by $C_{D,i}$, where $C_{D,i}$ now has the expanded interpretation as the coefficient of *drag due to lift*, including both the contributions due to induced drag and the increment in parasite drag due to angle of attack different than $\alpha_L = 0$. For compactness, we will designate $C_{D,0}$ as simply the *parasite drag coefficient*, although we recognize it to be more precisely the parasite drag coefficient at zero lift, i.e., the value of $C_{D,e}$ when $\alpha = \alpha_{L=0}$.

To study the performance of an airplane, the fundamental equations which govern its translational motion through the air must first be established, as follows.

6.2 EQUATIONS OF MOTION

Consider an airplane in flight, as sketched in Figure 6.1. The flight path (direction of motion of the airplane) is inclined at an angle θ with respect to the horizontal. In terms of the definitions in Chap. 5, the flight path direction and the relative wind are along the same line. The mean chord line is at a geometric angle of attack α with respect to the flight path direction. There are four physical forces

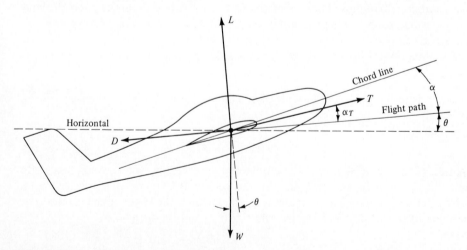

Figure 6.1 Force diagram for an airplane in flight.

acting on the airplane:

1. Lift L, which is perpendicular to the flight path direction
2. Drag D, which is parallel to the flight path direction
3. Weight W, which acts vertically toward the center of the earth (and hence is inclined at the angle θ with respect to the lift direction)
4. Thrust T, which in general is inclined at the angle α_T with respect to the flight path direction

The force diagram shown in Figure 6.1 is important. Study it carefully until you feel comfortable with it.

The flight path shown in Figure 6.1 is drawn as a straight line. This is the picture we see by focusing locally on the airplane itself. However, if we stand back and take a wider view of the space in which the airplane is traveling, the flight path is generally curved. This is obviously true if the airplane is maneuvering; even if the airplane is flying "straight and level" with respect to the ground, it is still executing a curved flight path with a radius of curvature equal to the absolute altitude h_a (as defined in Sec. 3.1).

When an object moves along a curved path, the motion is called curvilinear, as opposed to motion along a straight line, which is rectilinear. Newton's second law, which is a physical statement that force = mass × acceleration, holds in either case. Consider a curvilinear path. At a given point on the path, set up two mutually perpendicular axes, one along the direction of the flight path and the other normal to the flight path. Applying Newton's law along the flight path,

$$\sum F_{\parallel} = ma = m\frac{dV}{dt} \tag{6.2}$$

where $\sum F_{\parallel}$ is the summation of all forces parallel to the flight path, $a = dV/dt$ is the acceleration along the flight path, and V is the instantaneous value of the airplane's flight velocity. (Velocity V is always along the flight path direction, by definition.) Now, applying Newton's law perpendicular to the flight path, we have

$$\sum F_{\perp} = m\frac{V^2}{r_c} \tag{6.3}$$

where $\sum F_{\perp}$ is the summation of all forces perpendicular to the flight path and V^2/r_c is the acceleration normal to a curved path with radius of curvature r_c. This normal acceleration V^2/r_c should be familiar from basic physics. The right-hand side of Eq. (6.3) is nothing other than the "centrifugal force."

Examining Figure 6.1, we see that the forces parallel to the flight path (positive to the right, negative to the left) are

$$\sum F_{\parallel} = T\cos\alpha_T - D - W\sin\theta \tag{6.4}$$

and the forces perpendicular to the flight path (positive upward and negative downward) are

$$\sum F_{\perp} = L + T\sin\alpha_T - W\cos\theta \tag{6.5}$$

Combining Eq. (6.2) with (6.4), and (6.3) with (6.5) yields

$$T\cos\alpha_T - D - W\sin\theta = m\frac{dV}{dt} \tag{6.6}$$

$$L + T\sin\alpha_T - W\cos\theta = m\frac{V^2}{r_c} \tag{6.7}$$

Equations (6.6) and (6.7) are the *equations of motion* for an airplane in translational flight. (Note that an airplane can also rotate about its axes; this will be discussed in Chap. 7. Also note that we are not considering the possible sidewise motion of the airplane perpendicular to the page of Figure 6.1.)

Equations (6.6) and (6.7) describe the general two-dimensional translational motion of an airplane in accelerated flight. However, in the first part of this chapter we are interested in a specialized application of these equations, namely, the case where the acceleration is zero. The performance of an airplane for such unaccelerated flight conditions is called *static performance*. This may, at first thought, seem unduly restrictive; however, static performance analyses lead to reasonable calculations of maximum velocity, maximum rate of climb, maximum range, etc.—parameters of vital interest in airplane design and operation.

With this in mind, consider level, unaccelerated flight. Referring to Figure 6.1, level flight means that the flight path is along the horizontal; that is, $\theta = 0$. Unaccelerated flight means that the right-hand sides of Eqs. (6.6) and (6.7) are zero. Therefore, these equations reduce to

$$T\cos\alpha_T = D \tag{6.8}$$

$$L + T\sin\alpha_T = W \tag{6.9}$$

Furthermore, for most conventional airplanes, α_T is small enough that $\cos\alpha_T \approx 1$ and $\sin\alpha_T \approx 0$. Thus, from Eqs. (6.8) and (6.9),

$$T = D \tag{6.10}$$

$$L = W \tag{6.11}$$

Equations (6.10) and (6.11) are the equations of motion for level, unaccelerated flight. They can also be obtained directly from Figure 6.1, by inspection. In level, unaccelerated flight, the aerodynamic drag is balanced by the thrust of the engine, and the aerodynamic lift is balanced by the weight of the airplane—almost trivial, but very useful, results.

Let us now apply these results to the static performance analysis of an airplane. The following sections constitute the building blocks for such an analysis, which ultimately yields answers to such questions as how fast, how far, how long, and how high a given airplane can fly. Also, the discussion in these sections relies heavily on a graphical approach to the calculation of airplane performance. In modern aerospace engineering, such calculations are made directly on high-speed digital computers. However, the graphical illustrations in the following sections are essential to the programming and understanding of such computer solutions; moreover, they help to clarify and explain the concepts being presented.

6.3 THRUST REQUIRED FOR LEVEL, UNACCELERATED FLIGHT

Consider an airplane in steady level flight at a given altitude and a given velocity. For flight at this velocity, the airplane's power plant (e.g., turbojet engine or reciprocating engine–propeller combination) must produce a net thrust which is equal to the drag. The thrust required to obtain a certain steady velocity is easily calculated as follows. From Eqs. (6.10) and (5.18)

$$T = D = q_\infty S C_D \tag{6.12}$$

and from Eqs. (6.11) and (5.15)

$$L = W = q_\infty S C_L \tag{6.13}$$

Dividing Eq. (6.12) by (6.13) yields

$$\frac{T}{W} = \frac{C_D}{C_L} \tag{6.14}$$

Thus, from Eq. (6.14), the thrust required for an airplane to fly at a given velocity in level, unaccelerated flight is

$$\boxed{T_R = \frac{W}{C_L/C_D} = \frac{W}{L/D}} \tag{6.15}$$

(Note that a subscript R has been added to thrust to emphasize that it is thrust required.)

Thrust required T_R for a given airplane at a given altitude, varies with velocity V_∞. The *thrust-required curve* is a plot of this variation and has the general shape illustrated in Figure 6.2. To calculate a point on this curve, proceed as follows:

1. Choose a value of V_∞.
2. For this V_∞, calculate the lift coefficient from Eq. (6.13):

$$C_L = \frac{W}{\frac{1}{2}\rho_\infty V_\infty^2 S} \tag{6.16}$$

Note that ρ_∞ is known from the given altitude and S is known from the given airplane. The C_L calculated from Eq. (6.16) is that value necessary for the lift to balance the known weight W of the airplane.

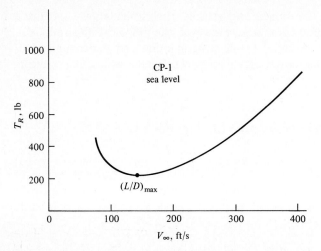

Figure 6.2 Thrust-required curve. The results on this and subsequent figures correspond to answers for some of the sample problems in Chap. 6.

3. Calculate C_D from the known drag polar for the airplane:

$$C_D = C_{D,0} + \frac{C_L^{\,2}}{\pi e \text{AR}}$$

where C_L is the value obtained from Eq. (6.16).
4. Form the ratio C_L/C_D.
5. Calculate thrust required from Eq. (6.15).

The value of T_R obtained from step 5 is that thrust required to fly at the specific velocity chosen in step 1. In turn, the curve in Figure 6.2 is the locus of all such points taken for all velocities in the flight range of the airplane. The reader should study Example 6.1 at the end of this section in order to become familiar with the above steps.

Note from Eq. (6.15) that T_R varies inversely as L/D. Hence, minimum thrust required will be obtained when the airplane is flying at a velocity where L/D is maximum. This condition is shown in Figure 6.2.

The lift-to-drag ratio L/D is a measure of the aerodynamic efficiency of an airplane; it only makes sense that maximum aerodynamic efficiency should lead to minimum thrust required. Consequently, lift-to-drag ratio is an important aerodynamic consideration in airplane design. Also note that L/D is a function of angle of attack, as sketched in Figure 6.3. For most conventional subsonic airplanes, L/D reaches a maximum at some specific value of α, usually on the order of 2 to 5°. Hence, when an airplane is flying at the velocity for minimum T_R as shown in Figure 6.2, it is simultaneously flying at the angle of attack for maximum L/D as shown in Figure 6.3.

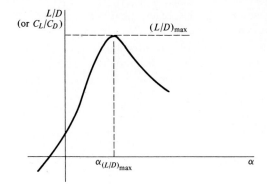

Figure 6.3 Lift-to-drag ratio vs. angle of attack.

As a corollary to this discussion, note that different points on the thrust-required curve correspond to different angles of attack. This is emphasized in Figure 6.4, which shows that, as we move from right to left on the thrust-required curve, the airplane angle of attack increases. This also helps to explain physically why T_R goes through a minimum. Recall that $L = W = q_\infty S C_L$. At high velocities (point a in Figure 6.4), most of the required lift is obtained from high dynamic pressure q_∞; hence C_L and therefore α are small. Also, under the same conditions, drag ($D = q_\infty S C_D$) is relatively large because q_∞ is large. As we move to the left on the thrust-required curve, q_∞ decreases; hence C_L and therefore α must increase to support the given airplane weight. Because q_∞ decreases, D and hence T_R initially decrease. However, recall that drag due to lift is a component of total drag and that $C_{D,i}$ varies as C_L^2. At low velocities, such as at point b in Figure 6.4, q_∞ is low, and hence C_L is large. At these conditions, $C_{D,i}$ increases rapidly, more rapidly than q_∞ decreases, and D, hence T_R, increases. This is why,

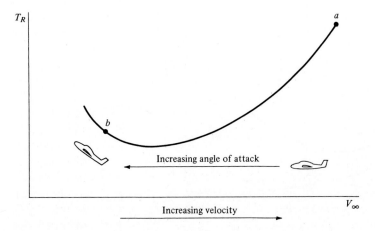

Figure 6.4 Thrust-required curve with associated angle-of-attack variation.

starting at point a, T_R first decreases as V_∞ decreases, then goes through a minimum and starts to increase, as shown at point b.

Recall from Eq. (6.1c) that the total drag of the airplane is the sum of parasite drag and drag due to lift. The corresponding drag coefficients are $C_{D,0}$ and $C_{D,i} = C_L^2/\pi e \mathrm{AR}$, respectively. At the condition for minimum T_R, there exists an interesting relation between $C_{D,0}$ and $C_{D,i}$, as follows. From Eq. (6.10)

$$T_R = D = q_\infty S C_D = q_\infty S (C_{D,0} + C_{D,i}) = \frac{C_D}{C_L} W = C_D \frac{1}{2}\rho v^2 \cdots$$

$$= q_\infty S \left(C_{D,0} + \frac{C_L^2}{\pi e \mathrm{AR}} \right)$$

$$T_R = q_\infty S C_{D,0} + q_\infty S \frac{C_L^2}{\pi e \mathrm{AR}} \qquad (6.17)$$

$$\underbrace{\qquad}_{\text{Parasite } T_R} \qquad \underbrace{\qquad}_{\text{Induced } T_R}$$

Note that, as identified in Eq. (6.17), the thrust required can be considered the sum of *parasite thrust required* (thrust required to balance parasite drag) and *induced thrust required* (thrust required to balance drag due to lift). Examining Figure 6.5, we find induced T_R decreases but parasite T_R increases as the velocity is increased. (Why?)

Recall that $C_L = W/q_\infty S$. From Eq. (6.17),

$$T_R = q_\infty S C_{D,0} + \frac{W^2}{q_\infty S \pi e \mathrm{AR}} \qquad (6.18)$$

Also

$$\frac{dT_R}{dq_\infty} = \frac{dT_R}{dV_\infty} \frac{dV_\infty}{dq_\infty} \qquad (6.19)$$

From calculus, we find that the point of minimum T_R in Figure 6.2 corresponds to $dT_R/dV_\infty = 0$. Hence, from Eq. (6.19), minimum T_R also corresponds to

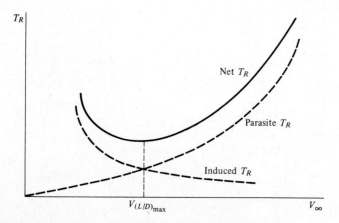

Figure 6.5 Comparison of induced and parasite thrust required.

$dT_R/dq_\infty = 0$. Differentiating Eq. (6.18) with respect to q_∞ and setting the derivative equal to zero, we have

$$\frac{dT_R}{dq_\infty} = SC_{D,0} - \frac{W^2}{q_\infty^2 S \pi e \text{AR}} = 0$$

Thus

$$C_{D,0} = \frac{W^2}{q_\infty^2 S^2 \pi e \text{AR}} \tag{6.20}$$

However

$$\frac{W^2}{q_\infty^2 S^2} = \left(\frac{W}{q_\infty S}\right)^2 = C_L^2$$

Hence, Eq. (6.20) becomes

$$\boxed{C_{D,0} = \frac{C_L^2}{\pi e \text{AR}} = C_{D,i}}$$

Parasite drag = drag due to lift

$$\tag{6.21}$$

Equation (6.21) yields the interesting aerodynamic result that at minimum thrust required, parasite drag equals drag due to lift. Hence, the curves for parasite and induced T_R intersect at the velocity for minimum T_R (i.e., for maximum L/D), as shown in Figure 6.5. We will return to this result in Sec. 6.13.

Example 6.1 For all the examples given in this chapter, two types of airplanes will be considered:
(a) A light, single-engine, propeller-driven, private airplane, approximately modeled after the Cessna Skylane shown in Figure 6.6. For convenience, we will designate our hypothetical

Figure 6.6 The hypothetical CP-1 studied in Chap. 6 sample problems is modeled after the Cessna Skylane shown here. (*Cessna Aircraft Corporation.*)

airplane as the CP-1, having the following characteristics:

Wingspan = 35.8 ft
Wing area = 174 ft^2
Normal gross weight = 2950 lb
Normal capacity: 65 gallons of aviation gasoline
Power plant: one-piston engine of 230 hp at sea level
Specific fuel consumption = 0.45 lb/(hp)(h)
Parasite drag coefficient = $C_{D,0} = 0.025$
Oswald efficiency factor = $e = 0.8$
Propeller efficiency = 0.8

(b) A jet-powered executive aircraft, approximately modeled after the Cessna Citation 3 shown in Figure 6.7. For convenience, we will designate our hypothetical jet as the CJ-1, having the following characteristics:

Wingspan = 53.3 ft
Wing area = 318 ft^2
Normal gross weight = 19,815 lb
Fuel capacity: 1119 gallons of kerosene
Power plant: two turbofan engines of 3650 lb of thrust each at sea level
Specific fuel consumption = 0.6 lb of fuel/(lb thrust)(h)
Parasite drag coefficient = $C_{D,0} = 0.02$
Oswald efficiency factor = $e = 0.81$

By the end of this chapter, all the examples taken together will represent a basic performance analysis of these two aircraft.

Figure 6.7 The hypothetical CJ-1 studied in Chap. 6 sample problems is modeled after the Cessna Citation 3 shown here. (*Cessna Aircraft Corp.*)

In this example, only the thrust required is considered. Calculate the T_R curves at sea level for both the CP-1 and the CJ-1.

SOLUTION

(a) For the CP-1 assume $V_\infty = 200$ ft/s $= 136.4$ mi/h. From Eq. (6.16),

$$C_L = \frac{W}{\frac{1}{2}\rho_\infty V_\infty^2 S} = \frac{2950}{\frac{1}{2}(0.002377)(200)^2(174)} = 0.357$$

The aspect ratio is

$$AR = \frac{b^2}{S} = \frac{(35.8)^2}{174} = 7.37$$

Thus, from Eq. (6.1c),

$$C_D = C_{D,0} + \frac{C_L^2}{\pi e AR} = 0.025 + \frac{(0.357)^2}{\pi(0.8)(7.37)}$$

$$= 0.0319$$

Hence

$$\frac{L}{D} = \frac{C_L}{C_D} = \frac{0.357}{0.0319} = 11.2$$

Finally, from Eq. (6.15),

$$T_R = \frac{W}{L/D} = \frac{2950}{11.2} = \boxed{263 \text{ lb}}$$

To obtain the thrust-required curve, the above calculation is repeated for many different values of V_∞. Some sample results are tabulated below.

V_∞, ft/s	C_L	C_D	L/D	T_R, lb
100	1.43	0.135	10.6	279
150	0.634	0.047	13.6	217
250	0.228	0.028	8.21	359
300	0.159	0.026	6.01	491
350	0.116	0.026	4.53	652

The above tabulation is given so that the reader can try such calculations and compare the results. Such tabulations will be given throughout this chapter. They are taken from a computer calculation where 100 different velocities are used to generate the data. The T_R curve obtained from these calculations is given in Figure 6.2.

(b) For the CJ-1 assume $V_\infty = 500$ ft/s $= 341$ mi/h. From Eq. (6.16),

$$C_L = \frac{W}{\frac{1}{2}\rho_\infty V_\infty^2 S} = \frac{19,815}{\frac{1}{2}(0.002377)(500)^2(318)}$$

$$= 0.210$$

The aspect ratio is

$$AR = \frac{b^2}{S} = \frac{(53.3)^2}{318} = 8.93$$

Thus, from Eq. (6.1c),

$$C_D = C_{D,0} + \frac{C_L^2}{\pi e AR} = 0.02 + \frac{(0.21)^2}{\pi(0.81)(8.93)}$$

$$= 0.022$$

Hence

$$\frac{L}{D} = \frac{C_L}{C_D} = \frac{0.21}{0.022} = 9.55$$

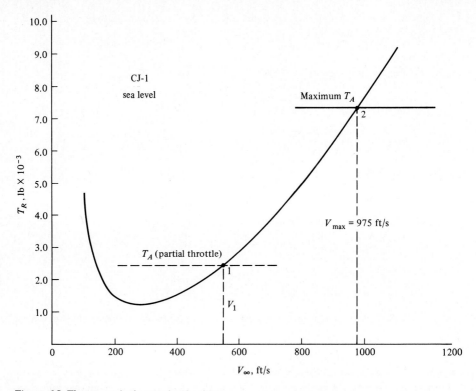

Figure 6.8 Thrust-required curve for the CJ-1.

Finally, from Eq. (6.15)

$$T_R = \frac{W}{L/D} = \frac{19,815}{9.55} = \boxed{2075 \text{ lb}}$$

A tabulation for a few different velocities follows.

V_∞, ft/s	C_L	C_D	L/D	T_R, lb
300	0.583	0.035	16.7	1188
600	0.146	0.021	6.96	2848
700	0.107	0.021	5.23	3797
850	0.073	0.020	3.59	5525
1000	0.052	0.020	2.61	7605

The thrust-required curve is given in Figure 6.8.

6.4 THRUST AVAILABLE AND MAXIMUM VELOCITY

Thrust required T_R, described in the previous section, is dictated by the aerodynamics and weight of the airplane itself; it is an *airframe-associated*

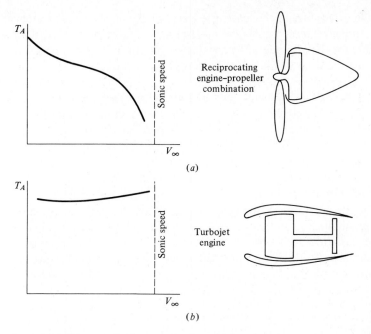

Figure 6.9 Thrust-available curves for piston engine–propeller combination and for a turbojet engine.

phenomenon. In contrast, the *thrust available* T_A is strictly associated with the engine of the airplane; it is the propulsive thrust provided by an engine-propeller combination, a turbojet, a rocket, etc. Propulsion is the subject of Chap. 9. Suffice it to say here that reciprocating piston engines with propellers exhibit a variation of thrust with velocity, as sketched in Figure 6.9a. Thrust at zero velocity (static thrust) is a maximum and decreases with forward velocity. At near-sonic flight speeds, the tips of the propeller blades encounter the same compressibility problems discussed in Chap. 5, and the thrust available rapidly deteriorates. On the other hand, the thrust of a turbojet engine is relatively constant with velocity, as sketched in Figure 6.9b. These two power plants are quite common in aviation today; reciprocating engine–propeller combinations power the average light, general aviation aircraft, whereas the jet engine is used by almost all large commercial transports and military combat aircraft. For these reasons, the performance analyses of the present chapter will consider only the above two propulsive mechanisms.

Consider a jet airplane flying in level, unaccelerated flight at a given altitude and with the velocity V_1, as shown in Figure 6.8. Point 1 on the thrust-required curve gives the value of T_R for the airplane to fly at velocity V_1. The pilot has adjusted the throttle such that the jet engine provides thrust available just equal to the thrust required at this point; $T_A = T_R$. This partial-throttle T_A is illustrated by the dashed curve in Figure 6.8. If the pilot now pushes the throttle forward and increases the engine thrust to a higher value of T_A, the airplane will accelerate to a higher velocity. If the throttle is increased to full position, maximum T_A will be produced by the jet engine. In this case, the speed of the airplane will further increase until the thrust required equals the maximum T_A (point 2 in Figure 6.8). It is now impossible for the airplane to fly any faster than the velocity at point 2; otherwise, the thrust required would exceed the maximum thrust available from the power plant. Hence, *the intersection of the T_R curve (dependent on the airframe) and the maximum T_A curve (dependent on the engine) defines the maximum velocity V_{max} of the airplane at the given altitude*, as shown in Figure 6.8. Calculating the maximum velocity is an important aspect of the airplane design process.

Conventional jet engines are rated in terms of thrust (usually in pounds). Hence, the thrust curves in Figure 6.8 are useful for the performance analysis of a jet-powered aircraft. On the other hand, piston engines are rated in terms of power (usually horsepower). Hence, the concepts of T_A and T_R are inconvenient for propeller-driven aircraft. In this case, power required and power available are the more relevant quantities. Moreover, considerations of power lead to results such as rate of climb and maximum altitude for both jet and propeller-driven airplanes. Therefore, for the remainder of this chapter, emphasis will be placed on power rather than thrust, as introduced in the next section.

Example 6.2 Calculate the maximum velocity of the CJ-1 at sea level (see Example 6.1).

SOLUTION The information given in Example 6.1 states that the power plant for the CJ-1 consists of two turbofan engines of 3650 lb of thrust each at sea level. Hence,

$$T_A = 2(3650) = 7300 \text{ lb}$$

Examining the results of Example 6.1, we see $T_R = T_A = 7300$ lb occurs when $V_\infty = 975$ ft/s (see Figure 6.8). Hence

$$\boxed{V_{max} = 975 \text{ ft/s} = 665 \text{ mi/h}}$$

It is interesting to note that, since the sea-level speed of sound is 1117 ft/s, the maximum sea-level Mach number is

$$M_{max} = \frac{V_{max}}{a} = \frac{975}{1117} = 0.87$$

In the present examples, $C_{D,0}$ is assumed constant; hence, the drag polar does not include drag-divergence effects, as discussed in Chap. 5. Because the drag-divergence Mach number for this type of airplane is normally on the order of 0.82 to 0.85, the above calculation indicates that M_{max} is above drag divergence, and our assumption of constant $C_{D,0}$ becomes inaccurate at this high a Mach number.

6.5 POWER REQUIRED FOR LEVEL, UNACCELERATED FLIGHT

Power is a precisely defined mechanical term; it is energy per unit time. The power associated with a moving object can be illustrated by a block moving at constant velocity V under the influence of the constant force F, as shown in Figure 6.10. The block moves from left to right through the distance d in a time

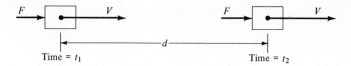

Figure 6.10 Force, velocity, and power of a moving body.

interval $t_2 - t_1$. (We assume that an opposing equal force not shown in Figure 6.10, say due to friction, keeps the block from accelerating.) Work is another precisely defined mechanical term; it is force multiplied by the distance through which the force moves. Moreover, work is energy, having the same units as energy. Hence

$$\text{Power} = \frac{\text{energy}}{\text{time}} = \frac{\text{force} \times \text{distance}}{\text{time}} = \text{force} \times \frac{\text{distance}}{\text{time}}$$

Applied to the moving block in Figure 6.10, this becomes

$$\text{Power} = F\left(\frac{d}{t_2 - t_1}\right) = FV \tag{6.22}$$

where $d/(t_2 - t_1)$ is the velocity V of the object. Thus, Eq. (6.22) demonstrates that the power associated with a force exerted on a moving object is force $\times$ velocity, an important result.

Consider an airplane in level, unaccelerated flight at a given altitude and with velocity V_∞. The thrust required is T_R. From Eq. (6.22), the *power required* P_R is therefore

$$\boxed{P_R = T_R V_\infty} \tag{6.23}$$

The effect of the airplane aerodynamics (C_L and C_D) on P_R is readily obtained by combining Eqs. (6.15) and (6.23):

$$P_R = T_R V_\infty = \frac{W}{C_L/C_D} V_\infty \tag{6.24}$$

From Eq. (6.11),

$$L = W = q_\infty S C_L = \tfrac{1}{2}\rho_\infty V_\infty^2 S C_L$$

Hence

$$V_\infty = \sqrt{\frac{2W}{\rho_\infty S C_L}} \tag{6.25}$$

Substituting Eq. (6.25) into (6.24), we obtain

$$P_R = \frac{W}{C_L/C_D} \sqrt{\frac{2W}{\rho_\infty S C_L}}$$

$$\boxed{P_R = \sqrt{\frac{2W^3 C_D^2}{\rho_\infty S C_L^3}} \propto \frac{1}{C_L^{3/2}/C_D}} \tag{6.26}$$

$$= \frac{C_D}{C_L} wV = C_D \tfrac{1}{2}\rho v^3 S$$

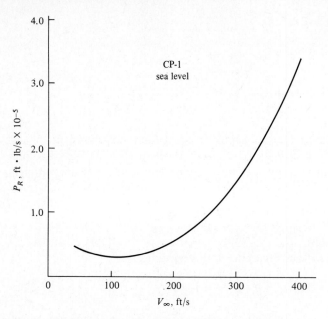

Figure 6.11 Power-required curve for the CP-1 at sea level.

In contrast to thrust required, which varies inversely as C_L/C_D [see Eq. (6.15)], power required varies inversely as $C_L^{3/2}/C_D$.

The power-required curve is defined as a plot of P_R versus V_∞, as sketched in Figure 6.11; note that it qualitatively resembles the thrust-required curve of Figure 6.2. As the airplane velocity increases, P_R first decreases, goes through a minimum, and then increases. At the velocity for minimum power required, the airplane is flying at the angle of attack which corresponds to a maximum $C_L^{3/2}/C_D$.

In Sec. 6.3, we demonstrated that minimum T_R aerodynamically corresponds to equal parasite and induced drag. An analogous but different relation holds at minimum P_R. From Eqs. (6.10) and (6.23),

$$P_R = T_R V_\infty = D V_\infty = q_\infty S \left(C_{D,0} + \frac{C_L^2}{\pi e \mathrm{AR}} \right) V_\infty$$

$$P_R = q_\infty S C_{D,0} V_\infty + q_\infty S V_\infty \frac{C_L^2}{\pi e \mathrm{AR}} \tag{6.27}$$

Parasite power   Induced power
required   required

Therefore, as in the earlier case of T_R, the power required can be split into the respective contributions needed to overcome parasite drag and drag due to lift. These contributions are sketched in Figure 6.12. Also as before, the aerodynamic conditions associated with minimum P_R can be obtained from Eq. (6.27) by setting $dP_R/dV_\infty = 0$. To do this, first obtain Eq. (6.27) explicitly in terms of V_∞,

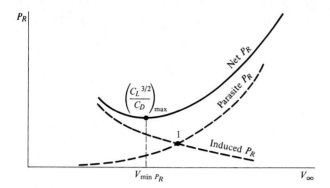

Figure 6.12 Comparison of induced, parasite, and net power required.

recalling that $q_\infty = \frac{1}{2}\rho V_\infty^2$ and $C_L = W/\frac{1}{2}\rho_\infty V_\infty^2 S$.

$$P_R = \frac{1}{2}\rho_\infty V_\infty^3 S C_{D,0} + \frac{1}{2}\rho_\infty V_\infty^3 S \frac{\left(W/\frac{1}{2}\rho_\infty V_\infty^2 S\right)^2}{\pi e \mathrm{AR}}$$

$$P_R = \frac{1}{2}\rho_\infty V_\infty^3 S C_{D,0} + \frac{W^2/\frac{1}{2}\rho_\infty V_\infty S}{\pi e \mathrm{AR}} \qquad (6.28)$$

For minimum power required, $dP_R/dV_\infty = 0$. Differentiating Eq. (6.28) yields

$$\frac{dP_R}{dV_\infty} = \frac{3}{2}\rho_\infty V_\infty^2 S C_{D,0} - \frac{W^2/\frac{1}{2}\rho_\infty V_\infty^2 S}{\pi e \mathrm{AR}}$$

$$= \frac{3}{2}\rho_\infty V_\infty^2 S \left(C_{D,0} - \frac{W^2/\frac{3}{4}\rho_\infty^2 S^2 V_\infty^4}{\pi e \mathrm{AR}} \right)$$

$$= \frac{3}{2}\rho_\infty V_\infty^2 S \left(C_{D,0} - \frac{\frac{1}{3}C_L^2}{\pi e \mathrm{AR}} \right)$$

$$= \frac{3}{2}\rho_\infty V_\infty^2 S \left(C_{D,0} - \frac{1}{3}C_{D,i} \right) = 0 \qquad \text{for minimum } P_R$$

Hence, the aerodynamic condition that holds at minimum power required is

$$\boxed{C_{D,0} = \tfrac{1}{3}C_{D,i}} \qquad (6.29)$$

The fact that parasite drag is one-third the drag due to lift at minimum P_R is reinforced by examination of Figure 6.12. Also note that point 1 in Figure 6.12 corresponds to $C_{D,0} = C_{D,i}$, that is, minimum T_R; hence, V_∞ for minimum P_R is less than that for minimum T_R.

The point on the power-required curve that corresponds to minimum T_R is easily obtained by drawing a line through the origin and tangent to the P_R curve, as shown in Figure 6.13. The point of tangency corresponds to minimum T_R (hence maximum L/D). To prove this, consider any line through the origin and intersecting the P_R curve, such as the dashed line in Figure 6.13. The slope of this

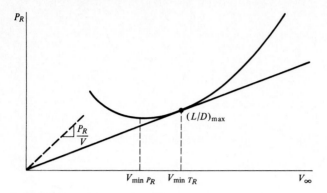

Figure 6.13 The tangent to the power-required curve locates the point of minimum thrust required (and hence the point of maximum L/D).

line is P_R/V_∞. As we move to the right along the P_R curve, the slope of an intersecting line will first decrease, then reach a minimum (at the tangent point), and then again increase. This is clearly seen simply by inspection of the geometry of Figure 6.13. Thus, the point of tangency corresponds to a minimum slope, hence a minimum value of P_R/V_∞. In turn, from calculus this corresponds to

$$\frac{d(P_R/V_\infty)}{dV_\infty} = \frac{d(T_R V_\infty/V_\infty)}{dV_\infty} = \frac{dT_R}{dV_\infty} = 0$$

The above result yields $dT_R/dV_\infty = 0$ at the tangent point, which is precisely the mathematical criterion for minimum T_R. Correspondingly, L/D is maximum at the tangent point.

Example 6.3 Calculate the power-required curves for (a) the CP-1 at sea level and for (b) the CJ-1 at an altitude of 22,000 ft.

SOLUTION

(a) For the CP-1 the values of T_R at sea level have already been tabulated and graphed in Example 6.1. Hence, from Eq. (6.23),

$$P_R = T_R V_\infty$$

we obtain the following tabulation:

V, ft/s	T_R, lb	P_R, ft · lb/s
100	279	27,860
150	217	32,580
250	359	89,860
300	491	147,200
350	652	228,100

The power-required curve is given in Figure 6.11.

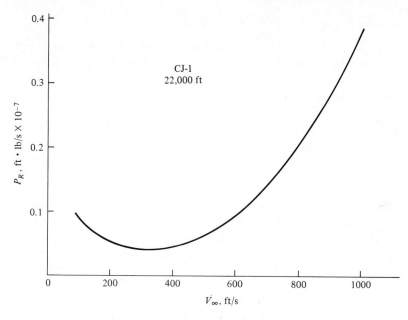

Figure 6.14 Power-required curve for the CJ-1 at 22,000 ft.

(b) For the CJ-1 at 22,000 ft, $\rho_\infty = 0.001184$ slug/ft³. The calculation of T_R is done with the same method as given in Example 6.1, and P_R is obtained from Eq. (6.23). Some results are tabulated below.

V_∞, ft/s	C_L	C_D	L/D	T_R, lb	P_R, ft · lb/s
300	1.17	0.081	14.6	1358	0.041×10^7
500	0.421	0.028	15.2	1308	0.065×10^7
600	0.292	0.024	12.3	1610	0.097×10^7
800	0.165	0.021	7.76	2553	0.204×10^7
1000	0.105	0.020	5.14	3857	0.386×10^7

The reader should attempt to reproduce these results.
The power-required curve is given in Figure 6.14.

6.6 POWER AVAILABLE AND MAXIMUM VELOCITY

Note again that P_R is a characteristic of the aerodynamic design and weight of the aircraft itself. In contrast, the *power available* P_A is a characteristic of the power plant. A detailed discussion on propulsion is deferred until Chap. 9; however, the following comments are made to expedite our performance analyses.

A Reciprocating Engine–Propeller Combination

A piston engine generates power by burning fuel in confined cylinders and using this energy to move pistons, which in turn deliver power to the rotating crankshaft, as schematically shown in Figure 6.15. The power delivered to the propeller by the crankshaft is defined as the *shaft brake* power P (the word "brake" stems from a method of laboratory testing which measures the power of an engine by loading it with a calibrated brake mechanism). However, not all of P is available to drive the airplane; some of it is dissipated by inefficiencies of the propeller itself (to be discussed in Chap. 9). Hence, the power available to propel the airplane, P_A, is given by

$$P_A = \eta P \tag{6.30}$$

where η is the propeller efficiency, $\eta < 1$. Propeller efficiency is an important quantity and is a direct product of the aerodynamics of the propeller. It is always less than unity. For our discussions here, both η and P are assumed to be known quantities for a given airplane.

A remark on units is necessary. In the engineering system, power is in ft · lb/s; in the SI system, power is in watts [which are equivalent to N · m/s]. However, the historical evolution of engineering has left us with a horrendously inconsistent (but very convenient) unit of power which is widely used, namely, horsepower. All reciprocating engines are rated in terms of horsepower, and it is important to note that

$$\begin{aligned} 1 \text{ horsepower} &= 550 \text{ ft} \cdot \text{lb/s} \\ &= 746 \text{ W} \end{aligned}$$

Therefore, it is common to use *shaft brake horsepower* bhp in place of P, and horsepower available hp_A in place of P_A. Equation (6.30) still holds in the form

$$\text{hp}_A = (\eta)(\text{bhp}) \tag{6.30'}$$

However, be cautious. As always in dealing with fundamental physical relations,

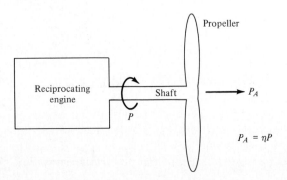

Propeller

$P_A = \eta P$

Figure 6.15 Relation between shaft brake power and power available.

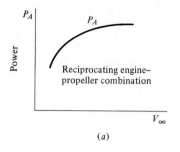
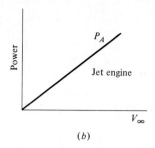

(a)   (b)

Figure 6.16 Power available for piston engine–propeller combination and for the jet engine.

units must be consistent; therefore, a good habit is to immediately convert horsepower to foot-pound per seconds or watts before starting an analysis. This approach will be used here.

The power-available curve for a typical piston engine–propeller combination is sketched in Figure 6.16a.

B Jet Engine

The jet engine (see Chap. 9) derives its thrust by combustion heating an incoming stream of air and then exhausting this hot air at high velocities through a nozzle. The power available from a jet engine is obtained from Eq. (6.22) as

$$P_A = T_A V_\infty \qquad (6.31)$$

Recall from Figure 6.9b that T_A for a jet engine is reasonably constant with velocity. Thus, the power-available curve varies essentially linearly with V_∞, as sketched in Figure 6.16b.

For both the propeller- and jet-powered aircraft, the maximum flight velocity is determined by the high-speed intersection of the maximum P_A and the P_R curves. This is illustrated in Figure 6.17. Because of their utility in determining other performance characteristics of an airplane, these power curves are essential to any performance analysis.

Example 6.4 Calculate the maximum velocity for (a) the CP-1 at sea level and (b) the CJ-1 at 22,000 ft.

SOLUTION

(a) For the CP-1 the information in Example 6.1 gave the horsepower rating of the power plant at sea level as 230 hp. Hence,

$$hp_A = (\eta)(bhp) = 0.80(230) = 184 \text{ hp}$$

The results of Example 6.3 for power required are replotted in Figure 6.17a in terms of horsepower. The horsepower available is also shown, and V_{max} is determined by the intersection

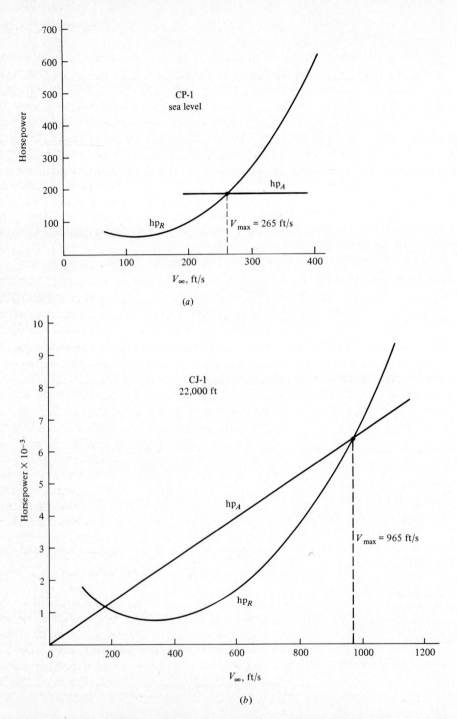

Figure 6.17 Power-available and power-required curves, and the determination of maximum velocity. (*a*) Propeller-driven airplane. (*b*) Jet-propelled airplane.

of the curves as

$$V_{max} = 265 \text{ ft/s} = 181 \text{ mi/h}$$

(b) For the CJ-1, again from the information given in Example 6.1, the sea-level static thrust for each engine is 3650 lb. There are two engines, hence $T_A = 2(3650) = 7300$ lb. From Eq. (6.31), $P_A = T_A V_\infty$, and in terms of horsepower, where T_A is in pounds and V_∞ in feet per seconds,

$$\text{hp}_A = \frac{T_A V_\infty}{550}$$

Let $\text{hp}_{A,0}$ be the horsepower at sea level. As we will see in Chap. 9, the thrust of a jet engine is, to a first approximation, proportional to the air density. If we make this approximation here, the thrust at altitude becomes

$$T_{A,\,alt} = \frac{\rho}{\rho_0} T_{A,0}$$

and hence

$$\text{hp}_{A,\,alt} = \frac{\rho}{\rho_0} \text{hp}_{A,0}$$

For the CJ-1 at 22,000 ft, where $\rho = 0.001184$ slug/ft^3,

$$\text{hp}_{A,\,alt} = \frac{(\rho/\rho_0) T_A V_\infty}{550} = \frac{0.001184/0.002377 (7300) V_\infty}{550} = 6.61 \, V_\infty$$

The results of Example 6.3 for power required are replotted in Figure 6.17b in terms of horsepower. The horsepower available, obtained from the above equation, is also shown, and V_{max} is determined by the intersection of the curves as

$$V_{max} = 965 \text{ ft/s} = 658 \text{ mi/h}$$

6.7 ALTITUDE EFFECTS ON POWER REQUIRED AND AVAILABLE

With regard to P_R, curves at altitude could be generated by repeating the calculations of the previous sections, with ρ_∞ appropriate to the given altitude. However, once the sea-level P_R curve is calculated by means of this process, the curves at altitude can be more quickly obtained by simple ratios, as follows. Let the subscript "0" designate sea-level conditions. From Eqs. (6.25) and (6.26),

$$V_0 = \sqrt{\frac{2W}{\rho_0 S C_L}} \qquad\qquad V_2 = V_1 \sqrt{\frac{\rho_1}{\rho_2}} \qquad (6.32)$$

$$P_{R,0} = \sqrt{\frac{2W^3 C_D^{\,2}}{\rho_0 S C_L^{\,3}}} \qquad\qquad P_{R_2} = P_{R_1} \frac{V_2}{V_1} \qquad (6.33)$$

$$V_2 = V_1 \sqrt{\frac{w_2}{w_1}}$$

$$P_{R_2} = P_{R_1} \frac{w_2}{w_1} \frac{V_2}{V_1} = P_{R_1} \left(\frac{w_2}{w_1}\right)^{\frac{3}{2}}$$

where V_0, $P_{R,0}$, and ρ_0 are velocity, power, and density at sea level. At altitude, where the density is ρ, these relations are

$$V_{alt} = \sqrt{\frac{2W}{\rho S C_L}} \tag{6.34}$$

$$P_{R,alt} = \sqrt{\frac{2W^3 C_D^2}{\rho S C_L^3}} \tag{6.35}$$

Now, strictly for the purposes of calculation, Let C_L remain fixed between sea level and altitude. Hence, because $C_D = C_{D,0} + C_L^2/\pi e AR$, C_D also remains fixed. Dividing Eq. (6.34) by (6.32), and (6.35) by (6.33), we obtain

$$V_{alt} = V_0 \left(\frac{\rho_0}{\rho} \right)^{1/2} \tag{6.36}$$

and

$$P_{R,alt} = P_{R,0} \left(\frac{\rho_0}{\rho} \right)^{1/2} \tag{6.37}$$

Geometrically, these equations allow us to plot a point on the P_R curve at altitude from a given point on the sea-level curve. For example, consider point 1 on the sea-level P_R curve sketched in Figure 6.18. By multiplying both the velocity and power at point 1 by $(\rho_0/\rho)^{1/2}$, a new point is obtained, point 2 in Figure 6.18. Point 2 is guaranteed to fall on the curve at altitude because of our analysis above. In this fashion, the complete P_R curve at altitude can be readily obtained from the sea-level curve. The results are qualitatively given in Figure 6.19, where the altitude curves tend to experience an upward and rightward translation as well as a slight clockwise rotation.

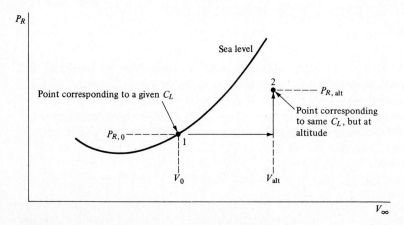

Figure 6.18 Correspondence of points on sea level and altitude power-required curves.

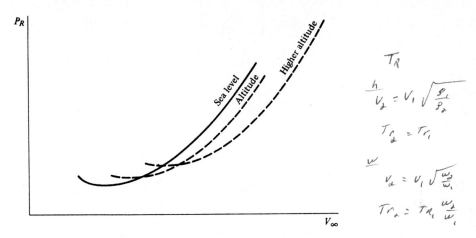

P_R

V_∞

T_R

$$\frac{h}{V_2} = V_1 \sqrt{\frac{\rho_1}{\rho_2}}$$

$$T_{a_2} = T_{r_1}$$

$$\frac{w}{V_2} = V_1 \sqrt{\frac{w_2}{w_1}}$$

$$T_{r_2} = T_{R_1} \frac{w_2}{w_1}$$

Figure 6.19 Effect of altitude on power required.

With regard to P_A, the lower air density at altitude invariably causes a reduction in power for both the reciprocating and jet engines. In this book we will assume P_A and T_A to be proportional to ambient density, as in Example 6.4. Reasons for this will be made clear in Chap. 9. For the reciprocating engine, the loss in power can be delayed by using a supercharger. Nevertheless, the impact on airplane performance due to altitude effects is illustrated in Figures 6.20a and 6.20b for the propeller- and jet-powered airplanes, respectively. Both P_R and maximum P_A are shown; the solid curves correspond to sea level, and the dashed curves to altitude. From these curves, note that $V_{\max}$ varies with altitude. Also note that at high enough altitude, the low-speed limit, which is usually dictated by V_{stall}, may instead be determined by maximum P_A. This effect is emphasized in Figure 6.21, where maximum P_A has been reduced to the extent that, at velocities just above stalling, P_R exceeds P_A. For this case, we make the interesting conclusion that stalling speed cannot be reached in level, steady flight.

To this point in our discussion, only the horizontal velocity performance—both maximum and minimum speeds in steady, level flight—has been emphasized. We have seen that maximum velocity of an airplane is determined by the high-speed intersection of the P_A and P_R curves and that the minimum velocity is determined either by stalling or by the low-speed intersection of the power curves. These velocity considerations are an important part of airplane performance; indeed, for some airplanes, such as many military fighter planes, squeezing the maximum velocity out of the aircraft is the pivotal design feature. However, this is just the beginning of the performance story; we will proceed to examine other important characteristics in the remaining sections of this chapter.

Example 6.5 Using the method of this section, from the CJ-1 power-required curve at 22,000 ft in Example 6.4, obtain the CJ-1 power-required curve at sea level. Compare the maximum velocities at both altitudes.

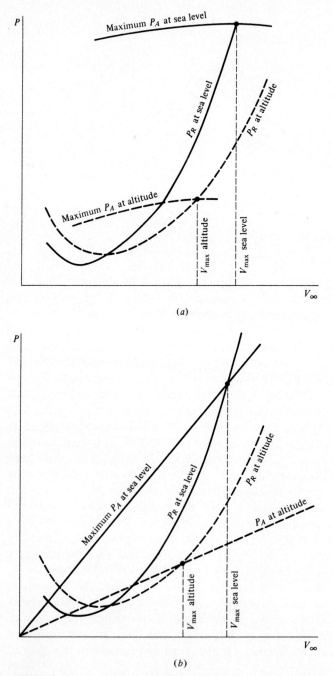

Figure 6.20 Effect of altitude on maximum velocity. (*a*) Propeller-driven airplane. (*b*) Jet-propelled airplane.

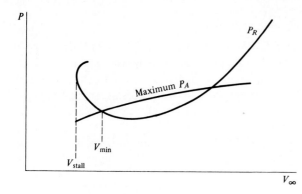

Figure 6.21 Situation when minimum velocity at altitude is greater than stalling velocity.

SOLUTION From Eqs. (6.36) and (6.37), corresponding points on the power-required curves for sea level and altitude are

$$V_0 = V_{\text{alt}}\left(\frac{\rho}{\rho_0}\right)^{1/2}$$

and

$$\text{hp}_{R,0} = \text{hp}_{R,\text{alt}}\left(\frac{\rho}{\rho_0}\right)^{1/2}$$

We are given V_{alt} and $\text{hp}_{R,\text{alt}}$ for 22,000 ft from the CJ-1 curve in Example 6.4. Using the above formulas, we can generate V_0 and $\text{hp}_{R,0}$ as in the following table, noting that

$$\left(\frac{\rho}{\rho_0}\right)^{1/2} = \left(\frac{0.001184}{0.002377}\right)^{1/2} = 0.706$$

Given Points			Generated Points	
V_{alt}, ft/s	$\text{hp}_{R,\text{alt}}$	$\left(\frac{\rho}{\rho_0}\right)^{1/2}$	V_0, ft/s	$\text{hp}_{R,0}$
200	889	0.706	141	628
300	741		212	523
500	1190		353	840
800	3713		565	2621
1,000	7012		706	4950

These results, along with the hp_A curves for sea level and 22,000 ft, are plotted in Figure 6.22. Looking closely at Figure 6.22, note that point 1 on the hp_R curve at 22,000 ft is used to generate point 2 on the hp_R curve at sea level. This illustrates the idea of this section. Also, note that V_{max} at sea level is 975 ft/s = 665 mi/h. This is slightly larger than V_{max} at 22,000 ft, which is 965 ft/s = 658 mi/h.

6.8 RATE OF CLIMB

Visualize a McDonnell-Douglas DC-9 transport powering itself to takeoff speed on an airport runway. It gently lifts off at about 160 mi/h, the nose rotates upward, and the airplane rapidly climbs out of sight. In a matter of minutes, it is cruising at 30,000 ft. This picture prompts the following questions: How fast can the airplane climb? How long does it take to reach a certain altitude? The next two sections provide some answers.

Consider an airplane in steady, unaccelerated, climbing flight, as shown in Figure 6.23. The velocity along the flight path is V_∞, and the flight path itself is

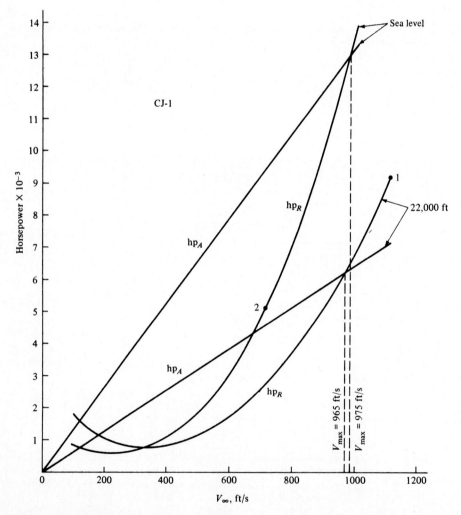

Figure 6.22 Altitude effects on V_{max} for the CJ-1.

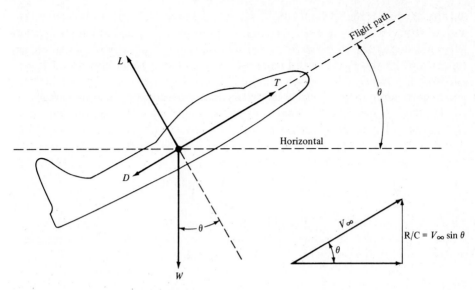

Figure 6.23 Airplane in climbing flight.

inclined to the horizontal at the angle θ. As always, lift and drag are perpendicular and parallel to V_∞, and the weight is perpendicular to the horizontal. Thrust T is assumed to be aligned with the flight path. Here, the physical difference from our previous discussion on level flight is that T is not only working to overcome the drag, but for climbing flight it is also supporting a component of weight. Summing forces parallel to the flight path, we get

$$T = D + W \sin \theta \qquad (6.38)$$

and perpendicular to the flight path, we have

$$L = W \cos \theta \qquad (6.39)$$

Note from Eq. (6.39) that the lift is now smaller than the weight. Equations (6.38) and (6.39) represent the equations of motion for steady, climbing flight and are analogous to Eqs. (6.10) and (6.11) obtained earlier for steady, horizontal flight.

Multiply Eq. (6.38) by V_∞:

$$TV_\infty = DV_\infty + WV_\infty \sin \theta$$

$$\frac{TV_\infty - DV_\infty}{W} = V_\infty \sin \theta \qquad (6.40)$$

Examine Eq. (6.40) closely. The right-hand side, $V_\infty \sin \theta$, is the airplane's *vertical velocity*, as illustrated in Figure 6.23. This vertical velocity is called the *rate of climb* R/C:

$$\boxed{R/C \equiv V_\infty \sin \theta} \qquad (6.41)$$

On the left-hand side of Eq. (6.40), TV_∞ is the power available, from Eq. (6.31), and is represented by the P_A curves in Figure 6.16. The second term on the left-hand side of Eq. (6.40) is DV_∞, which for level flight is the power required, as represented by the P_R curve in Figure 6.11. For climbing flight, however, DV_∞ is no longer precisely the power required, because power must be applied to overcome a component of weight as well as drag. Nevertheless, for small angles of climb, say $\theta < 20°$, it is reasonable to neglect this fact and to assume that the DV_∞ term in Eq. (6.40) is given by the level-flight P_R curve in Figure 6.11. With this,

$$\boxed{TV_\infty - DV_\infty = \text{excess power}} \tag{6.42}$$

where the excess power is the difference between power available and power required, as shown in Figures 6.24a and 6.24b, for propeller-driven and jet-powered aircraft, respectively. Combining Eqs. (6.40) to (6.42), we obtain

$$\boxed{\text{R/C} = \frac{\text{excess power}}{W}} \tag{6.43}$$

where the excess power is clearly illustrated in Figure 6.24.

Again, emphasis is made that the P_R curves in Figures 6.24a and 6.24b are taken, for convenience, as those already calculated for level flight. Hence, *in conjunction with these curves*, Eq. (6.43) is an *approximation* to the rate of climb, good only for small θ. To be more specific, a plot of DV_∞ versus V_∞ for climbing flight [which is exactly called for in Eq. (6.40)] is different from a plot of DV_∞ versus V_∞ for level flight [which is the curve assumed in Figure 6.24 and used in Eq. (6.43)] simply because D *is smaller for climbing than for level flight at the same* V_∞. To see this more clearly, consider an airplane with $W = 5000$ lb, $S = 100$ ft^2, $C_{D,0} = 0.015$, $e = 0.6$, and AR = 6. If the velocity is $V_\infty = 500$ ft/s at sea level, and if the airplane is in *level* flight, then $C_L = L/q_\infty S = W/\frac{1}{2}\rho_\infty V_\infty^2 S = 0.168$.

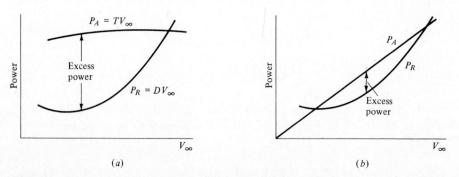

Figure 6.24 Illustration of excess power. (*a*) Propeller-driven airplane. (*b*) Jet-propelled airplane.

In turn

$$C_D = C_{D,0} + \frac{C_L{}^2}{\pi e \mathrm{AR}} = 0.015 + 0.0025 = 0.0175$$

Now, consider the same airplane in a 30° climb at sea level, with the same velocity $V_\infty = 500$ ft/s. Here the lift is smaller than the weight, $L = W\cos\theta$, and therefore $C_L = W\cos 30° / \frac{1}{2}\rho_\infty V_\infty{}^2 S = 0.145$. In turn $C_D = C_{D,0} + C_L{}^2/\pi e\mathrm{AR} = 0.015 + 0.0019 = 0.0169$. This should be compared with the higher value of 0.0175 obtained above for level flight. As seen in this example, for steady climbing flight, L (hence C_L) is smaller, and thus induced drag is smaller. Consequently, total drag for climbing flight is smaller than that for level flight at the same velocity.

Return again to Figure 6.24, which corresponds to a given altitude. Note that the excess power is different at different values of V_∞. Indeed, for both the propeller- and jet-powered aircraft there is some V_∞ at which the excess power is maximum. At this point, from Eq. (6.43), R/C will be maximum.

$$\max \mathrm{R/C} = \frac{\text{maximum excess power}}{W} \qquad (6.44)$$

This situation is sketched in Figure 6.25a, where the power available is that at full throttle, i.e., maximum P_A. The maximum excess power, shown in Figure 6.25a, via Eq. (6.44) yields the maximum rate of climb that can be generated by the airplane at the given altitude. A convenient graphical method of determining maximum R/C is to plot R/C versus V_∞, as shown in Figure 6.25b. A horizontal tangent defines the point of maximum R/C. Another useful construction is the *hodograph* diagram, which is a plot of the airplane's vertical velocity V_v versus its horizontal velocity V_h. Such a hodograph is sketched in Figure 6.26. Remember that R/C is defined as the vertical velocity, $\mathrm{R/C} \equiv V_v$; hence a horizontal tangent to the hodograph defines the point of maximum R/C (point 1 in Figure 6.26). Also, any line through the origin and intersecting the hodograph (say, at point 2) has the slope V_v/V_h; hence, from the geometry of the velocity components, such a line makes the climb angle θ with respect to the horizontal axis, as shown in Figure 6.26. Moreover, the length of the line is equal to V_∞. As this line is rotated counterclockwise, R/C first increases, then goes through its maximum, and then decreases. Finally, the line becomes tangent to the hodograph at point 3. This tangent line gives the maximum climb angle for which the airplane can maintain steady flight, shown as $\theta_{\max}$ in Figure 6.26. It is interesting that maximum R/C does *not* occur at maximum climb angle.

The large excess power and high thrust available in modern aircraft allow climbing flight at virtually any angle. Indeed, modern high-performance military aircraft (such as the F-14, F-15, and F-16) can accelerate to supersonic speeds flying straight up! For such large climb angles, the previous analysis is not valid. Instead, to deal with large θ, the original equations of motion [Eqs. (6.38) and (6.39)] must be solved algebraically, leading to an exact solution valid for any value of θ. The details of this approach can be found in the books by Dommasch,

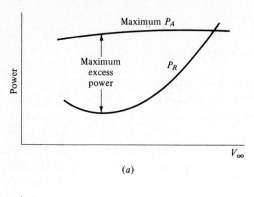

(a)

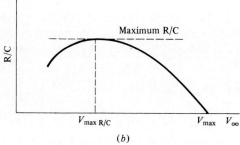

(b)

Figure 6.25 Determination of maximum rate of climb for a given altitude.

Sherbey, and Connolly and by Perkins and Hage (see Bibliography at the end of this chapter).

Returning briefly to Figures 6.24*a* and 6.24*b* for the propeller-driven and jet-powered aircraft, respectively, an important difference in the low-speed rate-of-climb performance can be seen between the two types. Due to the power-available characteristics of a piston engine–propeller combination, large excess powers are available at low values of V_∞, just above the stall. For an airplane on its

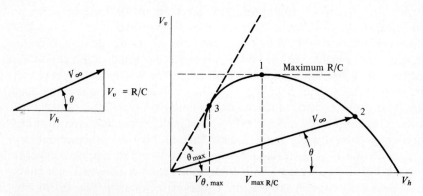

Figure 6.26 Hodograph for climb performance at a given altitude.

landing approach, this gives a comfortable margin of safety in case of a sudden wave-off (particularly important for landings on aircraft carriers). In contrast, the excess power available to jet aircraft at low V_∞ is small, with a correspondingly reduced rate-of-climb capability.

Figures 6.25b and 6.26 give R/C at a given altitude. In the next section we will ask how R/C varies with altitude. In pursuit of an answer, we will also find the answer to another question, namely, how high can the airplane fly.

Example 6.6 Calculate the rate of climb vs. velocity at sea level for (a) the CP-1 and (b) the CJ-1.

SOLUTION

(a) For the CP-1, from Eq. (6.43)

$$R/C = \frac{\text{excess power}}{W} = \frac{P_A - P_R}{W}$$

With power in foot-pounds per second and W in pounds, for the CP-1, this equation becomes

$$R/C = \frac{P_A - P_R}{2950}$$

From Example 6.3, at $V_\infty = 150$ ft/s, $P_R = 0.326 \times 10^5$ ft·lb/s. From Example 6.4, $P_A = 550(\text{hp}_A) = 550(184) = 1.012 \times 10^5$ ft·lb/s. Hence,

$$R/C = \frac{(1.012 - 0.326) \times 10^5}{2950} = 23.3 \text{ ft/s}$$

In terms of feet per minute,

$$R/C = 23.3(60) = \boxed{1395 \text{ ft/min}} \qquad \text{at } V_\infty = 150 \text{ ft/s}$$

This calculation can be repeated at different velocities, with the following results:

V_∞, ft/s	R/C, ft/min
100	1492
130	1472
180	1189
220	729
260	32.6

These results are plotted in Figure 6.27.

(b) For the CJ-1, from Eq. (6.43),

$$R/C = \frac{P_A - P_R}{W} = \frac{550(\text{hp}_A - \text{hp}_R)}{19,815}$$

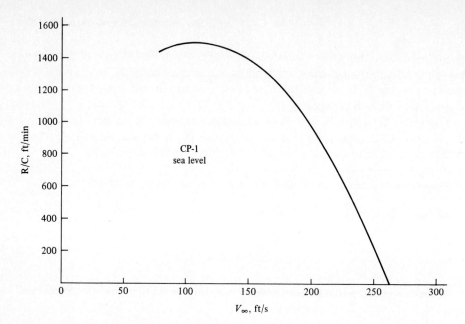

Figure 6.27 Sea-level rate of climb for the CP-1.

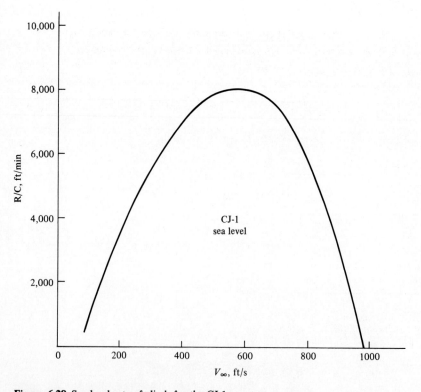

Figure 6.28 Sea-level rate of climb for the CJ-1.

From the results and curves of Example 6.5, at $V_\infty = 500$ ft/s, $hp_R = 1884$, and $hp_A = 6636$.

Hence
$$R/C = 550\frac{6636 - 1884}{19{,}815} = 132 \text{ ft/s}$$

or
$$R/C = 132(60) = \boxed{7914 \text{ ft/min}} \text{ at } V_\infty = 500 \text{ ft/s}$$

Again, a short tabulation for other velocities is given below for the reader to check.

V_∞, ft/s	R/C, ft/min
200	3546
400	7031
600	8088
800	5792
950	1230

These results are plotted in Figure 6.28

6.9 ABSOLUTE AND SERVICE CEILINGS

The effects of altitude on P_A and P_R were discussed in Sec. 6.7 and illustrated in Figures 6.20a and 6.20b. For the sake of discussion, consider a propeller-driven airplane; the results of this section will be qualitatively the same for a jet. As altitude increases, the maximum excess power decreases, as shown in Figure 6.29. In turn, maximum R/C decreases. This is illustrated by Figure 6.30, which is a plot of maximum R/C versus altitude, but with R/C as the abscissa.

There is some altitude high enough at which the P_A curve becomes tangent to the P_R curve (point 1 in Figure 6.31). The velocity at this point is the only value at which steady, level flight is possible; moreover, there is zero excess power, hence

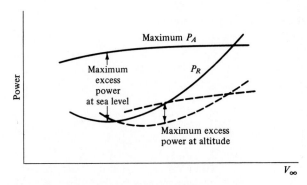

Figure 6.29 Variation of excess power with altitude.

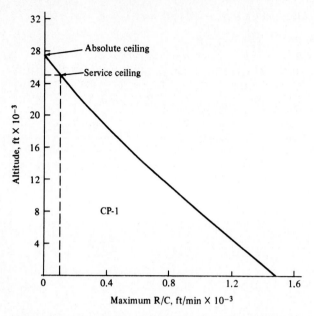

Figure 6.30 Determination of absolute and service ceilings for the CP-1.

zero maximum rate of climb, at this point. The altitude at which maximum R/C = 0 is defined as the *absolute ceiling* of the airplane. A more useful quantity is the *service ceiling*, defined as that altitude where the maximum R/C = 100 ft/min. The service ceiling represents the practical upper limit of steady, level flight.

The absolute and service ceilings can be determined as follows:

1. Using the technique of Sec. 6.8, calculate values of maximum R/C for a number of different altitudes.
2. Plot maximum rate of climb vs. altitude, as shown in Figure 6.30.
3. Extrapolate the curve to 100 ft/min and 0 ft/min to find the service and absolute ceilings, respectively, as also shown in Figure 6.30.

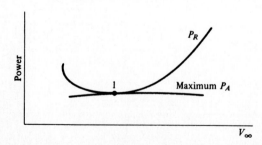

Figure 6.31 Power-required and power-available curves at the absolute ceiling.

Example 6.7 Calculate the absolute and service ceilings for (a) the CP-1 and (b) the CJ-1.

SOLUTION

(a) For the CP-1, as stated in Example 6.1, all the results presented in all the examples of this chapter are taken from a computer program which deals with 100 different velocities, each at different altitudes, beginning at sea level and increasing in 2000-ft increments. In modern engineering, using the computer to take the drudgery out of extensive and repeated calculations is an everyday practice. For example, note from Example 6.6 that the maximum rate of climb at sea level for the CP-1 is 1500 ft/min. In essence, this result is the product of all the work performed in Examples 6.1 to 6.6. Now, to obtain the absolute and service ceilings, these calculations must be repeated at several different altitudes in order to find where $R/C = 0$ and 100 ft/min, respectively. Some results are tabulated and plotted below; the reader should take the time to check a few of the numbers.

Altitude, ft	Maximum R/C, ft/min
0	1500
4,000	1234
8,000	987
12,000	755
16,000	537
20,000	331
24,000	135
26,000	40

These results are plotted in Figure 6.30. From these numbers, we find:

Absolute ceiling $(R/C = 0)$ is $\boxed{27,000 \text{ ft}}$

Service ceiling $(R/C = 100 \text{ ft/min})$ is $\boxed{25,000 \text{ ft}}$

(b) For the CJ-1, utilizing the results from Examples 6.1 to 6.6 and making similar calculations at various altitudes, we tabulate the following results:

Altitude, ft	Maximum R/C, ft/min
0	8118
6,000	6699
12,000	5448
18,000	4344
24,000	3369
30,000	2502
36,000	1718

These results are plotted in Figure 6.32.

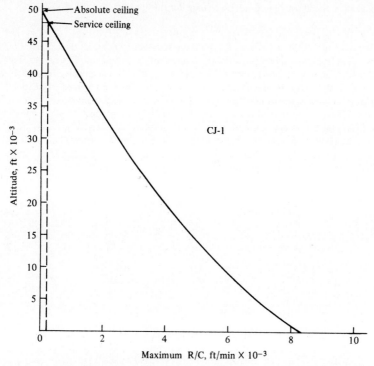

Figure 6.32 Determination of absolute and service ceilings for the CJ-1.

From these results, we find:

Absolute ceiling (R/C = 0) is $\boxed{49,000 \text{ ft}}$

Service ceiling (R/C = 100 ft/min) is $\boxed{48,000 \text{ ft}}$

6.10 TIME TO CLIMB

To carry out its defensive role adequately, a fighter airplane must be able to climb from sea level to the altitude of advancing enemy aircraft in as short a time as possible. In another case, a commercial aircraft must be able to rapidly climb to high altitudes to minimize the discomfort and risks of inclement weather and to minimize air traffic problems. As a result, the time for an airplane to climb to a given altitude can become an important design consideration. The calculation of the time to climb follows directly from our previous discussions, as described below.

The rate of climb was defined in Sec. 6.8 as the vertical velocity of the airplane. Velocity is simply the time rate of change of distance, the distance here being the altitude h. Hence, $R/C = dh/dt$. Therefore,

$$dt = \frac{dh}{R/C} \tag{6.45}$$

In Eq. (6.45), dt is the small increment in time required to climb a small increment dh in altitude. Therefore, from calculus, the time to climb from one altitude, h_1, to another, h_2, is obtained by integrating Eq. (6.45):

$$t = \int_{h_1}^{h_2} \frac{dh}{R/C}$$

Normally, time to climb is considered from sea level, where $h_1 = 0$. Hence, the time to climb to any given altitude h_2 is

$$t = \int_0^{h_2} \frac{dh}{R/C} \tag{6.46}$$

To calculate t graphically, first plot $(R/C)^{-1}$ versus h, as shown in Figure 6.33. The area under the curve from $h = 0$ to $h = h_2$ is the time to climb to altitude h_2.

Example 6.8 Calculate and compare the time required for (a) the CP-1 and (b) the CJ-1 to climb to 20,000 ft.

SOLUTION
    (a) For the CP-1, from Eq. (6.46), the time to climb is equal to the shaded area under the curve shown in Figure 6.33. The resulting area gives time to climb as 27.0 min.
    (b) For the CJ-1, Eq. (6.46) is plotted in Figure 6.34. The resulting area gives time to climb as 3.5 min.

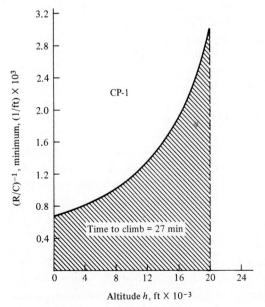

Figure 6.33 Determination of time to climb for the CP-1.

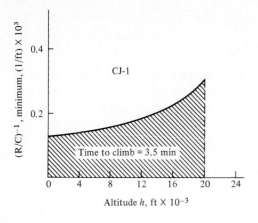

Figure **6.34** Determination of time to climb for the CJ-1.

Note that the CJ-1 climbs to 20,000 ft in one-eighth of the time required by the CP-1; this is to be expected for a high-performance executive jet transport in comparison to its propeller-driven piston engine counterpart.

6.11 RANGE AND ENDURANCE—PROPELLER-DRIVEN AIRPLANE

When Charles Lindbergh made his spectacular solo flight across the Atlantic Ocean on May 20–21, 1927, he could not have cared less about maximum velocity, rate of climb, or time to climb. Uppermost in his mind was the maximum distance he could fly on the fuel supply carried by the Spirit of St. Louis. Therefore, *range* was the all-pervasive consideration during the design and construction of Lindbergh's airplane. Indeed, throughout all of twentieth-century aviation, range has been an important design feature, especially for transcontinental and transoceanic transports and for strategic bombers for the military.

Range is technically defined as the total distance (measured with respect to the ground) traversed by the airplane on a tank of fuel. A related quantity is *endurance*, which is defined as the total time that an airplane stays in the air on a tank of fuel. In different applications, it may be desirable to maximize one or the other of these characteristics. The parameters which maximize range are different from those which maximize endurance; they are also different for propeller- and jet-powered aircraft. The purpose of this section is to discuss these variations for the case of a propeller-driven airplane; jet airplanes will be considered in the next section.

A Physical Considerations

One of the critical factors influencing range and endurance is the *specific fuel consumption*, a characteristic of the engine. For a reciprocating engine, specific fuel consumption (commonly abbreviated as SFC) is defined as the *weight of fuel consumed per unit power per unit time.* As mentioned earlier, reciprocating engines are rated in terms of horsepower, and the common units (although inconsistent) of specific fuel consumption are

$$\text{SFC} = \frac{\text{lb of fuel}}{(\text{bhp})(\text{h})}$$

where bhp signifies shaft brake horsepower, discussed in Sec. 6.6.

First, consider endurance. On a qualitative basis, in order to stay in the air for the longest period of time, common sense says that we must use the *minimum* number of pounds of fuel per hour. On a dimensional basis this quantity is proportional to the horsepower required by the airplane and to the SFC:

$$\frac{\text{lb of fuel}}{\text{h}} \propto (\text{SFC})(\text{hp}_R)$$

Therefore, minimum pounds of fuel per hour is obtained with minimum hp_R. Since minimum pounds of fuel per hour gives maximum endurance, we quickly conclude that

Maximum endurance for a propeller-driven airplane occurs when the airplane is flying at minimum power required.

This condition is sketched in Figure 6.35. Furthermore, in Sec. 6.5 we have already proven that minimum power required corresponds to a maximum value of $C_L^{3/2}/C_D$ [see Eq. (6.26)]. Thus

Maximum endurance for a propeller-driven airplane occurs when the airplane is flying at a velocity such that $C_L^{3/2}/C_D$ is maximum.

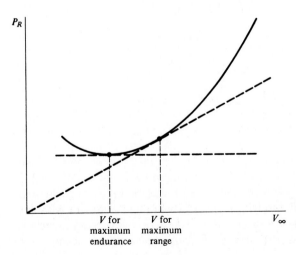

Figure 6.35 Points of maximum range and endurance on the power-required curve for a propeller-driven airplane.

Now, consider range. In order to cover the longest distance (say, in miles), common sense says that we must use the minimum number of pounds of fuel per mile. On a dimensional basis, we can state the proportionality

$$\frac{\text{lb of fuel}}{\text{mi}} \propto \frac{(\text{SFC})(\text{hp}_R)}{V_\infty}$$

(Check the units yourself, assuming V_∞ is in miles per hour.) As a result, minimum pounds of fuel per mile is obtained with a minimum hp_R/V_∞. This minimum value of hp_R/V_∞ precisely corresponds to the tangent point in Figure 6.13, which also corresponds to maximum L/D, as proved in Sec. 6.5. Thus

Maximum range for a propeller-driven airplane occurs when the airplane is flying at a velocity such that C_L/C_D is maximum.

This condition is also sketched in Figure 6.35.

B Quantitative Formulation

The important conclusions written above in italics were obtained from purely physical reasoning. We will develop quantitative formulas which substantiate these conclusions and which allow the direct calculation of range and endurance for given conditions.

In this development, the specific fuel consumption is couched in units that are consistent, i.e.,

$$\frac{\text{lb of fuel}}{[(\text{ft} \cdot \text{lb})/\text{s}](\text{s})} \quad \text{or} \quad \frac{\text{N of fuel}}{(\text{J/s})(\text{s})}$$

For convenience and clarification, c will designate the specific fuel consumption with consistent units.

Consider the product $cP\,dt$, where P is engine power and dt is a small increment in time. The units of this product are (in the English engineering system)

$$cP\,dt = \left(\frac{\text{lb of fuel}}{[(\text{ft} \cdot \text{lb})/\text{s}]\,\text{s}}\right)\left(\frac{\text{ft} \cdot \text{lb}}{\text{s}}\right)(\text{s}) = \text{lb of fuel}$$

Therefore, $cP\,dt$ represents the differential change in the weight of the fuel due to consumption over the short time period dt. The total weight of the airplane, W, is the sum of the fixed structural and payload weights, along with the changing fuel weight. Hence, any change in W is assumed to be due to the change in fuel weight. Recall that W denotes the weight of the airplane at any instant. Also, let $W_0 = $ gross weight of the airplane (weight with full fuel and payload), $W_f = $ weight of the fuel load, and $W_1 = $ weight of the airplane *without* fuel. With these considerations, we have

$$W_1 = W_0 - W_f$$

and

$$dW_f = dW = -cP\,dt$$

or

$$dt = -\frac{dW}{cP} \qquad (6.47)$$

The minus sign in Eq. (6.47) is necessary because dt is physically positive (time cannot move backward, except in science fiction novels), while at the same time W is decreasing (hence dW is negative). Integrating Eq. (6.47) between time $t = 0$, where $W = W_0$ (fuel tanks full), and time $t = E$, where $W = W_1$ (fuel tanks empty), we find

$$\int_0^E dt = -\int_{W_0}^{W_1} \frac{dW}{cP}$$

$$\boxed{E = \int_{W_1}^{W_0} \frac{dW}{cP}} \qquad (6.48)$$

In Eq. (6.48), E is the endurance in seconds.

To obtain an analogous expression for range, multiply Eq. (6.47) by V_∞:

$$V_\infty\,dt = -\frac{V_\infty\,dW}{cP} \qquad (6.49)$$

In Eq. (6.49), $V_\infty\,dt$ is the incremental distance ds covered in time dt.

$$ds = -\frac{V_\infty\,dW}{cP} \qquad (6.50)$$

The total distance covered throughout the flight is equal to the integral of Eq. (6.50) from $s = 0$, where $W = W_0$ (full fuel tank), to $s = R$, where $W = W_1$ (empty fuel tank):

$$\int_0^R ds = -\int_{W_0}^{W_1} \frac{V_\infty\,dW}{cP}$$

or

$$\boxed{R = \int_{W_1}^{W_0} \frac{V_\infty\,dW}{cP}} \qquad (6.51)$$

In Eq. (6.51), R is the range in consistent units, such as feet or meters.

Equations (6.48) and (6.51) can be evaluated graphically, as shown in Figures 6.36a and 6.36b for range and endurance, respectively. Range can be calculated accurately by plotting V_∞/cP versus W and taking the area under the curve from W_1 to W_0, as shown in Figure 6.36a. Analogously, the endurance can be calculated accurately by plotting $(cP)^{-1}$ versus W and taking the area under the curve from W_1 to W_0, as shown in Figure 6.36b.

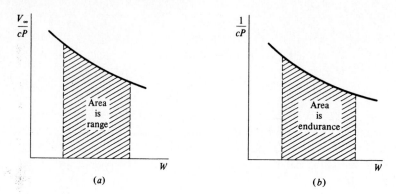

Figure 6.36 Determination of range and endurance.

Equations (6.48) and (6.51) are accurate formulations for endurance and range. In principle, they can include the entire flight—takeoff, climb, cruise, and landing—as long as the instantaneous values of W, V_∞, c, and P are known at each point along the flight path. However, Eqs. (6.48) and (6.51), although accurate, are also long and tedious to evaluate by the method discussed above. Therefore, simpler but approximate analytic expressions for R and E are useful. Such formulas are developed below.

C Breguet Formulas (Propeller-Driven Airplane)

For level, unaccelerated flight, we demonstrated in Sec. 6.5 that $P_R = DV_\infty$. Also, to maintain steady conditions, the pilot has adjusted the throttle such that power available from the engine-propeller combination is just equal to the power required: $P_A = P_R = DV_\infty$. In Eq. (6.47), P is the brake power output of the engine itself. Recall from Eq. (6.30) that $P_A = \eta P$, where η is the propeller efficiency. Thus

$$P = \frac{P_A}{\eta} = \frac{DV_\infty}{\eta} \tag{6.52}$$

Substitute Eq. (6.52) into (6.51):

$$R = \int_{W_1}^{W_0} \frac{V_\infty \, dW}{cP} = \int_{W_1}^{W_0} \frac{V_\infty \eta \, dW}{cDV_\infty} = \int_{W_1}^{W_0} \frac{\eta \, dW}{cD} \tag{6.53}$$

Multiplying Eq. (6.53) by W/W and noting that for steady, level flight, $W = L$, we obtain

$$R = \int_{W_1}^{W_0} \frac{\eta}{cD} \frac{W}{W} \, dW = \int_{W_1}^{W_0} \frac{\eta}{c} \frac{L}{D} \frac{dW}{W} \tag{6.54}$$

Unlike Eq. (6.51), which is exact, Eq. (6.54) now contains the direct assumption of level, unaccelerated flight. However, for practical use, it will be further simplified by assuming that η, $L/D = C_L/C_D$, and c are constant throughout the flight. This is a reasonable approximation for cruising flight conditions. Thus, Eq. (6.54) becomes

$$R = \frac{\eta}{c} \frac{C_L}{C_D} \int_{W_1}^{W_0} \frac{dW}{W}$$

$$\boxed{R = \frac{\eta}{c} \frac{C_L}{C_D} \ln \frac{W_0}{W_1}} \tag{6.55}$$

Equation (6.55) is a classic formula in aeronautical engineering; it is called the *Breguet range formula*, and it gives a quick, practical estimate for range which is generally accurate to within 10 to 20 percent. Also, keep in mind that, as with all proper physical derivations, Eq. (6.55) deals with consistent units. Hence, R is in feet or meters when c is in pounds of fuel per foot-pound per second per second or newtons of fuel per joule per second per second, respectively, as discussed in part B above. If c is given in terms of brake horsepower and if R is desired in miles, the proper conversions to consistent units should be made before using Eq. (6.55).

Look at Eq. (6.55). It says all the things that common sense would expect, namely, to maximize range for a reciprocating-engine, propeller-driven airplane, we want:

1. The largest possible propeller efficiency η.
2. The lowest possible specific fuel consumption c.
3. The highest ratio of W_0/W_1, which is obtained with the largest fuel weight W_F.
4. Most importantly, flight at *maximum L/D*. This confirms our above argument in part A that, for *maximum range*, we must fly at *maximum L/D*. Indeed, the Breguet range formula shows that range is directly proportional to L/D. This clearly explains why high values of L/D (high aerodynamic efficiency) have always been of importance in the design of airplanes. This importance was underscored in the 1970s due to the increasing awareness of the need to conserve energy (hence fuel).

A similar formula can be obtained for endurance. Recalling that $P = DV_\infty/\eta$ and that $W = L$, Eq. (6.48) becomes

$$E = \int_{W_1}^{W_0} \frac{dW}{cP} = \int_{W_1}^{W_0} \frac{\eta}{c} \frac{dW}{DV_\infty} = \int_{W_1}^{W_0} \frac{\eta}{c} \frac{L}{DV_\infty} \frac{dW}{W}$$

Since $L = W = \frac{1}{2}\rho_\infty V_\infty^2 SC_L$, then $V_\infty = \sqrt{2W/\rho_\infty SC_L}$.

Thus

$$E = \int_{W_1}^{W_0} \frac{\eta}{c} \frac{C_L}{C_D} \sqrt{\frac{\rho_\infty SC_L}{2}} \frac{dW}{W^{3/2}}$$

Assuming that C_L, C_D, η, c, and ρ_∞ (constant altitude) are all constant, this equation becomes

$$E = -2\frac{\eta}{c}\frac{C_L^{3/2}}{C_D}\left(\frac{\rho_\infty S}{2}\right)^{1/2}[W^{-1/2}]_{W_1}^{W_0}$$

or

$$\boxed{E = \frac{\eta}{c}\frac{C_L^{3/2}}{C_D}(2\rho_\infty S)^{1/2}\left(W_1^{-1/2} - W_0^{-1/2}\right)} \tag{6.56}$$

Equation (6.56) is the *Breguet endurance formula*, where E is in seconds (consistent units).

Look at Eq. (6.56). It says that to maximize endurance for a reciprocating-engine, propeller-driven airplane, we want:

1. The highest propeller efficiency η.
2. The lowest specific fuel consumption c.
3. The highest fuel weight W_f, where $W_0 = W_1 + W_f$.
4. Flight at maximum $C_L^{3/2}/C_D$. This confirms our above argument in part A that, for *maximum endurance*, we must fly at maximum $C_L^{3/2}/C_D$.
5. Flight at sea level, because $E \propto \rho_\infty^{1/2}$, and ρ_∞ is largest at sea level.

It is interesting to note that, subject to our approximations, endurance depends on altitude, whereas range [see Eq. (6.55)] is independent of altitude.

Emphasis is made that all the discussion in this section pertains to a piston engine–propeller combination only. For a jet-powered airplane, the picture changes, as will be discussed in the next section.

Example 6.9 Estimate the maximum range and maximum endurance for the CP-1.

SOLUTION The Breguet range formula is given by Eq. (6.55) for a propeller-driven airplane. This equation is

$$R = \frac{\eta}{c}\frac{C_L}{C_D}\ln\frac{W_0}{W_1}$$

with the specific fuel consumption c in consistent units, say pounds of fuel per foot-pound per second per second or simply per foot. However, in Example 6.1, the SFC is given as 0.45 lb of fuel/(hp)(h). This can be changed to consistent units, as

$$c = 0.45\frac{\text{lb}}{(\text{hp})(\text{h})}\frac{1\ \text{hp}}{550\ \text{ft·lb/s}}\frac{1\ \text{h}}{3600\ \text{s}}$$

$$= 2.27 \times 10^{-7}\ \text{ft}^{-1}$$

In Example 6.1, the variation of $C_L/C_D = L/D$ was calculated versus velocity. The variation of $C_L^{3/2}/C_D$ can be obtained in the same fashion. The results are plotted in Figure 6.37.

From these curves,

$$\max\left(\frac{C_L}{C_D}\right) = 13.62 \qquad \max\left(\frac{C_L^{3/2}}{C_D}\right) = 12.81$$

These are results pertaining to the aerodynamics of the airplane; even though the above plots

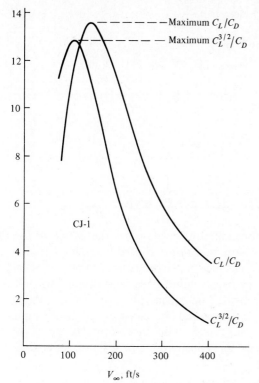

Figure 6.37 Aerodynamic ratios for the CP-1 at sea level.

were calculated at sea level (from Example 6.1), the *maximum* values of C_L/C_D and $C_L^{3/2}/C_D$ are independent of altitude, velocity, etc. They depend only on the aerodynamic design of the aircraft.

The gross weight of the CP-1 is $W_0 = 2950$ lb. The fuel capacity given in Example 6.1 is 65 gallons of aviation gasoline, which weighs 5.64 lb/gallon. Hence, the weight of the fuel, $W_p = 65(5.64) = 367$ lb. Thus, the empty weight $W_1 = 2950 - 367 = 2583$ lb.

Returning to Eq. (6.55)

$$ R = \frac{\eta}{c} \frac{C_L}{C_D} \ln \frac{W_0}{W_1} = \frac{0.8}{2.27 \times 10^{-7}} (13.62) \ln \frac{2950}{2583} $$

$$ \boxed{R = 6.38 \times 10^6 \text{ ft}} $$

Since 1 mi = 5280 ft, then

$$ R = \frac{6.38 \times 10^6}{5280} = \boxed{1207 \text{ mi}} $$

The endurance is given by Eq. (6.56):

$$ E = \frac{\eta}{c} \frac{C_L^{3/2}}{C_D} (2\rho_\infty S)^{1/2} \left(W_1^{-1/2} - W_0^{-1/2} \right) $$

Because of the explicit appearance of ρ_∞ in the endurance equation, maximum endurance will occur at sea level, $\rho_\infty = 0.002377$ slug/ft^3. Hence,

$$E = \frac{0.8}{2.27 \times 10^{-7}} (12.81)[2(0.002377)(174)]^{1/2} \left(\frac{1}{2583^{1/2}} - \frac{1}{2950^{1/2}} \right)$$

$$\boxed{E = 5.19 \times 10^4 \text{ s}}$$

Since 3600 s = 1 h,

$$E = \frac{5.19 \times 10^4}{3600} = \boxed{14.4 \text{ h}}$$

6.12 RANGE AND ENDURANCE—JET AIRPLANE

For a jet airplane, the specific fuel consumption is defined as the *weight of fuel consumed per unit thrust per unit time*. Note that *thrust* is used here, in contradistinction to power, as in the previous case for a reciprocating-engine–propeller combination. The fuel consumption of a jet engine physically depends on the thrust produced by the engine, whereas the fuel consumption of a reciprocating engine physically depends on the brake power produced. It is this simple difference which leads to different range and endurance formulas for a jet airplane. In the literature, thrust-specific fuel consumption (TSFC) for jet engines is commonly given as

$$\text{TSFC} = \frac{\text{lb of fuel}}{(\text{lb of thrust})(\text{h})}$$

(Note the inconsistent unit of time.)

A Physical Considerations

The maximum endurance of a jet airplane occurs for minimum pounds of fuel per hour, the same as for propeller-driven aircraft. However, for a jet,

$$\frac{\text{lb of fuel}}{\text{h}} = (\text{TSFC}) T_A$$

where T_A is the thrust available produced by the jet engine. Recall that in steady, level, unaccelerated flight, the pilot has adjusted the throttle such that thrust available T_A just equals the thrust required T_R: $T_A = T_R$. Therefore, minimum pounds of fuel per hour corresponds to minimum thrust required. Hence, we conclude that

Maximum endurance for a jet airplane occurs when the airplane is flying at minimum thrust required.

This condition is sketched in Figure 6.38. Furthermore, in Sec. 6.3, minimum

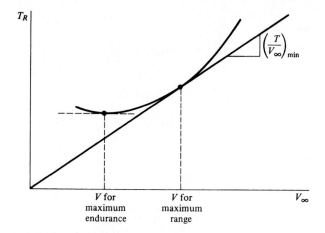

Figure 6.38 Points of maximum range and endurance on the thrust-required curve.

thrust required was shown to correspond to maximum L/D. Thus

> *Maximum endurance for a jet airplane occurs when the airplane is flying at a velocity such that C_L/C_D is maximum.*

Now, consider range. As before, maximum range occurs for a minimum pounds of fuel per mile. For a jet, on a dimensional basis,

$$\frac{\text{lb of fuel}}{\text{mi}} = \frac{(\text{TSFC})T_A}{V_\infty}$$

Recalling that for steady, level flight, $T_A = T_R$, we note that minimum pounds of fuel per mile corresponds to a minimum T_R/V_∞. In turn, T_R/V_∞ is the slope of a line through the origin and intersecting the thrust-required curve; its minimum value occurs when the line becomes tangent to the thrust-required curve, as sketched in Figure 6.38. The aerodynamic condition holding at this tangent point is obtained as follows. Recalling that for steady, level flight, $T_R = D$, then

$$\frac{T_R}{V_\infty} = \frac{D}{V_\infty} = \frac{\frac{1}{2}\rho_\infty V_\infty^2 S C_D}{V_\infty} = \frac{1}{2}\rho_\infty V_\infty S C_D$$

Recalling that $V_\infty = \sqrt{2W/\rho_\infty S C_L}$, we have

$$\frac{T_R}{V_\infty} = \frac{1}{2}\rho_\infty \sqrt{\frac{2W}{\rho_\infty S C_L}}\, C_D \propto \frac{1}{C_L^{1/2}/C_D}$$

Hence, minimum T_R/V_∞ corresponds to maximum $C_L^{1/2}/C_D$. In turn, we conclude that

> *Maximum range for a jet airplane occurs when the airplane is flying at a velocity such that $C_L^{1/2}/C_D$ is maximum.*

B Quantitative Formulation

Let c_t be the thrust-specific fuel consumption in consistent units, e.g.,

$$\frac{\text{lb of fuel}}{(\text{lb of thrust})(\text{s})} \quad \text{or} \quad \frac{\text{N of fuel}}{(\text{N of thrust})(\text{s})}$$

Let dW be the elemental change in weight of the airplane due to fuel consumption over a time increment dt. Then

$$dW = -c_t T_A \, dt$$

or
$$dt = \frac{-dW}{c_t T_A} \tag{6.57}$$

Integrating Eq. (6.57) between $t = 0$, where $W = W_0$, and $t = E$, where $W = W_1$, we obtain

$$E = -\int_{W_0}^{W_1} \frac{dW}{c_t T_A} \tag{6.58}$$

Recalling that $T_A = T_R = D$ and $W = L$, we have

$$E = \int_{W_1}^{W_0} \frac{1}{c_t} \frac{L}{D} \frac{dW}{W} \tag{6.59}$$

With the assumption of constant c_t and $C_L/C_D = L/D$, Eq. (6.59) becomes

$$\boxed{E = \frac{1}{c_t} \frac{C_L}{C_D} \ln \frac{W_0}{W_1}} \tag{6.60}$$

Note from Eq. (6.60) that for maximum endurance for a jet airplane, we want:

1. Minimum thrust-specific fuel consumption c_t.
2. Maximum fuel weight W_f.
3. Flight at maximum L/D. This confirms our above argument in part A that, for maximum endurance for a jet, we must fly such that L/D is maximum.

Note that, subject to our assumptions, E for a jet does not depend on ρ_∞, that is, it is independent of altitude.

Now consider range. Returning to Eq. (6.57) and multiplying by V_∞, we get

$$ds = V_\infty \, dt = -\frac{V_\infty \, dW}{c_t T_A} \tag{6.61}$$

where ds is the increment in distance traversed by the jet over the time increment dt. Integrating Eq. (6.61) from $s = 0$, where $W = W_0$, to $s = R$, where $W = W_1$, we have

$$R = \int_0^R ds = -\int_{W_0}^{W_1} \frac{V_\infty \, dW}{c_t T_A} \tag{6.62}$$

However, again noting that for steady, level flight, the engine throttle has been adjusted such that $T_A = T_R$ and recalling from Eq. (6.15) that $T_R = W/(C_L/C_D)$, we rewrite Eq. (6.62) as

$$R = \int_{W_1}^{W_0} \frac{V_\infty}{c_t} \frac{C_L}{C_D} \frac{dW}{W} \tag{6.63}$$

Since $V_\infty = \sqrt{2W/\rho_\infty S C_L}$, Eq. (6.63) becomes

$$R = \int_{W_1}^{W_0} \sqrt{\frac{2}{\rho_\infty S}} \frac{C_L^{1/2}/C_D}{c_t} \frac{dW}{W^{1/2}} \tag{6.64}$$

Again, assuming constant c_t, C_L, C_D, and ρ_∞ (constant altitude), we rewrite Eq. (6.64) as

$$R = \sqrt{\frac{2}{\rho_\infty S}} \frac{C_L^{1/2}}{C_D} \frac{1}{c_t} \int_{W_1}^{W_0} \frac{dW}{W^{1/2}}$$

$$\boxed{R = 2\sqrt{\frac{2}{\rho_\infty S}} \frac{1}{c_t} \frac{C_L^{1/2}}{C_D} \left(W_0^{1/2} - W_1^{1/2} \right)} \tag{6.65}$$

Note from Eq. (6.65) that to obtain maximum range for a jet airplane, we want:

1. Minimum thrust-specific fuel consumption c_t.
2. Maximum fuel weight W_f.
3. Flight at maximum $C_L^{1/2}/C_D$. This confirms our above argument in part A that for maximum range, a jet must fly at a velocity such that $C_L^{1/2}/C_D$ is maximum.
4. Flight at high altitudes, i.e., low ρ_∞. Of course, Eq. (6.65) says that R becomes infinite as ρ_∞ decreases to zero, i.e., as we approach outer space. This is physically ridiculous, however, because an airplane requires the atmosphere to generate lift and thrust. Long before outer space is reached, the assumptions behind Eq. (6.65) break down. Moreover, at extremely high altitudes, ordinary turbojet performance deteriorates and c_t begins to increase. All we can conclude from Eq. (6.65) is that range for a jet is poorest at sea level and increases with altitude, up to a point. Typical cruising altitudes for subsonic commercial jet transports are from 30,000 to 40,000 ft; for supersonic transports they are from 50,000 to 60,000 ft.

Example 6.10 Estimate the maximum range and endurance for the CJ-1.

SOLUTION From the calculations of Example 6.1, the variation of C_L/C_D and $C_L^{1/2}/C_D$ can be plotted vs. velocity, as given in Figure 6.39. From these curves, for the CJ-1,

$$\max\frac{C_L^{1/2}}{C_D} = 23.4$$

$$\max\frac{C_L}{C_D} = 16.9$$

In Example 6.1, the specific fuel consumption is given as TSFC = 0.6 lb of fuel per pound of thrust per hour. In consistent units,

$$c_t = 0.6\frac{\text{lb}}{\text{(lb)(h)}}\frac{1\text{ h}}{3600\text{ s}}$$

$$= 1.667\times10^{-4}\text{s}^{-1}$$

Also, the gross weight is 19,815 lb. The fuel capacity is 1119 gallons of kerosene, where 1 gallon of kerosene weighs 6.67 lb. Thus, $W_f = 1119(6.67) = 7463$ lb. Hence, the empty weight is $W_1 = W_0 - W_f = 19{,}815 - 7463 = 12{,}352$ lb.

The range of a jet depends on altitude, as shown by Eq. (6.65). Assume the cruising altitude is 22,000 ft. where $\rho_\infty = 0.00184$ slug/ft^3. From Eq. (6.65), using information from Example 6.1,

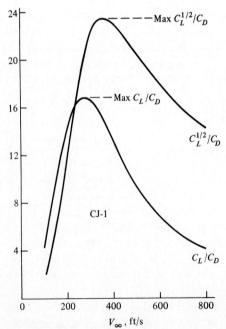

Figure 6.39 Aerodynamic ratios for the CJ-1 at sea level.

we obtain

$$R = 2\sqrt{\frac{2}{\rho_\infty S}} \frac{1}{c_t} \frac{C_L^{1/2}}{C_D} \left(W_0^{1/2} - W_1^{1/2} \right)$$

$$= 2\sqrt{\frac{2}{(0.001184)(318)}} \left(\frac{1}{1.667 \times 10^{-4}} \right) (23.4)(19{,}815^{1/2} - 12{,}352^{1/2})$$

$$\boxed{R = 19.2 \times 10^6 \text{ ft}}$$

In miles $\quad R = \dfrac{19.2 \times 10^6}{5280} = \boxed{3630 \text{ mi}}$

The endurance can be found from Eq. (6.60):

$$E = \frac{1}{c_t} \frac{C_L}{C_D} \ln \frac{W_0}{W_1}$$

$$= \frac{1}{1.667 \times 10^{-4}} (16.9) \ln \frac{19{,}815}{12{,}352}$$

or in hours $\quad \boxed{E = 4.79 \times 10^4 \text{ s}}$

$$E = \frac{4.79 \times 10^4}{3600} = \boxed{13.3 \text{ h}}$$

6.13 RELATIONS BETWEEN $C_{D,0}$ AND $C_{D,i}$

In the previous sections, we have observed that various aspects of the performance of different types of airplanes depend on the aerodynamic ratios $C_L^{1/2}/C_D$, C_L/C_D, or $C_L^{3/2}/C_D$. Moreover, in Sec. 6.3, we proved that at minimum T_R, drag due to lift equals parasite drag, that is, $C_{D,0} = C_{D,i}$. Analogously, for minimum P_R, we proved in Sec. 6.5 that $C_{D,0} = \frac{1}{3}C_{D,i}$. In this section, such results will be obtained strictly from aerodynamic considerations. The relations between $C_{D,0}$ and $C_{D,i}$ depend purely on the conditions for maximum $C_L^{1/2}/C_D$, C_L/C_D, or $C_L^{3/2}/C_D$; their derivations do not have to be associated with minimum T_R or P_R, as they were in Secs. 6.3 and 6.5.

For example, consider maximum L/D. Recalling that $C_D = C_{D,0} + C_L^2/\pi e \text{AR}$, we can write

$$\frac{C_L}{C_D} = \frac{C_L}{C_{D,0} + C_L^2/\pi e \text{AR}} \tag{6.66}$$

For maximum C_L/C_D, differentiate Eq. (6.66) with respect to C_L and set the result equal to zero:

$$\frac{d(C_L/C_D)}{dC_L} = \frac{C_{D,0} + C_L^2/\pi e\text{AR} - C_L(2C_L/\pi e\text{AR})}{\left(C_{D,0} + C_L^2/\pi e\text{AR}\right)^2} = 0$$

Thus,
$$C_{D,0} + \frac{C_L^2}{\pi e\text{AR}} - \frac{2C_L^2}{\pi e\text{AR}} = 0$$

or
$$C_{D,0} = \frac{C_L^2}{\pi e\text{AR}}$$

$$\boxed{C_{D,0} = C_{D,i}} \qquad \text{for } \left(\frac{C_L}{C_D}\right)_{\max} \tag{6.67}$$

Hence, Eq. (6.67), which is identical to Eq. (6.21), simply stems from the fact that L/D is maximum. The fact that it also corresponds to minimum T_R is only because T_R happens to be minimum when L/D is maximum.

Now consider maximum $C_L^{3/2}/C_D$. By setting $d(C_L^{3/2})/dC_L = 0$, a derivation similar to that above yields

$$\boxed{C_{D,0} = \tfrac{1}{3}C_{D,i}} \qquad \text{for } \left(\frac{C_L^{3/2}}{C_D}\right)_{\max} \tag{6.68}$$

Again, Eq. (6.68), which is identical to Eq. (6.29), simply stems from the fact that $C_L^{3/2}/C_D$ is maximum. The fact that it also corresponds to minimum P_R is only because P_R happens to be minimum when $C_L^{3/2}/C_D$ is maximum.

Similarly, when $C_L^{1/2}/C_D$ is maximum, setting $d(C_L^{1/2})/dC_L = 0$ yields

$$\boxed{C_{D,0} = 3C_{D,i}} \qquad \text{for } \left(\frac{C_L^{1/2}}{C_D}\right)_{\max} \tag{6.69}$$

You should not take Eqs. (6.68) and (6.69) for granted; derive them yourself.

We stated in Example 6.9 that the *maximum values* of $C_L^{1/2}/C_D$, C_L/C_D, and $C_L^{3/2}/C_D$ are independent of altitude, velocity, etc.; rather, they depend only on the aerodynamic design of the aircraft. The results of this section allow us to prove this statement, as follows.

First, consider again the case of maximum C_L/C_D. From Eq. (6.67),

$$C_{D,0} = C_{D,i} = \frac{C_L^2}{\pi e\text{AR}} \tag{6.70}$$

Thus,
$$C_L = \sqrt{\pi e\text{AR}C_{D,0}} \tag{6.71}$$

Substituting Eqs. (6.70) and (6.71) into Eq. (6.66), we obtain

$$\frac{C_L}{C_D} = \frac{C_L}{2C_L^2/\pi e\text{AR}} = \frac{\pi e\text{AR}}{2C_L} = \frac{\pi e\text{AR}}{2\sqrt{\pi e\text{AR}C_{D,0}}} \tag{6.72}$$

Hence, the value of the maximum C_L/C_D is obtained from Eq. (6.72) as

$$\left(\frac{C_L}{C_D}\right)_{\text{max}} = \frac{\left(\pi e \text{AR} C_{D,0}\right)^{1/2}}{2C_{D,0}} \tag{6.73}$$

Note from Eq. (6.73) that $(C_L/C_D)_{\text{max}}$ depends only on e, AR, and $C_{D,0}$, which are aerodynamic design parameters of the airplane. In particular, $(C_L/C_D)_{\text{max}}$ does not depend on altitude. However, note from Figures 6.37 and 6.39 that maximum C_L/C_D occurs at a certain velocity, and the velocity at which $(C_L/C_D)_{\text{max}}$ is obtained *does* change with altitude.

In the same vein, it is easily shown that

$$\left(\frac{C_L^{1/2}}{C_D}\right)_{\text{max}} = \frac{\left(\frac{1}{3}C_{D,0}\pi e \text{AR}\right)^{1/4}}{\frac{4}{3}C_{D,0}} \tag{6.74}$$

and

$$\left(\frac{C_L^{3/2}}{C_D}\right)_{\text{max}} = \frac{\left(3C_{D,0}\pi e \text{AR}\right)^{3/4}}{4C_{D,0}} \tag{6.75}$$

Prove this yourself.

Example 6.11 From the equations given in this section, directly calculate $(C_L/C_D)_{\text{max}}$ and $(C_L^{3/2}/C_D)_{\text{max}}$ for the CP-1.

SOLUTION From Eq. (6.73),

$$\left(\frac{C_L}{C_D}\right)_{\text{max}} = \frac{\left(\pi e \text{AR} C_{D,0}\right)^{1/2}}{2C_{D,0}}$$

$$= \frac{\left[\pi(0.8)(7.37)(0.025)\right]^{1/2}}{(2)(0.025)} = \boxed{13.6}$$

From Eq. (6.75),

$$\left(\frac{C_L^{3/2}}{C_D}\right)_{\text{max}} = \frac{\left(3C_{D,0}\pi e \text{AR}\right)^{3/4}}{4C_{D,0}}$$

$$= \frac{\left[(3)(0.025)\pi(0.8)(7.37)\right]^{3/4}}{(4)(0.025)} = \boxed{12.8}$$

Return to Example 6.9, where the values of $(C_L/C_D)_{\text{max}}$ and $(C_L^{3/2}/C_D)_{\text{max}}$ were obtained graphically, i.e., by plotting C_L/C_D and $C_L^{3/2}/C_D$ and finding their peak values. Note that the results obtained from Eqs. (6.73) and (6.75) agree with the graphical values obtained in Example 6.9 (as they should); however, the use of Eqs. (6.73) and (6.75) is much easier and quicker than plotting a series of numbers and finding the maximum.

Example 6.12 From the equations given in this section, directly calculate $(C_L^{1/2}/C_D)_{\text{max}}$ and $(C_L/C_D)_{\text{max}}$ for the CJ-1.

SOLUTION From Eq. (6.74),

$$\left(\frac{C_L^{1/2}}{C_D}\right)_{\max} = \frac{\left(\frac{1}{3}C_{D,0}\pi e\mathrm{AR}\right)^{1/4}}{\frac{4}{3}C_{D,0}}$$

$$= \frac{\left[\frac{1}{3}(0.02)\pi(0.81)(8.93)\right]^{1/4}}{\frac{4}{3}(0.02)} = \boxed{23.4}$$

From Eq. (6.73)

$$\left(\frac{C_L}{C_D}\right)_{\max} = \frac{(\pi e\mathrm{AR}C_{D,0})^{1/2}}{2C_{D,0}}$$

$$= \frac{\left[\pi(0.81)(8.93)(0.02)\right]^{1/2}}{(2)(0.02)} = \boxed{16.9}$$

These values agree with the graphically obtained maximums in Example 6.10.

6.14 TAKEOFF PERFORMANCE

To this point in our discussion of airplane performance, we have assumed that all accelerations are zero, i.e., we have dealt with aspects of *static* performance as defined in Sec. 6.2. For the remainder of this chapter, we will relax this restriction and consider several aspects of airplane performance that involve finite acceleration, such as takeoff and landing runs, turning flight, and accelerated rate of climb.

To begin with, we ask the question, what is the running length along the ground required by an airplane, starting from zero velocity, to gain flight speed and lift from the ground? This length is defined as the ground roll, or lift-off distance, s_{LO}.

To address this question, let us first consider the accelerated rectilinear motion of a body of mass m experiencing a constant force F, as sketched in Figure 6.40. From Newton's second law,

$$F = ma = m\frac{dV}{dt}$$

or

$$dV = \frac{F}{m}\,dt \tag{6.76}$$

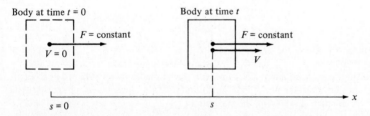

Body at time $t = 0$ Body at time t

F = constant $V = 0$ F = constant V

$s = 0$ s x

Figure 6.40 Sketch of a body moving under the influence of a constant force F, starting from rest ($V = 0$) at $s = 0$ and accelerating to velocity V at distance s.

Assume that the body starts from rest ($V = 0$) at location $s = 0$ at time $t = 0$ and is accelerated to velocity V over the distance s at time t. Integrating Eq. (6.76) between these two points, and remembering that both F and m are constant, we have

$$\int_0^V dV = \frac{F}{m} \int_0^t dt$$

or

$$V = \frac{F}{m} t \tag{6.77}$$

Solving for t,

$$t = \frac{Vm}{F} \tag{6.78}$$

Considering an instant when the velocity is V, the incremental distance ds covered during an incremental time dt is $ds = V\, dt$. From Eq. (6.77), we have

$$ds = V\, dt = \frac{F}{m} t\, dt \tag{6.79}$$

Integrating Eq. (6.79),

$$\int_0^s ds = \frac{F}{m} \int_0^t t\, dt$$

or

$$s = \frac{F}{m} \frac{t^2}{2} \tag{6.80}$$

Substituting Eq. (6.78) into (6.80), we obtain

$$\boxed{s = \frac{V^2 m}{2F}} \tag{6.81}$$

Equation (6.81) gives the distance required for a body of mass m to accelerate to velocity V under the action of a constant force F.

Now consider the force diagram for an airplane during its ground roll, as illustrated in Figure 6.41. In addition to the familiar forces of lift, drag, thrust,

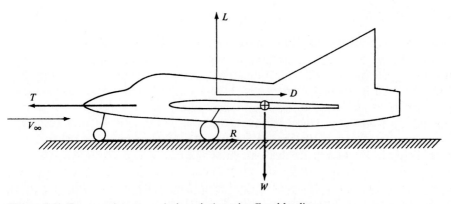

Figure 6.41 Forces acting on an airplane during takeoff and landing.

and weight, the airplane also experiences a resistance force R due to rolling friction between the tires and the ground. This resistance force is given by

$$R = \mu_r(W - L) \tag{6.82}$$

where $W - L$ is the net normal force exerted between the tires and the ground and μ_r is the coefficient of rolling friction. Summing forces parallel to the ground, and employing Newton's second law, we have

$$F = T - D - R = T - D - \mu_r(W - L) = m\frac{dV}{dt} \tag{6.83}$$

Let us examine Eq. (8.63) more closely. It gives the local instantaneous acceleration of the airplane, dV/dt, as a function of T, D, W, and L. For takeoff, over most of the ground roll, T is reasonably constant (this is particularly true for a jet-powered airplane). Also, W is constant. However, both L and D vary with velocity, since

$$L = \tfrac{1}{2}\rho_\infty V_\infty^2 S C_L \tag{6.84}$$

and

$$D = \tfrac{1}{2}\rho_\infty V_\infty^2 S\left(C_{D,0} + \phi\frac{C_L^2}{\pi e \text{AR}}\right) \tag{6.85}$$

The quantity of ϕ in Eq. (6.85) requires some explanation. When an airplane is flying close to the ground, the strength of the wingtip vortices is somewhat diminished due to interaction with the ground. Since these tip vortices induce downwash at the wing (see Sec. 5.13), which in turn generates induced drag (see Sec. 5.14), the downwash and hence induced drag are reduced when the airplane is flying close to the ground. This phenomenon is called *ground effect* and is the cause of the tendency for an airplane to flare, or "float," above the ground near the instant of landing. The reduced drag in the presence of ground effect is accounted for by ϕ in Eq. (6.85), where $\phi \leq 1$. An approximate expression for ϕ, based on aerodynamic theory, is given by McCormick (see Bibliography at the end of this chapter) as

$$\phi = \frac{(16\,h/b)^2}{1 + (16\,h/b)^2} \tag{6.86}$$

where h is the height of the wing above the ground and b is the wingspan.

In light of the above, to accurately calculate the variation of velocity with time during the ground roll, and ultimately the distance required for lift-off, Eq. (6.83) must be integrated numerically, taking into account the proper velocity variations of L and D from Eqs. (6.84) and (6.85), respectively, as well as any velocity effect on T. A typical variation of these forces with distance along the ground during takeoff is sketched in Figure 6.42. Note from Eq. (6.81) that s is proportional to V^2, and hence the horizontal axis in Figure 6.42 could just as well be V^2. Since both D and L are proportional to the dynamic pressure, $q_\infty = \tfrac{1}{2}\rho_\infty V_\infty^2$, they appear as linear variations in Figure 6.42. Also, Figure 6.42 is drawn for a jet-propelled airplane; hence T is relatively constant.

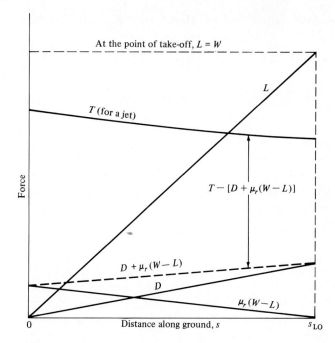

At the point of take-off, $L = W$

T (for a jet)

L

$T - [D + \mu_r(W - L)]$

$D + \mu_r(W - L)$

D

$\mu_r(W - L)$

Force

0   Distance along ground, s s_{LO}

Figure 6.42 Schematic of a typical variation of forces acting on an airplane during takeoff.

A simple but approximate expression for the lift-off distance s_{LO} can be obtained as follows. Assume that T is constant. Also, assume an *average value* for the sum of drag and resistance force, $[D + \mu_r(W - L)]_{ave}$, such that this average value, taken as a constant force, produces the proper lift-off distance s_{LO}. Then, we consider an effective constant force acting on the airplane during its take-off ground roll as

$$F_{eff} = T - [D + \mu_r(W - L)]_{ave} = \text{const} \qquad (6.87)$$

These assumptions are fairly reasonable, as seen from Figure 6.42. Note that the sum of $D + \mu_r(W - L)$ versus distance (or V^2) is reasonably constant, as shown by the dashed line in Figure 6.42. Hence, the accelerating force, $T - [D + \mu_r(W - L)]$, which is illustrated by the difference between the thrust curve and the dashed line in Figure 6.42, is also reasonably constant. Now return to Eq. (6.81). Considering F given by Eq. (6.87), $V = V_{LO}$ (the lift-off velocity), and $m = W/g$, where g is the acceleration of gravity, Eq. (6.81) yields

$$s_{LO} = \frac{(V^2_{LO})(W/g)}{2\{T - [D + \mu_r(W - L)]_{ave}\}} \qquad (6.88)$$

In order to ensure a margin of safety during takeoff, the lift-off velocity is typically 20 percent higher than stalling velocity. Hence, from Eq. (5.56), we have

$$V_{LO} = 1.2V_{stall} = 1.2\sqrt{\frac{2W}{\rho_\infty S C_{L,max}}} \tag{6.89}$$

Substituting Eq. (6.89) into (6.88), we obtain

$$s_{LO} = \frac{1.44W^2}{g\rho_\infty S C_{L,max}\{T - [D + \mu_r(W - L)]_{ave}\}} \tag{6.90}$$

In order to make a calculation using Eq. (6.90), Shevell (see Bibliography at the end of this chapter) suggests that the average force in Eq. (6.90) be set equal to its instantaneous value at a velocity equal to 0.7 V_{LO}, that is,

$$[D + \mu_r(W - L)]_{ave} = [D + \mu_r(W - L)]_{0.7 V_{LO}}$$

Also, experience has shown that the coefficient of rolling friction, μ_r, in Eq. (6.90) varies from 0.02 for a relatively smooth paved surface to 0.10 for a grass field.

A further simplification can be obtained by assuming that thrust is much larger than either D or R during takeoff. Referring to the case shown in Figure 6.42, this simplification is not unreasonable. Hence, ignoring D and R compared to T, Eq. (6.90) becomes simply

$$s_{LO} = \frac{1.44W^2}{g\rho_\infty S C_{L,max}T} \tag{6.91}$$

Equation (6.91) illustrates some important physical trends, as follows:

1. Lift-off distance is very sensitive to the weight of the airplane, varying directly as W^2. If the weight is doubled, the ground roll of the airplane is quadrupled.
2. Lift-off distance is dependent on the ambient density ρ_∞. If we assume that thrust is directly proportional to ρ_∞, as stated in Sec. 6.7, that is, $T \propto \rho_\infty$, then Eq. (6.91) demonstrates that

$$s_{LO} \propto \frac{1}{\rho_\infty^2}$$

This is why on hot summer days, when the air density is less than on cooler days, a given airplane requires a longer ground roll to get off the ground. Also, longer lift-off distances are required at airports which are located at higher altitudes (such as at Denver, Colorado, a mile above sea level).
3. The lift-off distance can be decreased by increasing the wing area, increasing $C_{L,max}$, and increasing the thrust, all of which simply make common sense.

The total takeoff distance, as defined in the Federal Aviation Requirements (FAR), is the sum of the ground roll distance s_{LO} and the distance (measured along the ground) to clear a 35-ft height (for jet-powered civilian transports) or a 50-ft height (for all other airplanes). A discussion of these requirements, as well as more details regarding the total takeoff distance, is beyond the scope of this book. See the books by Shevell and McCormick listed in the Bibliography at the end of this chapter for more information on this topic.

Example 6.13 Estimate the lift-off distance for the CJ-1 at sea level. Assume a paved runway, hence $\mu_r = 0.02$. Also, during the ground roll, the angle of attack of the airplane is restricted by the requirement that the tail not drag the ground, and therefore assume that $C_{L,\text{max}}$ during ground roll is limited to 1.0. Also, when the airplane is on the ground, the wings are 6 ft above the ground.

SOLUTION Use Eq. (6.90). In order to evaluate the average force in Eq. (6.90), first obtain the ground effect factor from Eq. (6.86), where $h/b = 6/53.3 = 0.113$.

$$\phi = \frac{(16\, h/b)^2}{1 + (16\, h/b)^2} = 0.764$$

Also, from Eq. (6.89),

$$V_{LO} = 1.2 V_{\text{stall}} = 1.2 \sqrt{\frac{2W}{\rho_\infty S C_{L,\text{max}}}}$$

$$= 1.2 \sqrt{\frac{2(19,815)}{0.002377(318)(1.0)}} = 230 \text{ ft/s}$$

Hence, $0.7\, V_{LO} = 160.3$ ft/s. The average force in Eq. (6.90) should be evaluated at a velocity of 160.3 ft/s. To do this, from Eq. (6.84) we get

$$L = \tfrac{1}{2} \rho_\infty V_\infty^2 S C_L$$

$$= \tfrac{1}{2}(0.002377)(160.3)^2(318)(1.0) = 9712 \text{ lb}$$

Equation (6.85) yields

$$D = \tfrac{1}{2} \rho_\infty V_\infty^2 S \left(C_{D,0} + \phi \frac{C_L^2}{\pi e \text{AR}} \right)$$

$$= \tfrac{1}{2}(0.002377)(160.3)^2(318) \left[0.02 + 0.764 \left(\frac{1.0^2}{\pi(0.81)(8.93)} \right) \right]$$

$$= 520.7 \text{ lb}$$

Finally, from Eq. (6.90),

$$s_{LO} = \frac{1.44 W^2}{g \rho_\infty S C_{L,\text{max}} \left\{ T - [D + \mu_r (W - L)]_{\text{ave}} \right\}}$$

$$= \frac{1.44(19815)^2}{32.2(0.002377)(318)(1.0)\{7300 - [520.7 + (0.02)(19,815 - 9712)]\}}$$

$$= \boxed{3532 \text{ ft}}$$

Note that $[D + \mu_r(W - L)]_{\text{ave}} = 722.8$ lb, which is about 10 percent of the thrust. Hence, the assumption leading to Eq. (6.91) is fairly reasonable, i.e., that D and R can sometimes be ignored compared with T.

6.15 LANDING PERFORMANCE

Consider an airplane during landing. After the airplane has touched the ground, the force diagram during the ground roll is exactly the same as that given in Figure 6.41, and the instantaneous acceleration (negative in this case) is given by Eq. (6.83). However, we assume that in order to minimize the distance required to come to a complete stop, the pilot has decreased the thrust to zero at touchdown, and therefore the equation of motion for the landing ground roll is obtained from Eq. (6.83) with $T = 0$.

$$-D - \mu_r(W - L) = m\frac{dV}{dt} \tag{6.92}$$

A typical variation of the forces on the airplane during landing is sketched in Figure 6.43. Designate the ground roll distance between touchdown at velocity V_T and a complete stop by s_L. An accurate calculation of s_L can be obtained by numerically integrating Eq. (6.92) along with Eqs. (6.84) and (6.85).

However, let us develop an approximate expression for s_L which parallels the philosophy used in Sec. 6.14. Assume an average constant value for $D + \mu_r(W - L)$ which effectively yields the correct ground roll distance at landing, s_L. Once again, we can assume that $[D + \mu_r(W - L)]_{ave}$ is equal to its instantaneous value evaluated at $0.7V_T$.

$$F = -[D + \mu_r(W - L)]_{ave} = -[D + \mu_r(W - L)]_{0.7V_T} \tag{6.93}$$

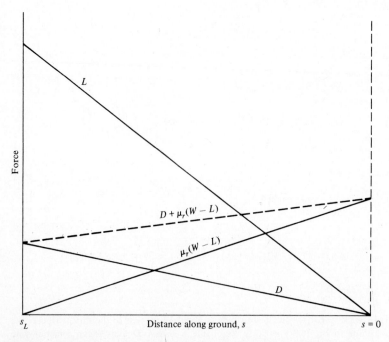

Figure 6.43 Schematic of a typical variation of forces acting on an airplane during landing.

(Note from Figure 6.43 that the net decelerating force, $D + \mu_r(W - L)$, can vary considerably with distance, as shown by the dashed line. Hence, our assumption here for landing is more tenuous than for takeoff.) Returning to Eq. (6.79), integrate between the touchdown point, where $s = s_L$ and $t = 0$, and the point where the airplane's motion stops, where $s = 0$ and time equals t.

$$\int_{s_L}^{0} ds = \frac{F}{m} \int_{0}^{t} t \, dt$$

or
$$s_L = -\frac{F}{m} \frac{t^2}{2} \tag{6.94}$$

Note that, from Eq. (6.93), F is a negative value; hence s_L in Eq. (6.94) is positive.

Combining Eqs. (6.78) and (6.94), we obtain

$$s_L = -\frac{V^2 m}{2F} \tag{6.95}$$

Equation (6.95) gives the distance required to decelerate from an initial velocity V to zero velocity under the action of a constant force F. In Eq. (6.95), F is given by Eq. (6.93), and V is V_T. Thus, Eq. (6.95) becomes

$$s_L = \frac{V_T^2(W/g)}{2[D + \mu_r(W - L)]_{0.7V_T}} \tag{6.96}$$

In order to maintain a factor of safety,

$$V_T = 1.3V_{\text{stall}} = 1.3\sqrt{\frac{2W}{\rho_\infty S C_{L,\text{max}}}} \tag{6.97}$$

Substituting Eq. (6.97) into (6.96), we obtain

$$s_L = \frac{1.69W^2}{g\rho_\infty S C_{L,\text{max}}[D + \mu_r(W - L)]_{0.7V_T}} \tag{6.98}$$

During the landing ground roll, the pilot is applying brakes; hence, in Eq. (6.98) the coefficient of rolling friction is that during braking, which is approximately $\mu_r = 0.4$ for a paved surface.

Modern jet transports utilize thrust reversal during the landing ground roll. Thrust reversal is created by ducting air from the jet engines and blowing it in the upstream direction, opposite to the usual downstream direction when normal thrust is produced. As a result, with thrust reversal, the thrust vector in Figure 6.41 is reversed and points in the drag direction, thus aiding the deceleration and shortening the ground roll. Designating the reversed thrust by T_R, Eq. (6.92) becomes

$$-T_R - D - \mu_r(W - L) = m\frac{dV}{dt} \tag{6.99}$$

Assuming that T_R is constant, Eq. (6.98) becomes

$$s_L = \frac{1.69W^2}{g\rho_\infty SC_{L,\max}\{T_R + [D + \mu_r(W - L)]_{0.7V_T}\}} \tag{6.100}$$

Another ploy to shorten the ground roll is to decrease the lift to near zero, hence impose the full weight of the airplane between the tires and the ground and increase the resistance force due to friction. The lift on an airplane wing can be destroyed by spoilers, which are simply long, narrow surfaces along the span of the wing, deflected directly into the flow, thus causing massive flow separation and a striking decrease in lift.

The total landing distance, as defined in the FAR, is the sum of the ground roll distance plus the distance (measured along the ground) to achieve touchdown in a glide from a 50-ft height. Such details are beyond the scope of this book; see the books by Shevell and McCormick (listed in the Bibliography at the end of this chapter) for more information.

Example 6.14 Estimate the landing ground roll distance at sea level for the CJ-1. No thrust reversal is used; however, spoilers are employed such that $L = 0$. The spoilers increase the parasite drag coefficient by 10 percent. The fuel tanks are essentially empty, so neglect the weight of any fuel carried by the airplane. The maximum lift coefficient, with flaps fully employed at touchdown, is 2.5.

SOLUTION The empty weight of the CJ-1 is 12,352 lb. Hence,

$$V_T = 1.3V_{\text{stall}} = 1.3\sqrt{\frac{2W}{\rho_\infty SC_{L,\max}}}$$

$$= 1.3\sqrt{\frac{2(12,352)}{0.002377(318)(2.5)}} = 148.6 \text{ ft/s}$$

Thus, $0.7V_T = 104$ ft/s. Also, $C_{D,0} = 0.02 + 0.1(0.02) = 0.022$. From Eq. (6.85), with $C_L = 0$ (remember, spoilers are employed, destroying the lift),

$$D = \tfrac{1}{2}\rho_\infty V_\infty^2 SC_{D,0} = \tfrac{1}{2}(0.002377)(104)^2(318)(0.022) = 89.9 \text{ lb}$$

From Eq. (6.98), with $L = 0$,

$$s_L = \frac{1.69W^2}{g\rho_\infty SC_{L,\max}(D + \mu_r W)_{0.7V_T}}$$

$$= \frac{1.69(12,352)^2}{32.2(0.002377)(318)(2.5)[89.9 + (0.4)(12352)]}$$

$$= \boxed{842 \text{ ft}}$$

6.16 TURNING FLIGHT AND THE V-n DIAGRAM

To this point in our discussion of airplane performance, we have considered rectilinear motion. Our static performance analyses dealt with zero acceleration leading to constant velocity along straight-line paths. Our discussion of takeoff and landing performance involved rectilinear acceleration, also leading to motion along a straight-line path. Let us now consider some cases involving *radial* acceleration, which leads to *curved* flight paths; i.e., let us consider the turning flight of an airplane. In particular, we will examine three specialized cases: (1) a level turn, (2) a pullup, and (3) a pulldown. A study of the generalized motion of an airplane along a three-dimensional flight path is beyond the scope of this book.

A level turn is illustrated in Figure 6.44. Here, the wings of the airplane are banked through the angle ϕ; hence the lift vector is inclined at the angle ϕ to the vertical. The bank angle ϕ and the lift L are such that the component of the lift in the vertical direction exactly equals the weight:

$$L \cos \phi = W$$

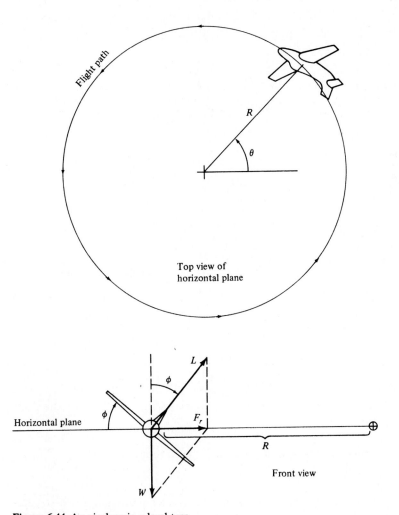

Figure 6.44 An airplane in a level turn.

and therefore the airplane maintains a constant altitude, moving in the same horizontal plane. However, the resultant of L and W leads to a resultant force F_r, which acts in the horizontal plane. This resultant force is perpendicular to the flight path, causing the airplane to turn in a circular path with a radius of curvature equal to R. We wish to study this turn radius R, as well as the turn rate $d\theta/dt$.

From the force diagram in Figure 6.44, the magnitude of the resultant force is

$$F_r = \sqrt{L^2 - W^2} \tag{6.101}$$

We introduce a new term, the *load factor n*, defined as

$$n \equiv \frac{L}{W} \tag{6.102}$$

The load factor is usually quoted in terms of "g's"; for example, an airplane with lift equal to five times the weight is said to be experiencing a load factor of 5 g's. Hence, Eq. (6.101) can be written as

$$F_r = W\sqrt{n^2 - 1} \tag{6.103}$$

The airplane is moving in a circular path at the velocity V_∞. Therefore, the radial acceleration is given by V_∞^2/R. From Newton's second law

$$F_r = m\frac{V_\infty^2}{R} = \frac{W}{g}\frac{V_\infty^2}{R} \tag{6.104}$$

Combining Eqs. (6.103) and (6.104) and solving for R, we have

$$\boxed{R = \frac{V_\infty^2}{g\sqrt{n^2 - 1}}} \tag{6.105}$$

The angular velocity, denoted by $\omega \equiv d\theta/dt$, is called the turn rate and is given by V_∞/R. Thus, from Eq. (6.105), we have

$$\boxed{\omega = \frac{g\sqrt{n^2 - 1}}{V_\infty}} \tag{6.106}$$

For the maneuvering performance of an airplane, both military and civil, it is frequently advantageous to have the smallest possible R and the largest possible ω. Equations (6.105) and (6.106) show that to obtain both a small turn radius and a large turn rate, we want:

1. The highest possible load factor (i.e., the highest possible L/W)
2. The lowest possible velocity

Consider another case of turning flight, where an airplane initially in straight level flight (where $L = W$) suddenly experiences an increase in lift. Since $L > W$,

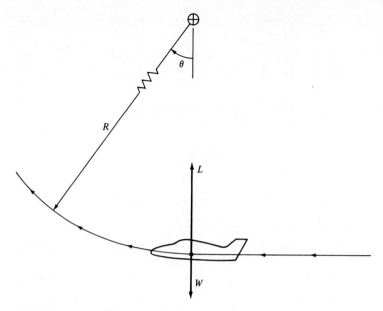

Figure 6.45 The pullup maneuver.

the airplane will begin to turn upward, as sketched in Figure 6.45. For this pullup maneuver, the flight path becomes curved in the vertical plane, with a turn rate $\omega = d\theta/dt$. From the force diagram in Figure 6.45, the resultant force F_r is vertical and is given by

$$F_r = L - W = W(n - 1) \tag{6.107}$$

From Newton's second law

$$F_r = m\frac{V_\infty^2}{R} = \frac{W}{g}\frac{V_\infty^2}{R} \tag{6.108}$$

Combining Eqs. (6.107) and (6.108) and solving for R,

$$R = \frac{V_\infty^2}{g(n - 1)} \tag{6.109}$$

and since $\omega = V_\infty/R$,

$$\omega = \frac{g(n - 1)}{V_\infty} \tag{6.110}$$

A related case is the pulldown maneuver, illustrated in Figure 6.46. Here, an airplane in initially level flight suddenly rolls to an inverted position, such that both L and W are pointing downward. The airplane will begin to turn downward

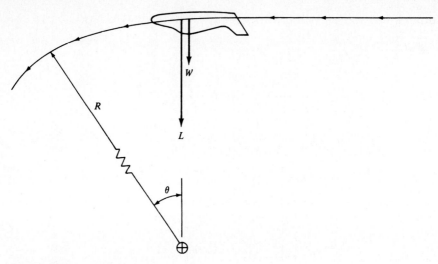

Figure 6.46 The pulldown maneuver.

in a circular flight path with a turn radius R and turn rate $\omega = d\theta/dt$. By an analysis similar to those above, the following results are easily obtained:

$$R = \frac{V_\infty^2}{g(n+1)} \tag{6.111}$$

$$\omega = \frac{g(n+1)}{V_\infty} \tag{6.112}$$

Prove this to yourself.

Considerations of turn radius and turn rate are particularly important to military fighter aircraft; everything else being equal, those airplanes with the smallest R and largest ω will have definite advantages in air combat. High-performance fighter aircraft are designed to operate at high load factors, typically from 3 to 10. When n is large, then $n + 1 \approx n$ and $n - 1 \approx n$; for such cases, Eqs. (6.105), (6.106), and (6.109) to (6.112) reduce to

$$R = \frac{V_\infty^2}{gn} \tag{6.113}$$

and

$$\omega = \frac{gn}{V_\infty} \tag{6.114}$$

Let us work with these equations further. Since

$$L = \tfrac{1}{2}\rho_\infty V_\infty^2 S C_L$$

then

$$V_\infty^2 = \frac{2L}{\rho_\infty S C_L} \tag{6.115}$$

Substituting Eqs. (6.115) and (6.102) into Eqs. (6.113) and (6.114), we obtain

$$R = \frac{2L}{\rho_\infty S C_L g(L/W)} = \frac{2}{\rho_\infty C_L g}\frac{W}{S} \tag{6.116}$$

and

$$\omega = \frac{gn}{\sqrt{(2L/\rho_\infty S C_L)(W/S)}}$$

$$= \frac{gn}{\sqrt{(2n/\rho_\infty C_L)(W/S)}} = g\sqrt{\frac{\rho_\infty C_L n}{2(W/S)}} \tag{6.117}$$

Note that in Eqs. (6.116) and (6.117) the factor W/S appears. This factor occurs frequently in airplane performance analyses and is labeled as

$$\frac{W}{S} \equiv \text{wing loading}$$

Equations (6.116) and (6.117) clearly show that airplanes with lower wing loadings will have smaller turn radii and larger turn rates, everything else being equal. However, the design wing loading of an airplane is usually determined by factors other than maneuvering, such as payload, range, and maximum velocity. As a result, wing loadings for light, general aviation aircraft are relatively low, but those for high-performance military aircraft are relatively large. Wing loadings for some typical airplanes are listed below.

Airplane	W/S, lb/ft^2
Wright Flyer (1903)	1.2
Beechcraft Bonanza	18.8
McDonnell Douglas F-15	66
General Dynamics F-16	74

From the above, we conclude that a small, light aircraft such as the Beechcraft Bonanza can outmaneuver a larger, heavier aircraft such as the F-16 because of smaller turn radius and larger turn rate. However, this is really comparing apples and oranges. Instead, let us examine Eqs. (6.116) and (6.117) for a *given* airplane with a given wing loading and ask the question, for this specific airplane, under what conditions will R be minimum and ω maximum? From these equations, clearly R will be minimum and ω will be maximum when both C_L and n are maximum. That is,

$$R_{\min} = \frac{2}{\rho_\infty g C_{L,\max}}\frac{W}{S} \tag{6.118}$$

$$\omega_{\max} = g\sqrt{\frac{\rho_\infty C_{L,\max} n_{\max}}{2(W/S)}} \tag{6.119}$$

Also note from Eqs. (6.118) and (6.119) that best performance will occur at sea level, where ρ_∞ is maximum.

There are some practical constraints on the above considerations. First, at low speeds, $n_{\max}$ is a function of $C_{L,\max}$ itself, because

$$n = \frac{L}{W} = \frac{\frac{1}{2}\rho_\infty V_\infty^2 S C_L}{W}$$

and hence $\qquad\qquad n_{\max} = \frac{1}{2}\rho_\infty V_\infty^2 \frac{C_{L,\max}}{W/S} \qquad\qquad\qquad (6.120)$

At higher speeds, $n_{\max}$ is limited by the structural design of the airplane. These considerations are best understood by examining Figure 6.47, which is a diagram showing load factor vs. velocity for a given airplane—*the V-n diagram.* Here, curve AB is given by Eq. (6.120). Consider an airplane flying at velocity V_1, where V_1 is shown in Figure 6.47. Assume that the airplane is at an angle of attack such that $C_L < C_{L,\max}$. This flight condition is represented by point 1 in Figure 6.47. Now assume that the angle of attack is increased to that for obtaining $C_{L,\max}$, keeping the velocity constant at V_1. The lift increases to its maximum value for the given V_1, and hence the load factor, $n = L/W$, reaches its maximum value, $n_{\max}$, for the given V_1. This value of $n_{\max}$ is given by Eq. (6.120), and the corresponding flight condition is given by point 2 in Figure 6.47. If the angle of attack is increased further, the wing stalls, and the load factor drops. Therefore, point 3 in Figure 6.46 is unobtainable in flight. Point 3 is in the "stall region" of the V-n diagram. Consequently, point 2 represents the highest possible load factor that can be obtained at the given velocity V_1. Now, as V_1 is increased, say to a value of V_4, then the maximum possible load factor $n_{\max}$ also increases, as given by point 4 in Figure 6.47 and as calculated from Eq. (6.120). However, $n_{\max}$ cannot be allowed to increase indefinitely. Beyond a certain value of load factor, defined as the positive limit load factor and shown as the horizontal line BC in Figure 6.47, structural damage may occur to the aircraft. The velocity corresponding to point B is designated as V^*. At velocities higher than V^*, say V_5, the airplane must fly at values of C_L less than $C_{L,\max}$ so that the positive limit load factor is not exceeded. If flight at $C_{L,\max}$ is obtained at velocity V_5, corresponding to point 5 in Figure 6.47, then structural damage will occur. The right-hand side of the V-n diagram, line CD, is a high-speed limit. At velocities greater than this, the dynamic pressure becomes so large that again structural damage may occur to the airplane. (This maximum velocity limit is, by design, much larger than the level-flight $V_{\max}$ calculated in Secs. 6.4 to 6.6. In fact, the structural design of most airplanes is such that the maximum velocity allowed by the V-n diagram is sufficiently greater than maximum diving velocity for the airplane.) Finally, the bottom part of the V-n diagram, given by curves AE and ED in Figure 6.47, corresponds to negative absolute angles of attack, i.e., negative load factors. Curve AE defines the stall limit. (At absolute angles of attack less than zero, the lift is negative and acts in the downward direction. If the wing is pitched downward to a large enough negative angle of attack, the flow will separate from the bottom surface of the wing and the downward-acting lift will decrease in magnitude; i.e., the wing "stalls.") Line ED gives the negative limit load factor, beyond which structural damage will occur.

As a final note concerning the V-n diagram, consider point B in Figure 6.47. This point is called the *maneuver point*. At this point, both C_L and n are simultaneously at their highest possible values that can be obtained anywhere throughout the allowable flight envelope of the aircraft. Consequently, from Eqs. (6.118) and (6.119), this point corresponds simultaneously to the smallest possible turn radius and the largest possible turn rate for the airplane. The velocity corresponding to point B is called the *corner velocity* and is designated by V^* in Figure 6.47. The corner velocity can be obtained by solving Eq. (6.120) for velocity, yielding

$$V^* = \sqrt{\frac{2n_{\max}}{\rho_\infty C_{L,\max}} \frac{W}{S}} \tag{6.121}$$

In Eq. (6.121), the value of $n_{\max}$ corresponds to that at point B in Figure 6.47. The corner velocity is an interesting dividing line. At flight velocities less than V^*, it is not possible to structurally damage the airplane due to the generation of too much lift. In contrast, at velocities greater than V^*, lift can be obtained that can structurally damage the aircraft (for example, point 5 in Figure 6.47), and the pilot must make certain to avoid such a case.

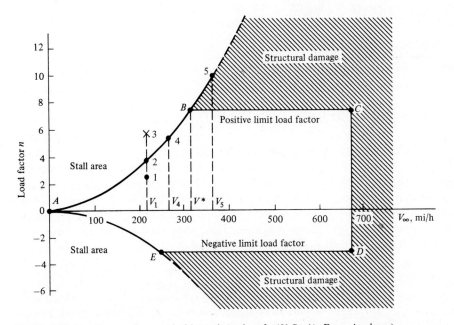

Figure 6.47 V-n diagram for a typical jet trainer aircraft. (*U.S. Air Force Academy.*)

Figure 6.48 General Dynamics F-16 in a 90° vertical accelerated climb. (*U.S. Air Force.*)

6.17 ACCELERATED RATE OF CLIMB (ENERGY METHOD)*

Modern high-performance airplanes, such as the supersonic General Dynamics F-16 shown in Figure 6.48, are capable of highly accelerated rates of climb. Therefore, the performance analysis of such airplanes requires methods that go beyond the static rate of climb considerations given in Secs. 6.8 to 6.10. The purpose of the present section is to introduce one such method, namely a method dealing with the *energy* of an airplane. This is in contrast to our previous discussions that have dealt explicitly with forces on the airplane.

*This section is based in part on material presented by the faculty of the department of aeronautics at the U.S. Air Force Academy at its annual aerodynamics workshop, held each July at Colorado Springs. This author has had the distinct privilege to participate in this workshop since its inception in 1979. Special thanks for this material go to Col. James D. Lang, Major Thomas Parrot, and Col. Daniel Daley.

Consider an airplane of mass m in flight at some altitude h and with some velocity V. Due to its altitude, the airplane has *potential energy* PE equal to mgh. Due to its velocity, the airplane has *kinetic energy* KE equal to $\frac{1}{2}mV^2$. The total energy of the airplane is the sum of these energies; i.e.,

$$\text{Total aircraft energy} = \text{PE} + \text{KE} = mgh + \tfrac{1}{2}mV^2 \qquad (6.122)$$

The energy per unit weight of the airplane is obtained by dividing Eq. (6.122) by $W = mg$. This yields the *specific energy*, denoted by H_e, as

$$H_e \equiv \frac{\text{PE} + \text{KE}}{W} = \frac{mgh + \frac{1}{2}mV^2}{mg}$$

or

$$\boxed{H_e = h + \frac{V^2}{2g}} \qquad (6.123)$$

The specific energy H_e has units of height and is therefore also called the *energy height* of the aircraft. Thus, let us become accustomed to quoting the energy of an airplane in terms of the energy height H_e, which is simply the sum of the potential and kinetic energies of the airplane per unit weight. Contours of constant H_e are illustrated in Figure 6.49, which is an "altitude–Mach number map." Here, the ordinate and abscissa are altitude h and Mach number M, respectively, and the dashed curves are lines of constant energy height.

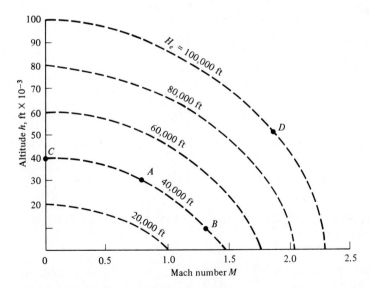

Figure 6.49 Altitude–Mach number map showing curves of constant-energy height. These are universal curves that represent the variation of kinetic and potential energies per unit weight. They do not depend on the specific design factors of a given airplane.

To obtain a feeling for the significance of Figure 6.49, consider two airplanes, one flying at an altitude of 30,000 ft at Mach 0.81 (point A in Figure 6.49) and the other flying at an altitude of 10,000 ft at Mach 1.3 (point B). Both airplanes have the same energy height, 40,000 ft (check this yourself by calculation). However, airplane A has more potential energy and less kinetic energy (per unit weight) than airplane B. If both airplanes maintain their same states of total energy, then both are capable of "zooming" to an altitude of 40,000 ft at zero velocity (point C) simply by trading all their kinetic energy for potential energy. Consider another airplane, flying at an altitude of 50,000 ft at Mach 1.85, denoted by point D in Figure 6.49. This airplane will have an energy height of 100,000 ft and is indeed capable of zooming to an actual altitude of 100,000 ft by trading all of its kinetic energy for potential energy. Airplane D is in a much higher energy state ($H_e = 100,000$ ft) than airplanes A and B (which have $H_e = 40,000$ ft). Therefore, airplane D has a much greater capability for speed and altitude performance than airplanes A and B. In air combat, everything else being equal, it is advantageous to be in a higher energy state (have a larger H_e) than your adversary.

How does an airplane change its energy state; e.g., in Figure 6.49, how could airplanes A and B increase their energy height to equal that of D? To answer this question, return to the force diagram in Figure 6.1 and the resulting equation of motion along the flight path, given by Eq. (6.6). Assuming that α_T is small, Eq. (6.6) becomes

$$T - D - W\sin\theta = m\frac{dV}{dt} \tag{6.124}$$

Recalling that $m = W/g$, Eq. (6.24) can be rearranged as

$$T - D = W\left(\sin\theta + \frac{1}{g}\frac{dV}{dt}\right)$$

Multiplying by V/W, we obtain

$$\frac{TV - DV}{W} = V\sin\theta + \frac{V}{g}\frac{dV}{dt} \tag{6.125}$$

Examining Eq. (6.125) and recalling some of the definitions from Sec. 6.8, we observe that $V\sin\theta = R/C = dh/dt$ and that

$$\frac{TV - DV}{W} = \frac{\text{excess power}}{W} \equiv P_s$$

where the excess power per unit weight is defined as the *specific excess power* and is denoted by P_s. Hence, Eq. (6.125) can be written as

$$\boxed{P_s = \frac{dh}{dt} + \frac{V}{g}\frac{dV}{dt}} \tag{6.126}$$

Equation (6.126) states that an airplane with excess power can use this excess for rate of climb (dh/dt) or to accelerate along its flight path (dV/dt) or for a combination of both. For example, consider an airplane in level flight at a velocity of 800 ft/s. Assume that when the pilot pushes the throttle all the way forward, an excess power is generated in the amount $P_s = 300$ ft/s. Equation (6.126) shows that the pilot can choose to use all this excess power to obtain a maximum unaccelerated rate of climb of 300 ft/s ($dV/dt = 0$, hence $P_s = dh/dt = R/C$). In this case, the velocity along the flight path stays constant at 800 ft/s. Alternatively, the pilot may choose to maintain level flight ($dh/dt = 0$) and to use all this excess power to accelerate at the rate of $dV/dt = gP_s/V = 32.2(300)/800 = 12.1$ ft/s^2. On the other hand, some combination could be achieved, such as a rate of climb $dh/dt = 100$ ft/s along with an acceleration along the flight path of $dV/dt = 32.2(200)/800 = 8.1$ ft/s^2. [Note that Eqs. (6.125) and (6.126) are generalizations of Eq. (6.43). In Sec. 6.8, we assumed that $dV/dt = 0$, which resulted in Eq. (6.43) for a steady climb. In the present section, we are treating the more general case of climb with a finite acceleration.] Now return to Eq. (6.123) for the energy height. Differentiating with respect to time, we have

$$\frac{dH_e}{dt} = \frac{dh}{dt} + \frac{V}{g}\frac{dV}{dt} \qquad (6.127)$$

The right-hand sides of Eqs. (6.126) and (6.127) are identical, hence we see that

$$\boxed{P_s = \frac{dH_e}{dt}} \qquad (6.128)$$

That is, the *time rate of change of energy height is equal to the specific excess power.* This is the answer to the question at the beginning of this paragraph. An airplane can increase its energy state simply by the application of excess power. In Figure 6.49, airplanes A and B can reach the energy state of airplane D *if* they have enough excess power to do so.

This immediately leads to the next question, namely, how can we ascertain whether or not a given airplane has enough P_s to reach a certain energy height? To address this question, recall the definition of excess power as illustrated in Figure 6.24, i.e., the difference between power available and power required. For a given altitude, say h, the excess power (hence P_s) can be plotted vs. velocity (or Mach number). For a subsonic airplane below the drag-divergence Mach number, the resulting curve will resemble the sketch shown in Figure 6.50a. At a given altitude h_1, P_s will be an inverted, U-shaped curve. (This is essentially the same type of plot as shown in Figures 6.27 and 6.28.) For progressively higher altitudes, such as h_2 and h_3, P_s becomes smaller, as also shown in Figure 6.50a. Hence, Figure 6.50a is simply a plot of P_s versus Mach number with altitude as a parameter. These results can be cross-plotted on an altitude–Mach number map using P_s as a parameter, as illustrated in Figure 6.50b. For example, consider all the points on Figure 6.50a where $P_s = 0$; these correspond to points along a horizontal axis through $P_s = 0$, such as points a, b, c, d, e, and f in Figure 6.50a. Now replot these points on the altitude–Mach number map in Figure 6.50b. Here, points a, b, c, d, e, and f form a bell-shaped curve, along which

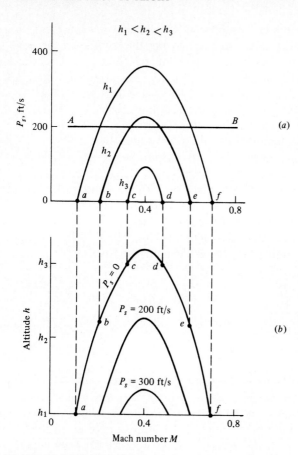

$h_1 < h_2 < h_3$

(a)

(b)

Figure 6.50 Construction of the specific excess power contours in the altitude–Mach number map for a subsonic airplane below the drag-divergence Mach number. These contours are constructed for a fixed load factor; if the load factor is changed, the P_s contours will shift.

$P_s = 0$. This curve is called the P_s contour for $P_s = 0$. Similarly, all points with $P_s = 200$ ft/s are on the horizontal line AB in Figure 6.50a, and these points can be cross-plotted to generate the $P_s = 200$ ft/s contour in Figure 6.50b. In this fashion, an entire series of P_s contours can be generated in the altitude–Mach number map. For a supersonic airplane, the P_s versus Mach number curves at different altitudes will appear as sketched in Figure 6.51a. The "dent" in the U-shaped curves around Mach 1 is due to the large drag increase in the transonic flight regime (see Sec. 5.10). In turn, these curves can be cross-plotted on the altitude–Mach number map, producing the P_s contours as illustrated in Figure 6.51b. Due to the double-humped shape of the P_s curves in Figure 6.51a, the P_s contours in Figure 6.51b have different shapes in the subsonic and supersonic regions. The shape of the P_s contours shown in Figure 6.51b is characteristic of most supersonic aircraft. Now, we are close to the answer to our question at the beginning of this paragraph. Let us overlay the P_s contours, say from Figure 6.51b, and the energy states illustrated in Figure 6.49—all on an altitude–Mach number map. We obtain a diagram, as illustrated in Figure 6.52. In this figure,

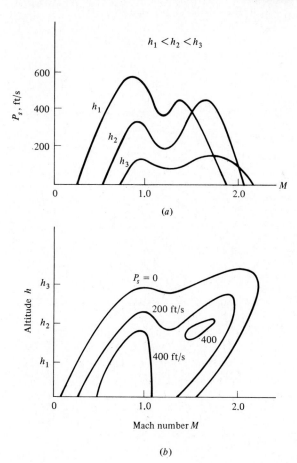

Figure 6.51 Specific excess power contours for a supersonic airplane.

(b)

note that the P_s contours always correspond to a given airplane at a given load factor, whereas the H_e lines are universal fundamental physical curves that have nothing to do with any given airplane. The usefulness of Figure 6.52 is that it clearly establishes what energy states are obtainable by a given airplane. The regime of sustained flight for the airplane lies *inside* the envelope formed by the $P_s = 0$ contour. Hence, all values of H_e inside this envelope are obtainable by the airplane. A comparison of figures like Figure 6.52 for different airplanes will clearly show in what regions of altitude and Mach number an airplane has maneuver advantages over another.

Figure 6.52 is also useful for representing the proper flight path to achieve minimum time to climb. For example, consider two energy heights, $H_{e,1}$ and $H_{e,2}$, where $H_{e,2} > H_{e,1}$. The time to move between these energy states can be obtained from Eq. (6.128), written as

$$dt = \frac{dH_e}{P_s}$$

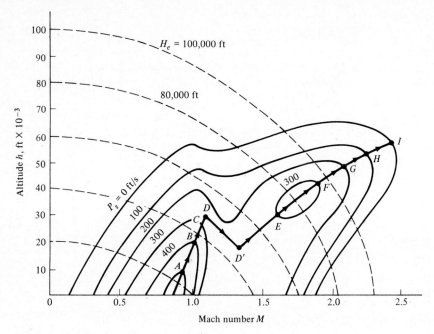

Figure 6.52 Overlay of P_s contours and specific energy states on an altitude–Mach number map. The P_s values shown here approximately correspond to a Lockheed F-104G supersonic fighter. Load factor $n = 1$. $W = 18,000$ lb. Airplane is at maximum thrust. The path given by points A through I is the flight path for minimum time to climb.

Integrating between $H_{e,1}$ and $H_{e,2}$, we have

$$t_2 - t_1 = \int_{H_{e,1}}^{H_{e,2}} \frac{dH_e}{P_s} \tag{6.129}$$

From Eq. (6.129), the time to climb will be a minimum when P_s is a maximum. Looking at Figure 6.52, for each H_e curve, there is a point where P_s is a maximum. Indeed, at this point, the P_s curve is tangent to the H_e curve. Such points are illustrated by points A to I in Figure 6.52. The heavy curve through these points illustrates the variation of altitude and Mach number along the flight path for minimum time to climb. The segment of the flight path between D and D' represents a constant energy dive to accelerate through the drag-divergence region near Mach 1.

As a final note, analyses of modern high-performance airplanes make extensive use of energy concepts such as those described above. Indeed, military pilots fly with P_s diagrams in the cockpit. Our purpose here has been to simply introduce some of the definitions and basic ideas involving these concepts. A more extensive treatment is beyond the scope of this book.

6.18 A COMMENT

We end the technical portion of this chapter by noting that detailed computer programs now exist within NASA and the aerospace industry for the accurate estimation of airplane performance. These programs are usually geared to specific types of airplanes, e.g., general aviation aircraft (light single- or twin-engine private airplanes), military fighter aircraft, and commercial transports. Such considerations are beyond the scope of this book. However, the principles developed in this chapter are stepping-stones to more-advanced studies of airplane performance; the Bibliography at the end of this chapter provides some suggestions for such studies.

6.19 HISTORICAL NOTE: DRAG REDUCTION—THE NACA COWLING AND THE FILLET

The radial piston engine came into wide use in aviation during and after World War I. As described in Chap. 9, a radial engine has its pistons arranged in a circular fashion about the crankshaft, and the cylinders themselves are cooled by airflow over the outer finned surfaces. Until 1927, these cylinders were generally directly exposed to the main airstream of the airplane, as sketched in Figure 6.53. As a result, the drag on the engine-fuselage combination was inordinately high. The problem was severe enough that a group of aircraft manufacturers met at Langley Field on May 24, 1927, to urge NACA to undertake an investigation of means to reduce this drag. Subsequently, under the direction of Fred E. Weick, an extensive series of tests was conducted in the Langley 20-ft propeller research tunnel using a Wright Whirlwind J-5 radial engine mounted to a conventional fuselage. In these tests, various types of aerodynamic surfaces, called cowlings, were used to cover, partly or completely, the engine cylinders, directly guiding part of the airflow over these cylinders for cooling purposes but at the same time not interfering with the smooth primary aerodynamic flow over the fuselage. The

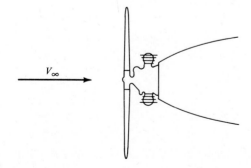

V_∞

Figure 6.53 Engine mounted with no cowling.

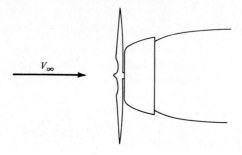

Figure 6.54 Engine mounted with full cowling.

best cowling, illustrated in Figure 6.54, completely covered the engine. The results were dramatic! Compared with the uncowled fuselage, a full cowling reduced the drag by a stunning 60 percent! This is illustrated in Figure 6.55, taken directly from Weick's report entitled "Drag and Cooling with Various Forms of Cowling for a Whirlwind Radial Air-Cooled Engine," NACA technical report no. 313, published in 1928. After this work, virtually all radial engine–equipped airplanes since 1928 have been designed with a full NACA cowling. The development of this cowling was one of the most important aerodynamic advancements of the 1920s; it led the way to a major increase in aircraft speed and efficiency.

A few years later, a second major advancement was made, but by a completely different group and on a completely different part of the airplane. In the early 1930s, the California Institute of Technology at Pasadena, California, established a program in aeronautics under the direction of Theodore von Karman. Von Karman, a student of Ludwig Prandtl, became probably the leading aerodynamicist of the 1920–1960 time period. At Caltech, von Karman established an aeronautical laboratory of high quality, which included a large

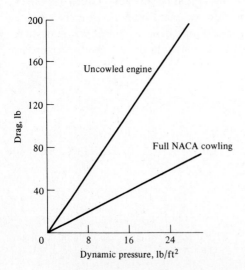

Figure 6.55 Reduction in drag due to a full cowling.

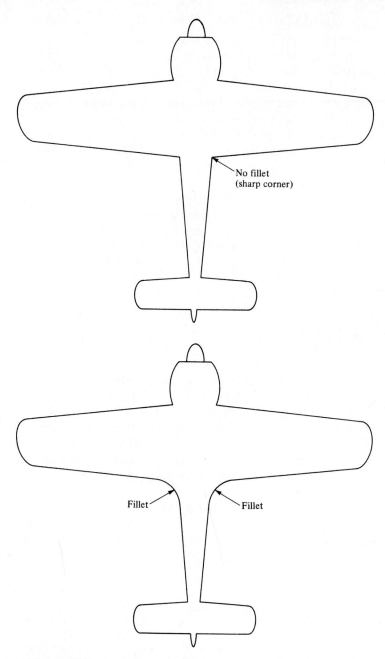

Figure 6.56 Illustration of the wing fillet.

subsonic wind tunnel funded by a grant from the Guggenheim Foundation. The first major experimental program in this tunnel was a commercial project for the Douglas Aircraft Company. Douglas was designing the DC-1, the forerunner of a series of highly successful transports (including the famous DC-3, which revolutionized commercial aviation in the 1930s). The DC-1 was plagued by unusual buffeting in the region where the wing joined the fuselage. The sharp corner at the juncture caused severe flow field separation, which resulted in high drag as well as shed vortices which buffeted the tail. The Caltech solution, which was new and pioneering, was to fair the trailing edge of the wing smoothly into the fuselage. These fairings, called fillets, were empirically designed and were modeled in clay on the DC-1 wind-tunnel models. The best shape was found by trial and error. The addition of a fillet (see Figure 6.56) solved the buffeting problem by smoothing out the separated flow and hence also reduced the interference drag. Since that time, fillets have become a standard airplane design feature. Moreover, the fillet is an excellent example of how university laboratory research in the 1930s contributed directly to the advancement of practical airplane design.

6.20 HISTORICAL NOTE: EARLY PREDICTIONS OF AIRPLANE PERFORMANCE

The airplane of today is a modern work of art and engineering. In turn, the prediction of airplane performance as described in this chapter is sometimes viewed as a relatively modern discipline. However, contrary to intuition, some of the basic concepts have roots deep in history; indeed, some of the very techniques detailed in previous sections were being used in practice only a few years after the Wright brothers' successful first flight in 1903. This section traces a few historic paths for some of the basic ideas of airplane performance, as follows:

1. Some understanding of the *power required* P_R for an airplane was held by George Cayley. He understood that the rate of energy lost by an airplane in a steady glide under gravitational attraction must be essentially the power that must be supplied by an engine to maintain steady, level flight. In 1853, Cayley wrote:

   The whole apparatus when loaded by a weight equal to that of the man intended ultimately to try the experiment, and with the horizontal rudder [the elevator] described on the essay before sent, adjusted so as to regulate the oblique descent from some elevated point, to its proper pitch, it may be expected to skim down, with no force but its own gravitation, in an angle of about 11 degrees with the horizon; or possibly, if well executed, as to direct resistance something less, at a speed of about 36 feet per second, if loaded 1 pound to each square foot of surface. This having by repeated experiments, in perfectly calm weather, been ascertained, for both the safety of the man, and the datum required, let the wings be plied with the man's utmost strength; and let the angle measured by the greater extent of horizontal range of flight be noted; when this point, by repeated experiments, has been accurately found, we shall have ascertained a sound practical basis for calculating what engine power is necessary under the same circumstances as to weight and surface to produce horizontal flight....

2. The *drag polar*, a concept introduced in Sec. 5.14, sketched in Figure 5.33, and embodied in Eq. (6.1*a*), represents simply a plot of C_D versus C_L, illustrating that C_D varies as the square of C_L. A knowledge of the drag polar is essential to the calculation of airplane performance. It is interesting that the concept of the drag polar was first introduced by the Frenchman M. Eiffel about 1890. Eiffel was interested in determining the laws of resistance (drag) on bodies of various shapes, and he conducted such drag measurements by dropping bodies from the Eiffel Tower and measuring their terminal velocity.

3. Some understanding of the requirements for *rate of climb* existed as far back as 1913, when in an address by Granville E. Bradshaw before the Scottish Aeronautical Society in Glasgow in December, the following comment was made: "Among the essential features of all successful aeroplanes [is that] it shall climb very quickly. This depends almost entirely on the weight efficiency of the engine. The rate of climb varies directly as the power developed and indirectly as the weight to be lifted." This is essentially a partial statement of Eq. (6.43).

4. No general understanding of the prediction of *airplane performance* existed before the twentieth century. The excellent summary of aeronautics written by Octave Chanute in 1894, *Progress in Flying Machines*, does not contain any calculational technique even remotely resembling the procedures set forth in this chapter. At best, it was understood by that time that lift and drag varied as the first power of the area and as the second power of velocity, but this does not constitute a performance calculation. However, this picture radically changed in 1911. In that year, the Frenchman Duchène received the Monthyon Prize from the Paris Academy of Sciences for his book entitled *The Mechanics of the Airplane: A Study of the Principles of Flight*. Captain Duchène was a French engineering officer, born in Paris on December 27, 1869, educated at the famous Ecole Polytechnique, and later assigned to the fortress at Toul, one of the centers of "aerostation" in France. It was in this capacity that Captain Duchène wrote his book during 1910–1911. In this book, the basic elements of airplane performance, as discussed in this chapter, are put forth for the first time. Duchène gives curves of power required and power available, as we illustrated in Figure 6.17*a*; he discusses airplane maximum velocity; he also gives the same relation as Eq. (6.43) for rate of climb. Thus, some of our current concepts for the calculation of airplane performance date back as far as 1910–1911—four years before the beginning of World War I, and only seven years after the Wright brothers' first flight in 1903. Later, in 1917, Duchène's book was translated into English by John Ledeboer and T. O'B. Hubbard (see Bibliography at the end of this chapter). Finally, during 1918–1920, three additional books on airplane

performance were written (again, see Bibliography), the most famous being the authoritative *Applied Aerodynamics* by Leonard Bairstow. By this time the foundations discussed in this chapter had been well set.

6.21 HISTORICAL NOTE: BREGUET AND THE RANGE FORMULA

Louis-Charles Breguet was a famous French aviator, airplane designer, and industrialist. Born in Paris on January 2, 1880, he was educated in electrical engineering at the Lycée Condorcet, the Lycée Carnot, and the École Superieure d'Electricité. After graduation, he joined the electrical engineering firm of his father, Maison Breguet. However, in 1909 Breguet built his first airplane and then plunged his life completely into aviation. During World War I, his airplanes were mass-produced for the French air force. In 1919, he founded a commercial airline company which later grew into Air France. His airplanes set several long-range records during the 1920s and 1930s. Indeed, Breguet was active in his own aircraft company until his death on May 4, 1955, in Paris. His name is associated with a substantial part of French aviation history.

The formula for range of a propeller-driven airplane given by Eq. (6.55) has also become associated with Breguet's name; indeed, it is commonly called the Breguet range equation. However, the reason for this association is historically obscure. In fact, the historical research of the present author can find no substance to Breguet's association with Eq. (6.55). On one hand, we find absolutely no reference to airplane range or endurance in any of the airplane performance literature before 1919, least of all a reference to Breguet. The authoritative books by Cowley and Levy (1918), Judge (1919), and Bairstow (1920) (see Bibliography at the end of this chapter) amazingly enough do not discuss this subject. On the other hand, in 1919, NACA report no. 69, entitled "A Study of Airplane Ranges and Useful Loads," by J. G. Coffin, gives a complete derivation of the formulas for range, Eq. (6.55), and endurance, Eq. (6.56). However, Coffin, who was director of research for Curtiss Engineering Corporation at that time, gives absolutely no references to *anybody*. Coffin's work appears to be original and clearly seems to be the first presentation of the range and endurance formulas in the literature. However, to confuse matters, we find a few years later, in NACA report no. 173, entitled "Reliable Formulae for Estimating Airplane Performance and the Effects of Changes in Weight, Wing Area or Power," by Walter S. Diehl (we have met Diehl before, in Sec. 3.6), the following statement: "The common formula for range, usually credited to Breguet, is easily derived." Diehl's report then goes on to use Eq. (6.55), with no further reference to Breguet. This report was published in 1923, four years after Coffin's work.

Consequently, to say the least, the proprietorship of Eq. (6.55) is not clear. It appears to this author that, in the United States at least, there is plenty of documentation to justify calling Eq. (6.55) the Coffin-Breguet range equation. However, it has come down to us through the ages simply as Breguet's equation, apparently without documented substance.

6.22 CHAPTER SUMMARY

A few of the important aspects of this chapter are listed below.

1. For a complete airplane, the drag polar is given as

$$C_D = C_{D,0} + \frac{C_L^2}{\pi e \text{AR}} \qquad (6.1c)$$

where $C_{D,0}$ is the parasite drag coefficient at zero lift and the term $C_L^2/\pi e \text{AR}$ includes both induced drag and the contribution to parasite drag due to lift.

2. Thrust required for level, unaccelerated flight is

$$T_R = \frac{W}{L/D} \qquad (6.15)$$

Thrust required is a minimum when L/D is maximum.

3. Power required for level, unaccelerated flight is

$$P_R = \sqrt{\frac{2W^3 C_D^2}{\rho_\infty S C_L^3}} \qquad (6.26)$$

Power required is a minimum when $C_L^{3/2}/C_D$ is a maximum.

4. The rate of climb, $R/C = dh/dt$, is given by

$$\frac{dh}{dt} = \frac{TV - DV}{W} - \frac{V}{g}\frac{dV}{dt} \qquad (6.126)$$

where $(TV - DV)/W = P_s$, the specific excess power. For an unaccelerated climb, $dV/dt = 0$, and hence

$$R/C = \frac{dh}{dt} = \frac{TV - DV}{W} \qquad (6.43)$$

5. The absolute ceiling is defined as that altitude where maximum $R/C = 0$. The service ceiling is that altitude where maximum $R/C = 100$ ft/min.

6. For a propeller-driven airplane, range R and endurance E are given by

$$R = \frac{\eta}{c} \frac{C_L}{C_D} \ln \frac{W_0}{W_1} \qquad (6.55)$$

and

$$E = \frac{\eta}{c} \frac{C_L^{3/2}}{C_D} (2\rho_\infty S)^{1/2} \left(W_1^{-1/2} - W_0^{-1/2} \right) \qquad (6.56)$$

Maximum range occurs at maximum C_L/C_D. Maximum endurance occurs at sea level with maximum $C_L^{3/2}/C_D$.

7. For a jet-propelled airplane, range and endurance are given by

$$R = 2\sqrt{\frac{2}{\rho_\infty S}} \frac{1}{c_t} \frac{C_L^{1/2}}{C_D} \left(W_0^{1/2} - W_1^{1/2} \right) \qquad (6.65)$$

and

$$E = \frac{1}{c_t} \frac{C_L}{C_D} \ln \frac{W_0}{W_1} \qquad (6.60)$$

8. At *maximum* $C_L^{3/2}/C_D$, $C_{D,0} = \frac{1}{3}C_{D,i}$. For this case,

$$\left(\frac{C_L^{3/2}}{C_D}\right)_{\max} = \frac{(3C_{D,0}\pi e\mathrm{AR})^{3/4}}{4C_{D,0}} \tag{6.75}$$

At *maximum* C_L/C_D, $C_{D,0} = C_{D,i}$. For this case,

$$\left(\frac{C_L}{C_D}\right)_{\max} = \frac{(C_{D,0}\pi e\mathrm{AR})^{1/2}}{2C_{D,0}} \tag{6.73}$$

At *maximum* $C_L^{1/2}/C_D$, $C_{D,0} = 3C_{D,i}$. For this case

$$\left(\frac{C_L^{1/2}}{C_D}\right)_{\max} = \frac{\left(\frac{1}{3}C_{D,0}\pi e\mathrm{AR}\right)^{1/4}}{\frac{4}{3}C_{D,0}} \tag{6.74}$$

9. Takeoff ground roll is given by

$$s_{\mathrm{LO}} = \frac{1.44W^2}{g\rho_\infty S C_{L,\max}\left\{T - \left[D + \mu_r(W - L)\right]_{\mathrm{ave}}\right\}} \tag{6.90}$$

10. The landing ground roll is

$$s_L = \frac{1.69W^2}{g\rho_\infty S C_{L,\max}\left[D + \mu_r(W - L)\right]_{\mathrm{ave}}} \tag{6.98}$$

11. The load factor is defined as

$$n = L/W \tag{6.102}$$

12. In *level turning flight*, the turn radius is

$$R = \frac{V_\infty^2}{g\sqrt{n^2 - 1}} \tag{6.105}$$

and the turn rate is

$$\omega = \frac{g\sqrt{n^2 - 1}}{V_\infty} \tag{6.106}$$

13. The *V-n* diagram is illustrated in Figure 6.47. It is a diagram showing load factor vs. velocity for a given airplane, along with the constraints on both n and V due to structural limitations. The *V-n* diagram illustrates some particularly important aspects of overall airplane performance.

14. The energy height (specific energy) of an airplane is given by

$$H_e = h + \frac{V^2}{2g} \tag{6.123}$$

This, in combination with the specific excess power,

$$P_s = \frac{TV - DV}{W}$$

leads to the analysis of accelerated-climb performance using energy considerations only.

BIBLIOGRAPHY

Bairstow, L., *Applied Aerodynamics*, Longmans, London, 1920.

Cowley, W. L., and Levy, H., *Aeronautics in Theory and Experiment*, E. Arnold, London, 1918.

Dommasch, D. O., Sherbey, S. S., and Connolly, T. F., *Airplane Aerodynamics*, 3rd ed., Pitman, New York, 1961.

Duchène, Captain, *The Mechanics of the Airplane: A Study of the Principles of Flight* (transl. by J. H. Ledeboer and T. O'B. Hubbard), Longmans, London, 1917.

Hale, F. J., *Introduction to Aircraft Performance, Selection and Design*, Wiley, New York, 1984.

Judge, A. W., *Handbook of Modern Aeronautics*, Appleton, London, 1919.

McCormick, B. W., *Aerodynamics, Aeronautics and Flight Mechanics*, Wiley, New York, 1979.

Perkins, C. D., and Hage, R. E., *Airplane Performance, Stability and Control*, Wiley, New York, 1949.

Shevell, R. S., *Fundamentals of Flight*, Prentice-Hall, Englewood Cliffs, NJ, 1983.

PROBLEMS

6.1 Consider an airplane patterned after the twin-engine Beechcraft Queen Air executive transport. The airplane weight is 38,220 N, wing area is 27.3 m^2, aspect ratio is 7.5, Oswald efficiency factor is 0.9, and parasite drag coefficient $C_{D,0} = 0.03$. Calculate the thrust required to fly at a velocity of 350 km/h at (a) standard sea level and (b) an altitude of 4.5 km.

6.2 An airplane weighing 5000 lb is flying at standard sea level with a velocity of 200 mi/h. At this velocity, the L/D ratio is a maximum. The wing area and aspect ratio are 200 ft^2 and 8.5, respectively. The Oswald efficiency factor is 0.93. Calculate the total drag on the airplane.

6.3 Consider an airplane patterned after the Fairchild Republic A-10, a twin-jet attack aircraft. The airplane has the following characteristics: wing area = 47 m^2, aspect ratio = 6.5, Oswald efficiency factor = 0.87, weight = 103,047 N, and parasite drag coefficient = 0.032. The airplane is equipped with two jet engines with 40,298 N of static thrust *each* at sea level.

    (a) Calculate and plot the power-required curve at sea level.

    (b) Calculate the maximum velocity at sea level.

    (c) Calculate and plot the power-required curve at 5-km altitude.

    (d) Calculate the maximum velocity at 5-km altitude. (Assume the engine thrust varies directly with freestream density.)

6.4 Consider an airplane patterned after the Beechcraft Bonanza V-tailed, single-engine light private airplane. The characteristics of the airplane are as follows: aspect ratio = 6.2, wing area = 181 ft^2, Oswald efficiency factor = 0.91, weight = 3000 lb, and parasite drag coefficient = 0.027. The airplane is powered by a single piston engine of 285 hp maximum at sea level. Assume the power of the engine is proportional to freestream density. The two-bladed propeller has an efficiency of 0.83.

    (a) Calculate the power required at sea level.

    (b) Calculate the maximum velocity at sea level.

    (c) Calculate the power required at 12,000-ft altitude.

    (d) Calculate the maximum velocity at 12,000-ft altitude.

6.5 From the information generated in Prob. 6.3, calculate the maximum rate of climb for the twin-jet aircraft at sea level and at an altitude of 5 km.

6.6 From the information generated in Prob. 6.4, calculate the maximum rate of climb for the single-engine light plane at sea level and at 12,000-ft altitude.

6.7 From the rate of climb information for the twin-jet aircraft in Prob. 6.5, estimate the absolute ceiling of the airplane. (Note: Assume maximum R/C varies linearly with altitude—not a precise assumption, but not bad, either.)

6.8 From the rate of climb information for the single-engine light plane in Prob. 6.6, estimate the absolute ceiling of the airplane. (Again, make the linear assumption described in Prob. 6.7.)

6.9 Consider an airplane with a parasite drag coefficient of 0.025, an aspect ratio of 6.72, and an Oswald efficiency factor of 0.9. Calculate the value of $(L/D)_{max}$.

6.10 Consider the single-engine light plane described in Prob. 6.4. If the specific fuel consumption is 0.42 lb of fuel per horsepower per hour, the fuel capacity is 44 gallons, and the maximum gross weight is 3400 lb, calculate the range and endurance at standard sea level.

6.11 Consider the twin-jet airplane described in Prob. 6.3. The thrust-specific fuel consumption is 1.0 N of fuel per Newton of thrust per hour, the fuel capacity is 1900 gallons, and the maximum gross weight is 136,960 N. Calculate the range and endurance at a standard altitude of 8 km.

6.12 Derive Eqs. 6.68 and 6.69.

6.13 Derive Eqs. 6.74 and 6.75.

6.14 Estimate the sea-level lift-off distance for the airplane in Prob. 6.3. Assume a paved runway. Also, during the ground roll, the angle of attack is restricted by the requirement that the tail not drag the ground. Hence, assume $C_{L,max}$ during the ground roll is limited to 0.8. Also, when the airplane is on the ground, the wings are 5 ft above the ground.

6.15 Estimate the sea-level lift-off distance for the airplane in Prob. 6.4. Assume a paved runway, and $C_{L,max} = 1.1$ during the ground roll. When the airplane is on the ground, the wings are 4 ft above the ground.

6.16 Estimate the sea-level landing ground roll distance for the airplane in Prob. 6.3. Assume the airplane is landing at full gross weight. The maximum lift coefficient with flaps fully employed at touchdown is 2.8.

6.17 Estimate the sea-level landing ground roll distance for the airplane in Prob. 6.4. Assume the airplane is landing with a weight of 2900 lb. The maximum lift coefficient with flaps at touchdown is 1.8.

6.18 For the airplane in Prob. 6.3, the sea-level corner velocity is 250 mph, and the maximum lift coefficient with no flap deflection is 1.2. Calculate the minimum turn radius and maximum turn rate at sea level.

6.19 The airplane in Prob. 6.3 is flying at 15,000 ft with a velocity of 375 mi/h. Calculate its specific energy at this condition.

SEVEN

PRINCIPLES OF STABILITY
AND CONTROL

An important problem to aviation is ... improvement in the form of the aeroplane leading toward natural inherent stability to such a degree as to relieve largely the attention of the pilot while still retaining sufficient flexibility and control to maintain any desired path, without seriously impairing the efficiency of the design.

From the First Annual Report of the NACA, 1915

7.1 INTRODUCTION

The scene: A French army drill field at Issy-les-Moulineaux just outside Paris. *The time:* The morning of January 13, 1908. *The character:* Henri Farman, a bearded, English-born but French-speaking aviator, who had flown for his first time just four months earlier. *The action:* A delicately constructed Voisin-Farman I-bis biplane (see Figure 7.1) is poised, ready for the takeoff in the brisk Parisian wind, with Farman seated squarely in front of the 50-hp Antoinette engine. The winds ripple the fabric on the Voisin's box kite–shaped tail as Farman powers to a bumpy lift-off. Fighting against a head wind, he manipulates his aircraft to a marker 1000 m from his takeoff point. In a struggling circular turn, Farman deflects the rudder and mushes the biplane around the marker, the wings remaining essentially level to the ground. Continuing in its rather wide and tenuous circular arc, the airplane heads back. Finally, Farman lands at his original takeoff point, amid cheers from the crowd that had gathered for the occasion. Farman has been in the air for 1 min and 28 s—the longest flight in Europe to that date—and has just performed the first circular flight of 1-km extent. For this, he is awarded the Grand Prix d'Aviation. (Coincidentally, in the crowd is a young Hungarian engineer, Theodore von Karman, who is present only due to the insistence of his female companion—waking at 5 A.M. in order to see history be made. However, von Karman is mesmerized by the flight, and his

Figure 7.1 The Voisin-Farman I-bis plane. *(National Air and Space Museum.)*

interest in aeronautical science is catalyzed. Von Karman will go on to become a leading aerodynamic genius of the first half century of powered flight.)

The scene shifts to a small race track near Le Mans, France. *The time:* Just seven months later, August 8, 1908. *The character:* Wilbur Wright, intense, reserved, and fully confident. *The action:* A new Wright type A biplane (see Figure 1.22), shipped to France in crates and assembled in a friend's factory near Le Mans, is ready for flight. A crowd is present, enticed to the field by much advance publicity and an intense curiosity to see if the "rumors" about the Wright brothers' reported success were really true. Wilbur takes off. Using the Wrights' patented concept of twisting the wingtips ("wing warping"), Wilbur is able to bank and turn at will. He makes two graceful circles and then effortlessly lands after 1 min and 45 s of flight. The crowds cheer. The French press is almost speechless but then heralds the flight as epoch-making. The European aviators who witness this demonstration gaze in amazement and then quickly admit that the Wrights' airplane is far advanced over the best European machines of that day. Wilbur goes on to make 104 flights in France before the end of the year and in the process transforms the direction of aviation in Europe.

The distinction between the two scenes above, and the reason for Wilbur's mastery of the air in comparison to Farman's struggling circular flight, involves stability and control. The Voisin aircraft of Farman, which represented the best European state of the art, had only rudder control and could make only a laborious, flat turn by simply swinging the tail around. In contrast, the Wright airplane's wing-twisting mechanism provided control of roll, which when combined with rudder control, allowed effortless turning and banking flight, figure eights, etc. Indeed, the Wright brothers were "airmen" (see Chap. 1) who concentrated on designing total control into their aircraft before adding an engine for powered flight. Since those early days, airplane stability and control has been a dominant aspect of airplane design. This is the subject of the present chapter.

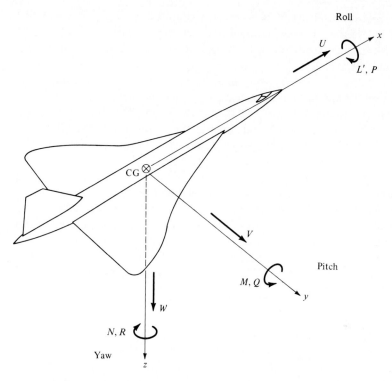

Figure 7.2 Definition of the airplane's axes along with the translational and rotational motion along and about these axes.

Airplane performance, as discussed in Chap. 6, is governed by forces (along and perpendicular to the flight path), with the translational motion of the airplane as a response to these forces. In contrast, airplane stability and control, discussed in the present chapter, are governed by moments about the center of gravity, with the rotational motion of the airplane as a response to these moments. Therefore, moments and rotational motion are the main focus of this chapter.

Consider an airplane in flight, as sketched in Figure 7.2. The center of gravity (the point through which the weight of the complete airplane effectively acts) is denoted as cg. The xyz orthogonal axis system is fixed relative to the airplane; the x axis is along the fuselage, the y axis is along the wingspan perpendicular to the x axis, and the z axis is directed downward, perpendicular to the xy plane. The origin is at the center of gravity. The translational motion of the airplane is given by the velocity components U, V, and W along the x, y, and z directions, respectively. (Note that the resultant freestream velocity V_∞ is the vector sum of U, V, and W.) The rotational motion is given by the angular velocity components P, Q, and R about the x, y, z axes, respectively. These rotational velocities are due to the moments L', M, and N about the x, y, and z axes, respectively. (The prime is put over the symbol "L" so that the reader avoids confusing it with lift.) Rotational motion about the x axis is called *roll*; L' and P are the *rolling* moment and velocity, respectively. Rotational motion about the y axis is called *pitch*; M and Q are the *pitching* moment and velocity, respectively. Rotational motion about the z axis is called *yaw*; N and R are the *yawing* moment and velocity, respectively.

There are three basic controls on an airplane—the ailerons, elevator, and rudder—which are designed to change and control the moments about the x, y, and z axes. These control surfaces are shown in Figure 7.3; they are flaplike surfaces that can be deflected back and forth at the command of the pilot. The ailerons are mounted at the trailing edge of the wing, near the wingtips. The elevators are located on the horizontal stabilizer. In some modern aircraft, the complete horizontal stabilizer is rotated instead of just the elevator (so-called flying tails). The rudder is located on the vertical stabilizer, at the trailing edge. Just as in the case of wing flaps discussed in Sec. 5.17, a downward deflection of the control surface will increase the lift of the wing or tail. In turn, the moments will be changed, as sketched in Figure 7.4. Consider Figure 7.4a. One aileron is deflected up, the other down, creating a differential lifting force on the wings, thus contributing to the rolling moment L'. In Figure 7.4b, the elevator is deflected

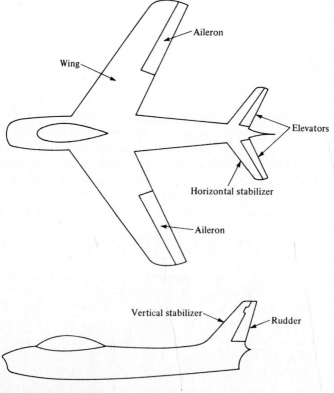

Figure 7.3 Some airplane nomenclature.

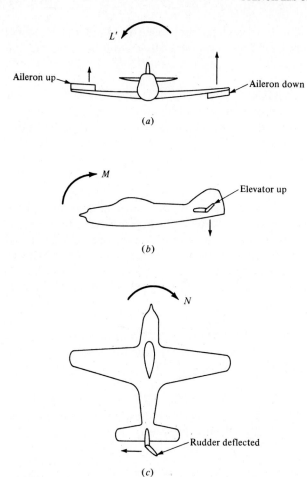

Figure 7.4 Effect of control deflections on roll, pitch, and yaw. (*a*) Effect of aileron deflection; lateral control. (*b*) Effect of elevator deflection; longitudinal control. (*c*) Effect of rudder deflection; directional control.

upward, creating a negative lift at the tail, thus contributing to the pitching moment M. In Figure 7.4*c*, the rudder is deflected to the right, creating a leftward aerodynamic force on the tail, thus contributing to the yawing moment N.

Rolling (about the x axis) is also called *lateral motion*. Referring to Figure 7.4*a*, we see that ailerons control roll; hence they are known as *lateral controls*. Pitching (about the y axis) is also called *longitudinal motion*. In Figure 7.4*b*, we see that elevators control pitch; hence they are known as *longitudinal controls*. Yawing (about the z axis) is also called *directional motion*. Figure 7.4*c* shows that the rudder controls yaw; hence it is known as the *directional control*.

All the above definitions and concepts are part of the basic language of airplane stability and control; they should be studied carefully. Also, in the process, the following question becomes apparent: what is meant by the words "stability and control" themselves? This question is answered in the next section.

7.2 DEFINITION OF STABILITY AND CONTROL

There are two types of stability: static and dynamic. They can be visualized as follows.

A Static Stability

Consider a marble on a curved surface, such as a bowl. Imagine that the bowl is upright and the marble is resting inside, as shown in Figure 7.5a. The marble is stationary; it is in a state of *equilibrium*, which means that the moments acting on the marble are zero. If the marble is now disturbed (moved to one side, as shown by the dotted circle in Figure 7.5a) and then released, it will roll back toward the bottom of the bowl, i.e., toward its original equilibrium position. Such a system is *statically* stable. In general, we can state that

> *If the forces and moments on the body caused by a disturbance tend initially to return the body toward its equilibrium position, the body is statically stable. The body has positive static stability.*

Now, imagine the bowl is upside-down, with the marble at the crest, as shown in Figure 7.5b. If the marble is placed precisely at the crest, the moments will be zero and the marble will be in equilibrium. However, if the marble is now

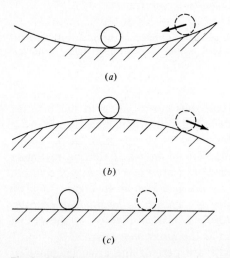

(a)

(b)

(c)

Figure 7.5 Illustration of static stability. (a) Statically stable system. (b) Statically unstable system. (c) Statically neutral system.

disturbed (as shown by the dotted circle in Figure 7.5*b*), it will tend to roll down the side, away from its equilibrium position. Such a system is *statically unstable*. In general, we can state that

> *If the forces and moments are such that the body continues to move* away *from its equilibrium position after being disturbed, the body is* statically unstable. *The body has* negative *static stability.*

Finally, imagine the marble on a flat horizontal surface, as shown in Figure 7.5*c*. Its moments are zero; it is in equilibrium. If the marble is now disturbed to another location, the moments will still be zero, and it will still be in equilibrium. Such a system is *neutrally stable*. This situation is rare in flight vehicles, and we will not be concerned with it here.

Emphasis is made that static stability (or the lack of it) deals with the *initial* tendency of a vehicle to return to equilibrium (or to diverge from equilibrium) after being disturbed. It says nothing about whether it ever reaches its equilibrium position, nor how it gets there. Such matters are the realm of dynamic stability, as follows.

B Dynamic Stability

Dynamic stability deals with the *time history* of the vehicle's motion after it initially responds to its static stability. For example, consider an airplane flying at an angle of attack α_e such that its moments about the center of gravity are zero. The airplane is therefore in equilibrium at α_e; in this situation, it is *trimmed*, and α_e is called the trim angle of attack. Now assume that the airplane is disturbed (say, by encountering a wind gust) to a new angle of attack α as shown in Figure 7.6. The airplane has been pitched through a *displacement* $\alpha - \alpha_e$. Now, let us observe the subsequent pitching motion after the airplane has been disturbed by the gust. We can describe this motion by plotting the instantaneous displacement vs. time, as shown in Figure 7.7. Here $\alpha - \alpha_e$ is given as a function of time t. At $t = 0$, the displacement is equal to that produced by the gust. If the airplane is

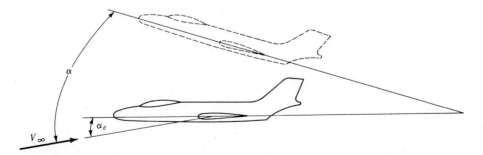

Figure 7.6 Disturbance from the equilibrium angle of attack.

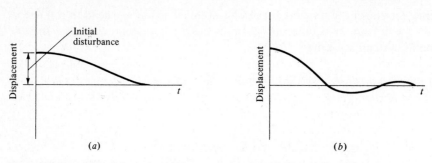

Figure 7.7 Examples of dynamic stability. (*a*) Aperiodic; (*b*) Damped oscillations.

statically stable, it will *initially* tend to move back toward its equilibrium position, that is, $\alpha - \alpha_e$ will initially decrease. Over a lapse of time, the vehicle may monotonically "home-in" to its equilibrium position, as shown in Figure 7.7a. Such motion is called aperiodic. Alternately, it may first overshoot the equilibrium position and approach α_e after a series of oscillations with decreasing amplitude, as shown in Figure 7.7b. Such motion is described as damped oscillations. In both situations, Figures 7.7a and 7.7b, the airplane eventually returns to its equilibrium position after some interval of time. These two situations are examples of *dynamic stability* in an airplane. Thus, we can state that

> *A body is dynamically stable if, out of its own accord, it eventually returns to and remains at its equilibrium position over a period of time.*

On the other hand, after initially responding to its static stability, the airplane may oscillate with increasing amplitude, as shown in Figure 7.8. Here, the equilibrium position is never maintained for any period of time; the airplane in this case is *dynamically unstable* (even though it is statically stable). Also, it is theoretically possible for the airplane to pitch back and forth with constant-

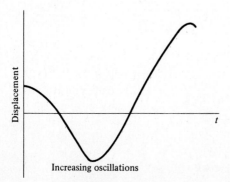

Increasing oscillations

Figure 7.8 An example of dynamic instability.

amplitude oscillations. This is an example of a *dynamically neutral* body; such a case is of little practical interest here.

It is important to observe from the above examples that a dynamically stable airplane must always be statically stable. On the other hand, static stability is *not* sufficient to ensure dynamic stability. Nevertheless, static stability is usually the first stability characteristic to be designed into an airplane. (There are some exceptions, to be discussed later.) Such considerations are of paramount importance in conventional airplanes, and therefore most of the present chapter will deal with *static* stability and control. A study of dynamic stability, although of great importance, requires rather advanced analytical techniques beyond the scope of this book.

C Control

The conventional control surfaces (elevators, ailerons, and rudder) on an airplane were discussed in Sec. 7.1 and sketched in Figures 7.3 and 7.4. Their function is usually (1) to change the airplane from one equilibrium position to another and (2) to produce nonequilibrium accelerated motions such as maneuvers. The study of the *deflections* of the ailerons, elevators, and rudder necessary to make the airplane do what we want and of the amount of *force* that must be exerted by the pilot (or the hydraulic boost system) to deflect these controls is part of a discipline called "airplane control," to be discussed later in this chapter.

D The Partial Derivative

Some physical definitions associated with stability and control have been given above. In addition, a mathematical definition, namely, that of the partial derivative, will be useful in the equations developed later, not only in this chapter but in our discussion of astronautics (Chap. 8) as well. For those readers with only a nodding acquaintance of calculus, hopefully this section will be self-explanatory; for those with a deeper calculus background, this should serve as a brief review.

Consider a function, say $f(x)$, of a single variable x. The derivative of $f(x)$ is defined from elementary calculus as

$$\frac{df}{dx} \equiv \lim_{\Delta x \to 0} \left(\frac{f(x + \Delta x) - f(x)}{\Delta x} \right)$$

Physically, this limit represents the instantaneous rate of change of $f(x)$ with respect to x.

Now consider a function which depends on more than one variable, say, for example, the function $g(x, y, z)$, which depends on the three independent variables x, y, and z. Let x vary while y and z are held constant. Then, the instantaneous rate of change of g with respect to x is given by

$$\frac{\partial g}{\partial x} \equiv \lim_{\Delta x \to 0} \left(\frac{g(x + \Delta x, y, z) - g(x, y, z)}{\Delta x} \right)$$

Here, $\partial g/\partial x$ is the *partial derivative* of g with respect to x. Now, let y vary while holding x and z constant. Then the instantaneous rate of change of g with respect to y is given by

$$\frac{\partial g}{\partial y} \equiv \lim_{\Delta y \to 0} \left(\frac{g(x, y + \Delta y, z) - g(x, y, z)}{\Delta y} \right)$$

Here, $\partial g/\partial y$ is the *partial derivative* of g with respect to y. An analogous definition holds for the partial derivative with respect to z, $\partial g/\partial z$.

In this book, we will use the concept of the partial derivative as a definition only. The calculus of partial derivatives is essential to the advanced study of virtually any field of engineering, but such considerations are beyond the scope of this book.

Example 7.1 If $g = x^2 + y^2 + z^2$, calculate $\partial g/\partial z$.

SOLUTION From the definition given above, the partial derivative is taken with respect to z holding x and y constant.

$$\frac{\partial g}{\partial z} = \frac{\partial (x^2 + y^2 + z^2)}{\partial z} = \frac{\partial x^2}{\partial z} + \frac{\partial y^2}{\partial z} + \frac{\partial z^2}{\partial z}$$

$$= 0 + 0 + 2z = 2z$$

7.3 MOMENTS ON THE AIRPLANE

A study of stability and control is focused on moments: moments on the airplane and moments on the control surfaces. At this stage, it would be well for the reader to review the discussion of aerodynamically produced moments in Sec. 5.2. Recall that the pressure and shear stress distributions over a wing produce a pitching moment. This moment can be taken about any arbitrary point (the leading edge, the trailing edge, the quarter chord, etc.). However, there exists a particular point about which the moments are independent of angle of attack. This point is defined as the *aerodynamic center* for the wing. The moment and its coefficient about the aerodynamic center are denoted by M_{ac} and $C_{M,ac}$, respectively, where $C_{M,ac} \equiv M_{ac}/q_\infty Sc$.

Reflecting again on Sec. 5.2, consider the force diagram of Figure 5.4. Assume the wing is flying at zero lift; hence F_1 and F_2 are equal and opposite forces. Thus, the moment established by these forces is a pure couple, which we know from elementary physics can be translated anywhere on the body at constant value. Therefore, at *zero lift*, $M_{ac} = M_{c/4} = M_{\text{any point}}$. In turn

$$C_{M,ac} = \left(C_{M,c/4} \right)_{L=0} = \left(C_{M,\text{any point}} \right)_{L=0}$$

This says that the value of $C_{M,ac}$ (which is constant for angles of attack) can be obtained from the value of the moment coefficient about any point when the wing is at the zero-lift angle of attack $\alpha_{L=0}$. For this reason, M_{ac} is sometimes called the *zero-lift moment*.

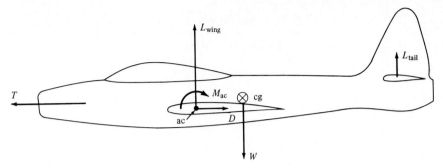

Figure 7.9 Contributions to the moment about the center of gravity of the airplane.

The aerodynamic center is a useful concept for the study of stability and control. In fact, the force and moment system on a wing can be completely specified by the lift and drag acting through the aerodynamic center, plus the moment about the aerodynamic center, as sketched in Figure 7.9. We will adopt this convention for the remainder of the present chapter.

Now consider the complete airplane, as sketched in Figure 7.9. Here, we are most concerned with the pitching moment about the center of gravity of the airplane, M_{cg}. Clearly, by examination of Figure 7.9, M_{cg} is created by (1) L, D, and M_{ac} of the wing, (2) lift of the tail, (3) thrust, and (4) aerodynamic forces and moments on other parts of the airplane, such as the fuselage and engine nacelles. (Note that weight does not contribute, since it acts through the center of gravity.) These contributions to M_{cg} will be treated in detail later. The purpose of Figure 7.9 is simply to illustrate the important conclusion that a moment does exist about the center of gravity of an airplane, and it is this moment which is fundamental to the stability and control of the airplane.

The moment coefficient about the center of gravity is defined as

$$C_{M,cg} = \frac{M_{cg}}{q_\infty S c} \tag{7.1}$$

Combining the above concept with the discussion of Sec. 7.2, we find an airplane is in equilibrium (in pitch) when the moment about the center of gravity is zero; i.e., when $M_{cg} = C_{M,cg} = 0$, the airplane is said to be *trimmed*.

7.4 ABSOLUTE ANGLE OF ATTACK

Continuing with our collection of tools to analyze stability and control, consider a wing at an angle of attack such that lift is zero; i.e., the wing is at the zero-lift angle of attack $\alpha_{L=0}$, as shown in Figure 7.10a. With the wing in this orientation, draw a line through the trailing edge parallel to the relative wind V_∞. This line is

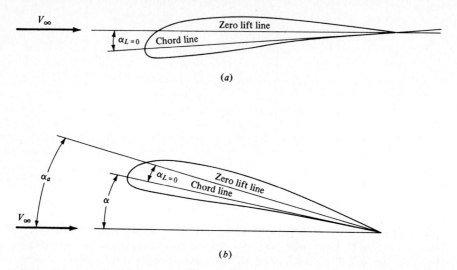

Figure 7.10 Illustration of the zero-lift line and absolute angle of attack. (*a*) No lift; (*b*) with lift.

defined as the *zero-lift line* for the airfoil. It is a fixed line; visualize it frozen into the geometry of the airfoil, as sketched in Figure 7.10*a*. As discussed in Chap. 5, conventional cambered airfoils have slightly negative zero-lift angles; therefore the zero-lift line lies slightly above the chord line, as shown (with overemphasis) in Figure 7.10*a*.

Now consider the wing pitched to the geometric angle of attack α such that lift is generated, as shown in Figure 7.10*b*. (Recall from Chap. 5 that the geometric angle of attack is the angle between the freestream relative wind and the chord line.) In the same configuration, Figure 7.10*b* demonstrates that the angle between the zero-lift line and the relative wind is equal to the sum of α plus the absolute value of α_{L-0}. This angle is defined as the *absolute angle of attack* α_a. From Figure 7.10*b*, $\alpha_a = \alpha + \alpha_{L-0}$ (using α_{L-0} in an absolute sense). The geometry of Figures 7.10*a* and 7.10*b* should be studied carefully until the concept of α_a is clearly understood.

The definition of the absolute angle of attack has a major advantage. When $\alpha_a = 0$, then $L = 0$, no matter what the camber of the airfoil. To further illustrate, consider the lift curves sketched in Figure 7.11. The conventional plot (as discussed in detail in Chap. 5), C_L versus α, is shown in Figure 7.11*a*. Here, the lift curve does not go through the origin, and of course α_{L-0} is different for different airfoils. In contrast, when C_L is plotted versus α_a, as sketched in Figure 7.11*b*, the curve always goes through the origin (by definition of α_a). The curve in Figure 7.11*b* is identical to that in Figure 7.11*a*; only the abscissa has been translated by the value α_{L-0}.

The use of α_a in lieu of α is common in studies of stability and control. We will adopt this convention for the remainder of this chapter.

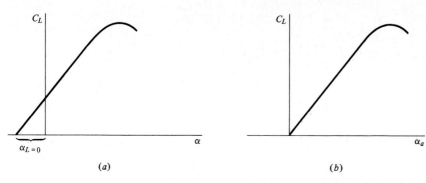

Figure 7.11 Lift coefficient vs. (*a*) geometric angle of attack and (*b*) absolute angle of attack.

7.5 CRITERIA FOR LONGITUDINAL STATIC STABILITY

Static stability and control about all three axes shown in Figure 7.2 is usually a necessity in the design of conventional airplanes. However, a complete description of all three types—lateral, longitudinal, and directional static stability and control (see Figure 7.4)—is beyond the scope of this book. Rather, the intent here is to provide only the flavor of stability and control concepts, and to this end only the airplane's longitudinal motion (pitching motion about the *y* axis) will be considered. This pitching motion is illustrated in Figure 7.4*b*. It takes place in the plane of symmetry of the airplane. Longitudinal stability is also the most important static stability mode; in airplane design, wind-tunnel testing, and flight research, it usually earns more attention than lateral or directional stability.

Consider a rigid airplane with fixed controls, e.g., the elevator in some fixed position. Assume the airplane has been tested in a wind tunnel or free flight and that its variation of M_{cg} with angle of attack has been measured. This variation is illustrated in Figure 7.12, where $C_{M,cg}$ is sketched versus α_a. For many conventional airplanes, the curve is nearly linear, as shown in Figure 7.12. The value of $C_{M,cg}$ at zero lift (where $\alpha_a = 0$) is denoted by $C_{M,0}$. The value of α_a where

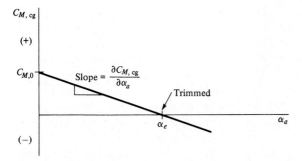

Figure 7.12 Moment coefficient curve with a negative slope.

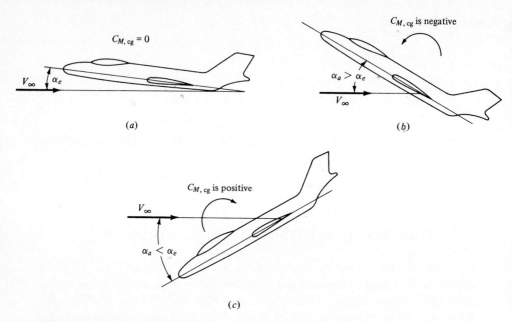

(a)

(b)

(c)

Figure 7.13 Illustration of static stability: (a) Equilibrium position (trimmed). (b) Pitched upward by disturbance. (c) Pitched downward by disturbance. In both b and c, the airplane has the initial tendency to return to its equilibrium position.

$M_{cg} = 0$ is denoted by α_e; as stated in Sec. 7.3, this is the equilibrium, or trim, angle of attack.

Consider the airplane in steady, equilibrium flight at its trim angle of attack α_e as shown in Figure 7.13a. Suddenly, the airplane is disturbed by hitting a wind gust, and the angle of attack is momentarily changed. There are two possibilities: an increase or a decrease in α_a. If the airplane is pitched upward, as shown in Figure 7.13b, then $\alpha_a > \alpha_e$. From Figure 7.12, if $\alpha_a > \alpha_e$, the moment about the center of gravity is negative. As discussed in Sec. 5.4, a negative moment (by convention) is counterclockwise, tending to pitch the nose downward. Hence, in Figure 7.13b, the airplane will initially tend to move back toward its equilibrium position after being disturbed. On the other hand, if the plane is pitched downward by the gust, as shown in Figure 7.13c, then $\alpha_a < \alpha_e$. From Figure 7.12, the resulting moment about the center of gravity will be positive (clockwise) and will tend to pitch the nose upward. Thus, again we have the situation where the airplane will initially tend to move back toward its equilibrium position after being disturbed. From Sec. 7.2, this is precisely the definition of static stability. Therefore, we conclude that an airplane which has a $C_{M,cg}$ versus α_a variation like that shown in Figure 7.12 is *statically stable*. Note from Figure 7.12 that $C_{M,0}$ is positive and that the slope of the curve, $\partial C_{M,cg}/\partial \alpha_a$, is negative. Here, the partial derivative, defined in Sec. 7.2 D, is used for the slope of the moment coefficient curve. This is because (as we shall see) $C_{M,cg}$ depends on a number of

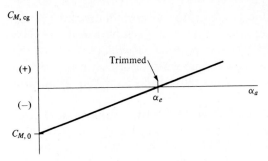

Figure 7.14 Moment coefficient curve with a positive slope.

other variables in addition to α_a, and therefore it is mathematically proper to use $\partial C_{M,cg}/\partial \alpha_a$ rather than $dC_{M,cg}/d\alpha_a$ to represent the slope of the line in Figure 7.12. As defined in Sec. 7.2 D, $\partial C_{M,cg}/\partial \alpha_a$ symbolizes the instantaneous rate of change of $C_{M,cg}$ with respect to α_a, with all other variables held constant.

Consider now a different airplane, with a measured $C_{M,cg}$ variation as shown in Figure 7.14. Imagine the airplane is flying at its trim angle of attack α_e as shown in Figure 7.15a. If it is disturbed by a gust, pitching the nose upward, as shown in Figure 7.15b, then $\alpha_a > \alpha_e$. From Figure 7.14 this results in a positive (clockwise) moment, which tends to pitch the nose even further away from its

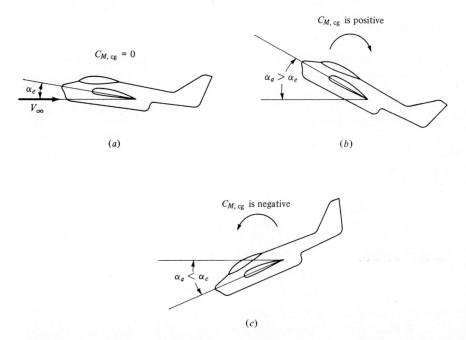

Figure 7.15 Illustration of static instability. (a) Equilibrium position (trimmed). (b) Pitched upward by disturbance. (c) Pitched downward by disturbance. In both b and c, the airplane has the initial tendency to diverge farther away from its equilibrium position.

equilibrium position. Similarly, if the gust pitches the nose downward (Figure 7.15c), a negative (counterclockwise) moment results, which also tends to pitch the nose further away from its equilibrium position. Therefore, because the airplane always tends to diverge from equilibrium when disturbed, it is *statically unstable*. Note from Figure 7.14 that $C_{M,0}$ is negative and $\partial C_{M,\mathrm{cg}}/\partial \alpha_a$ is positive for this airplane.

For both airplanes, Figures 7.12 and 7.14 show a positive value of α_e. Recall from Figure 6.4 that an airplane moves through a range of angle of attack as it flies through its velocity range from V_{stall} (where α_a is the largest) to V_{max} (where α_a is the smallest). The value of α_e must fall within this flight range of angle of attack, or else the airplane cannot be trimmed for steady flight. (Remember that we are assuming a fixed elevator position: we are discussing "stick-fixed" stability.) When α_e does fall within this range, the airplane is *longitudinally balanced*.

From the above considerations, we conclude the following.

The necessary criteria for longitudinal balance and static stability are:

1. $C_{M,0}$ *must be positive*.
2. $\partial C_{M,\mathrm{cg}}/\partial \alpha_a$ *must be negative*.

That is, the $C_{M,\mathrm{cg}}$ curve must look like Figure 7.12.

Of course, implicit in the above criteria is that α_e must also fall within the flight range of angle of attack for the airplane.

We are now in a position to explain why a conventional airplane has a horizontal tail (the horizontal stabilizer shown in Figure 7.3). First, consider an ordinary wing (by itself) with a conventional airfoil, say an NACA 2412 section. Note from the airfoil data in Appendix D that the moment coefficient about aerodynamic center is negative. This is characteristic of all airfoils with positive camber. Now, assume that the wing is at zero lift. In this case, the only moment on the wing is a pure couple, as explained in Sec. 7.3; hence, at zero lift the moment about one point is equal to the moment about any other point. In particular,

$$C_{M,\mathrm{ac}} = C_{M,\mathrm{cg}} \qquad \text{for zero lift (wing only)} \qquad (7.2)$$

On the other hand, examination of Figure 7.12 shows that $C_{M,0}$ is by definition, the moment coefficient about the center of gravity at zero lift (when $\alpha_a = 0$). Hence, from Eq. (7.2)

$$C_{M,0} = C_{M,\mathrm{ac}} \qquad \text{(wing only)} \qquad (7.3)$$

Equation (7.3) demonstrates that, for a wing with positive camber ($C_{M,\mathrm{ac}}$ negative) $C_{M,0}$ is also negative. Hence, such a wing by itself is *unbalanced*. To rectify this situation, a horizontal tail must be added to the airplane, as shown in Figures 7.16a and 7.16b. If the tail is mounted behind the wing, as shown in Figure 7.16a, and if it is inclined downward to produce a negative tail lift as shown, then a

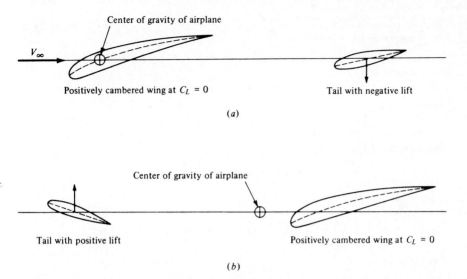

Figure 7.16 (*a*) Conventional wing-tail combination. The tail is set at such an angle as to produce negative lift, thus providing a positive $C_{M,0}$. (*b*) Canard wing-tail combination. The tail is set at such an angle as to produce positive lift, thus providing a positive $C_{M,0}$.

clockwise moment about the center of gravity will be created. If this clockwise moment is strong enough, it will overcome the negative $C_{M,ac}$, and $C_{M,0}$ for the wing-tail combination will become positive. The airplane will then be balanced.

The arrangement shown in Figure 7.16*a* is characteristic of most conventional airplanes. However, the tail can also be placed ahead of the wing, as shown in Figure 7.16*b*; this is called a *canard configuration*. For a canard, the tail is inclined upward to produce a positive lift, hence creating a clockwise moment about the center of gravity. If this moment is strong enough, $C_{M,0}$ for the wing-tail combination will become positive and again the airplane will be balanced. Unfortunately, the forward-located tail of a canard interferes with the smooth aerodynamic flow over the wing. For this and other reasons, canard configurations have not been popular. Of course, a notable exception were the Wright Flyers, which were canards. In fact, it was not until 1910 that the Wright brothers went to a conventional arrangement. Using the word "rudder" to mean elevator, Orville wrote to Wilbur in 1909 that "the difficulty in handling our machine is due to the rudder being in front, which makes it hard to keep on a level course.... I do not think it is necessary to lengthen the machine, but to simply put the rudder behind instead of before." Originally, the Wrights thought the forward-located elevator would help to protect them from the type of fatal crash encountered by Lilienthal. This rationale persisted until the design of their model B in 1910. Finally, a modern example of a canard is the North American XB-70, an experimental supersonic bomber developed for the Air Force in the 1960s. The canard surfaces ahead of the wing are clearly evident in the photograph shown in Figure 7.17.

Figure 7.17 The North American XB-70. Note the canard surfaces immediately behind the cockpit. *(Rockwell International Corp.)*

In retrospect, using essentially qualitative arguments based on physical reasoning and without resort to complicated mathematical formulas, we have developed some fundamental results for longitudinal static stability. Indeed, it is somewhat amazing how far our discussion has progressed on such a qualitative basis. However, we now turn to some quantitative questions. For a given airplane, how far should the wing and tail be separated in order to obtain stability? How large should the tail be made? How do we design for a desired trim angle α_e? These and other such questions are addressed in the remainder of this chapter.

7.6 QUANTITATIVE DISCUSSION: CONTRIBUTION OF THE WING TO M_{cg}

The calculation of moments about the center of gravity of the airplane, M_{cg}, is critical to a study of longitudinal static stability. The previous sections have already underscored this fact. Therefore, we now proceed to consider individually the contributions of the wing, fuselage, and tail to moments about the center of gravity of the airplane, in the end combining them to obtain the total M_{cg}.

Consider the forces and moments on the wing only, as shown in Figure 7.18. Here, the zero-lift line is drawn horizontally for convenience; hence, the relative wind is inclined at the angle α_w with respect to the zero-lift line, where α_w is the absolute angle of attack of the wing. Let c denote the mean zero-lift chord of the wing (the chord measured along the zero-lift line). The difference between

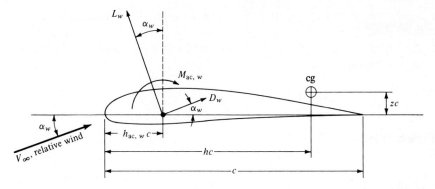

Figure 7.18 Airfoil nomenclature and geometry.

the zero-lift chord and the geometric chord (as defined in Chap. 5) is usually insignificant and will be ignored here. The center of gravity for the airplane is located a distance hc behind the leading edge, and zc above the zero-lift line, as shown. Hence, h and z are coordinates of the center of gravity in fractions of chord length. The aerodynamic center is a distance $h_{ac_w}c$ from the leading edge. The moment of the wing about the aerodynamic center of the wing is denoted by M_{ac_w}, and the wing lift and drag are L_w and D_w, respectively, as shown. As usual, L_w and D_w are perpendicular and parallel, respectively, to the relative wind.

We wish to take moments about the center of gravity with pitch-up moments positive as usual. Clearly, from Figure 7.18, L_w, D_w, and M_{ac_w} all contribute to moments about the center of gravity. For convenience, split L_w and D_w into components perpendicular and parallel to the chord. Then, referring to Figure 7.18, we find the moments about the center of gravity of the airplane due to the wing are

$$M_{cg_w} = M_{ac_w} + L_w \cos \alpha_w \left(hc - h_{ac_w}c \right) + D_w \sin \alpha_w \left(hc - h_{ac_w}c \right)$$
$$+ L_w \sin \alpha_w zc - D_w \cos \alpha_w zc \qquad (7.4)$$

[Study Eq. (7.4) and Figure 7.18 carefully, and make certain that you understand each term before progressing further.] For the normal-flight range of a conventional airplane, α_w is small; hence, the approximation is made that $\cos \alpha_w \approx 1$ and $\sin \alpha_w \approx \alpha_w$ (where α_w is in radians). Then Eq. (7.4) becomes

$$M_{cg_w} = M_{ac_w} + \left(L_w + D_w \alpha_w \right)\left(h - h_{ac_w} \right)c + \left(L_w \alpha_w - D_w \right)zc \qquad (7.5)$$

Dividing Eq. (7.5) by $q_\infty Sc$ and recalling that $C_M = M/q_\infty Sc$, we obtain the moment coefficient about the center of gravity as

$$C_{M, cg_w} = C_{M, ac_w} + \left(C_{L,w} + C_{D,w}\alpha_w \right)\left(h - h_{ac_w} \right)$$
$$+ \left(C_{L,w}\alpha_w - C_{D,w} \right)z \qquad (7.6)$$

For most airplanes, the center of gravity is located close to the zero-lift line;

hence z is usually small ($z \approx 0$) and will be neglected. Furthermore, α_w (in radians) is usually much less than unity, and $C_{D,w}$ is usually less than $C_{L,w}$; hence, the product $C_{D,w}\alpha_w$ is small in comparison to $C_{L,w}$. With these assumptions, Eq. (7.6) simplifies to

$$C_{M,\mathrm{cg}_w} = C_{M,\mathrm{ac}_w} + C_{L,w}\left(h - h_{\mathrm{ac}_w}\right) \tag{7.7}$$

Referring to Figure 7.11b, we find $C_{L,w} = (dC_{L,w}/d\alpha)\alpha_w = a_w\alpha_w$, where a_w is the *lift slope of the wing*. Thus, (Eq. (7.7) can be written as

$$C_{M,\mathrm{cg}_w} = C_{M,\mathrm{ac}_w} + a_w\alpha_w\left(h - h_{\mathrm{ac}_w}\right) \tag{7.8}$$

Equations (7.7) and (7.8) give the contribution of the wing to moments about the center of gravity of the airplane, subject of course to the above assumptions. Closely examine Eqs. (7.7) and (7.8) along with Figure 7.18. On a physical basis, they state that the wing's contribution to M_{cg} is essentially due to two factors: the moment about the aerodynamic center, M_{ac_w}, and the lift acting through the moment arm $(h - h_{\mathrm{ac}_w})c$.

The above results are slightly modified if a fuselage is added to the wing. Consider a cigar-shaped body at an angle of attack to an airstream. This fuselage-type body experiences a moment about its aerodynamic center, plus some lift and drag due to the airflow around it. Now consider the fuselage and wing joined together: a *wing-body combination*. The airflow about this wing-body combination is different from that over the wing and body separately; aerodynamic interference occurs where the flow over the wing affects the fuselage flow, and vice versa. Due to this interference, the moment due to the wing-body combination is *not* simply the sum of the separate wing and fuselage moments. Similarly, the lift and drag of the wing-body combination are affected by aerodynamic interference. Such interference effects are extremely difficult to predict theoretically. Consequently, the lift, drag, and moments of a wing-body combination are usually obtained from wind-tunnel measurements. Let $C_{L_{wb}}$ and $C_{M,\mathrm{ac}_{wb}}$ be the lift coefficient and moment coefficient about the aerodynamic center, respectively, for the wing-body combination. Analogous to Eqs. (7.7) and (7.8) for the wing only, the contribution of the wing-body combination to M_{cg} is

$$C_{M,\mathrm{cg}_{wb}} = C_{M,\mathrm{ac}_{wb}} + C_{L_{wb}}\left(h - h_{\mathrm{ac}_{wb}}\right) \tag{7.9}$$

$$C_{M,\mathrm{cg}_{wb}} = C_{M,\mathrm{ac}_{wb}} + a_{wb}\alpha_{wb}\left(h - h_{\mathrm{ac}_{wb}}\right) \tag{7.10}$$

where a_{wb} and α_{wb} are the slope of the lift curve and absolute angle of attack, respectively, for the wing-body combination. In general, adding a fuselage to a wing shifts the aerodynamic center forward, increases the lift curve slope, and contributes a negative increment to the moment about the aerodynamic center.

Emphasis is again made that the aerodynamic coefficients in Eqs. (7.9) and (7.10) are almost always obtained from wind-tunnel data.

Example 7.2 For a given wing-body combination, the aerodynamic center lies 0.05 chord length ahead of the center of gravity. The moment coefficient about the aerodynamic center is -0.016. If the lift coefficient is 0.45, calculate the moment coefficient about the center of gravity.

SOLUTION From Eq. (7.9),

$$C_{M,\text{cg}_{wb}} = C_{M,\text{ac}_{wb}} + C_{L_{wb}}\left(h - h_{\text{ac}_{wb}}\right)$$

where

$$h - h_{\text{ac}_{wb}} = 0.05$$

$$C_{L_{wb}} = 0.45$$

$$C_{M,\text{ac}_{wb}} = -0.016$$

Thus

$$C_{M,\text{cg}_{wb}} = -0.016 + 0.45(0.05) = \boxed{0.0065}$$

Example 7.3 A wing-body model is tested in a subsonic wind tunnel. The lift is found to be zero at a geometric angle of attack $\alpha = -1.5°$. At $\alpha = 5°$, the lift coefficient is measured as 0.52. Also, at $\alpha = 1.0°$ and $7.88°$, the moment coefficients about the center of gravity are measured as -0.01 and 0.05, respectively. The center of gravity is located at $0.35c$. Calculate the location of the aerodynamic center and the value of $C_{M,\text{ac}_{wb}}$.

SOLUTION First, calculate the lift slope:

$$a_{wb} \equiv \frac{dC_L}{d\alpha} = \frac{0.52 - 0}{5 - (-1.5)} = \frac{0.52}{6.5} = 0.08 \text{ per degree}$$

Write Eq. (7.10),

$$C_{M,\text{cg}_{wb}} = C_{M,\text{ac}_{wb}} + a_{wb}\alpha_{wb}\left(h - h_{\text{ac}_{wb}}\right)$$

evaluated at $\alpha = 1.0°$ [remember that α is the geometric angle of attack, whereas in Eq. (7.10) α_{wb} is the absolute angle of attack]:

$$-0.01 = C_{M,\text{ac}_{wb}} + 0.08(1 + 1.5)\left(h - h_{\text{ac}_{wb}}\right)$$

Then evaluate it at $\alpha = 7.88°$:

$$0.05 = C_{M,\text{ac}_{wb}} + 0.08(7.88 + 1.5)\left(h - h_{\text{ac}_{wb}}\right)$$

The above two equations have two unknowns, $C_{M,\text{ac}_{wb}}$ and $h - h_{\text{ac}_{wb}}$. They can be solved simultaneously.

Subtracting the second equation from the first, we get

$$-0.06 = 0 - 0.55\left(h - h_{\text{ac}_{wb}}\right)$$

$$h - h_{\text{ac}_{wb}} = \frac{-0.06}{-0.55} = 0.11$$

The value of h is given: $h = 0.35$. Thus

$$h_{\text{ac}_{wb}} = 0.35 - 0.11 = \boxed{0.24}$$

In turn $$-0.01 = C_{M, ac_{wb}} + 0.08(1 - 1.5)(0.11)$$

$$C_{M, ac_{wb}} = \boxed{-0.032}$$

7.7 CONTRIBUTION OF THE TAIL TO M_{cg}

An analysis of moments due to an isolated tail taken independently of the airplane would be the same as that given for the isolated wing above. However, in real life the tail is obviously connected to the airplane itself; it is not isolated. Moreover, the tail is generally mounted behind the wing; hence it feels the wake of the airflow over the wing. As a result, there are two interference effects that influence the tail aerodynamics:

1. The airflow at the tail is deflected downward by the *downwash* due to the finite wing (see Secs. 5.13 and 5.14); i.e., the relative wind seen by the tail is not in the same direction as the relative wind V_∞ seen by the wing.
2. Due to the retarding force of skin friction and pressure drag over the wing, the airflow reaching the tail has been slowed. Therefore, the velocity of the relative wind seen by the tail is less than V_∞. In turn, the dynamic pressure seen by the tail is less than q_∞.

These effects are illustrated in Figure 7.19. Here V_∞ is the relative wind as seen by the wing, and V' is the relative wind at the tail, inclined below V_∞ by the downwash angle ε. The tail lift and drag, L_t and D_t, are (by definition) perpendicular and parallel, respectively, to V'. In contrast, the lift and drag of the complete airplane are always (by definition) perpendicular and parallel, respectively, to V_∞. Therefore, considering components of L_t and D_t perpendicular to V_∞, we demonstrate in Figure 7.19 that the tail contribution to the total airplane lift is $L_t \cos \varepsilon - D_t \sin \varepsilon$. In many cases, ε is very small, and thus $L_t \cos \varepsilon - $

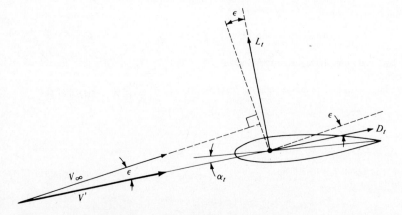

Figure 7.19 Flow and force diagram in the vicinity of the tail.

$D_t \sin \varepsilon \approx L_t$. Hence, for all practical purposes, it is sufficient to add the tail lift directly to the wing-body lift to obtain the lift of the complete airplane.

Consider the tail in relation to the wing-body zero-lift line, as illustrated in Figure 7.20. It is useful to pause and study this figure. The wing-body combination is at an absolute angle of attack α_{wb}. The tail is twisted downward in order to provide a positive $C_{M,0}$, as discussed at the end of Sec. 7.5. Thus the zero-lift line of the tail is intentionally inclined to the zero-lift line of the wing-body combination at the *tail-setting angle* i_t. (The airfoil section of the tail is generally symmetric, for which the tail zero-lift line and the tail chord line are the same.) The absolute angle of attack of the tail, α_t is measured between the local relative wind V' and the tail zero-lift line. The tail has an aerodynamic center, about which there is a moment M_{ac_t} and through which L_t and D_t act perpendicular and parallel, respectively, to V'. As before, V' is inclined below V_∞ by the downwash angle ε; hence L_t makes an angle $\alpha_{wb} - \varepsilon$ with the vertical. The tail aerodynamic center is located a distance l_t behind and z_t below the center of gravity of the airplane. Make certain to carefully study the geometry shown in Figure 7.20; it is fundamental to the derivation which follows.

Split L_t and D_t into their vertical components, $L_t \cos(\alpha_{wb} - \varepsilon)$ and $D_t \sin(\alpha_{wb} - \varepsilon)$, and their horizontal components, $L_t \sin(\alpha_{wb} - \varepsilon)$ and $D_t \cos(\alpha_{wb} - \varepsilon)$. By inspection of Figure 7.20, the sum of moments about the center of gravity due to L_t, D_t, and M_{ac_t} of the tail is

$$M_{cg_t} = -l_t \left[L_t \cos(\alpha_{wb} - \varepsilon) + D_t \sin(\alpha_{wb} - \varepsilon) \right]$$

$$+ z_t L_t \sin(\alpha_{wb} - \varepsilon) - z_t D_t \cos(\alpha_{wb} - \varepsilon) + M_{ac_t} \qquad (7.11)$$

Here, M_{cg_t} denotes the contribution to moments about the airplane's center of gravity due to the horizontal tail.

In Eq. (7.11), the first term on the right-hand side, $l_t L_t \cos(\alpha_{wb} - \varepsilon)$, is by far the largest in magnitude. In fact, for conventional airplanes, the following simplifications are reasonable:

1. $z_t \ll l_t$
2. $D_t \ll L_t$
3. The angle $\alpha_{wb} - \varepsilon$ is small, hence $\sin(\alpha_{wb} - \varepsilon) \approx 0$ and $\cos(\alpha_{wb} - \varepsilon) \approx 1$.
4. M_{ac_t} is small in magnitude.

With the above approximations, which are based on experience, Eq. (7.11) is dramatically simplified to

$$M_{cg_t} = -l_t L_t \qquad (7.12)$$

Define the *tail lift coefficient*, based on freestream dynamic pressure $q_\infty = \frac{1}{2}\rho_\infty V_\infty^2$ and the tail planform area S_t, as

$$C_{L,t} = \frac{L_t}{q_\infty S_t} \qquad (7.13)$$

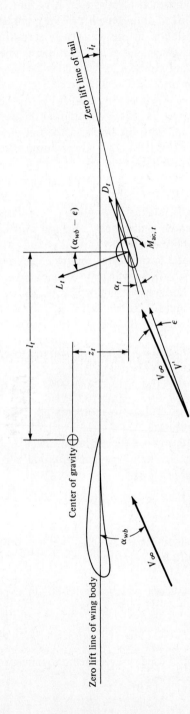

Figure 7.20 Geometry of wing-tail combination.

Combining Eqs. (7.12) and (7.13), we obtain

$$M_{cg_t} = -l_t q_\infty S_t C_{L,t} \tag{7.14}$$

Dividing Eq. (7.14) by $q_\infty Sc$, where c is the wing chord and S is the wing planform area,

$$\frac{M_{cg_t}}{q_\infty Sc} \equiv C_{M,cg_t} = -\frac{l_t S_t}{cS} C_{L,t} \tag{7.15}$$

Examining the right-hand side of Eq. (7.15), note that $l_t S_t$ is a *volume* characteristic of the size and location of the tail and that cS is a *volume* characteristic of the size of the wing. The ratio of these two volumes is called the *tail volume ratio* V_H, where

$$V_H \equiv \frac{l_t S_t}{cS} \tag{7.16}$$

Thus, Eq. (7.15) becomes

$$\boxed{C_{M,cg_t} = -V_H C_{L,t}} \tag{7.17}$$

The simple relation in Eq. (7.17) gives the total contribution of the tail to moments about the airplane's center of gravity. With the simplifications above and referring to Figure 7.20, Eqs. (7.12) and (7.17) say that the moment is equal to tail lift operating through the moment arm l_t.

It will be useful to couch Eq. (7.17) in terms of angle of attack, as was done in Eq. (7.10) for the wing-body combination. Keep in mind that the stability criterion in Figure 7.12 involves $\partial C_{M,cg}/\partial \alpha_a$; hence equations in terms of α_a are directly useful. Specifically, referring to the geometry of Figure 7.20, we see that the angle of attack of the tail is

$$\alpha_t = \alpha_{wb} - i_t - \varepsilon \tag{7.18}$$

Let a_t denote the lift slope of the tail. Thus, from Eq. (7.18),

$$C_{L,t} = a_t \alpha_t = a_t(\alpha_{wb} - i_t - \varepsilon) \tag{7.19}$$

The downwash angle ε is difficult to predict theoretically and is usually obtained from experiment. It can be written as

$$\varepsilon = \varepsilon_0 + \left(\frac{\partial \varepsilon}{\partial \alpha}\right) \alpha_{wb} \tag{7.20}$$

where ε_0 is the downwash angle when the wing-body combination is at zero lift. Both ε_0 and $\partial \varepsilon/\partial \alpha$ are usually obtained from wind-tunnel data. Thus, combining Eqs. (7.19) and (7.20) yields

$$C_{L,t} = a_t \alpha_{wb}\left(1 - \frac{\partial \varepsilon}{\partial \alpha}\right) - a_t(i_t + \varepsilon_0) \tag{7.21}$$

Substituting Eq. (7.21) into (7.17), we obtain

$$C_{M,\text{cg}_t} = -a_t V_H \alpha_{\text{wb}}\left(1 - \frac{\partial \varepsilon}{\partial \alpha}\right) + a_t V_H (\varepsilon_0 + i_t) \tag{7.22}$$

Equation (7.22), although lengthier than Eq. (7.17), contains the explicit dependence on angle of attack and will be useful for our subsequent discussions.

7.8 TOTAL PITCHING MOMENT ABOUT THE CENTER OF GRAVITY

Consider the airplane as a whole. The total M_{cg} is due to the contribution of the wing-body combination, plus that of the tail:

$$C_{M,\text{cg}} = C_{M,\text{cg}_{\text{wb}}} + C_{M,\text{cg}_t} \tag{7.23}$$

Here, $C_{M\,\text{cg}}$ is the total moment coefficient about the center of gravity for the complete airplane. Substituting Eqs. (7.9) and (7.17) into (7.23),

$$C_{M,\text{cg}} = C_{M,\text{ac}_{\text{wb}}} + C_{L_{\text{wb}}}\left(h - h_{\text{ac}_{\text{wb}}}\right) - V_H C_{L,t} \tag{7.24}$$

In terms of angle of attack, an alternate expression can be obtained by substituting Eqs. (7.10) and (7.22) into (7.23):

$$C_{M,\text{cg}} = C_{M,\text{ac}_{\text{wb}}} + a_{\text{wb}}\alpha_{\text{wb}}\left[\left(h - h_{\text{ac}_{\text{wb}}}\right) - V_H \frac{a_t}{a_{\text{wb}}}\left(1 - \frac{\partial \varepsilon}{\partial \alpha}\right)\right] + V_H a_t (i_t + \varepsilon_0) \tag{7.25}$$

The angle of attack needs further clarification. Referring again to Figure 7.12, we find the moment coefficient curve is usually obtained from wind-tunnel data, preferably on a model of the complete airplane. Hence, α_a in Figure 7.12 should be interpreted as the absolute angle of attack referenced to the zero-lift line of the *complete airplane*, which is not necessarily the same as the zero-lift line for the wing-body combination. This comparison is sketched in Figure 7.21. However, for many conventional aircraft, the difference is small. Therefore, in the remainder of this chapter, we will assume the two zero-lift lines in Figure 7.21 to be the same. Thus, α_{wb} becomes the angle of attack of the complete airplane, α_a. Consistent with this assumption, the total lift of the airplane is due to the wing-body combination, with the tail lift neglected. Hence, $C_{L_{\text{wb}}} = C_L$ and the lift slope $a_{\text{wb}} = a$, where C_L and a are for the complete airplane. With these interpretations, Eq. (7.25) can be rewritten as

$$C_{M,\text{cg}} = C_{M,\text{ac}_{\text{wb}}} + a\alpha_a\left[\left(h - h_{\text{ac}_{\text{wb}}}\right) - V_H \frac{a_t}{a}\left(1 - \frac{\partial \varepsilon}{\partial \alpha}\right)\right] + V_H a_t (i_t + \varepsilon_0) \tag{7.26}$$

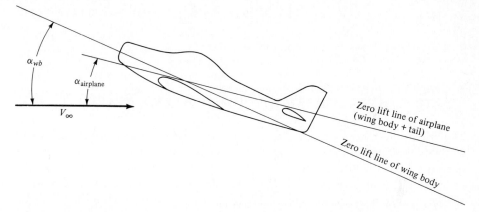

Figure 7.21 Zero-lift line of the wing-body compared with that of the complete airplane.

Equation (7.26) is the same as Eq. (7.25), except that the subscript "wb" on some terms has been dropped in deference to properties for the whole airplane.

Example 7.4 Consider the wing-body model in Example 7.3. The area and chord of the wing are 0.1 m² and 0.1 m, respectively. Now assume that a horizontal tail is added to this model. The distance from the airplane's center of gravity to the tail's aerodynamic center is 0.17 m, the tail area is 0.02 m², the tail-setting angle is 2.7°, the tail lift slope is 0.1 per degree, and from experimental measurement, $\varepsilon_0 = 0$ and $\partial\varepsilon/\partial\alpha = 0.35$. If $\alpha = 7.88°$, calculate $C_{M,\text{cg}}$ for the airplane model.

SOLUTION From Eq. (7.26)

$$C_{M.\text{cg}} = C_{M.\text{ac}_{\text{wb}}} + a\alpha_a\left[\left(h - h_{\text{ac}_{\text{wb}}}\right) - V_H\frac{a_t}{a}\left(1 - \frac{\partial\varepsilon}{\partial\alpha}\right)\right] + V_H a_t\left(i_t + \varepsilon_0\right)$$

where $C_{M.\text{ac}_{\text{wb}}} = -0.032$ (from Example 7.3)

$a = 0.08$ (from Example 7.3)

$\alpha_a = 7.88 + 1.5 = 9.38°$ (from Example 7.3)

$\left(h - h_{\text{ac}_{\text{wb}}}\right) = 0.11$ (from Example 7.3)

$V_H = \dfrac{l_t S_t}{cS} = \dfrac{0.17(0.02)}{0.1(0.1)} = 0.34$

$a_t = 0.1$ per degree

$\dfrac{\partial\varepsilon}{\partial\alpha} = 0.35$

$i_t = 2.7°$

$\varepsilon_0 = 0$

Thus $\qquad C_{M,\text{cg}} = -0.032 + (0.08)(9.38)\left[0.11 - 0.34\dfrac{0.1}{0.08}(1-0.35)\right] + 0.34(0.1)(2.7+0)$

$$= -0.032 - 0.125 + 0.092 = \boxed{-0.065}$$

7.9 EQUATIONS FOR LONGITUDINAL STATIC STABILITY

The criteria necessary for longitudinal balance and static stability were developed in Sec. 7.5. They are (1) $C_{M,0}$ must be positive and (2) $\partial C_{M,\text{cg}}/\partial \alpha_a$ must be negative, both conditions with the implicit assumption that α_e falls within the practical flight range of angle of attack; i.e., the moment coefficient curve must be similar to that sketched in Figure 7.12. In turn, the ensuing sections developed a quantitative formalism for static stability culminating in Eq. (7.26) for $C_{M,\text{cg}}$. The purpose of the present section is to combine the above results in order to obtain formulas for the direct calculation of $C_{M,0}$ and $\partial C_{M,\text{cg}}/\partial \alpha_a$. In this manner, we will be able to make a quantitative assessment of the longitudinal static stability of a given airplane, as well as point out some basic philosophy of airplane design.

Recall that, by definition, $C_{M,0}$ is the value of $C_{M,\text{cg}}$ when $\alpha_a = 0$, that is, when the lift is zero. Substituting $\alpha_a = 0$ into Eq. (7.26), we directly obtain

$$\boxed{C_{M,0} \equiv \left(C_{M,\text{cg}}\right)_{L=0} = C_{M,\text{ac}_{\text{wb}}} + V_H a_t(i_t + \varepsilon_0)} \qquad (7.27)$$

Examine Eq. (7.27). We know that $C_{M,0}$ must be positive in order to balance the airplane. However, the previous sections have pointed out that $C_{M,\text{ac}_{\text{wb}}}$ is negative for conventional airplanes. Therefore, $V_H a_t(i_t + \varepsilon_0)$ must be positive and large enough to more than counterbalance the negative $C_{M,\text{ac}}$. Both V_H and a_t are positive quantities, and ε_0 is usually so small that it exerts only a minor effect. Thus, i_t *must be a positive quantity.* This verifies our previous physical arguments that the tail must be set at an angle relative to the wing in the manner shown in Figures 7.16a and 7.20. This allows the tail to generate enough negative lift to produce a positive $C_{M,0}$.

Consider now the slope of the moment coefficient curve. Differentiating Eq. (7.26) with respect to α_a, we obtain

$$\boxed{\dfrac{\partial C_{M,\text{cg}}}{\partial \alpha_a} = a\left[(h - h_{\text{ac}_{\text{wb}}}) - V_H\dfrac{a_t}{a}\left(1 - \dfrac{\partial \varepsilon}{\partial \alpha}\right)\right]} \qquad (7.28)$$

This equation clearly shows the powerful influence of the location h of the center of gravity and the tail volume ratio V_H in determining longitudinal static stability.

Equations (7.27) and (7.28) allow us to check the static stability of a given airplane, assuming we have some wind-tunnel data for a, a_t, $C_{M,\text{ac}_{\text{wb}}}$, ε_0, and $\partial \varepsilon/\partial \alpha$. They also establish a certain philosophy in the design of an airplane. For example, consider an airplane where the location h of the center of gravity is essentially dictated by payload or other mission requirements. Then, the desired amount of static stability can be obtained simply by designing V_H large enough, via Eq. (7.28). Once V_H is fixed in this manner, then the desired $C_{M,0}$ (or the desired α_e) can be obtained by designing i_t appropriately, via Eq. (7.27). Thus, the values of $C_{M,0}$ and $\partial C_{M,cg}/\partial\alpha_a$ basically dictate the design values of i_t and V_H, respectively (for a fixed center of gravity location).

Example 7.5 Consider the wing-body-tail wind-tunnel model of Example 7.4. Does this model have longitudinal static stability and balance?

SOLUTION From Eq. (7.28)

$$\frac{\partial C_{M,cg}}{\partial\alpha_a} = a\left[\left(h - h_{ac_{wb}}\right) - V_H\frac{a_t}{a}\left(1 - \frac{\partial\varepsilon}{\partial\alpha}\right)\right]$$

where, from Examples 7.3 and 7.4,

$$a = 0.08$$

$$h - h_{ac_{wb}} = 0.11$$

$$V_H = 0.34$$

$$a_t = 0.1 \text{ per degree}$$

$$\frac{\partial\varepsilon}{\partial\alpha} = 0.35$$

Thus

$$\frac{\partial C_{M,cg}}{\partial\alpha_a} = 0.08\left[0.11 - 0.34\frac{0.1}{0.08}(1 - 0.35)\right]$$

$$= \boxed{-0.0133}$$

The slope of the moment coefficient curve is negative, hence the airplane model is statically stable.

However, is the model longitudinally balanced? To answer this, we must find $C_{M,0}$, which in combination with the above result for $\partial C_{M,cg}/\partial\alpha$, will yield the equilibrium angle of attack α_e. From Eq. (7.27),

$$C_{M,0} = C_{M,ac_{wb}} + V_H a_t (i_t + \varepsilon_0)$$

where from Examples 7.3 and 7.4,

$$C_{M,ac_{wb}} = -0.032$$

$$i_t = 2.7°$$

Thus

$$C_{M,0} = -0.032 + 0.34(0.1)(2.7)$$

$$= 0.06$$

From Figure 7.12, the equilibrium angle of attack is obtained from

$$0 = 0.06 - 0.0133\alpha_e$$

Thus

$$\alpha_e = 4.5°$$

Clearly, this angle of attack falls within the reasonable flight range. Therefore, the airplane is longitudinally balanced as well as statically stable.

7.10 THE NEUTRAL POINT

Consider the situation where the location h of the center of gravity is allowed to move, with everything else remaining fixed. In fact, Eq. (7.28) indicates that static stability is a strong function of h. Indeed, the value of $\partial C_{M,\text{cg}}/\partial \alpha_a$ can always be made negative by properly locating the center of gravity. In the same vein, there is one specific location of the center of gravity such that $\partial C_{M,\text{cg}}/\partial \alpha_a = 0$. The value of h when this condition holds is defined as the neutral point, denoted by h_n. When $h = h_n$, the slope of the moment coefficient curve is zero, as illustrated in Figure 7.22.

The location of the neutral point is readily obtained from Eq. (7.28) by setting $h = h_n$ and $\partial C_{M,\text{cg}}/\partial \alpha_a = 0$, as follows.

$$0 = a\left[h_n - h_{\text{ac}_{\text{wb}}} - V_H \frac{a_t}{a}\left(1 - \frac{\partial \varepsilon}{\partial \alpha}\right)\right] \tag{7.29}$$

Solving Eq. (7.29) for h_n,

$$\boxed{h_n = h_{\text{ac}_{\text{wb}}} + V_H \frac{a_t}{a}\left(1 - \frac{\partial \varepsilon}{\partial \alpha}\right)} \tag{7.30}$$

Examine Eq. (7.30). The quantities on the right-hand side are, for all practical purposes, established by the design configuration of the airplane. Thus, for a *given* airplane design, the neutral point is a *fixed quantity*, i.e., a point that is frozen somewhere on the airplane. It is quite independent of the actual location h of the center of gravity.

The concept of the neutral point is introduced as an alternative stability criterion. For example, inspection of Eqs. (7.28) and (7.30) show that $\partial C_{M,\text{cg}}/\partial \alpha_a$ is negative, zero, or positive depending on whether h is less than, equal to, or greater than h_n. These situations are sketched in Figure 7.22. Remember that h is measured from the leading edge of the wing, as shown in Figure 7.18. Hence,

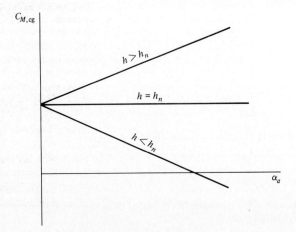

Figure 7.22 Effect of the location of the center of gravity, relative to the neutral point, on static stability.

$h < h_n$ means that the center of gravity location is *forward* of the neutral point. Thus, an alternative stability criterion is:

> For longitudinal static stability, the position of the center of gravity must always be forward of the neutral point.

Recall that the definition of the aerodynamic center for a wing is that point about which moments are independent of angle of attack. This concept can now be extrapolated to the whole airplane by considering again Figure 7.22. Clearly, when $h = h_n$, $C_{M,\mathrm{cg}}$ is independent of angle of attack. Therefore, the neutral point might be considered the aerodynamic center of the complete airplane.

Again examining Eq. (7.30), we see that the tail strongly influences the location of the neutral point. *By proper selection of the tail parameters, principally V_H, h_n can be located at will by the designer.*

Example 7.6 For the wind-tunnel model of Examples 7.3 to 7.5, calculate the neutral point location.

SOLUTION From Eq. (7.30)

$$h_n = h_{ac_{wb}} + V_H \frac{a_t}{a} \left(1 - \frac{\partial \varepsilon}{\partial \alpha}\right)$$

where $h_{ac_{wb}} = 0.24$ (from Example 7.3).

Thus
$$h_n = 0.24 + 0.34 \frac{0.1}{0.08} (1 - 0.35)$$

$$\boxed{h_n = 0.516}$$

Note from Example 7.3 that $h = 0.35$. Compare this center of gravity location with the neutral point location of 0.516. The center of gravity is comfortably *forward* of the neutral point; this again confirms the results of Example 7.5 that the airplane is statically stable.

7.11 THE STATIC MARGIN

A corollary to the above discussion can be obtained, as follows. Solve Eq. (7.30) for $h_{ac_{wb}}$.

$$h_{ac_{wb}} = h_n - V_H \frac{a_t}{a} \left(1 - \frac{\partial \varepsilon}{\partial \alpha}\right) \tag{7.31}$$

Note that in Eqs. (7.29) to (7.31), the value of V_H is not precisely the same number as in Eq. (7.28). Indeed, in Eq. (7.28), V_H is based on the moment arm l_t measured from the center of gravity location, as shown in Figure 7.20. In contrast, in Eq. (7.29), the center of gravity location has been moved to the neutral point, and V_H is therefore based on the moment arm measured from the neutral point location. However, the difference is usually small, and this effect will be ignored here. Therefore, substituting Eq. (7.31) into Eq. (7.28) and cancelling

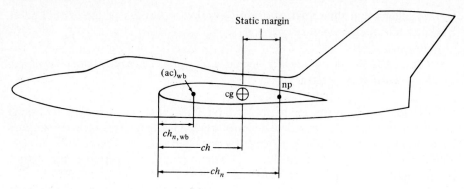

Figure 7.23 Illustration of the static margin.

the terms involving V_H, we obtain

$$\frac{\partial C_{M,cg}}{\partial \alpha_a} = a(h - h_n) \tag{7.32}$$

The distance, $h_n - h$, is defined as the *static margin* and is illustrated in Figure 7.23. Thus, from Eq. (7.32),

$$\frac{\partial C_{M,cg}}{\partial \alpha_a} = -a(h_n - h) = -a \times \text{static margin} \tag{7.33}$$

Equation (7.33) shows that the static margin is a direct measure of longitudinal static stability. For static stability, the static margin must be positive. Moreover, the larger the static margin, the more stable is the airplane.

Example 7.7 For the wind-tunnel model of the previous examples, calculate the static margin.

SOLUTION From Example 7.6, $h_n = 0.516$ and $h = 0.35$. Thus, by definition,

$$\text{static margin} \equiv h_n - h = 0.516 - 0.35 = \boxed{0.166}$$

For a check on the consistency of our calculations, consider Eq. (7.33).

$$\frac{\partial C_{M,cg}}{\partial \alpha_a} = -a \times \text{static margin}$$

$$= -0.08(0.166)$$

$$= -0.0133 \text{ per degree}$$

This is the same value calculated in Example 7.5; our calculations are indeed consistent.

7.12 THE CONCEPT OF STATIC LONGITUDINAL CONTROL

A study of stability and control is double-barrelled. The first aspect—that of stability itself—has been the subject of the preceding sections. However, for the remainder of this chapter, the focus will turn to the second aspect—control.

Consider a statically stable airplane in trimmed (equilibrium) flight. Recalling Figure 7.12, the airplane must therefore be flying at the trim angle of attack α_e. In turn, this value of α_e corresponds to a definite value of lift coefficient, namely, the trim lift coefficient $C_{L_{trim}}$. For steady, level flight, this corresponds to a definite velocity, which from Eq. (6.25) is

$$V_{trim} = \sqrt{\frac{2W}{\rho_\infty S C_{L_{trim}}}} \tag{7.34}$$

Now assume that the pilot wishes to fly at a lower velocity $V_\infty < V_{trim}$. At a lower velocity, the lift coefficient, hence the angle of attack, must be increased in order to offset the decrease in dynamic pressure (remember from Chap. 6 that the lift must always balance the weight for steady, level flight). However, from Figure 7.12, if α is increased, $C_{M,cg}$ becomes negative (i.e., the moment about the center of gravity is no longer zero), and the airplane is no longer trimmed. Consequently, if nothing else is changed about the airplane, it cannot achieve steady, level, equilibrium flight at any other velocity than V_{trim} or at any other angle of attack than α_e.

Obviously, this is an intolerable situation—an airplane must be able to change its velocity at the will of the pilot and still remain balanced. The only way to accomplish this is to effectively change the moment coefficient curve for the airplane. Say that the pilot wishes to fly at a *faster* velocity but still remain in steady, level, balanced flight. The lift coefficient must decrease, hence a new angle of attack α_n must be obtained, where $\alpha_n < \alpha_e$. At the same time, the moment coefficient curve must be changed such that $C_{M,cg} = 0$ at α_n. Figures 7.24 and 7.25 demonstrate two methods of achieving this change. In Figure 7.24, the slope is made more negative, such that $C_{M,cg}$ goes through zero at α_n. From Eq. (7.28) or (7.32), the slope can be changed by shifting the center of gravity. In our example, the center of gravity must be shifted forward. Otto Lilienthal (see Sec. 1.5) used this method in his gliding flights. Figure 1.15 shows Lilienthal hanging loosely below his glider; by simply swinging his hips, he was able to shift the center of gravity and change the stability of the aircraft. This principle is carried over today to the modern hang gliders for sports use.

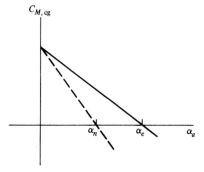

Figure 7.24 Change in trim angle of attack due to change in slope of moment coefficient curve.

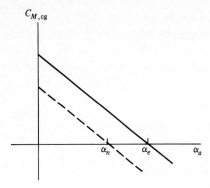

Figure 7.25 Change in trim angle of attack due to change in $C_{M,0}$.

However, for a conventional airplane, shifting the center of gravity is highly impractical. Therefore, another method for changing the moment curve is employed, as shown in Figure 7.25. Here, the slope remains the same, but $C_{M,0}$ is changed such that $C_{M,cg} = 0$ at α_n. This is accomplished by deflecting the elevator on the horizontal tail. Hence, we have arrived at a major concept of static, longitudinal control, namely, that the elevator deflection can be used to control the trim angle of attack, hence to control the equilibrium velocity of the airplane.

Consider Figure 7.25. We stated above, without proof, that a translation of the moment curve without a change in slope can be obtained simply by deflecting the elevator. But *how* and *to what extent* does the elevator deflection change $C_{M,cg}$? To provide some answers, first consider the horizontal tail with the elevator fixed in the neutral position, i.e., no elevator deflection, as shown in Figure 7.26. The absolute angle of attack of the tail is α_t, as defined earlier. The variation of tail lift coefficient with α_t is also sketched in Figure 7.26; note that it has the same general shape as the airfoil and wing lift curves discussed in Chap. 5. Now,

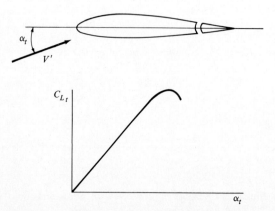

Figure 7.26 Tail-lift coefficient curve with no elevator deflection.

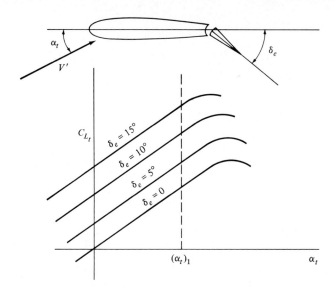

Figure 7.27 Tail-lift coefficient with elevator deflection.

assume that the elevator is deflected downward through the angle δ_e, as shown in Figure 7.27. This is the same picture as a wing with a deflected flap, as was discussed in Sec. 5.17. Consequently, just as in the case of a deflected flap, the deflected elevator causes the tail lift coefficient curve to shift to the left, as shown in Figure 7.27. By convention (and for convenience later on), a downward elevator deflection is positive. Therefore, if the elevator is deflected by an angle of, say, 5° and then held fixed as the complete tail is pitched through a range of α_t, the tail lift curve is translated to the left. If the elevator is then deflected farther, say to 10°, the lift curve is shifted even farther to the left. This behavior is clearly illustrated in Figure 7.27. Note that for all the lift curves, the slope, $\partial C_{L,t}/\partial \alpha_t$, is the same.

With the above in mind, now consider the tail at a fixed angle of attack, say $(\alpha_t)_1$. If the elevator is deflected from, say, 0 to 15°, $C_{L,t}$ will increase along the vertical dashed line in Figure 7.27. This variation can be cross-plotted as $C_{L,t}$ versus δ_e, as shown in Figure 7.28. For most conventional airplanes, the curve in Figure 7.28 is essentially linear, and its slope, $\partial C_{L,t}/\partial \delta_e$, is called the *elevator control effectiveness*. This quantity is a direct measure of the "strength" of the elevator as a control; because δ_e has been defined as positive for downward deflections, the $\partial C_{L,t}/\partial \delta_e$ is *always positive*.

Consequently, the tail lift coefficient is a function of *both* α_t and δ_e (hence, the partial derivative notation is used, as discussed earlier). Keep in mind that, physically, $\partial C_{L,t}/\partial \alpha_t$ is the rate of change of $C_{L,t}$ with respect to α_t, keeping δ_e constant; similarly, $\partial C_{L,t}/\partial \delta_e$ is the rate of change of $C_{L,t}$ with respect to δ_e,

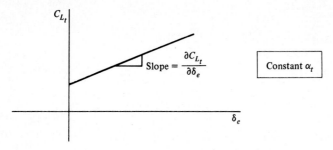

Figure 7.28 Tail-lift coefficient vs. elevator deflection at constant angle of attack; a cross plot of Figure 7.27.

keeping α_t constant. Hence, on a physical basis,

$$C_{L,t} = \frac{\partial C_{L,t}}{\partial \alpha_t}\alpha_t + \frac{\partial C_{L,t}}{\partial \delta_e}\delta_e \qquad (7.35)$$

Recalling that the tail lift slope is $a_t = \partial C_{L,t}/\partial \alpha_t$, Eq. (7.35) can be written as

$$C_{L,t} = a_t\alpha_t + \frac{\partial C_{L,t}}{\partial \delta_e}\delta_e \qquad (7.36)$$

Substituting Eq. (7.36) into (7.24), we have for the pitching moment about the center of gravity,

$$\boxed{C_{M,\text{cg}} = C_{M,\text{ac}_{wb}} + C_{L,\text{wb}}(h - h_{\text{ac}}) - V_H\left(a_t\alpha_t + \frac{\partial C_{L,t}}{\partial \delta_e}\delta_e\right)} \qquad (7.37)$$

Equation (7.37) gives explicitly the effect of elevator deflection on moments about the center of gravity of the airplane.

The rate of change of $C_{M,\text{cg}}$ due *only* to elevator deflection is, by definition, $\partial C_{M,\text{cg}}/\partial \delta_e$. This partial derivative can be found by differentiating Eq. (7.37) with respect to δ_e, keeping everything else constant.

$$\frac{\partial C_{M,\text{cg}}}{\partial \delta_e} = -V_H\frac{\partial C_{L,t}}{\partial \delta_e} \qquad (7.38)$$

Note that, from Figure 7.28, $\partial C_{L,t}/\partial \delta_e$ is constant; moreover, V_H is a specific value for the given airplane. Thus, the right-hand side of Eq. (7.38) is a constant. Therefore, on a physical basis, the increment in $C_{M,\text{cg}}$ due *only* to a given elevator deflection δ_e is

$$\Delta C_{M,\text{cg}} = -V_H\frac{\partial C_{L,t}}{\partial \delta_e}\delta_e \qquad (7.39)$$

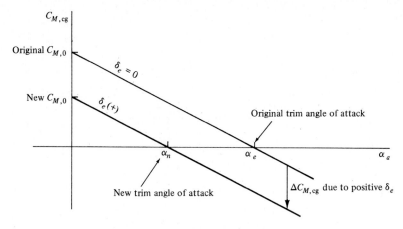

Figure 7.29 Effect of elevator deflection on moment coefficient.

Equation (7.39) now answers the questions asked earlier: how and to what extent the elevator deflection changes $C_{M,cg}$. Consider the moment curve labeled $\delta_e = 0$ in Figure 7.29. This is the curve with the elevator fixed in the neutral position; it is the curve we originally introduced in Figure 7.12. If the elevator is now deflected through a positive angle (downward), Eq. (7.39) states that all points on this curve will now be shifted down by the constant amount $\Delta C_{M,cg}$. Hence, the slope of the moment curve is preserved; only the value of $C_{M,0}$ is changed by elevator deflection. This now proves our earlier statement made in conjunction with Figure 7.25.

For emphasis, we repeat the main thrust of this section. The elevator can be used to change and control the trim of the airplane. In essence, this controls the equilibrium velocity of the airplane. For example, by a downward deflection of the elevator, a new trim angle α_n, smaller than the original trim angle α_e, can be obtained. (This is illustrated in Figure 7.29.) This corresponds to an increase in velocity of the airplane.

As another example, consider the two velocity extremes—stalling velocity and maximum velocity. Figure 7.30 illustrates the elevator deflection necessary to trim the airplane at these two extremes. First consider Figure 7.30a, which corresponds to an airplane flying at $V_\infty \approx V_{stall}$. This would be the situation on a landing approach, for example. The airplane is flying at $C_{L_{max}}$, hence the angle of attack is large. Therefore, from our previous discussion, the airplane must be trimmed by an *up-elevator* position, i.e., by a negative δ_e. On the other hand, consider Figure 7.30b, which corresponds to an airplane flying at $V_\infty \approx V_{max}$ (near full throttle). Because q_∞ is large, the airplane requires only a small C_L to generate the required lift force; hence the angle of attack is small. Thus, the airplane must be trimmed by a *down-elevator* position, i.e., by a positive δ_e.

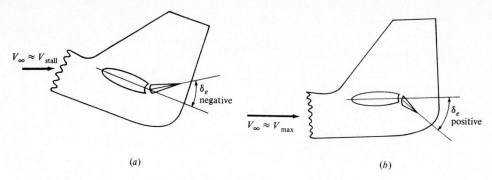

$$V_\infty \approx V_{\text{stall}}$$

δ_e
negative

$$V_\infty \approx V_{\text{max}}$$

δ_e
positive

(a)   (b)

Figure 7.30 Elevator deflection required for trim at (*a*) low flight velocity, and (*b*) high flight velocity.

7.13 CALCULATION OF ELEVATOR ANGLE TO TRIM

The concepts and relations developed in Sec. 7.12 allow us now to calculate the precise elevator deflection necessary to trim the airplane at a given angle of attack. Consider an airplane with its moment coefficient curve given as in Figure 7.31. The equilibrium angle of attack with no elevator deflection is α_e. We wish to trim the airplane at a new angle of attack α_n. What value of δ_e is required for this purpose?

To answer this question, first write the equation for the moment curve with $\delta_e = 0$ (the solid line in Figure 7.31). This is a straight line with a constant slope equal to $\partial C_{M,\text{cg}}/\partial\alpha_a$ and intercepting the ordinate at $C_{M,0}$. Hence, from analytic geometry, the equation of this line is

$$C_{M,\text{cg}} = C_{M,0} + \frac{\partial C_{M,\text{cg}}}{\partial\alpha_a}\alpha_a \tag{7.40}$$

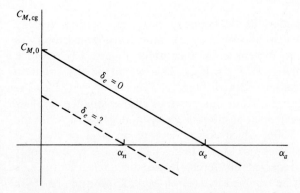

Figure 7.31 Given the equilibrium angle of attack at zero elevator deflection, what elevator deflection is necessary to establish a given new equilibrium angle of attack?

Now assume the elevator is deflected through an angle δ_e. The value of $C_{M,\text{cg}}$ will now change by the increment $\Delta C_{M,\text{cg}}$, and the moment equation given by Eq. (7.40) is now modified as

$$C_{M,\text{cg}} = C_{M,0} + \frac{\partial C_{M,\text{cg}}}{\partial \alpha_a}\alpha_a + \Delta C_{M,\text{cg}} \qquad (7.41)$$

The value of $\Delta C_{M,\text{cg}}$ was obtained earlier as Eq. (7.39). Hence, substituting Eq. (7.39) into (7.41), we obtain

$$C_{M,\text{cg}} = C_{M,0} + \frac{\partial C_{M,\text{cg}}}{\partial \alpha_a}\alpha_a - V_H \frac{\partial C_{L,t}}{\partial \delta_e}\delta_e \qquad (7.42)$$

Equation (7.42) allows us to calculate $C_{M,\text{cg}}$ for any arbitrary angle of attack α_a and any arbitrary elevator deflection δ_e. However, we are interested in the specific situation where $C_{M,\text{cg}} = 0$ at $\alpha_a = \alpha_n$ and where the value of δ_e necessary to obtain this condition is $\delta_e = \delta_{\text{trim}}$. That is, we want to find the value of δ_e which gives the dashed line in Figure 7.31. Substituting the above values into Eq. (7.42),

$$0 = C_{M,0} + \frac{\partial C_{M,\text{cg}}}{\partial \alpha_a}\alpha_n - V_H \frac{\partial C_{L,t}}{\partial \delta_e}\delta_{\text{trim}}$$

and solving for δ_{trim}, we obtain

$$\boxed{\delta_{\text{trim}} = \frac{C_{M,0} + (\partial C_{M,\text{cg}}/\partial \alpha_a)\alpha_n}{V_H(\partial C_{L,t}/\partial \delta_e)}} \qquad (7.43)$$

Equation (7.43) is the desired result. It gives the elevator deflection necessary to trim the airplane at a given angle of attack α_n. In Eq. (7.43), V_H is a known value from the airplane design, and $C_{M,0}$, $\partial C_{M,\text{cg}}/\partial \alpha_a$, and $\partial C_{L,t}/\partial \delta_e$ are known values usually obtained from wind-tunnel or free-flight data.

Example 7.8 Consider a full-size airplane with the same aerodynamic and design characteristics as the wind-tunnel model of Examples 7.3 to 7.7. The airplane has a wing area of 19 m², a weight of 2.27×10^4 N, and an elevator control effectiveness of 0.04. Calculate the elevator deflection angle necessary to trim the airplane at a velocity of 61 m/s at sea level.

SOLUTION First, we must calculate the angle of attack for the airplane at $V_\infty = 61$ m/s. Recall that

$$C_L = \frac{2W}{\rho_\infty V_\infty^2 S} = \frac{2(2.27 \times 10^4)}{1.225(61)^2(19)} = 0.52$$

From Example 7.3, the lift slope is $a = 0.08$ per degree. Hence, the absolute angle of attack of the airplane is

$$\alpha_a = \frac{C_L}{a} = \frac{0.52}{0.08} = 6.5°$$

From Eq. (7.43), the elevator deflection angle required to trim the airplane at this angle of attack is

$$\delta_{\text{trim}} = \frac{C_{M,0} + \left(\partial C_{M,\text{cg}} / \partial \alpha_a \right) \alpha_n}{V_H \left(\partial C_{L,t} / \partial \delta_e \right)}$$

where

$$C_{M,0} = 0.06 \qquad \text{(from Example 7.5)}$$

$$\frac{C_{M,\text{cg}}}{\partial \alpha_a} = -0.0133 \qquad \text{(from Example 7.5)}$$

$$\alpha_n = 6.5° \qquad \text{(this is the } \alpha_a \text{ calculated above)}$$

$$V_H = 0.34 \qquad \text{(from Example 7.4)}$$

$$\frac{\partial C_{L,t}}{\partial \delta_e} = 0.04 \qquad \text{(given above)}$$

Thus, from Eq. (7.43),

$$\delta_{\text{trim}} = \frac{0.06 + (-0.0133)(6.5)}{0.34(0.04)} = \boxed{-1.94°}$$

Recall that positive δ is downward. Hence, to trim the airplane at an angle of attack of 6.5°, the elevator must be deflected *upward* by 1.94°.

7.14 STICK-FIXED VERSUS STICK-FREE STATIC STABILITY

The second paragraph of Sec. 7.5 initiated our study of a rigid airplane with *fixed controls*, e.g., the elevator *fixed* at a given deflection angle. The ensuing sections developed the static stability for such a case, always assuming that the elevator can be deflected to a desired angle δ_e but held fixed at that angle. This is the situation when the pilot (human or automatic) moves the control stick to a given position and then rigidly holds it there. Consequently, the static stability that we have discussed to this point is called *stick-fixed static stability*. Modern high-performance airplanes designed to fly near or beyond the speed of sound have hydraulically assisted power controls; therefore a stick-fixed static stability analysis is appropriate for such airplanes.

On the other hand, consider a control stick connected to the elevator via wire cables without a power boost of any sort. This was characteristic of most early airplanes until the 1940s and is representative of many light, general aviation, private aircraft of today. In this case, in order to hold the stick fixed in a given position, the pilot must continually exert a manual force. This is uncomfortable and impractical. Consequently, in steady, level flight, the control stick is left essentially free; in turn, the elevator is left free to float under the influence of the natural aerodynamic forces and moments at the tail. The static stability of such an airplane is therefore called *stick-free static stability*. This will be the subject of the remainder of this chapter.

7.15 ELEVATOR HINGE MOMENT

Consider a horizontal tail with an elevator which rotates about a hinge axis, as shown in Figure 7.32. Assume the airfoil section of the tail is symmetrical, which is almost always the case for both the horizontal and vertical tail. First, consider the tail at zero angle of attack, as shown in Figure 7.32a. The aerodynamic pressure distribution on the top and bottom surfaces of the elevator will be the same, i.e., symmetrical about the chord. Hence, there will be no moment exerted on the elevator about the hinge line. Now assume that the tail is pitched to the angle of attack α_t, but the elevator is not deflected, that is, $\delta_e = 0$. This is illustrated in Figure 7.32b. As discussed in Chap. 5, there will be a low pressure on the top surface of the airfoil and a high pressure on the bottom surface.

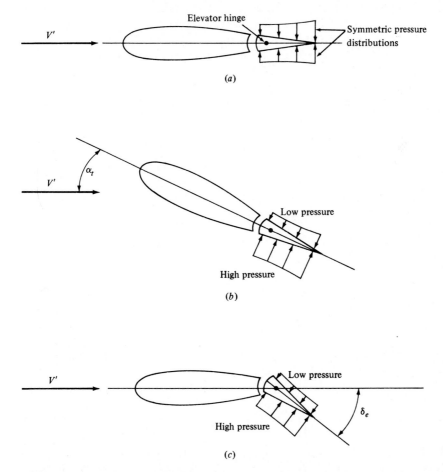

Figure 7.32 Illustration of the aerodynamic generation of elevator hinge moment. (a) No hinge moment; (b) hinge moment due to angle of attack; (c) hinge moment due to elevator deflection.

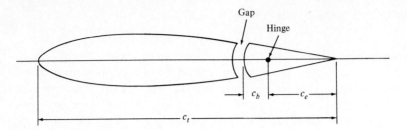

Figure 7.33 Nomenclature and geometry for hinge moment coefficient.

Consequently, the aerodynamic force on the elevator will not be balanced, and there will be a moment about the hinge axis tending to deflect the elevator upward. Finally, consider the horizontal tail at zero angle of attack but with the elevator deflected downward and held fixed at the angle δ_e, as shown in Figure 7.32c. Recall from Sec. 5.17 that a flap deflection effectively changes the camber of the airfoil and alters the pressure distribution. Therefore, in Figure 7.32c, there will be low and high pressures on the top and bottom elevator surfaces, respectively. As a result, a moment will again be exerted about the hinge line, tending to rotate the elevator upward. Thus, we see that both the tail angle of attack α_t and the elevator deflection δ_e result in a moment about the elevator hinge line—such a moment is defined as the *elevator hinge moment*. It is the governing factor in stick-free static stability, as discussed in the next section.

Let H_e denote the elevator hinge moment. Also, referring to Figure 7.33, the chord of the tail is c_t, the distance from the leading edge of the elevator to the hinge line is c_b, the distance from the hinge line to the trailing edge is c_e, and that portion of the elevator planform area that lies *behind* (aft of) the hinge line is S_e. The *elevator hinge moment* coefficient, C_{h_e}, is then defined as

$$C_{h_e} = \frac{H_e}{\frac{1}{2}\rho_\infty V_\infty^2 S_e c_e} \tag{7.44}$$

where V_∞ is the freestream velocity of the airplane.

Recall that the elevator hinge moment is due to the tail angle of attack and the elevator deflection. Hence, C_{h_e} is a function of both α_t and δ_e. Moreover, experience has shown that at both subsonic and supersonic speeds, C_{h_e} is approximately a linear function of α_t and δ_e. Thus, recalling the definition of the partial derivative in Sec. 7.2 D, the hinge moment coefficient can be written as

$$C_{h_e} = \frac{\partial C_{h_e}}{\partial \alpha_t}\alpha_t + \frac{\partial C_{h_e}}{\partial \delta_e}\delta_e \tag{7.45}$$

where $\partial C_{h_e}/\partial \alpha_t$ and $\partial C_{h_e}/\partial \delta_e$ are approximately constant. However, the actual magnitudes of these constant values depend in a complicated way on c_e/c_t, c_b/c_e, the elevator nose shape, gap, trailing edge angle, and planform. Moreover, H_e is very sensitive to local boundary layer separation. As a result, the values of the partial derivatives in Eq. (7.45) must almost always be obtained empirically, such as from wind-tunnel tests, for a given design.

Consistent with the convention that downward elevator deflections are positive, hinge moments which tend to deflect the elevator downward are also defined as positive. Note from Figure 7.32b that a positive α_t physically tends to produce a negative hinge moment (tending to deflect the elevator upward). Hence $\partial C_{h_e}/\partial \alpha_t$ is usually negative. (However, if the hinge axis is placed very far back, near the trailing edge, the sense of H_e may become positive. This is usually not done for conventional airplanes.) Also, note from Figure 7.32c that a positive δ_e usually produces a negative H_e, hence $\partial C_{h_e}/\partial \delta_e$ is also negative.

7.16 STICK-FREE LONGITUDINAL STATIC STABILITY

Let us return to the concept of stick-free static stability introduced in Sec. 7.14. If the elevator is left free to float, it will always seek some equilibrium deflection angle such that the hinge moment is zero, that is, $H_e = 0$. This is obvious, because as long as there is a moment on the free elevator, it will always rotate. It will come to rest (equilibrium) only for that position where the moment is zero.

Recall our qualitative discussion of longitudinal static stability in Sec. 7.5. Imagine that an airplane is flying in steady, level flight at the equilibrium angle of attack. Now assume the airplane is disturbed by a wind gust and is momentarily pitched to another angle of attack, as was sketched in Figure 7.13. If the airplane is statically stable, it will initially tend to return toward its equilibrium position. In subsequent sections, we saw that the design of the horizontal tail was a powerful mechanism governing this static stability. However, until now, the elevator was always considered fixed. On the other hand, if the elevator is allowed to float freely when the airplane is pitched by some disturbance, the elevator will seek some momentary equilibrium position different from its position before the disturbance. This deflection of the free elevator will change the static stability characteristics of the airplane. In fact, such stick-free stability is usually less than stick-fixed stability. For this reason, it is usually desirable to design an airplane such that the difference between stick-free and stick-fixed longitudinal stability is small.

With the above in mind, consider the equilibrium deflection angle of a free elevator. Denote this angle by δ_{free}, as sketched in Figure 7.34. At this angle, $H_e = 0$. Thus, from Eq. (7.45),

$$C_{h_e} = 0 = \frac{\partial C_{h_e}}{\partial \alpha_t} \alpha_t + \frac{\partial C_{h_e}}{\partial \delta_e} \delta_{\text{free}} \qquad (7.46)$$

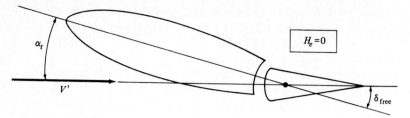

Figure 7.34 Illustration of free elevator deflection.

Solving Eq. (7.46) for δ_{free},

$$\boxed{\delta_{\text{free}} = -\frac{\partial C_{h_e}/\partial \alpha_t}{\partial C_{h_e}/\partial \delta_e}\alpha_t}$$ (7.47)

Equation (7.47) gives the equilibrium, free-floating angle of the elevator as a function of tail angle of attack. As stated earlier, both partial derivatives in Eq. (7.47) are usually negative; hence, a positive α_t yields a negative δ_{free} (an upward deflection). This is intuitively correct, as verified by Figure 7.34, which shows a negative δ_{free}.

Obviously, δ_{free} affects the tail lift coefficient, which in turn affects the static stability of the airplane. The tail lift coefficient for angle of attack α_t and fixed-elevator deflection δ_e was given in Eq. (7.36), repeated below.

$$C_{L,t} = a_t\alpha_t + \frac{\partial C_{L,t}}{\partial \delta_e}\delta_e$$

However, for a free elevator, $\delta_e = \delta_{\text{free}}$. Denoting the tail lift coefficient for a free elevator as $C'_{L,t}$, a substitution of Eq. (7.47) into (7.36) gives

$$C'_{L,t} = a_t\alpha_t + \frac{\partial C_{L,t}}{\partial \delta_e}\delta_{\text{free}}$$

$$C'_{L,t} = a_t\alpha_t - \frac{\partial C_{L,t}}{\partial \delta_e}\frac{\partial C_{h_e}/\partial \alpha_t}{\partial C_{h_e}/\partial \delta_e}\alpha_t$$

or

$$\boxed{C'_{L,t} = a_t\alpha_t F}$$ (7.48)

where F is the *free elevator factor*, defined as

$$F = 1 - \frac{1}{a_t}\frac{\partial C_{L,t}}{\partial \delta_e}\frac{\partial C_{h_e}/\partial \alpha_t}{\partial C_{h_e}/\partial \delta_e}$$

The free elevator factor is a number usually less than unity, and usually on the order of 0.7 to 0.8. It represents a reduction in the tail's contribution to static stability when the elevator is free. The magnitude of this reduction is developed below.

Consider now the moment about the center of gravity of the airplane. For a fixed elevator, the moment coefficient is given by Eq. (7.24),

$$C_{M,cg} = C_{M,ac_{wb}} + C_{L_{wb}}(h - h_{ac_{wb}}) - V_H C_{L,t}$$

For a free elevator, the tail lift coefficient is now changed to $C'_{L,t}$. Hence, the moment coefficient for a free elevator, $C'_{M,cg}$, is

$$C'_{M,cg} = C_{M,ac_{wb}} + C_{L_{wb}}(h - h_{ac_{wb}}) - V_H C'_{L,t} \qquad (7.49)$$

Substituting Eq. (7.48) into (7.49),

$$\boxed{C'_{M,cg} = C_{M,ac_{wb}} + C_{L_{wb}}(h - h_{ac_{wb}}) - V_H a_t \alpha_t F} \qquad (7.50)$$

Equation (7.50) gives the final form of the moment coefficient about the center of gravity of the airplane with a free elevator.

Using Eq. (7.50), the same analyses as given in Sec. 7.9 can be used to obtain equations for stick-free longitudinal static stability. The results are:

$$\boxed{\begin{aligned} C'_{M,0} &= C_{M,ac_{wb}} + F V_H a_t (i_t + \varepsilon_0) \end{aligned}} \qquad (7.51)$$

$$\boxed{h'_n = h_{ac_{wb}} + F V_H \frac{a_t}{a}\left(1 - \frac{\partial \varepsilon}{\partial \alpha}\right)} \qquad (7.52)$$

$$\boxed{\frac{\partial C'_{M,cg}}{\partial \alpha} = -a(h'_n - h)} \qquad (7.53)$$

Equations (7.51), (7.52), and (7.53) apply for stick-free conditions, denoted by the prime notation. They should be compared with Eqs. (7.27), (7.30), and (7.33), respectively, for stick-fixed stability. Note that $h'_n - h$ is the stick-free static margin; because $F < 1.0$, this is smaller than the stick-fixed static margin.

Hence, it is clear from Eqs. (7.51) to (7.53) that a free elevator usually decreases the static stability of the airplane.

Example 7.9 Consider the airplane of Example 7.8. Its elevator hinge moment derivatives are $\partial C_{h_e}/\partial \alpha_t = -0.008$ and $\partial C_{h_e}/\partial \alpha_e = -0.013$. Assess the *stick-free* static stability of this airplane.

SOLUTION First, obtain the free elevator factor F defined from Eq. (7.48),

$$F = 1 - \frac{1}{a_t} \frac{\partial C_{L,t}}{\partial \delta_e} \frac{\partial C_{h_e}/\partial \alpha_t}{\partial C_{h_e}/\partial \delta_e}$$

where $\qquad a_t = 0.1 \qquad$ (from Example 7.4)

$$\frac{\partial C_{L,t}}{\partial \delta_e} = 0.04 \qquad \text{(from Example 7.8)}$$

$$F = 1 - \frac{1}{0.1}(0.04)\frac{-0.008}{-0.013} = 0.754$$

The stick-free static stability characteristics are given by Eqs. (7.51) to (7.53). First, from Eq. (7.51),

$$C'_{M,0} = C_{M,ac_{wb}} + FV_H a_t (i_t + \varepsilon_0)$$

where $\qquad C_{M,ac_{wb}} = -0.032 \qquad$ (from Example 7.3)

$$V_H = 0.34 \qquad \text{(from Example 7.4)}$$

$$i_t = 2.7° \qquad \text{(from Example 7.4)}$$

$$\varepsilon_0 = 0 \qquad \text{(from Example 7.4)}$$

Thus, $\qquad C'_{M,0} = -0.032 + 0.754(0.34)(0.1)(2.7)$

$$\boxed{C'_{M,0} = 0.037}$$

This is to be compared with $C_{M,0} = 0.06$ obtained for stick-fixed conditions in Example 7.5. From Eq. (7.52),

$$h'_n = h_{ac_{wb}} + FV_H \frac{a_t}{a}\left(1 - \frac{\partial \varepsilon}{\partial \alpha}\right)$$

where $\qquad h_{ac_{wb}} = 0.24 \qquad$ (from Example 7.3)

$$\frac{\partial \varepsilon}{\partial \alpha} = 0.35 \qquad \text{(from Example 7.4)}$$

$$a = 0.08 \qquad \text{(from Example 7.4)}$$

$$h'_n = 0.24 + 0.754(0.34)\left(\frac{0.1}{0.08}\right)(1 - 0.35)$$

$$\boxed{h'_n = 0.448}$$

This is to be compared with $h_n = 0.516$ obtained for stick-fixed conditions in Example 7.6. Note that the neutral point has moved forward for stick-free conditions, hence decreasing the stability. In fact, the stick-free static margin is

$$h'_n - h = 0.448 - 0.35 = 0.098$$

This is a 41 percent decrease, in comparison with the stick-fixed static margin from Example 7.7. Finally, from Eq. (7.53),

$$\frac{\partial C'_{M,cg}}{\partial \alpha} = -a\left(h'_n - h\right) = -0.08(0.098) = \boxed{-0.0078}$$

Thus, as expected, the slope of the stick-free moment coefficient curve, although still negative, is small in absolute value.

In conclusion, this problem indicates that stick-free conditions cut the static stability of our hypothetical airplane by nearly one-half. This helps to dramatize the differences between stick-fixed and stick-free considerations.

7.17 A COMMENT

This brings to a close our technical discussion of stability and control. The preceding sections constitute an introduction to the subject; however, we have just scratched the surface. There are many other considerations: control forces, dynamic stability, lateral and directional stability, etc. Such matters are the subject of more advanced studies of stability and control and are beyond the scope of this book. However, this subject is one of the fundamental pillars of aeronautical engineering, and the interested reader can find extensive presentations in books such as those of Perkins and Hage, and Etkin (see Bibliography at the end of this chapter).

7.18 HISTORICAL NOTE: THE WRIGHT BROTHERS VERSUS THE EUROPEAN PHILOSOPHY ON STABILITY AND CONTROL

The two contrasting scenes depicted in Sec. 7.1—the lumbering, belabored flight of Farman vs. the relatively effortless maneuvering of Wilbur Wright—underscore two different schools of aeronautical thought during the first decade of powered flight. One school, consisting of virtually all early European and U.S. aeronautical engineers, espoused the concept of inherent stability (statically stable aircraft); the other, consisting solely of Wilbur and Orville Wright, practiced the design of statically unstable aircraft that had to be controlled every instant by the pilot. Both philosophies have their advantages and disadvantages, and because they have an impact on modern airplane design, let us examine their background more closely.

The basic principles of airplane stability and control began to evolve at the time of George Cayley. His glider of 1804, sketched in Figure 1.8, incorporated a vertical and horizontal tail that could be adjusted up and down. In this fashion, the complete tail unit acted as an elevator.

The next major advance in airplane stability was made by Alphonse Penaud, a brilliant French aeronautical engineer who committed suicide in 1880 at the age of 30. Penaud built small model airplanes powered by twisted rubber bands, a precursor of the flying balsa-and-tissue paper models of today. Penaud's design had a fixed wing and tail, like Cayley's even though at the time Penaud was not aware of Cayley's work. Of particular note was Penaud's horizontal tail design, which was set at a negative 8° with respect to the wing chord line. Here we find the first true understanding of the role of the tail-setting angle i_t (see Secs. 7.5 and 7.7) on the static stability of an airplane. Penaud flew his model in the Tuileries Gardens in Paris on August 18, 1871, before members of the Société de Navigation Aérienne. The aircraft flew for 11 s, covering 131 ft. This event, along with Penaud's theory for stability, remained branded on future aeronautical designs right down to the present day.

After Penaud's work, the attainment of "inherent" (static) stability became a dominant feature in aeronautical design. Lilienthal, Pilcher, Chanute, and Langley all strived for it. However, static stability has one disadvantage: the more stable the airplane, the harder it is to maneuver. An airplane that is highly stable is also sluggish in the air; its natural tendency to return to equilibrium somewhat defeats the purpose of the pilot to change its direction by means of control deflections. The Wright brothers recognized this problem in 1900. Since Wilbur and Orville were "airmen" in the strictest meaning of the word, they aspired for quick and easy maneuverability. Therefore, they discarded the idea of inherent stability that was entrenched by Cayley and Penaud. Wilbur wrote that "we... resolved to try a fundamentally different principle. We would arrange the machine so that it would not tend to right itself." The Wright brothers designed their aircraft to be statically unstable! This feature, along with their development of lateral control through wing warping, is primarily responsible for the fantastic aerial performance of all their airplanes from 1903 to 1912 (when Wilbur died). Of course, this design feature heavily taxed the pilot, who had to keep the airplane under control at every instant, continuously operating the controls to compensate for the unstable characteristics of the airplane. Thus, the Wright airplanes were difficult to fly, and long periods were required to train pilots for these aircraft. In the same vein, such unstable aircraft were more dangerous.

These undesirable characteristics were soon to be compelling. After Wilbur's dramatic public demonstrations in France in 1908 (see Sec. 1.8), the European designers quickly adopted the Wrights' patented concept of combined lateral and directional control by coordinated wing warping (or by ailerons) and rudder deflection. But they rejected the Wrights' philosophy of static instability. By 1910, the Europeans were designing and flying aircraft that properly mated the Wrights' control ideas with the long-established static stability principles. On the other hand, the Wrights stubbornly clung to their basic unstable design. As a result, by 1910 the European designs began to surpass the Wrights' machines, and the lead in aeronautical engineering established in America in 1903 now swung to France, England, and Germany, where it remained for almost 20 years. In the process, static stability became an unquestioned design feature in all successful aircraft up to the 1970s.

It is interesting that very modern airplane design has returned full circle to the Wright brothers' original philosophy, at least in some cases. Recent lightweight military fighter designs, such as the F-16, F-17, and F-18, are statically unstable in order to obtain dramatic increases in maneuverability. At the same time, the airplane is instantaneously kept under control by computer-calculated and electrically adjusted positions of the control surfaces—the "fly by wire" concept. In this fashion, the maneuverability advantages of static instability can be realized without heavily taxing the pilot: the work is done by electronics! Even when maneuverability is not a prime feature, such as in civil transport airplanes, static instability has some advantages. For example, the tail surfaces for an unstable airplane can be smaller, with a subsequent savings in structural weight and reductions in aerodynamic drag. Hence, with the advent of the "fly by wire" system, the cardinal airplane design principle of static stability may be somewhat relaxed in the future. The Wright brothers may indeed ride again!

7.19 HISTORICAL NOTE: THE DEVELOPMENT OF FLIGHT CONTROLS

Figure 7.3 illustrates the basic aerodynamic control surfaces on an airplane, the ailerons, elevator, and rudder. They have been an integral part of airplane designs for most of the twentieth century, and we take them almost for granted. But where are their origins? When did such controls first come into practical use? Who had the first inspirations for such controls?

In the previous section, we have already mentioned that by 1809 George Cayley employed a movable tail in his designs, the first effort at some type of longitudinal control. Cayley's idea of moving the complete horizontal tail to obtain such control persisted through the first decade of the twentieth century. Henson, Stringfellow, Penaud, Lilienthal, the Wright brothers all envisioned or utilized movement of the complete horizontal tail surface for longitudinal control. It was not until 1908–1909 that the first "modern" tail control configuration was put into practice. This was achieved by the French designer Levavasseur on his famous Antoinette airplanes, which had fixed vertical and horizontal tail surfaces with movable, flaplike rudder and elevator surfaces at the trailing edges. So the configuration for elevators and rudders as shown in Figure 7.3 dates back to 1908, five years after the dawn of powered flight.

The origin of ailerons (a French word for the extremity of a bird's wing) is steeped in more history and controversy. It is known that the Englishman M. P. W. Boulton patented a concept for lateral control by ailerons in 1868. Of course, at that time no practical aircraft existed, so the concept could not be demonstrated and verified, and Boulton's invention quickly retreated to the background and was forgotten. Ideas of warping the wings or inserting vertical surfaces (spoilers) at the wingtips cropped up several times in Europe during the late nineteenth century and into the first decade of the twentieth century, but always in the context of a braking surface which would slow one wing down and pivot the airplane about a vertical axis. The true function of ailerons or wing warping, that for lateral control for banking and consequently turning an airplane, was not fully appreciated until Orville and Wilbur incorporated wing warping on their Flyers (see Chap. 1). The Wright brothers' claim that they were the first to invent wing warping may not be historically precise, but clearly they were the first to demonstrate its function and to obtain a legally enforced patent on its use (combined with simultaneous rudder action for total control in banking). The early European airplane designers did not appreciate the need for lateral control until Wilbur's dramatic public flights in France in 1908. This is in spite of the fact that Wilbur had fully described their wing warping concept in a paper at Chicago on September 1, 1901, and again on June 24, 1903; indeed, Octave Chanute clearly described the Wrights' concept in a lecture to the Aero Club de France in

Paris in April 1903. Other aeronautical engineers at that time, if they listened, did not pay much heed. As a result, European aircraft before 1908, even though they were making some sustained flights, were awkward to control.

However, the picture changed after 1908, when in the face of the indisputable superiority of the Wrights' control system, virtually everybody turned to some type of lateral control. Wing warping was quickly copied and was employed on numerous different designs. Moreover, the idea was refined to include movable surfaces near the wingtips. These were first separate "winglets" mounted either above, below, or between the wings. But, in 1909, Henri Farman (see Sec. 7.1) designed a biplane named the *Henri Farman III*, which included a flaplike aileron at the trailing edge of all four wingtips; this was the true ancestor of the conventional modern-day aileron, as sketched in Figure 7.3. Farman's design was soon adopted by most designers, and wing warping quickly became passé. Only the Wright brothers clung to their old concept; a Wright airplane did not incorporate ailerons until 1915, six years after Farman's development.

7.20 HISTORICAL NOTE: THE "TUCK-UNDER" PROBLEM

A quick examination of Figure 7.20, and the resulting stability equations such as Eqs. (7.26), (7.27), and (7.28), clearly underscores the importance of the downwash angle ε in determining longitudinal static stability. Downwash is a rather skittish aerodynamic phenomenon, very difficult to calculate accurately for real airplanes and therefore usually measured in wind-tunnel tests or in free flight. A classic example of the stability problems that can be caused by downwash, and how wind-tunnel testing can help, occurred during World War II, as described below.

In numerous flights during 1941 and 1942, the Lockheed P-38, a twin-engine, twin-boomed, high-performance fighter plane (see Figure 7.35), went into sudden dives from which recovery was exceptionally difficult. Indeed, several pilots were killed in this fashion. The problem occurred at high subsonic speeds, usually in a dive, where the airplane had a tendency to nose over, putting the plane in yet a steeper dive. Occasionally, the airplane would become locked in this position, and even with maximum elevator deflection, a pullout could not be achieved. This "tuck-under" tendency could not be tolerated in a fighter aircraft which was earmarked for a major combat role.

Therefore, with great urgency, NACA was asked to investigate the problem. Since the effect occurred only at high speeds, usually above Mach 0.6, compressibility appeared to be the culprit. Tests in the Langley 30 ft by 60 ft low-speed tunnel and in the 8-ft high-speed tunnel (see Sec. 4.22) correlated the tuck-under tendency with the simultaneous formation of shock waves on the wing surface. Such compressibility effects were discussed in Secs. 5.9 and 5.10, where it was pointed out that, beyond the critical Mach number for the wing, shock waves will form on the upper surface, encouraging flow separation far upstream of the trailing edge. The P-38 was apparently the first operational airplane to encounter

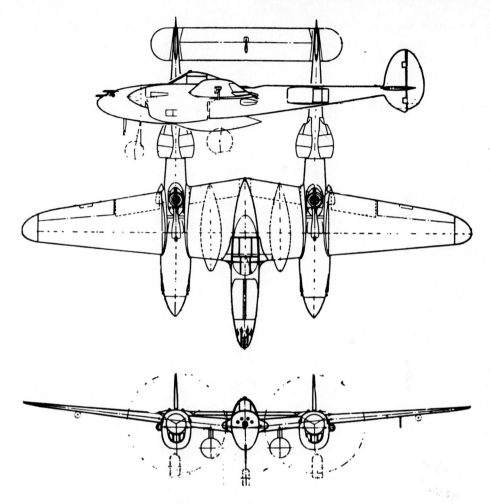

Figure 7.35 The Lockheed P-38 of World War II fame.

this problem. The test engineers at Langley made several suggestions to rectify the situation, but they all involved major modifications of the airplane. For a model already in production, a quicker fix was needed.

Next, the 16-ft high-speed wind tunnel at the NACA Ames Aeronautical Laboratory in California (see again Sec. 4.22) was pressed into service on the P-38 problem. Here, further tests indicated that the shock-induced separated flow over the wing was drastically reducing the lift. In turn, because the downwash is directly related to lift, as discussed in Secs. 5.13 and 5.14, the downwash angle ε was greatly reduced. Consequently (see Figure 7.20), the tail angle of attack α_t was markedly increased. This caused a sharp increase in the positive lift on the tail, creating a strong pitching moment, nosing the airplane into a steeper dive. After the series of Ames tests in April 1943, Al Erickson of NACA suggested the addition of flaps on the lower surface of the wing at the $0.33c$ point in order to increase the lift, hence increase the downwash. This was the quick fix that Lockheed was looking for, and it worked.

7.21 CHAPTER SUMMARY

Some of the important points of this chapter are given below.

1. If the forces and moments on a body caused by a disturbance tend *initially* to return the body *toward* its equilibrium position, the body is statically stable. In contrast, if these forces and moments tend *initially* to move the body *away from* its equilibrium position, the body is *statically unstable*.
2. The necessary criteria for longitudinal balance and static stability are (a) $C_{M,0}$ must be positive, (b) $\partial C_{M,cg}/\partial \alpha_a$ must be negative, and (c) the trim angle of attack α_e must fall within the flight range of angle of attack for the airplane. These criteria may be evaluated quantitatively for a given airplane from

$$C_{M,0} = C_{M,ac_{wb}} + V_H a_t(i_t + \varepsilon_0) \qquad (7.27)$$

and

$$\frac{\partial C_{M,cg}}{\partial \alpha_a} = a\left[(h - h_{ac_{wb}}) - V_H \frac{a_t}{a}\left(1 - \frac{\partial \varepsilon}{\partial \alpha}\right)\right] \qquad (7.28)$$

where the tail volume ratio is given by

$$V_H = \frac{l_t S_t}{cS}$$

3. The neutral point is that location of the center of gravity where $\partial C_{M,cg}/\partial \alpha_a = 0$. It can be calculated from

$$h_n = h_{ac_{wb}} + V_H \frac{a_t}{a}\left(1 - \frac{\partial \varepsilon}{\partial \alpha}\right) \qquad (7.30)$$

4. The static margin is defined as $h_n - h$. For static stability, the location of the center of gravity must be ahead of the neutral point; i.e., the static margin must be positive.
5. The effect of elevator deflection δ_e on the pitching moment about the center of gravity is given by

$$C_{M,cg} = C_{M,ac_{wb}} + C_{L,wb}(h - h_{ac}) - V_H\left(a_t\alpha_t + \frac{\partial C_{L,t}}{\partial \delta_e}\delta_e\right) \qquad (7.37)$$

6. The elevator deflection necessary to trim an airplane at a given angle of attack α_n is

$$\delta_{trim} = \frac{C_{M,0} + (\delta C_{M,cg}/\partial \alpha_a)\alpha_n}{V_H(\partial C_{L,t}/\partial \delta_e)} \qquad (7.43)$$

BIBLIOGRAPHY

Etkin, B., *Dynamics of Flight*, Wiley, New York, 1959.

Gibbs-Smith, C. H., *Aviation: An Historical Survey from its Origins to the End of World War II*, Her Majesty's Stationery Office, London, 1970.

Perkins, C. D., and Hage, R. E., *Airplane Performance, Stability, and Control*, Wiley, New York, 1949.

PROBLEMS

7.1 For a given wing-body combination, the aerodynamic center lies 0.03 chord length ahead of the center of gravity. The moment coefficient about the center of gravity is 0.0050, and the lift coefficient is 0.50. Calculate the moment coefficient about the aerodynamic center.

7.2 Consider a model of a wing-body shape mounted in a wind tunnel. The flow conditions in the test section are standard sea-level properties with a velocity of 100 m/s. The wing area and chord are 1.5 m^2 and 0.45 m respectively. Using the wind-tunnel force and moment-measuring balance, the moment about the center of gravity when the lift is zero is found to be -12.4 N $\cdot$ m. When the model is pitched to another angle of attack, the lift and moment about the center of gravity are measured to be 3675 N and 20.67 N $\cdot$ m, respectively. Calculate the value of the moment coefficient about the aerodynamic center and the location of the aerodynamic center.

7.3 Consider the model in Prob. 7.2. If a mass of lead is added to the rear of the model such that the center of gravity is shifted rearward by a length equal to 20 percent of the chord, calculate the moment about the center of gravity when the lift is 4000 N.

7.4 Consider the wing-body model in Prob. 7.2. Assume that a horizontal tail with no elevator is added to this model. The distance from the airplane's center of gravity to the tail's aerodynamic center is 1.0 m. The area of the tail is 0.4 m^2, and the tail-setting angle is $2.0°$. The lift slope of the tail is 0.12 per degree. From experimental measurement, $\varepsilon_0 = 0$ and $\partial \varepsilon / \partial \alpha = 0.42$. If the absolute angle of attack of the model is $5°$ and the lift at this angle of attack is 4134 N, calculate the moment about the center of gravity.

7.5 Consider the wing-body-tail model of Prob. 7.4. Does this model have longitudinal static stability and balance?

7.6 For the configuration of Prob. 7.4, calculate the neutral point and static margin.

7.7 Assume that an elevator is added to the horizontal tail of the configuration given in Prob. 7.4. The elevator control effectiveness is 0.04. Calculate the elevator deflection angle necessary to trim the configuration at an angle of attack of $8°$.

7.8 Consider the configuration of Prob. 7.7. The elevator hinge moment derivatives are $\partial C_{h_e} / \partial \alpha_t = -0.007$ and $\partial C_{h_e} / \partial \alpha_e = -0.012$. Assess the stick-free static stability of this configuration.

EIGHT

ASTRONAUTICS

It is difficult to say what is impossible, for the dream of yesterday is the hope of today and the reality of tomorrow.

Robert H. Goddard, at his high school graduation, 1904

Houston, Tranquillity Base here. The Eagle has landed.

Neil Armstrong, in a radio transmission to Mission Control, at the instant of the first manned landing on the moon, July 20, 1969

8.1 INTRODUCTION

Space—that last frontier, that limitless expanse which far outdistances the reach of our strongest telescopes, that region which may harbor other intelligent civilizations on countless planets; space—whose unknown secrets have attracted the imagination of humanity for centuries and whose technical conquest has labeled the latter half of the twentieth century as the "space age"; space—is the subject of this chapter.

To this point in our introduction of flight, emphasis has been placed on aeronautics, the science and engineering of vehicles which are designed to move within the atmosphere and which depend on the atmosphere for their lift and propulsion. However, as presented in Sec. 1.8, the driving force behind the advancement of aviation has always been the desire to fly higher and faster. The ultimate, of course, is to fly so high and so fast that you find yourself in outer space, beyond the limits of the sensible atmosphere. Here, motion of the vehicle takes place only under the influence of gravity and possibly some type of propulsive force; however, the mode of propulsion must be entirely independent of the air for its thrust. Therefore, the physical fundamentals and engineering principles associated with space vehicles are somewhat different from those

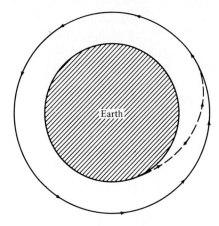

Figure 8.1 Earth orbit.

Figure 8.2 The Skylab—an earth satellite. (*NASA.*)

associated with airplanes. The purpose of this chapter is to introduce some of the basic concepts of space flight, i.e., to introduce the discipline of *astronautics*. In particular, the early sections of this chapter will emphasize the calculation and analysis of orbits and trajectories of space vehicles operating under the influence of gravitational forces only (such as in the vacuum of free space). In the later sections, we will consider several aspects of the reentry of a space vehicle into the earth's atmosphere, especially the reentry trajectory and aerodynamic heating of the vehicle.

The space age formally began on October 4, 1957, when the Soviet Union launched *Sputnik I*, the first artificial satellite to go into orbit about the earth.

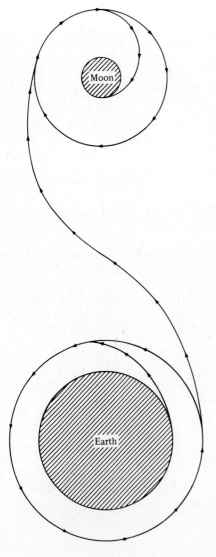

Figure 8.3 Earth-moon mission (not to scale).

Unlike the first flight of the Wright brothers in 1903, which took years to have any impact on society, the effect of *Sputnik I* on the world was immediate. Within 12 years, people had walked on the moon, and after another 7 years, unmanned probes were resting on the surfaces of Venus and Mars. A variety of different space vehicles designed for different missions have been launched since 1957. Most of these vehicles fall into three main categories, as follows:

1. Earth satellites, launched with enough velocity to go into orbit about the earth, as sketched in Figure 8.1. As we will show later, velocities on the order of 26,000 ft/s (7.9 km/s) are necessary to place a vehicle in orbit about the earth, and these orbits are generally elliptical. Figure 8.2 shows a photograph of an artificial earth satellite.
2. Lunar and interplanetary vehicles, launched with enough velocity to overcome the gravitational attraction of the earth and travel into deep space. Velocities of 36,000 ft/s (approximately 11 km/s) or larger are necessary for this purpose. Such trajectories are parabolic or hyperbolic. A typical path from the earth to the moon is sketched in Figure 8.3; here, the space vehicle is first placed in earth orbit, from which it is subsequently boosted by on-board rockets to an orbit about the moon, from which it finally makes a landing on the moon's surface. This is the mode employed by all the Apollo manned

Figure 8.4 The Apollo spacecraft. (*Smithsonian National Air and Space Museum.*)

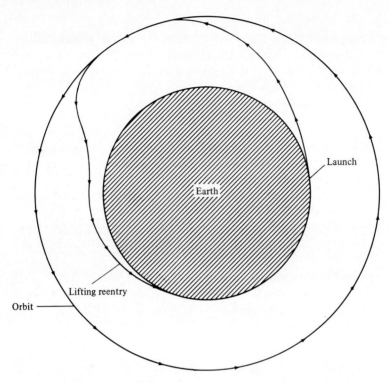

Figure 8.5 Earth orbit with lifting reentry.

lunar missions, beginning with the historic first moon landing on July 20, 1969. A photograph of the Apollo spacecraft is shown in Figure 8.4.

3. Space shuttles, designed to take off from the earth's surface, perform a mission in space, and then return and land on the earth's surface, all self-contained in the same vehicle. These are lifting reentry vehicles, designed with a reasonable L/D ratio to allow the pilot to land the craft just like an airplane. Earth orbit with a lifting reentry path is sketched in Figure 8.5. The first successful flight of a space shuttle into space, with a subsequent lifting reentry and landing, was carried out by NASA's *Columbia* during the period April 12–14, 1981. A photograph of the space shuttle is given in Figure 8.6.

Finally, a discussion of astronautics, even an introductory one, requires slightly more mathematics depth than just basic differential and integral calculus. Therefore, this chapter will incorporate more mathematical rigor than other parts of this book. In particular, some concepts from differential equations must be employed. However, it will be assumed that the reader has *not* had exposure to such mathematics, and therefore the necessary ideas will be introduced in a self-contained fashion.

Figure 8.6 The space shuttle. (*NASA*.)

8.2 DIFFERENTIAL EQUATIONS

Consider a dependent variable r which depends on an independent variable t. Thus, $r = f(t)$. The concept of the derivative of r with respect to t, dr/dt, has been used frequently in this book. The physical interpretation of dr/dt is simply the rate of change of r with respect to t. If r is a distance and t is time, then dr/dt is the rate of change of distance with respect to time, i.e., velocity. The second derivative of r with respect to t is simply

$$\frac{d(dr/dt)}{dt} \equiv \frac{d^2r}{dt^2}$$

This is the rate of change of the derivative itself with respect to t. If r and t are distance and time, respectively, then d^2r/dt^2 is the rate of change of velocity with respect to time, i.e., acceleration.

A differential equation is simply an equation which has derivatives in some of its terms. For example

$$\frac{d^2r}{dt^2} + r\frac{dr}{dt} - 2t^3 = 2 \tag{8.1}$$

is a differential equation; it contains derivatives along with the variables r and t

themselves. By comparison, the equation

$$r + \frac{t^2}{r} = 0$$

is an algebraic equation; it contains only r and t without any derivatives.

To find a *solution* of the differential equation, Eq. (8.1), means to find a functional relation $r = f(t)$ which satisfies the equation. For example, assume that $r = t^2$. Then, $dr/dt = 2t$ and $d^2r/dt^2 = 2$. Substitute into Eq. (8.1):

$$2 + t^2 2t - 2t^3 = 2$$
$$2 + 2t^3 - 2t^3 = 2$$
$$2 = 2$$

Hence, $r = t^2$ does indeed satisfy the differential equation, Eq. (8.1). Thus, $r = t^2$ is called a solution of that equation.

Calculations of space vehicle trajectories involve distance r and time t. Some of the fundamental equations involve first and second derivatives of r with respect to t. To simplify the notation in these equations, we now introduce the *dot notation* for time derivatives, i.e.,

$$\dot{r} \equiv \frac{dr}{dt}$$

$$\ddot{r} \equiv \frac{d^2r}{dt^2}$$

A single dot over the variable means the first time derivative of that variable; a double dot means the second derivative. For example, the differential equation, Eq. (8.1), can be written as

$$\ddot{r} + r\dot{r} - 2t^3 = 2 \tag{8.2}$$

Equations (8.1) and (8.2) are identical; only the notation is different. The dot notation for time derivatives is common in physical science; you will encounter it frequently in more-advanced studies of science and engineering.

8.3 LAGRANGE'S EQUATION

In physical science, a study of the forces and motion of bodies is called *mechanics*. If the body is motionless, this study is further identified as *statics*; if the body is moving, the study is one of *dynamics*. In this chapter, we are concerned with the *dynamics* of space vehicles.

Problems in dynamics usually involve the use of Newton's second law, $F = ma$, where F is force, m is mass, and a is acceleration. Perhaps the reader is familiar with various applications of $F = ma$ from basic physics; indeed, we applied this law in Chap. 4 to obtain the momentum equation in aerodynamics and again in Chap. 6 to obtain the equations of motion for an airplane. However, in this section we introduce an equation, Lagrange's equation, which is essentially

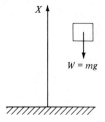

Figure 8.7 Falling body.

a corollary to Newton's second law. The use of Lagrange's equation represents an alternative approach to the solution of dynamics problems in lieu of $F = ma$; in the study of space vehicle orbits and trajectories, Lagrange's equation greatly simplifies the analysis. We will not derive Lagrange's equation; rather, we will simply introduce it by way of an example and then, in the next section, apply it to obtain the orbit equation. A rigorous derivation of Lagrange's equation is left to more-advanced studies of mechanics.

Consider the following example. A body of mass m is falling freely in the earth's gravitational field, as sketched in Figure 8.7. Let x be the vertical distance of the body from the ground. If we ignore drag, the only force on the body is its weight w directed downward. By definition, the weight of a body is equal to its mass m times the acceleration of gravity, g; $w = mg$. From Newton's second law,

$$F = ma \tag{8.3}$$

The force is weight, directed downward. Since the direction of positive x is upward, then a downward-acting force is negative. Hence,

$$F = -w = -mg \tag{8.4}$$

From the discussion in Sec. 8.2, the acceleration can be written as

$$a \equiv \frac{d^2x}{dt^2} \equiv \ddot{x} \tag{8.5}$$

Substituting Eqs. (8.4) and (8.5) into (8.3) yields

$$-mg = m\ddot{x}$$

$$\boxed{\ddot{x} = -g} \tag{8.6}$$

Equation (8.6) is the equation of motion for the body in our example. It is a differential equation whose solution will yield $x = f(t)$. Moreover, Eq. (8.6) was obtained by the application of Newton's second law.

Now consider an alternative formulation of this example, using Lagrange's equation. This will serve as an introduction to Lagrange's equation. Let T denote the *kinetic energy* of the body, where by definition,

$$T = \tfrac{1}{2}mV^2 = \tfrac{1}{2}m(\dot{x})^2 \tag{8.7}$$

Let Φ denote the potential energy of the body. By definition, potential energy of a body referenced to the earth's surface is the weight of the body times the distance above the surface:

$$\Phi = wx = mgx \tag{8.8}$$

Now define the *lagrangian function B* as the difference between kinetic and potential energy:

$$B \equiv T - \Phi \tag{8.9}$$

For our example, combining Eqs. (8.7) to (8.9), we get

$$B = \tfrac{1}{2}m(\dot{x})^2 - mgx \tag{8.10}$$

We now write down *Lagrange's equation*, which will have to be accepted without proof; it is simply a corollary to Newton's second law:

$$\frac{d}{dt}\left(\frac{\partial B}{\partial \dot{x}}\right) - \frac{\partial B}{\partial x} = 0 \tag{8.11}$$

In Lagrange's equation above, recall the definition of the partial derivative given in Sec. 7.2 D. For example, $\partial B/\partial \dot{x}$ means the derivative of B with respect to $\dot{x}$, holding everything else constant. Hence, from Eq. (8.10),

$$\frac{\partial B}{\partial \dot{x}} = m\dot{x} \tag{8.12}$$

and

$$\frac{\partial B}{\partial x} = -mg \tag{8.13}$$

Substituting Eqs. (8.12) and (8.13) into Eq. (8.11), we have

$$\frac{d}{dt}(m\dot{x}) - (-mg) = 0$$

or, because m is a constant,

$$m\frac{d}{dt}(\dot{x}) - (-mg) = 0$$

$$m\ddot{x} + mg = 0$$

$$\boxed{\ddot{x} = -g} \tag{8.14}$$

Compare Eqs. (8.14) and (8.6); they are identical equations of motion. Therefore, we induce that Lagrange's equation and Newton's second law are equivalent mechanical relations and lead to the same equations of motion for a mechanical system. In the above example, the use of Lagrange's equation resulted in a slightly more complicated formulation than the direct use of $F = ma$. However, in the analysis of space vehicle orbits and trajectories, Lagrange's equation is the most expedient formulation, as will be detailed in the next section.

With the above example in mind, a more general formulation of Lagrange's equation can be given. Again, no direct proof is given; the reader must be content with the "cookbook" recipe given below, using the above example as a basis for induction. Consider a body moving in three-dimensional space, described by some generalized spatial coordinates q_1, q_2, and q_3. (These may be r, θ, and ϕ for a spherical coordinate system; x, y, and z for rectangular coordinate system; etc.) Set up the expression for the *kinetic energy* of the body, which may depend on the coordinates q_1, q_2, and q_3 themselves as well as the velocities $\dot{q}_1$, $\dot{q}_2$, and $\dot{q}_3$:

$$T = T(q_1, q_2, q_3, \dot{q}_1, \dot{q}_2, \dot{q}_3) \qquad (8.15)$$

Then, set up the expression for the *potential energy* of the body, which depends only on spatial location:

$$\Phi = \Phi(q_1, q_2, q_3) \qquad (8.16)$$

Now form the *lagrangian function*:

$$B = T - \Phi \qquad (8.17)$$

Finally, obtain three equations of motion (one along each coordinate direction) by writing Lagrange's equation for each coordinate:

$$
\begin{array}{ll}
q_1 \text{ coordinate: } & \dfrac{d}{dt}\left(\dfrac{\partial B}{\partial \dot{q}_1}\right) - \dfrac{\partial B}{\partial q_1} = 0 \\[2em]
q_2 \text{ coordinate: } & \dfrac{d}{dt}\left(\dfrac{\partial B}{\partial \dot{q}_2}\right) - \dfrac{\partial B}{\partial q_2} = 0 \\[2em]
q_3 \text{ coordinate: } & \dfrac{d}{dt}\left(\dfrac{\partial B}{\partial \dot{q}_3}\right) - \dfrac{\partial B}{\partial q_3} = 0
\end{array}
\qquad (8.18)
$$

Let us now apply this formalism to obtain the orbit or trajectory equations for a space vehicle.

8.4 THE ORBIT EQUATION

Space vehicles are launched from a planet's surface by rocket boosters. The rocket engines driving these boosters are discussed in Chap. 9. Here, we are concerned with the motion of the vehicle after all stages of the booster have burned out and the satellite, interplanetary probe, etc., is smoothly moving through space under the influence of gravitational forces. At the instant the last booster stage burns out, the space vehicle is at a given distance from the center of the planet, moving in a specific direction at a specific velocity. Obviously, nature prescribes a specific path (a specific orbit about the planet or possibly a specific trajectory away from the planet) for these given conditions at burnout. The purpose of this section is to derive the equation which describes this path.

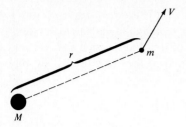

Figure 8.8 Movement of a small mass in the gravitational field of a large mass.

A Force and Energy

Consider a vehicle of mass m moving with velocity V in the vicinity of a planet of large mass M, as sketched in Figure 8.8. The distance between the centers of the two masses is r. In a stroke of genius during the last quarter of the seventeenth century, Isaac Newton uncovered the law of universal gravitation, which states that the gravitational force between two masses varies inversely as the square of the distance between their centers. In particular, this force is given by

$$F = \frac{GmM}{r^2} \tag{8.19}$$

where G is the universal gravitational constant, $G = 6.67 \times 10^{-11} m^3/(kg)(s)^2$.

Lagrange's equation deals with energy, both potential and kinetic. First consider the potential energy of the system shown in Figure 8.8. Potential energy is always based on some reference point, and for gravitational problems in astronautics, it is conventional to establish the potential energy as zero at r equal to infinity. Hence, the potential energy at a distance r is defined as the work done in moving the mass m from infinity to the location r. Let Φ be the potential energy. If the distance between M and m is changed by a small increment dr, then the work done in producing this change is $F\,dr$. This is also the change in potential energy, $d\Phi$. Using Eq. (8.19), we obtain

$$d\Phi = F\,dr = \frac{GmM}{r^2}\,dr$$

Integrating from r equals infinity, where Φ by definition is 0, to $r = r$, where the potential energy is $\Phi = \Phi$, we get

$$\int_0^{\Phi} d\Phi = \int_{\infty}^{r} \frac{GmM}{r^2}\,dr$$

or

$$\boxed{\Phi = \frac{-GmM}{r}} \tag{8.20}$$

Equation (8.20) gives the potential energy of small mass m in the gravitational field of large mass M at the distance r. The potential energy at r is a negative value due to our choice of $\Phi = 0$ at r going to infinity. However, if the idea of a negative energy is foreign to you, do not be concerned. In mechanical systems, we

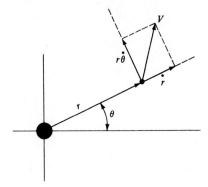

Figure 8.9 Polar coordinate system.

are usually concerned with *changes* in energy, and such changes are independent of our choice of reference for potential energy.

Now consider the kinetic energy. Here we need more precisely to establish our coordinate system. In more-advanced studies of mechanics, it can be proven that the motion of a body in a central force field (such as we are dealing with here) takes place in a plane. Hence, we need only two coordinates to designate the location of mass m. Polar coordinates are particularly useful in this case, as shown in Figure 8.9. Here, the origin is at the center of mass M, r is the distance between m and M, and θ is the angular orientation of r. The velocity of the vehicle of mass m is V. The velocity component parallel to r is $V_r = dr/dt = \dot{r}$. The velocity component perpendicular to r is equal to the radius vector r times the time rate of change of θ, that is, times the angular velocity; $V_\theta = r(d\theta/dt) = r\dot{\theta}$. Therefore, the kinetic energy of the vehicle is

$$T = \tfrac{1}{2}mV^2 = \tfrac{1}{2}\left[\dot{r}^2 + (r\dot{\theta})^2\right]m \tag{8.21}$$

B The Equation of Motion

From Eqs. (8.17), (8.20), and (8.21), the lagrangian function is

$$B = T - \Phi = \tfrac{1}{2}m\left[\dot{r}^2 + (r\dot{\theta})^2\right] + \frac{GmM}{r} \tag{8.22}$$

In orbital analysis, it is common to denote the product GM by k^2. If we are dealing with the earth, where $M = 5.98 \times 10^{24}$ kg, then

$$k^2 \equiv GM = 3.986 \times 10^{14} \text{ m}^3/\text{s}^2$$

Equation (8.22) then becomes

$$B = \tfrac{1}{2}m\left[\dot{r}^2 + (r\dot{\theta})^2\right] + \frac{mk^2}{r} \tag{8.23}$$

Now invoke Lagrange's equation, Eq. (8.18), where $q_1 = \theta$ and $q_2 = r$. First, the θ equation is

$$\frac{d}{dt}\frac{\partial B}{\partial \dot{\theta}} - \frac{\partial B}{\partial \theta} = 0 \tag{8.24}$$

From Eq. (8.23)

$$\frac{\partial B}{\partial \dot\theta} = mr^2\dot\theta \tag{8.25}$$

and

$$\frac{\partial B}{\partial \theta} = 0 \tag{8.26}$$

Substituting Eqs. (8.25) and (8.26) into (8.24), we obtain

$$\frac{d}{dt} mr^2\dot\theta = 0 \tag{8.27}$$

Equation (8.27) is the equation of motion of the space vehicle in the θ direction. It can be immediately integrated as

$$\boxed{mr^2\dot\theta = \text{const} = c_1} \tag{8.28}$$

From elementary physics, linear momentum is defined as mass times velocity. Analogously, for angular motion, *angular momentum* is defined as $I\dot\theta$, where I is the moment of inertia and $\dot\theta$ is the angular velocity. For a point mass m, $I = mr^2$. Hence, the product $mr^2\dot\theta$ is the *angular momentum of the space vehicle*, and from Eq. (8.28),

$$\boxed{mr^2\dot\theta = \text{angular momentum} = \text{const}}$$

For a central force field, Eq. (8.28) demonstrates that the angular momentum is constant.

Now consider the r equation. From Eq. (8.18), where $q_2 = r$,

$$\frac{d}{dt}\frac{\partial B}{\partial \dot r} - \frac{\partial B}{\partial r} = 0 \tag{8.29}$$

From Eq. (8.23)

$$\frac{\partial B}{\partial \dot r} = m\dot r \tag{8.30}$$

$$\frac{\partial B}{\partial r} = -\frac{mk^2}{r^2} + mr\dot\theta^2 \tag{8.31}$$

Substituting Eqs. (8.30) and (8.31) into (8.29), we get

$$\frac{d}{dt}m\dot r + \frac{mk^2}{r^2} - mr\dot\theta^2 = 0 \tag{8.32}$$

or

$$m\ddot r - mr\dot\theta^2 + \frac{mk^2}{r^2} = 0 \tag{8.33}$$

Equation (8.28) demonstrated that, since m is constant, $r^2\dot\theta$ is constant. Denote this quantity by h.

$$r^2\dot\theta \equiv h = \text{angular momentum per unit mass}$$

Multiplying and dividing the second term of Eq. (8.33) by r^3 and canceling m yields

$$m\ddot{r} - m\frac{r^4\dot{\theta}^2}{r^3} + \frac{mk^2}{r^2} = 0$$

or

$$\ddot{r} - \frac{h^2}{r^3} + \frac{k^2}{r^2} = 0 \qquad (8.34)$$

Equation (8.34) is the equation of motion for the space vehicle in the r direction. Note that both h^2 and k^2 are constants. Recalling our discussion in Sec. 8.2, we see that Eq. (8.34) is a differential equation. Its solution will provide a relation for r as the function of time, that is, $r = f(t)$.

However, examine Figure 8.9. The equation of the *path* of the vehicle in space should be geometrically given by $r = f(\theta)$, not $r = f(t)$. We are interested in this path, i.e., we want the equation of the space vehicle motion in terms of its geometric coordinates r and θ. Therefore, Eq. (8.34) must be reworked, as follows.

Let us transform Eq. (8.34) to a new dependent variable u, where

$$r = \frac{1}{u} \qquad (8.35)$$

Then

$$h \equiv r^2\dot{\theta} = \frac{\dot{\theta}}{u^2} \qquad (8.36)$$

Hence

$$\dot{r} \equiv \frac{dr}{dt} = \frac{d(1/u)}{dt} = -\frac{1}{u^2}\frac{du}{dt}$$

$$= -\frac{1}{u^2}\frac{du}{d\theta}\frac{d\theta}{dt} = -\frac{\dot{\theta}}{u^2}\frac{du}{d\theta} = -h\frac{du}{d\theta} \qquad (8.37)$$

Differentiating Eq. (8.37) with respect to t, we get

$$\ddot{r} = -h\frac{d}{dt}\frac{du}{d\theta} = -h\left(\frac{d}{d\theta}\frac{du}{d\theta}\right)\frac{d\theta}{dt}$$

$$= -h\left(\frac{d^2u}{d\theta^2}\right)\frac{d\theta}{dt} = -h\frac{d^2u}{d\theta^2}\dot{\theta} \qquad (8.38)$$

But from Eq. (8.36), $\dot{\theta} = u^2h$. Substituting into Eq. (8.38), we obtain

$$\ddot{r} = -h^2u^2\frac{d^2u}{d\theta^2} \qquad (8.39)$$

Substituting Eqs. (8.39) and (8.35) into Eq. (8.34) yields

$$-h^2u^2\frac{d^2u}{d\theta^2} - h^2u^3 + k^2u^2 = 0$$

or, dividing by h^2u^2,

$$\frac{d^2u}{d\theta^2} + u - \frac{k^2}{h^2} = 0 \tag{8.40}$$

Equation (8.40) is just as valid an equation of motion as is the original Eq. (8.34). Equation (8.40) is a differential equation, and its solution gives $u = f(\theta)$. Specifically, a solution of Eq. (8.40) is

$$u = \frac{k^2}{h^2} + A\cos(\theta - C) \tag{8.41}$$

where A and C are constants (essentially, constants of integration). You should satisfy yourself that Eq. (8.41) is indeed a solution of Eq. (8.40) by substitution of (8.41) into (8.40).

Return to the original transformation, Eq. (8.35). Substituting $u = 1/r$ into Eq. (8.41) yields

$$r = \frac{1}{k^2/h^2 + A\cos(\theta - C)} \tag{8.42}$$

Multiply and divide Eq. (8.42) by h^2/k^2:

$$r = \frac{h^2/k^2}{1 + A(h^2/k^2)\cos(\theta - C)} \tag{8.43}$$

Equation (8.43) is the desired equation of the path (the orbit, or trajectory) of the space vehicle. It is an algebraic equation for $r = f(\theta)$; it gives the geometric coordinates r and θ for a given path. The *specific* path is dictated by the values of the constants h^2, A, and C in Eq. (8.43). In turn, refer to Figure 8.10: these constants are fixed by conditions at the instant of burnout of the rocket booster. At burnout, the vehicle is at distance r_b from the center of the earth, and its velocity has a magnitude V_b and is in a direction β_b with respect to a perpendicular to r. These burnout conditions completely specify the vehicle's path; i.e., they determine the values of h^2, A, and C for Eq. (8.43).

Equation (8.43) is sometimes generically called the "orbit equation." However, it applies to the trajectory of a space vehicle escaping from the gravitational

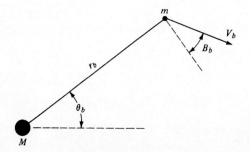

Figure 8.10 Conditions at the instant of burnout.

field of the earth as well as to an artificial satellite in orbit about the earth. In fact, what kind of orbit or trajectory is described by Eq. (8.43)? What type of mathematical curve is it? What physical conditions are necessary for a body to go into orbit or to escape from the earth? The answers can be found by further examination of Eq. (8.43), as discussed below.

8.5 SPACE VEHICLE TRAJECTORIES— SOME BASIC ASPECTS

Examine Eq. (8.43) closely. It has the general form

$$r = \frac{p}{1 + e\cos(\theta - C)} \tag{8.44}$$

where $p = h^2/k^2$, $e = A(h^2/k^2)$, and C is simply a phase angle. From analytic geometry, Eq. (8.44) is recognized as the standard form of a *conic section* in polar coordinates; i.e., Eq. (8.44) is the equation of a circle, ellipse, parabola, or hyperbola, depending on the value of e, where e is the *eccentricity* of the conic section. Specifically:

If $e = 0$, the path is a *circle*.
If $e < 1$, the path is an *ellipse*.
If $e = 1$, the path is a *parabola*.
If $e > 1$, the path is a *hyperbola*.

These possibilities are sketched in Figure 8.11. Note that point b on these sketches denotes the point of burnout and that θ is referenced to the dashed line through b, that is, θ is arbitrarily chosen as zero at burnout. Then C is simply a phase angle which orients the x and y axes with respect to the burnout point, where the x axis is a line of symmetry for the conic section. From inspection of Figure 8.11, circular and elliptical paths result in an orbit about the large mass M (the earth), whereas parabolic and hyperbolic paths result in escape from the earth.

On a physical basis, the eccentricity, hence the type of path for the space vehicle, is governed by the difference between the kinetic and potential energies of the vehicle. To prove this, consider first the kinetic energy, $T = \frac{1}{2}mV^2$. From Eq. (8.21),

$$T = \tfrac{1}{2}m\left[\dot{r}^2 + (r\dot{\theta})^2\right]$$

Differentiate Eq. (8.44) with respect to t:

$$\frac{dr}{dt} = \dot{r} = \frac{[re\sin(\theta - C)]\dot{\theta}}{1 + e\cos(\theta - C)} \tag{8.45}$$

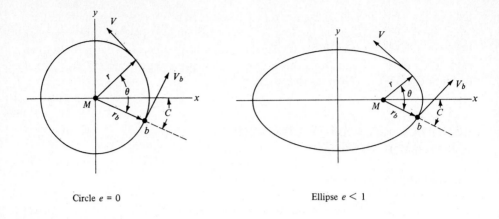

Circle $e = 0$ Ellipse $e < 1$

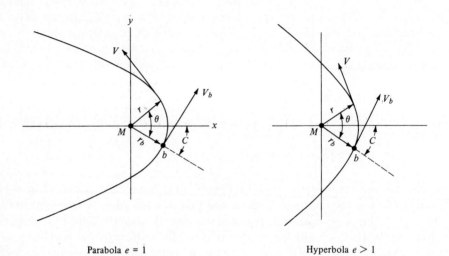

Parabola $e = 1$ Hyperbola $e > 1$

Figure 8.11 The four types of orbits and trajectories, illustrating the relation of the burnout point and phase angle with the axes of symmetry.

Substitute Eq. (8.45) into (8.21):

$$T = \frac{1}{2}m\left\{\frac{[r^2e^2\sin^2(\theta - C)]\dot{\theta}^2}{[1 + e\cos(\theta - C)]^2} + r^2\dot{\theta}^2\right\} \tag{8.46}$$

Recall that $r^2\dot{\theta} \equiv h$, hence $\dot{\theta}^2 = h^2/r^4$. Thus, Eq. (8.46) becomes

$$T = \frac{1}{2}m\left\{\frac{h^2e^2\sin^2(\theta - C)}{r^2[1 + e\cos(\theta - C)]^2} + \frac{h^2}{r^2}\right\} \tag{8.47}$$

Putting the right-hand side of Eq. (8.47) under the same common denominator and remembering from Eq. (8.44) that

$$r^2[1 + e\cos(\theta - C)]^2 = \left(\frac{h^2}{k^2}\right)^2$$

we transform Eq. (8.47) into

$$T = \frac{1}{2}m\frac{k^4}{h^2}\left[1 + 2e\cos(\theta - C) + e^2\right] \tag{8.48}$$

The reader should put in the few missing algebraic steps to obtain Eq. (8.48).

Consider now the absolute value of the potential energy, denoted as $|\Phi|$. From Eq. (8.20),

$$|\Phi| = \frac{GMm}{r} = \frac{k^2m}{r} \tag{8.49}$$

Substitute Eq. (8.44) into (8.49):

$$|\Phi| = \frac{k^4m}{h^2}\left[1 + e\cos(\theta - C)\right] \tag{8.50}$$

The difference between the kinetic and potential energies is obtained by subtracting Eq. (8.50) from (8.48):

$$T - |\Phi| = \frac{1}{2}m\frac{k^4}{h^2}\left[1 + 2e\cos(\theta - C) + e^2\right] - \frac{k^4m}{h^2}\left[1 + e\cos(\theta - C)\right] \tag{8.51}$$

Let H denote $T - |\Phi|$. Then, Eq. (8.51) becomes

$$H \equiv T - |\Phi| = -\frac{1}{2}m\frac{k^4}{h^2}(1 - e^2) \tag{8.52}$$

Solving Eq. (8.52) for e, we get

$$\boxed{e = \sqrt{1 + \frac{2h^2H}{mk^4}}} \tag{8.53}$$

Equation (8.53) is the desired result, giving the eccentricity e in terms of the difference between kinetic and potential energies, H.

Examine Eq. (8.53). If the kinetic energy is smaller than the potential energy, H will be negative, and hence $e < 1$. If the kinetic and potential energies are equal, $H = 0$ and $e = 1$. Similarly, if the kinetic energy is larger than the potential energy, H is postive and $e > 1$. Referring again to Figure 8.11, we can make the following tabulation:

Type of Trajectory	e	Energy Relation
Ellipse	< 1	$\frac{1}{2}mV^2 < \dfrac{GMm}{r}$
Parabola	$= 1$	$\frac{1}{2}mV^2 = \dfrac{GMm}{r}$
Hyperbola	> 1	$\frac{1}{2}mV^2 > \dfrac{GMm}{r}$

From this we make the important conclusion that a vehicle intended to escape the earth and travel into deep space (a parabolic or hyperbolic trajectory) must be launched such that its kinetic energy at burnout is equal to or greater than its potential energy, a conclusion that makes intuitive sense even without the above derivation.

Equation (8.53) tells us more. For example, what velocity is required for a circular orbit? To answer this question, recall that a circle has zero eccentricity. Putting $e = 0$ in Eq. (8.53), we get

$$0 = \sqrt{1 + \frac{2h^2 H}{mk^4}}$$

or
$$H = -\frac{mk^4}{2h^2} \tag{8.54}$$

Recall that $H = T - |\Phi| = \frac{1}{2}mV^2 - GMm/r$. Hence, Eq. (8.54) becomes

$$\frac{1}{2}mV^2 = -\frac{mk^4}{2h^2} + \frac{GMm}{r} \tag{8.55}$$

From Eq. (8.44), with $e = 0$,

$$r = \frac{h^2}{k^2} \tag{8.56}$$

Substitute Eq. (8.56) into (8.55) and solve for V:

$$\frac{1}{2}mV^2 = -\frac{m}{2}\frac{k^2}{r} + \frac{k^2 m}{r} = \frac{k^2 m}{2r}$$

Thus
$$\boxed{V = \sqrt{\frac{k^2}{r}} \qquad \text{circular velocity}} \tag{8.57}$$

Equation (8.57) gives the velocity required to obtain a circular orbit. Recall from Sec. 8.4 B that $k^2 = GM = 3.956 \times 10^{14}$ m^3/s^2. Assume that $r = 6.4 \times 10^6$ m, essentially the radius of the earth. Then

$$V = \sqrt{\frac{3.986 \times 10^{14}}{6.4 \times 10^6}} = 7.9 \times 10^3 \text{ m/s}$$

This is a convenient number to remember; *circular, or orbital, velocity is 7.9 km/s, or approximately 26,000 ft/s.*

The velocity required to escape the earth can be obtained in much the same fashion. We have demonstrated above that a vehicle will escape if it has a parabolic ($e = 1$) or a hyperbolic ($e > 1$) trajectory. Consider a parabolic trajectory. For this, we know that the kinetic and potential energies are equal; $T = |\Phi|$. Hence,

$$\tfrac{1}{2}mV^2 = \frac{GMm}{r} = \frac{k^2m}{r}$$

Solving for V, we get

$$\boxed{V = \sqrt{\frac{2k^2}{r}} \qquad \text{parabolic velocity}} \qquad (8.58)$$

Equation (8.58) gives the velocity required to obtain a parabolic trajectory. This is called the *escape velocity*; note by comparing Eqs. (8.57) and (8.58) that escape velocity is larger than orbital velocity by a factor of $\sqrt{2}$. Again assuming r is the radius of the earth, $r = 6.4 \times 10^6$ m, then *escape velocity is 11.2 km/s, or approximately 36,000 ft/s*. Return to Figure 8.10; if at burnout $V_b \geq 11.2$ km/s, then the vehicle will escape the earth, independent of the direction of motion, β_b.

Example 8.1 At the end of a rocket launch of a space vehicle, the burnout velocity is 9 km/s in a direction due north and 3° above the local horizontal. The altitude above sea level is 500 mi. The burnout point is located at the 27th parallel (27°) above the equator. Calculate and plot the trajectory of the space vehicle.

SOLUTION The burnout conditions are sketched in Figure 8.12. The altitude above sea level is

$$h_G = 500 \text{ mi} = 0.805 \times 10^6 \text{ m}$$

The distance from the center of the earth to the burnout point is (where the earth's radius is $r_e = 6.4 \times 10^6$ m)

$$r_b = r_e + h_G = 6.4 \times 10^6 + 0.805 \times 10^6 = 7.2 \times 10^6 \text{ m}$$

As given in Sec. 8.4 B,

$$k^2 \equiv GM = 3.986 \times 10^{14} \text{ m}^3/\text{s}^2$$

Also, as defined earlier,

$$h = r^2\dot{\theta} = r(r\dot{\theta}) = rV_\theta$$

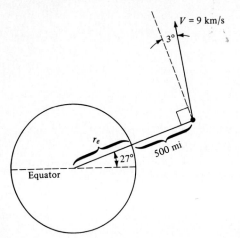

Figure 8.12 Burnout conditions for Example 8-1.

where V_θ is the velocity component perpendicular to the radius vector r. Thus,

$$h = rV_\theta = r_bV\cos\beta_b = (7.2\times10^6)(9\times10^3)\cos3°$$

$$= 6.47\times10^{10} \text{ m}^2/\text{s}$$

$$h^2 = 4.188\times10^{21} \text{ m}^4/\text{s}^2$$

Hence $\qquad p \equiv \dfrac{h^2}{k^2} = \dfrac{4.188\times10^{21}}{3.986\times10^{14}} = 1.0506\times10^7 \text{ m}$

The trajectory equation is given by Eq. (8.44), where the above value of p is the numerator of the right-hand side. To proceed further, we need the eccentricity e. This can be obtained from Eq. (8.53)

$$e = \sqrt{1 + \frac{2h^2H}{mk^4}}$$

where $H/m = (T - |\Phi|)/m$.

$$\frac{T}{m} = \frac{V^2}{2} = \frac{(9\times10^3)^2}{2} = 4.05\times10^7 \text{ m}^2/\text{s}^2$$

$$\left|\frac{\Phi}{m}\right| = \frac{GM}{r_b} = \frac{k^2}{r_b} = \frac{3.986\times10^{14}}{7.2\times10^6} = 5.536\times10^7 \text{ m}^2/\text{s}^2$$

Hence $\qquad \dfrac{H}{m} = (4.05 - 5.536)\times10^7 = -1.486\times10^7 \text{ m}^2/\text{s}^2$

Thus $\qquad e = \left[1 + \dfrac{2h^2}{k^4}\left(\dfrac{H}{m}\right)\right]^{1/2}$

$$= \left(1 + \frac{2(4.188\times10^{21})(-1.486\times10^7)}{(3.986\times10^{14})^2}\right)^{1/2}$$

$$= \sqrt{0.2166} = 0.4654$$

Immediately we recognize that the trajectory is an elliptical orbit, because $e < 1$ and also because

$T < |\Phi|$. From Eq. (8.44),

$$r = \frac{p}{1 + e\cos(\theta - C)}$$

$$= \frac{1.0506 \times 10^7}{1 + 0.4654\cos(\theta - C)}$$

To find the phase angle C, simply substitute the burnout location ($r_b = 7.2 \times 10^6$ m and $\theta = 0°$) into the above equation. (Note that $\theta = 0°$ at burnout, and hence θ is measured relative to the radius vector at burnout, with increasing θ taken in the direction of motion; this is sketched in Figure 8.11.)

$$r_b = \frac{p}{1 + e\cos(-C)}$$

$$7.2 \times 10^6 = \frac{1.0506 \times 10^7}{1 + 0.4654\cos(-C)}$$

Solve for $\cos(-C)$:

$$\cos(-C) = 0.9878$$

Thus

$$C = -8.96°$$

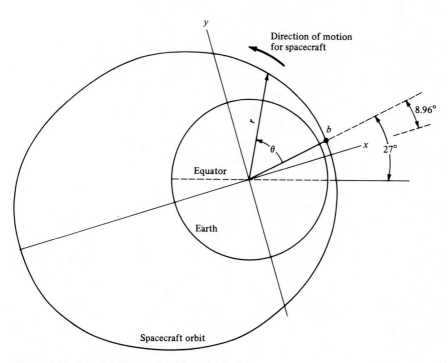

Figure 8.13 Orbit for the spacecraft in Example 8-1.

Finally, the complete equation of the orbit is

$$r = \frac{1.0506 \times 10^7}{1 + 0.4654 \cos(\theta + 8.96°)}$$

where θ is in degrees and r is in meters.

The orbit is drawn to scale in Figure 8.13. Note that b designates the burnout point, which is 27° above the equator. The x and y axes are the axes of symmetry for the elliptical orbit, and the phase angle orients the x axis at 8.96° *below* (because C is negative in this problem) the radius vector through point b. The angle θ is measured from the radius through b, with positive θ in the counterclockwise direction. The spacecraft is traveling in a counterclockwise direction, following an elliptical orbit. The perigee and apogee are 7.169×10^6 and 1.965×10^7 m, respectively.

8.6 KEPLER'S LAWS

To this point, our discussion has been couched in terms of an artificial space vehicle launched from the earth. However, most of the above analysis and results hold in general for orbits and trajectories of any mass in a central gravitational force field. The most familiar natural example of such motion is our solar system, i.e., the orbits of the planets about the sun. Such motion has held peoples' attention since the early days of civilization. Early observations and mapping of planetary motion evolved over millennia passing from the Babylonians to the Egyptians to the Greeks to the Romans, carried throughout the dark ages by the Arabians, and reaching the age of Copernicus in the fifteenth century (about the time Christopher Columbus was discovering America). However, by this time, astronomical observations were still inaccurate and uncertain. Then, during the period 1576 to 1597, Tycho Brahe, a Danish noble, made a large number of precise astronomical observations which improved the accuracy of existing tables by a factor of 50. Near the end of his life, Brahe was joined by Johannes Kepler, a young German astronomer and mathematician, who further improved these observations. Moreover, Kepler made some pioneering conclusions about the geometry of planetary motion. From 1609 to 1618, Kepler induced and published three laws of planetary motion, obtained strictly from an exhaustive examination of the astronomical data. Kepler did not have the advantage of Newton's law of universal gravitation or newtonian mechanics, which came three-quarters of a century later. Nevertheless, Kepler's inductions were essentially correct, and his classic three laws are as important today for the understanding of artificial satellite motion as they were in the seventeenth century for the understanding of planetary motion. Therefore, his conclusions will be discussed in this section. Moreover, we will take advantage of our previous derivations of orbital motion to derive Kepler's laws, a luxury Kepler did not have himself.

Kepler's first major conclusion was as follows.

Kepler's first law: *A satellite describes an* elliptical *path around its center of attraction.*

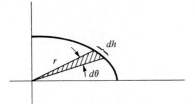

Figure 8.14 Area swept out by the radius vector in moving through the angle $d\theta$.

We have already proved this fact in Secs. 8.4 and 8.5; therefore, nothing more needs to be said.

To prove Kepler's second law, recall from Eq. (8.28) that angular momentum is constant; that is, $mr^2\dot\theta =$ const. Consider Figure 8.14, which shows the radius vector r sweeping through an infinitesimally small angle $d\theta$. The area of the small triangle swept out is $dA = \frac{1}{2}r\,dh$. However, $dh = r\,d\theta$. Thus $dA = \frac{1}{2}r^2\,d\theta$. The time rate of change of the area swept out by the radius is then

$$\frac{dA}{dt} = \frac{\frac{1}{2}r^2\,d\theta}{dt} = \frac{1}{2}r^2\dot\theta \tag{8.59}$$

However, from Eq. (8.28), $r^2\dot\theta$ is a constant. Hence, Eq. (8.59) shows that

$$\frac{dA}{dt} = \text{const} \tag{8.60}$$

which proves Kepler's second law.

Kepler's second law: *In equal times, the areas swept out by the radius vector of a satellite are the same.*

An obvious qualitative conclusion follows from this law, as illustrated in Figure 8.15. Here, the elliptical orbit of a small mass m is shown about a large mass M. In order for equal areas to be swept out in equal times, the satellite must have a larger velocity when it is near M and a smaller velocity when it is far away. This is characteristic of all satellite motion.

To derive Kepler's third law, consider the elliptical orbit shown in Figure 8.16. The point of closest approach, where r is minimum, is defined as the *perigee*; the point farthest away, where r is maximum, is defined as the *apogee*.

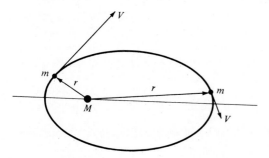

Figure 8.15 Illustration of the variation in velocity at different points along the orbit.

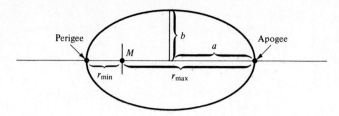

Figure 8.16 Illustration of apogee, perigee, and semimajor and semiminor axes.

The mass M (e.g., that of the earth or the sun) is at the focus of the ellipse. The major axis of the ellipse is the distance from the perigee to the apogee, and one-half this distance is defined as the *semimajor axis a*. The *semiminor axis b* is also shown in Figure 8.16. Let us assume for simplicity that the phase angle C of the orbit is zero. Thus, from Eq. (8.44), the maximum and minimum radii are, respectively,

$$r_{max} = \frac{h^2/k^2}{1 - e} \tag{8.61}$$

$$r_{min} = \frac{h^2/k^2}{1 + e} \tag{8.62}$$

From the definition of a, and using Eqs. (8.61) and (8.62), we obtain

$$a = \tfrac{1}{2}(r_{max} + r_{min})$$

$$= \frac{1}{2}\frac{h^2}{k^2}\left(\frac{1}{1 - e} + \frac{1}{1 + e}\right) = \frac{h^2/k^2}{1 - e^2} \tag{8.63}$$

The eccentricity e of the ellipse is geometrically related to the semimajor and semiminor axes; taking a result from analytic geometry, we get

$$e = \frac{(a^2 - b^2)^{1/2}}{a}$$

Solving for b,

$$b = a(1 - e^2)^{1/2} \tag{8.64}$$

If we lift another result from analytic geometry, we find the area of an ellipse is

$$A = \pi ab \tag{8.65}$$

Substituting Eq. (8.64) into (8.65) yields

$$A = \pi a\left[a(1 - e^2)^{1/2}\right] = \pi a^2(1 - e^2) \tag{8.66}$$

Now return to Eq. (8.59):

$$dA = \tfrac{1}{2}r^2\dot{\theta}\,dt = \tfrac{1}{2}h\,dt \tag{8.67}$$

Thus, the area of the ellipse can be obtained by integrating Eq. (8.67) around the complete orbit. That is, imagine the satellite starting at the perigee at time = 0. Now allow the satellite to move around one complete orbit, returning to the perigee. The area swept out by the radius vector is the whole area of the ellipse, A. The time taken by the satellite in executing the complete orbit is defined as the *period* and is denoted by τ. Thus, integrating Eq. (8.67) around the complete orbit, we get

$$\int_0^A dA = \int_0^\tau \tfrac{1}{2}h\,dt$$

or
$$A = \tfrac{1}{2}h\tau \tag{8.68}$$

We now have two independent results for A: Eq. (8.66) from analytic geometry and Eq. (8.68) from orbital mechanics. Equating these two relations, we have

$$\tfrac{1}{2}h\tau = \pi a^2(1 - e^2)^{1/2} \tag{8.69}$$

Solve Eq. (8.63) for h:

$$h = a^{1/2}k(1 - e^2)^{1/2} \tag{8.70}$$

Substitute Eq. (8.70) into (8.69):

$$\tfrac{1}{2}\tau a^{1/2}k(1 - e^2)^{1/2} = \pi a^2(1 - e^2)^{1/2}$$

or, squaring both sides,

$$\tfrac{1}{4}\tau^2 ak^2 = \pi^2 a^4$$

or
$$\tau^2 = \frac{4\pi^2}{k^2}a^3 \tag{8.71}$$

Examine Eq. (8.71). The factor $4\pi^2/k^2$ is a constant. Hence

$$\tau^2 = (\text{const})a^3 \tag{8.72}$$

i.e., the square of the period is proportional to the cube of the semimajor axis. If we have two satellites in orbit about the same planet, with values of τ_1, a_1 and τ_2, a_2, respectively, then Kepler's third law can be written as follows.

Kepler's third law: The periods of any two satellites about the same planet are related to their semimajor axes as

$$\frac{\tau_1^{\,2}}{\tau_2^{\,2}} = \frac{a_1^{\,3}}{a_2^{\,3}}$$

On this note, we conclude our discussion of space vehicle orbits and trajectories. Consistent with the scope of this book, we have only provided an introduction to the topic. Modern orbital and trajectory analysis is performed on high-speed digital computers, taking into account the gravitational attraction of several bodies simultaneously (e.g., gravitational attraction of the earth, sun, and moon on a lunar space vehicle), perturbations of the gravitational field due to the real nonspherical shape of the earth, trajectory corrections and orbital transfers due to in-flight propulsion, etc. Also, much attention is given to satellite altitude control, i.e., the proper orientation of the vehicle with respect to some given reference system. The reader is encouraged to look further into such matters in more-advanced studies of astronautics.

Example 8.2 The period of revolution of the earth about the sun is 365.256 days. The semimajor axis of the earth's orbit is 1.49527×10^{11} m. In turn, the semimajor axis of the orbit of Mars is 2.2783×10^{11} m. Calculate the period of Mars.

SOLUTION From Kepler's third law, we have

$$\tau_2 = \tau_1 \left(\frac{a_2}{a_1} \right)^{3/2}$$

where
$$a_1 = 1.49527 \times 10^{11} \text{ m} \qquad \text{for earth}$$
$$\tau_1 = 365.256 \text{ days}$$

and
$$a_2 = 2.2783 \times 10^{11} \text{ m} \qquad \text{for Mars}$$

Hence
$$\tau_2 = 365.256 \left(\frac{2.2783}{1.49527} \right)^{3/2}$$

$$\boxed{\tau_2 = 686.96 \text{ days for Mars}}$$

8.7 INTRODUCTION TO REENTRY

In all cases of contemporary manned space vehicles, and with many unmanned vehicles, it is necessary to terminate the orbit or trajectory at some time and return to the earth. Obviously, this necessitates negotiating the atmosphere at high velocities. Recall from Sec. 8.5 that an orbital vehicle will enter the outer regions of the atmosphere at a velocity near 26,000 ft/s; a vehicle returning from a moon mission (e.g., an Apollo vehicle) will enter at an even higher velocity, near 36,000 ft/s. These velocities correspond to flight Mach numbers of 30 or more! Such hypersonic flight conditions are associated with several uniquely difficult aerodynamic problems—so unique and difficult that they dominated the research efforts of aerodynamicists during the late 1950s and throughout the 1960s. Subsequently, the successful manned reentries of the Mercury, Gemini, and Apollo vehicles were striking testimonials to the success of this hypersonic research.

Consider a space vehicle in orbit about the earth, as shown in Figure 8.17. We wish to terminate this orbit and land the vehicle somewhere on the earth's surface. First, the path of the vehicle is changed by firing a retro-rocket, decreasing the vehicle's velocity. In terms of the orbit equation, Eq. (8.43) or (8.44), the retro-rocket's firing effectively changes the values of h, e, and C such that the vehicle curves toward the earth. When the vehicle encounters the outer region of the atmosphere (given by the dashed circle in Figure 8.17), three types of reentry

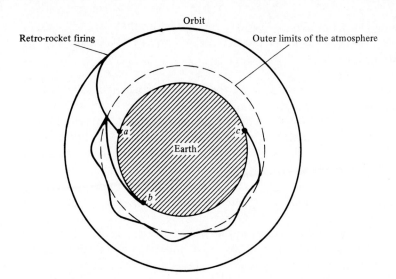

Figure 8.17 Three types of reentry paths. (*a*) Ballistic; (*b*) glide; (*c*) skip.

paths are possible:

1. *Ballistic reentry*—Here, the vehicle has little or no aerodynamic lift. It falls
   through the atmosphere under the influence of drag and gravity, impacting
   the surface at point *a* in Figure 8.17. The impact point is predetermined by
   the conditions at first entry to the atmosphere. The pilot has no control over
   his or her landing position during this ballistic trajectory. It literally is the
   same as *falling* to the surface. Before the space shuttle, virtually all reentries
   of existing space vehicles were ballistic. (A slight exception might be the
   Apollo capsule shown in Figure 8.4, which at an angle of attack can generate
   a small lift-to-drag ratio, $L/D < 1$. However, for all practical purposes, this
   is still a ballistic reentry vehicle.)
2. *Skip reentry*—Here, the vehicle generates a value of L/D between 1 and 4
   and uses this lifting ability to first graze the atmosphere, then slow down a
   bit, then pitch up such that the lift carries it back out of the atmosphere. This
   is repeated several times, much like a flat stone skipping over the surface of a
   pond, until finally the vehicle is slowed down appropriately and penetrates
   the atmosphere, landing at point *c* in Figure 8.17. Unfortunately, the
   aerodynamic heating of a skip reentry vehicle is inordinately large, and
   therefore such a reentry mode has never been used and is not contemplated in
   the future.
3. *Glide reentry*—Here, the vehicle is essentially an airplane, generating a
   lift-to-drag ratio of 4 or larger. The vehicle enters the atmosphere at a high
   angle of attack (30° or more) and flies to the surface, landing at point *b* in
   Figure 8.17. An example of such a lifting reentry vehicle is given in Figure

8.6. The compelling advantages of the space shuttle are that the pilot can, in principle, choose the landing site and that the vehicle can be landed intact, to be used again.

In all of the above reentry modes, there are two overriding technical concerns —maximum deceleration and aerodynamic heating. For the safety of the occupants of a manned reentry vehicle, the maximum deceleration should not exceed 10 times that of the acceleration of gravity, that is, 10 g's. Furthermore, the aerodynamic heating of the vehicle should be low enough to maintain tolerable temperatures inside the capsule; if the vehicle is unmanned, it still must be kept from burning up in the atmosphere. For these reasons, reentry trajectories, maximum deceleration, and aerodynamic heating will be the subject of the remainder of this chapter.

Finally, there is an extra consideration in regard to the reentry of manned space vehicles returning from lunar or planetary missions. Such vehicles will approach the earth with parabolic or hyperbolic trajectories, as shown in Figure 8.18. If the vehicle is traveling along path A in Figure 8.18, penetration of the atmosphere will be too rapid and the maximum deceleration will be too large. On the other hand, if the vehicle is traveling along path B, it will not penetrate the atmosphere enough; the drag will be too low, the velocity will not be decreased enough for the vehicle to be captured by the earth, and it will go shooting past, back into outer space, never to return again. Consequently, there is a narrow *reentry corridor* into which the vehicle must be guided for a successful return to

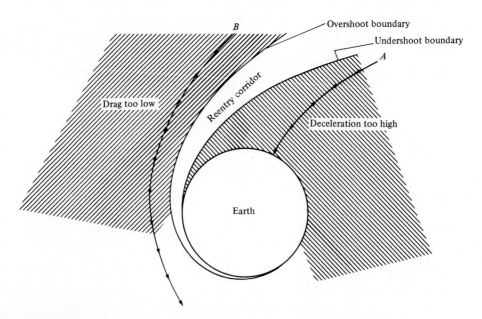

Figure 8.18 Illustration of the reentry corridor.

the earth's surface. This reentry corridor is shown in Figure 8.18, bounded above by the *overshoot* boundary and below by the *undershoot* boundary.

8.8 THE EXPONENTIAL ATMOSPHERE

Since reentry involves motion through the atmosphere, it is reasonable to expect reentry performance to depend on the physical properties of the atmosphere. Such properties have been discussed in Chap. 3, where the atmospheric temperature distribution is given in Figure 3.3. Detailed reentry trajectory calculations made on computers take into account the precise variation of the standard atmosphere as given in Chap. 3. However, for a first approximation, a completely isothermal atmosphere, with a constant temperature equal to some mean of the variation shown in Figure 3.3, can be assumed. In this case, the density variation with altitude is a simple exponential, as given by Eq. (3.10). [At this point, the reader should review the derivation of Eq. (3.10).] Writing Eq. (3.10) with point 1 at sea level, we obtain

$$\frac{\rho}{\rho_0} = e^{-g_0 h / RT} \qquad (8.73)$$

Equation (8.73) establishes the *exponential model atmosphere*. It agrees reasonably well with the actual density variation of the earth's standard atmosphere up to about 450,000 ft (about 140 km); above this height, the air is so thin that it has no meaningful influence on the reentry trajectory. The exponential model atmosphere was used by NASA and other laboratories in the early studies of reentry during the 1950s and early 1960s. We will adopt it here for the remainder of this chapter.

8.9 GENERAL EQUATIONS OF MOTION FOR ATMOSPHERIC REENTRY

Consider a space vehicle reentering the atmosphere, as sketched in Figure 8.19. At a given altitude h, the velocity of the vehicle is V, inclined at the angle θ below the local horizontal. The weight W is directed toward the center of the earth, and drag D and lift L are parallel and perpendicular, respectively, to the flight path, as usual. Summing forces parallel and perpendicular to the flight path and using Newton's second law, we obtain, respectively,

$$-D + W \sin \theta = m \frac{dV}{dt} \qquad (8.74)$$

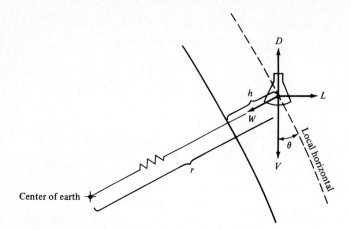

Figure 8.19 Geometry of reentry vehicle forces and motion.

and
$$L - W\cos\theta = m\frac{V^2}{r_c} \tag{8.75}$$

where r_c is the radius of curvature of the flight path. Equations (8.74) and (8.75) are identical to the equations of motion obtained in Chap. 6, specifically Eqs. (6.6) and (6.7), with $T = 0$ and θ measured below rather than above the horizontal.

We wish to establish an analysis which will yield velocity V as a function of altitude h. Dealing first with the drag equation, Eq. (8.74), we have

$$-D + W\sin\theta = m\frac{dV}{dt} = m\frac{dV}{ds}\frac{ds}{dt} = mV\frac{dV}{ds}$$

$$-D + W\sin\theta = \frac{1}{2}m\frac{dV^2}{ds} \tag{8.76}$$

where s denotes distance along the flight path. From the definition of drag coefficient,

$$D = \tfrac{1}{2}\rho V^2 S C_D \tag{8.77}$$

Also, from the geometry shown in Figure 8.20,

$$ds = -\frac{dh}{\sin\theta} \tag{8.78}$$

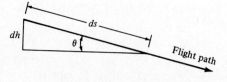

Figure 8.20 Flight path geometry.

Substitute Eqs. (8.77) and (8.78) into (8.76):

$$-\tfrac{1}{2}\rho V^2 S C_D + W\sin\theta = -\tfrac{1}{2}m\sin\theta\frac{dV^2}{dh} \qquad (8.79)$$

We are interested in obtaining V as a function of h. However, recall from Eq. (8.73) that $\rho = f(h)$, namely

$$\frac{\rho}{\rho_0} = e^{-g_0 h/RT} = e^{-Zh} \qquad (8.80)$$

where $Z \equiv g_0/RT$ for simplicity of notation. Therefore, if we instead had a relation between velocity and density, $V = f(\rho)$, we could still find the variation of V with h by using Eq. (8.80) as an intermediary. Let us take this approach and seek an equation relating V to ρ, as follows.

Differentiating Eq. (8.80), we obtain

$$\frac{d\rho}{\rho_0} = e^{-Zh}(-Z\,dh) = \frac{\rho}{\rho_0}(-Z\,dh)$$

or

$$dh = -\frac{d\rho}{Z\rho} \qquad (8.81)$$

Substitute Eq. (8.81) into Eq. (8.79):

$$-\tfrac{1}{2}\rho V^2 S C_D + W\sin\theta = -\tfrac{1}{2}m\sin\theta\frac{dV^2}{d\rho}(-Z\rho) \qquad (8.82)$$

Divide Eq. (8.82) by $-\tfrac{1}{2}\rho Z m\sin\theta$:

$$\frac{V^2 S C_D}{Zm\sin\theta} - \frac{2mg}{Z\rho m} = -\frac{dV^2}{d\rho}$$

or

$$\boxed{\frac{dV^2}{d\rho} + \frac{1}{m/C_D S}\frac{V^2}{Z\sin\theta} = \frac{2g}{Z\rho}} \qquad (8.83)$$

Equation (8.83) is an exact equation of motion for a vehicle reentering the atmosphere—the only approximation it contains is the exponential model atmosphere. Also note that the parameter $m/C_D S$, which appears in the second term in Eq. (8.83), is essentially a constant for a given space vehicle; it is identified as

$$\frac{m}{C_D S} \equiv \text{ballistic parameter}$$

The value of $m/C_D S$ strongly governs the reentry trajectory, as will be demonstrated later.

Equation (8.83) is also a differential equation, and in principle it can be solved to obtain $V = f(\rho)$ and hence $V = f(h)$ through Eq. (8.80). However, in general, the angle θ in Eq. (8.83) also varies with altitude h, and this variation must be obtained before Eq. (8.83) can be solved. This is the role of our second equation of motion, Eq. (8.75)—the lift equation. Equation (8.75) can be re-

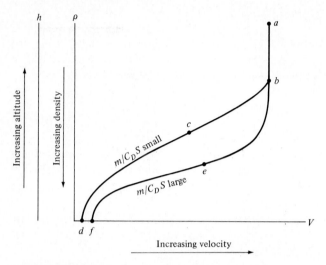

Figure 8.21 Reentry trajectory on a velocity-altitude map.

worked to obtain a differential equation in terms of $d\theta/d\rho$, which can then be solved simultaneously with Eq. (8.83) to obtain an explicit relation for V as a function of ρ for a vehicle with a given $m/C_D S$ and L/D. The details will not be given here; our intent has been to simply map out an approach to calculate a lifting reentry path, as given above. The reader can obtain more details from the NACA and NASA reports given in the Bibliography at the end of this chapter.

After the above analysis is carried to completion, what does the actual reentry path look like? An answer is given in Figure 8.21, which illustrates the variation of velocity (the abscissa) with density (the ordinate). Since ρ is a function of altitude through Eq. (8.80), h is also given on the ordinate. Thus, Figure 8.21 shows the reentry path in terms of velocity vs. altitude—a so-called *velocity-altitude map* for reentry. Such velocity-altitude maps are frequently used in reentry vehicle design and analysis. Examine Figure 8.21 more closely. Imagine a reentry vehicle just beginning to penetrate the atmosphere. It is at a very high altitude and velocity, such as point a in Figure 8.21. During the early portion of reentry, the atmospheric density is so low that the drag is virtually insignificant; the vehicle penetrates the upper region of the atmosphere with only a small decrease in velocity, as shown from point a to point b in Figure 8.21. However, below the altitude denoted by point b, the air density rapidly increases with an attendant marked increase in drag, causing the velocity to decrease rapidly. This is the situation at point c in Figure 8.21. Finally, the vehicle impacts the surface at point d. In Figure 8.21, the path a-b-c-d is for a given ballistic parameter. If $m/C_D S$ is made larger, the vehicle penetrates deeper into the atmosphere before slowing down, as illustrated by path a-b-e-f. Thus, as suspected from an examination of Eq. (8.83), the ballistic parameter is an important design aspect of reentry vehicles.

8.10 APPLICATION TO BALLISTIC REENTRY

A solution of the exact equations of motion, such as Eq. (8.83), must be performed numerically on a high-speed computer. That is, the curves in Figure 8.21 are obtained from numbers generated by a computer; they are not given by simple, closed-form analytic equations. However, such an analytic solution can be obtained for a purely ballistic reentry (no lift), with a few assumptions. This is the purpose of the present section.

Return to the picture of a vehicle reentering the atmosphere, as shown in Figure 8.19. If the path is purely ballistic, then $L = 0$ by definition. Also, recall that the initial reentry velocities are high—26,000 ft/s for circular orbits, 36,000 ft/s for parabolic space trajectories, etc. Therefore, the dynamic pressures associated with reentry velocities throughout most of the velocity-altitude map are large. As a result, drag is large—much larger, in fact, than the vehicle's weight; $D \gg W$. With this in mind, W can be ignored, and the original drag equation, Eq. (8.74), becomes

$$-D = m\frac{dV}{dt} \tag{8.84}$$

Following Eq. (8.84) with the same derivation that led to Eq. (8.83), we obtain

$$\frac{dV^2}{d\rho} + \frac{1}{m/C_DS}\frac{V^2}{Z\sin\theta} = 0 \tag{8.85}$$

(The reader should carry through this derivation to satisfy his or her own curiosity.) Equation (8.85) is the same as Eq. (8.83), with the right side now zero because W has been neglected.

Furthermore, assume that θ is constant in Eq. (8.85). Referring to Figure 8.19, this implies a straight-line reentry path through the atmosphere. This is a reasonable approximation for many actual ballistic reentry vehicles. If θ is constant, Eq. (8.85) can be integrated in closed form, as follows. First, rearrange Eq. (8.85):

$$\frac{dV^2}{V^2} = -\frac{d\rho}{(m/C_DS)Z\sin\theta} \tag{8.86}$$

Integrate Eq. (8.86) from the point of initial contact with the atmosphere, where $\rho = 0$ and $V = V_E$ (the initial reentry velocity), to some point in the atmosphere, where the density is ρ and the vehicle velocity is V.

$$\int_{V_E}^{V}\frac{dV^2}{V^2} = -\frac{1}{(m/C_DS)Z\sin\theta}\int_0^\rho d\rho$$

or

$$\ln\frac{V^2}{V_E^2} = 2\ln\frac{V}{V_E} = -\frac{\rho}{(m/C_DS)Z\sin\theta}$$

Thus,

$$\boxed{\frac{V}{V_E} = e^{-\rho/2(m/C_DS)Z\sin\theta}} \qquad (8.87)$$

Equation (8.87) is a closed-form expression for the variation of V with ρ, hence of V with h via Eq. (8.80). It is an explicit equation for the reentry trajectory on a velocity-altitude map, as sketched in Figure 8.21, except Eq. (8.87) now tells us precisely how the velocity changes; before, we had to take the shape of the curves in Figure 8.21 on faith. For example, examine Eq. (8.87). As ρ increases (i.e., as the altitude decreases), V decreases. This confirms the shape of the curves shown in Figure 8.21. Also, if m/C_DS is made larger, the exponential term in Eq. (8.87) does not have as strong an effect until ρ becomes larger (i.e., until the altitude is smaller). Hence, a vehicle with a large m/C_DS penetrates deeper into the atmosphere with a high velocity, as shown in Figure 8.21. Therefore, the variations shown in Figure 8.21 are directly verified by the form of Eq. (8.87).

In Sect. 8.7, maximum deceleration was identified as an important reentry consideration. We now have enough background to examine deceleration in more detail. First, consider the equation of motion, Eq. (8.84), which neglects the vehicle's weight. By definition, dV/dt in Eq. (8.84) is the acceleration, and from Eq. (8.84) it is negative for reentry.

$$\frac{dV}{dt} = -\frac{D}{m}$$

Also, by definition, a negative value of acceleration is *deceleration*, denoted by $|dV/dt|$. From the above equation,

$$\text{Deceleration} = \left|\frac{dV}{dt}\right| = \frac{D}{m} \qquad (8.88)$$

From the definition of drag coefficient, $D = \frac{1}{2}\rho V^2 S C_D$, Eq. (8.88) becomes

$$\left|\frac{dV}{dt}\right| = \frac{\rho V^2 S C_D}{2m} \qquad (8.89)$$

[In Eq. (8.89), the subscript "∞" has been dropped from ρ and V for convenience.] Note from Eq. (8.89) that $|dV/dt|$ increases as ρ increases, and decreases as V decreases. This allows us to qualitatively sketch the deceleration vs. altitude curve shown in Figure 8.22. At high altitudes, the velocity is large but relatively constant (see Figure 8.21, from points a to b), whereas ρ is beginning to increase. Therefore, from Eq. (8.89), deceleration will first increase as the vehicle reenters the atmosphere, as shown in Figure 8.22 at high altitude. However, at lower altitudes, Figure 8.21 shows that the velocity rapidly decreases. From Eq. (8.89), the velocity decrease now overshadows the increase in density, hence the deceleration will decrease in magnitude. This is shown in Figure 8.22 at low altitude. Consequently, the deceleration experienced by a vehicle throughout reentry first increases, then goes through a maximum, and then decreases; this variation is clearly illustrated in Figure 8.22.

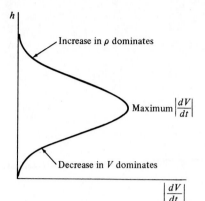

h

Increase in ρ dominates

Maximum $\left|\dfrac{dV}{dt}\right|$

Decrease in V dominates

$\left|\dfrac{dV}{dt}\right|$

Figure 8.22 The variation of deceleration with altitude for ballistic reentry.

The quantitative value of the *maximum* deceleration is of interest. It was stated in Sec. 8.7 that a manned reentry vehicle should not exceed a maximum deceleration of 10 g's; furthermore, even unmanned vehicles have limitations dictated by structural failure of the vehicle itself or its components. Therefore, let us derive an equation or maximum deceleration. To begin with, Eq. (8.89) gives an expression for deceleration which holds at any point along our straight-line ballistic trajectory. We wish to find the maximum deceleration. Therefore, from calculus, we wish to differentiate Eq. (8.89) and set the result equal to zero in order to find the conditions for maximum deceleration. Differentiating Eq. (8.89) with respect to time, and noting that both ρ and V vary along the trajectory,

$$\left|\frac{d^2V}{dt^2}\right| = \frac{SC_D}{2m}\left(2\rho V\frac{dV}{dt} + V^2\frac{d\rho}{dt}\right) \tag{8.90}$$

From Eq. (8.84),

$$\frac{dV}{dt} = -\frac{D}{m} = -\frac{1}{m}\left(\frac{1}{2}\rho V^2 SC_D\right) \tag{8.91}$$

Substitute Eq. (8.91) into Eq. (8.90):

$$\left|\frac{d^2V}{dt^2}\right| = \frac{SC_D}{2m}\left[2\rho V\left(-\frac{\rho V^2 SC_D}{2m}\right) + V^2\frac{d\rho}{dt}\right]$$

$$\left|\frac{d^2V}{dt^2}\right| = \frac{SC_D V^2}{2m}\left(-\frac{\rho^2 VSC_D}{m} + \frac{d\rho}{dt}\right) \tag{8.92}$$

Setting Eq. (8.92) equal to zero for conditions at maximum $|dV/dt|$, we find that

$$\frac{d\rho}{dt} = \frac{\rho^2 VSC_D}{m} \tag{8.93}$$

From the exponential model atmosphere, differentiating Eq. (8.80) with respect to time,

$$\frac{d\rho}{dt} = -\rho_0 Z e^{-Zh} \frac{dh}{dt} = -Z\rho \frac{dh}{dt} \tag{8.94}$$

However, from the geometric construction of Figure 8.20 and from Eq. (8.78),

$$\frac{dh}{dt} = -\frac{ds}{dt} \sin\theta = -V\sin\theta \tag{8.95}$$

Substitute Eq. (8.95) into (8.94):

$$\frac{d\rho}{dt} = \rho Z V \sin\theta \tag{8.96}$$

Substitute Eq. (8.96) into (8.93):

$$\rho Z V \sin\theta = \frac{\rho^2 V S C_D}{m} \tag{8.97}$$

Solve Eq. (8.97) for ρ:

$$\rho = \frac{m}{C_D S} Z \sin\theta \tag{8.98}$$

Equation (8.98) gives the value of density at the point of maximum deceleration. Substituting this into Eq. (8.89) in order to obtain maximum deceleration,

$$\left|\frac{dV}{dt}\right|_{\text{max}} = \frac{1}{2m} \frac{m}{C_D S} Z(\sin\theta) V^2 S C_D$$

$$\left|\frac{dV}{dt}\right|_{\text{max}} = \tfrac{1}{2} V^2 Z \sin\theta \tag{8.99}$$

The velocity at the point of maximum deceleration is obtained by combining Eqs. (8.98) and (8.87), yielding

$$V = V_E e^{-1/2} \tag{8.100}$$

Substituting Eq. (8.100) into (8.99), we find

$$\boxed{\left|\frac{dV}{dt}\right|_{\text{max}} = \frac{V_E^2 Z \sin\theta}{2e}} \tag{8.101}$$

Equation (8.101) is the desired result. It gives us a closed-form expression from which we can quickly calculate the maximum deceleration for a straight-line ballistic reentry trajectory. Note from Eq. (8.101) that

$$\left|\frac{dV}{dt}\right|_{\text{max}} \propto V_E^2 \quad \text{and} \quad \left|\frac{dV}{dt}\right|_{\text{max}} \propto \sin\theta$$

Hence, reentry from a parabolic or hyperbolic trajectory ($V_E \geq 11.2$ km/s) is much more severe than from a near-circular orbit ($V_E = 7.9$ km/s). However, for reentry, there is little we can do to adjust the value of V_E—it is primarily determined by the orbit or trajectory *before* reentry, which in turn is dictated by the desired mission in space. Therefore, Eq. (8.101) tells us that maximum deceleration must be primarily adjusted by the entry angle θ. *In fact, we conclude from Eq. (8.101) that to have reasonably low values of deceleration during reentry, the vehicle must enter the atmosphere at a shallow angle, i.e., at a small θ.*

Finally, Eq. (8.101) yields a rather startling result. Maximum deceleration depends only on V_E and θ. Note that the design of the vehicle, i.e., the ballistic parameter $m/C_D S$, does not influence the value of maximum deceleration. However, you might correctly suspect that $m/C_D S$ determines the altitude at which maximum deceleration occurs.

This concludes our discussion of deceleration, and of reentry trajectories in general. In the next section, we will move on to examine the second major problem of reentry as discussed in Sec. 8.7, namely, aerodynamic heating.

Example 8.3 Consider a solid iron sphere reentering the earth's atmosphere at 13 km/s (slightly above escape velocity) and at an angle of 15° below the local horizontal. The sphere diameter is 1 m. The drag coefficient for a sphere at hypersonic speeds is approximately 1. The density of iron is 6963 kg/m³. Calculate (a) the altitude at which maximum deceleration occurs, (b) the value of the maximum deceleration, and (c) the velocity at which the sphere would impact the earth's surface.

SOLUTION

First, calculate the ballistic parameter $m/C_D S$:

$$m = \rho v = \rho \tfrac{4}{3} \pi r^3$$

$$S = \pi r^2$$

where r = radius of sphere. Hence,

$$\frac{m}{C_D S} = \frac{4}{3} \frac{r\rho}{C_D} = \frac{4}{3} \frac{0.5(6963)}{1.0} = 4642 \text{ kg/m}^2$$

Also, by definition, $Z = g_0/RT$. For our exponential atmosphere, assume a constant temperature of 288 K (recall from Sec. 8.8 that the exponential atmosphere is just an approximation of the detailed standard atmosphere discussed in Chap. 3). Hence,

$$Z = \frac{g_0}{RT} = \frac{9.8}{287(288)} = 0.000118 \text{ m}^{-1}$$

To obtain the altitude for maximum deceleration, calculate the corresponding density from Eq. (8.98):

$$\rho = \frac{m}{C_D S} Z \sin \theta$$

$$\rho = 4642(.000118)\sin 15° = 0.1418 \text{ kg/m}^3$$

This can be translated to an altitude value via Eq. (8.73):

$$\frac{\rho}{\rho_0} = e^{-Zh}$$

or

$$h = -\frac{1}{Z} \ln \frac{\rho}{\rho_0}$$

$$= \frac{1}{0.000118} \ln \frac{0.1418}{1.225} = 18{,}275 \text{ m}$$

Thus, the altitude for maximum deceleration is

$$\boxed{h = 18.275 \text{ km}}$$

(b) The value of maximum deceleration is obtained from Eq. (8.101):

$$\left|\frac{dV}{dt}\right|_{\max} = \frac{V_E^2 Z \sin\theta}{2e}$$

$$= \frac{(13,000)^2(.000118)\sin 15°}{2e}$$

$$= 949.38 \text{ m/s}^2$$

Since 9.8 m/s^2 is the sea-level acceleration of gravity, then the maximum deceleration in terms of g's is:

$$\left|\frac{dV}{dt}\right|_{\max} = \frac{949.38}{9.8} = \boxed{96.87 \text{ } g\text{'s}}$$

This deceleration is very large; it is way beyond that which can be tolerated by humans.

(c) The velocity at impact on the earth's surface is obtained from Eq. (8.87):

$$\frac{V}{V_E} = e^{-\rho/2(m/C_D S)Z\sin\theta}$$

where the value used for ρ is the standard sea-level value, $\rho_0 = 1.225 \text{ kg/m}^3$. Hence,

$$\frac{V}{V_E} = e^{-1.225/2(4642)(.000118)\sin 15°}$$

$$= 0.01329$$

Thus,

$$V = 0.01329 V_E = 0.01329(13,000)$$

$$\boxed{V = 172.8 \text{ m/s}}$$

It is interesting to note that the sphere has slowed down to subsonic velocity before impact. At sea level, $a_s = 340.9 \text{ m/s}$, hence the Mach number at impact is

$$M = \frac{V}{a_s} = \frac{172.8}{340.9} = 0.507$$

Please note that, in reality, the iron sphere will encounter tremendous aerodynamic heating during reentry, especially at the large velocity of 13 km/s. Hence, it is likely that the sphere would vaporize in the atmosphere and never impact the surface; this is the fate of most meteors that enter the atmosphere from outer space. Aerodynamic heating is the subject of the next section.

8.11 REENTRY HEATING

Imagine a reentry body, e.g., the Apollo capsule, as it penetrates the atmosphere. For reasons to be developed later, this body has a very blunt nose, as shown in Figure 8.23. The reentry velocities are extremely high, and the corresponding Mach numbers are hypersonic. From the aerodynamic discussions in Chap. 4, we know there will be a shock wave in front of the vehicle—the bow shock wave

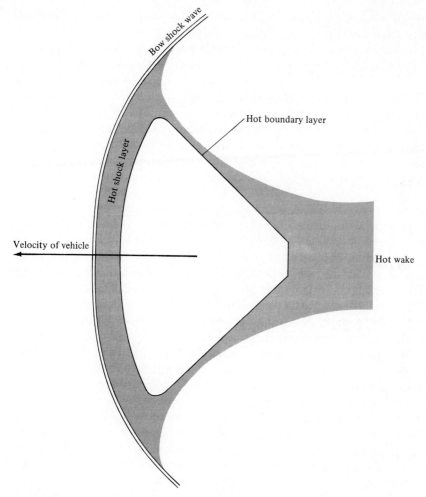

Figure 8.23 High temperature flow field around a blunt reentry vehicle.

shown in Figure 8.23. Because the reentry velocities are so large, this shock wave will be very strong. Consequently, the temperature of the air behind the shock will be extraordinarily high. For example, during the 11.2 km/s reentry of the Apollo, the air temperature behind the shock wave reaches 11,000 K—higher than on the surface of the sun! At these temperatures, the air itself breaks down; the O_2 and N_2 molecules dissociate into O and N atoms, and ionize into O^+ and N^+ ions and electrons. The air becomes a chemically reacting gas. Of more importance, however, is that such high temperatures result in large heat inputs to the reentry vehicle itself. As shown in Figure 8.23, the vehicle is sheathed in a layer of hot air, first from the hot shock layer at the nose, then from the hot boundary layer on the forward and rearward surfaces. These hot gases then flow downstream in the wake of the vehicle. A major objective of reentry vehicle design is to shield the vehicle from this severe aerodynamic heating.

An alternate way of looking at this problem is to consider the combined kinetic and potential energy of the reentry vehicle. At the beginning of reentry, where V_E and h are large, this combined energy is large. At the end of reentry, i.e., at impact, V and h are essentially zero, and the vehicle has no kinetic or potential energy. However, energy is conserved, so where did it go? The answer is that the kinetic and potential energies of the vehicle are ultimately dissipated as *heat*. Returning to Figure 8.23, some of this heat goes into the vehicle itself, and the remainder goes into the air. The object of successful reentry vehicle design is to minimize that heat which goes into the vehicle and maximize that which goes into the air.

The main physical mechanism of aerodynamic heating is related to the action of friction in the boundary layer, as discussed in reference to shear stress and drag in Chap. 4. Clearly, if you take the palm of your hand and rub it vigorously over the surface of a table, your skin will soon get hot. The same applies to the high-speed flow of a gas over an aerodynamic surface. The same frictional forces which create skin friction drag will also at the same time heat the air. The net result is heat transfer to the surface—*aerodynamic heating*.

Incidentally, aerodynamic heating becomes a problem at velocities far below reentry velocity. For example, even at Mach 2 at sea level, the temperature behind a normal shock, and also deep within a boundary layer, can be as high as 520 K. Therefore, aerodynamic heating of the surfaces of supersonic airplanes such as the Concord Supersonic Transport, the F-15, etc., is important and influences the type of materials used in their construction. For example, this is why titanium, rather than the more conventional aluminum, is extensively used on high-speed aircraft—titanium has more strength at high temperatures. However, with the advent of hypervelocity reentry vehicles in the space age, aerodynamic heating became an overriding problem with regard to the very survival of the vehicle itself. It even dictates the shape of the vehicle, as we will soon see.

For a quantitative analysis of aerodynamic heating, it is convenient to introduce a dimensionless heat transfer coefficient called the *Stanton number C_H*, defined as

$$C_H = \frac{dQ/dt}{\rho_\infty V_\infty (h_0 - h_w) S} \tag{8.102}$$

where ρ_∞ and V_∞ are the freestream density and velocity, respectively, h_0 is the total enthalpy (defined as the enthalpy of a fluid element which is slowed adiabatically to zero velocity, in the same spirit as the definition of T_0 in Chap. 4), h_w is the enthalpy at the aerodynamic surface (remember that the velocity is zero at the surface due to friction), S is a reference area (planform area of a wing, cross-sectional area of a spherical reentry vehicle, etc.), and dQ/dt is the heating rate (energy/second) going into the surface. Let us use Eq. (8.102) to obtain a quantitative expression for reentry vehicle heating.

Rewriting Eq. (8.102),

$$\frac{dQ}{dt} = \rho_\infty V_\infty (h_0 - h_w) S C_H \tag{8.103}$$

Considering the energy equation, Eq. (4.41), and the definition of h_0, we obtain

$$h_0 = h_\infty + \frac{V_\infty^2}{2} \tag{8.104}$$

For high-speed reentry conditions, V_∞ is very large. Also, the ambient air far ahead of the vehicle is relatively cool, hence $h_\infty = c_p T$ is relatively small. Thus, from Eq. (8.104)

$$h_0 \approx \frac{V_\infty^2}{2} \tag{8.105}$$

Also, the surface temperature, although hot by normal standards, still must remain less than a few thousand kelvin, below the melting or decomposition temperature of the surface. On the other hand, the temperatures associated with h_0 are large (11,000 K for the Apollo reentry, as stated earlier). Hence, we can easily make the assumption that

$$h_0 \gg h_w \approx 0 \tag{8.106}$$

Substituting Eq. (8.106) and (8.105) into (8.103),

$$\boxed{\frac{dQ}{dt} = \tfrac{1}{2}\rho_\infty V_\infty^3 S C_H} \tag{8.107}$$

Note that Eq. (8.107) states that the *aerodynamic heating rate varies as the cube of the velocity*. This is in contrast to aerodynamic drag, which varies only as the square of the velocity (as we have seen in Chaps. 4 and 5). For this reason, at very high velocities aerodynamic heating becomes a dominant aspect and drag retreats into the background. Also, recall the reasoning that led from Eq. (8.89) to the curve for deceleration vs. altitude in Figure 8.22. This same reasoning leads from Eq. (8.107) to the curve for heating rate vs. altitude sketched in Figure 8.24. During the early part of reentry, dQ/dt increases because of the increasing atmospheric density. In contrast, during the later portion of reentry, dQ/dt decreases because of the rapidly decreasing velocity. Hence, dQ/dt goes through a maximum, as shown in Figure 8.24.

In addition to the local heating rate dQ/dt, we are also concerned with the *total heating* Q, that is, the total amount of energy transferred to the vehicle from beginning to end of reentry. The result for Q will give us some vital information on the desired *shape* for reentry vehicles. First, we draw on a relation between aerodynamic heating and skin friction, called Reynold's analogy. Indeed, it makes sense that aerodynamic heating and skin friction should somehow be connected because both are influenced by friction in the boundary layer. Based on experi-

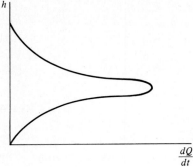

$\dfrac{dQ}{dt}$

Figure 8.24 The variation of heat transfer rate during ballistic reentry.

ment and theory, we approximate Reynold's analogy (without proof) as

$$C_H \approx \tfrac{1}{2} C_f \tag{8.108}$$

where C_f is the mean skin friction coefficient averaged over the complete surface. Substituting Eq. (8.108) into (8.107), we obtain

$$\frac{dQ}{dt} = \tfrac{1}{4} \rho_\infty V_\infty^{\;3} S C_f \tag{8.109}$$

Returning to the equation of motion, Eq. (8.84),

$$\frac{dV_\infty}{dt} = -\frac{D}{m} = -\frac{1}{2m} \rho_\infty V_\infty^{\;2} S C_D \tag{8.110}$$

Mathematically, we can write dQ/dt as $(dQ/dV_\infty)(dV_\infty/dt)$, where dV_∞/dt is given by Eq. (8.110):

$$\frac{dQ}{dt} = \frac{dQ}{dV_\infty} \frac{dV_\infty}{dt}$$

$$= \frac{dQ}{dV_\infty} \left(-\frac{1}{2m} \rho_\infty V_\infty^{\;2} S C_D \right) \tag{8.111}$$

Equating Eqs. (8.111) and (8.109),

$$\frac{dQ}{dV_\infty} \left(-\frac{1}{2m} \rho_\infty V_\infty^{\;2} S C_D \right) = \tfrac{1}{4} \rho_\infty V_\infty^{\;3} S C_f$$

or

$$\frac{dQ}{dV_\infty} = -\frac{1}{2} m V_\infty \frac{C_f}{C_D}$$

or

$$dQ = -\frac{1}{2} m \frac{C_f}{C_D} \frac{dV_\infty^{\;2}}{2} \tag{8.112}$$

Integrate Eq. (8.112) from the beginning of reentry, where $Q = 0$ and $V_\infty = V_E$, and the end of reentry, where $Q = Q_{total}$ and $V_\infty = 0$:

$$\int_0^{Q_{total}} dQ = -\frac{1}{2}\frac{C_f}{C_D}\int_{V_E}^0 d\left(m\frac{V_\infty^2}{2}\right)$$

$$\boxed{Q_{total} = \frac{1}{2}\frac{C_f}{C_D}\frac{1}{2}mV_E^2} \qquad (8.113)$$

Equation (8.113) is the desired result for total heat input to the reentry vehicle. It is an important relation—examine it closely. It reflects two important conclusions, as follows:

1. The quantity $\frac{1}{2}mV_E^2$ is the initial kinetic energy of the vehicle as it first enters the atmosphere. Equation (8.113) says that total heat input is directly proportional to this initial kinetic energy.
2. Total heat input is directly proportional to the ratio of skin friction drag to total drag, C_f/C_D.

The second conclusion above is of particular importance. Recall from Chap. 5 that total drag of a nonlifting body is pressure drag plus skin friction drag:

$$C_D = C_{D_P} + C_f$$

Equation (8.113) says that to minimize reentry heating, we need to minimize the ratio

$$\frac{C_f}{C_{D_P} + C_f}$$

Now consider two extremes of aerodynamic configurations; a sharp-nosed *slender body* such as the cone shown in Figure 8.25a, and the *blunt body* shown in Figure 8.25b. For a slender body, the skin friction drag is large in comparison to the pressure drag, hence $C_D \approx C_f$ and

$$\frac{C_f}{C_D} \approx 1 \qquad \text{(for a slender body)}$$

On the other hand, for a blunt body, the pressure drag is large in comparison to the skin friction drag, hence $C_D \approx C_{D_p}$ and

$$\frac{C_f}{C_D} \ll 1 \qquad \text{(for a blunt body)}$$

In light of Eq. (8.113), this leads to the following vital conclusion:

To minimize reentry heating, the vehicle must have a blunt nose.

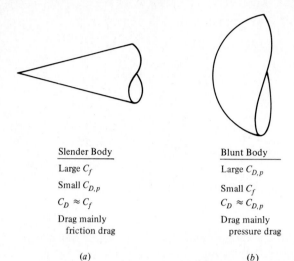

Slender Body

Large C_f

Small $C_{D,p}$

$C_D \approx C_f$

Drag mainly
  friction drag

(a)

Blunt Body

Large $C_{D,p}$

Small C_f

$C_D \approx C_{D,p}$

Drag mainly
  pressure drag

(b)

Figure 8.25 Comparison of blunt and slender bodies.

For this reason, all successful reentry vehicles in practice, from intercontinental ballistic missiles (ICBMs) to the Apollo, have utilized rounded nose shapes.

Returning to our qualitative discussion surrounding Figure 8.23, the advantage of a blunt body can also be reasoned on a purely physical basis. If the body is blunt, as shown in Figure 8.23, the bow shock wave will be strong; i.e., a substantial portion of the wave in the vicinity of the nose will be nearly normal. In this case, the temperature of extensive regions of the air will be high, and much of this high-temperature air will simply flow past the body without encountering the surface. Therefore, a blunt body will deposit much of its initial kinetic and potential energy into heating the air, and little into heating the body. In this fashion, a blunt body tends to minimize the total heat input to the vehicle, as proven quantitatively from Eq. (8.113).

The mechanism of aerodynamic heating discussed above is called *convective heating*. To conclude this section on reentry heat transfer, another mechanism is mentioned—*radiative heating* from the shock layer. Consider Figure 8.26, which shows a blunt reentry body at high velocity. It was mentioned earlier that at speeds associated with lunar missions (11.2 km/s, or 36,000 ft/s), the air temperature behind the shock wave is as high as 11,000 K. At this high temperature, the shock layer literally *radiates* energy in all directions, as illustrated in Figure 8.26, much as you feel the warmth radiated from a fireplace on a cold winter day. Some of this radiation is incident upon and absorbed by the vehicle itself, giving rise to an additional heat transfer component, Q_R. This radiative heat transfer rate is proportional to a power of velocity ranging from V_∞^5 to V_∞^{12}, depending on the nose radius, density, and velocity. For ICBM and orbital vehicles, radiative heating is not significant. But as sketched in Figure 8.27, because of its strong velocity dependence, radiative heating becomes dominant at very high velocities. For the Apollo mission from the moon ($V_E = 36,000$ ft/s),

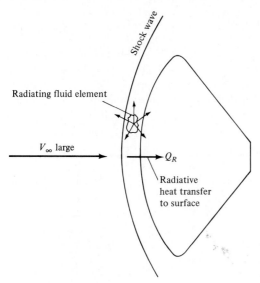

Figure 8.26 Mechanism of radiative heating from the high-temperature shock layer.

radiative heating was slightly less than convective heating. However, for future manned missions from the planets ($V_E \approx 50{,}000$ ft/s), radiative heating will swamp convective heating. This is illustrated schematically in Figure 8.27. Moreover, entry into the atmospheres of other large planets, especially Jupiter, will be overwhelmed by radiative heating. For these reasons, the designers of vehicles for advanced space missions will have to be vitally concerned about radiative heating from the shock layer during atmospheric entry. The interested reader can find more details on radiative heating in the AIAA paper by Anderson listed in the Bibliography at the end of this chapter.

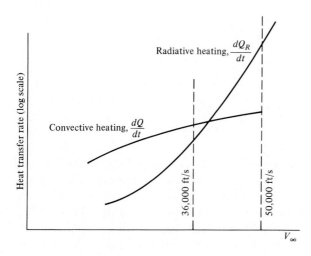

Figure 8.27 Comparison of convective and radiative heat transfer rates, illustrating dominance of radiative heating at high velocities.

8.12 HISTORICAL NOTE: KEPLER

The sixteenth century was a period of quandry for astronomy. The conservative line of scientific thought held the earth as the center of the universe, with the sun, planets, and stars revolving about it on various celestial spheres. This *geocentric* system was popular among the Greeks. Put into a somewhat rational form by Claudius Ptolemy in the second century A.D., this earth-centered system was adopted as the truth by the Church in western Europe and was carried through to the sixteenth century. However, about the time that Columbus was discovering America, a Polish scientist by the name of Nicolaus Copernicus was beginning to develop different ideas. Copernicus reasoned that the earth as well as all the other planets revolved around the sun, a *heliocentric* system. He established his line of thought in a main work entitled *Six Books Concerning the Revolutions of the Heavenly Spheres* published in the year of his death, 1543. Here, Copernicus was diplomatic with regard to the Church's dogma. He stated that his heliocentric theory was not new, having been held by a few early Greek astronomers, and also that he was just "postulating and theorizing," not necessarily speaking the absolute truth. However, it was clear that Copernicus personally believed in what he wrote. Another astronomer, Giordano Bruno, who evangelized Copernicus's theory, was not so diplomatic and was burned at the stake in 1600. Galileo Galilei took up the heliocentric banner in 1632 and was ultimately exiled under guard for his heresy. Finally, the Danish astronomer Tycho Brahe, while shunning a direct association with the controversial heliocentric theory, spent virtually his complete life from 1546 to 1601 making astronomical observations of planet and star movements, resulting in spectacular improvements in the precision of existing knowledge.

Into this tenuous time, Johannes Kepler was born in Württemberg, Germany, on December 27, 1571. By winning scholarships, he was able to finish elementary school and to go on to the University of Tübingen. There he was converted to the heliocentric theory by Michael Mastlin, a professor of astronomy. Later, Kepler became a teacher of mathematics and an ardent astronomer. Through his writings on celestial motion, Kepler came to the attention of Tycho Brahe, who was now living in Prague. In 1599, Kepler went to Prague to work under Brahe, who died just two years later. Kepler stayed in Prague, extending and improving the existing tables of celestial movement. In 1627, he published his *Rudolphine Tables*, which were much more accurate than any existing tables at that time.

However, Kepler was also thinking and theorizing about his observations, attempting to bring some reason and order to the movement of the heavenly bodies. For example, the heliocentric system of Copernicus assumed circular orbits of the planets about the sun; but Kepler's accurate observations did not precisely fit circular motion. In 1609 he found that elliptical orbits fit his measurements exactly, giving rise to *Kepler's first law* (see Sec. 8.6). In the same year, he induced that a line drawn from the sun to a planet sweeps out equal areas in equal times—*Kepler's second law*. His first and second laws were published in his book *New Astronomy* in 1609. It was not until nine years later that he discovered that the square of the period of planetary orbits was proportional to the cube of the semimajor axis of the elliptical orbit—*Kepler's third law*. This was published in 1618 in his book *Epitome of the Copernican Astronomy*.

Kepler's impact on astronomy was massive; in fact, his work was the founding of modern astronomy. His contributions are all the more stunning because his laws were induced from empirical observation. Kepler did not have the tools developed later by Newton. Therefore, he could not derive his laws with the same finesse as we did in Sec. 8.6.

It is interesting to note that Kepler also wrote science fiction. In his book *Somnium* (Dream), Kepler describes a trip from the earth to the moon. Recognizing that the void of space would not support flight by wings, he had to resort to demons as a supernatural mode of propulsion. These demons would carry along humans, suitably anesthetized to survive the rigors of space travel. He described the moon in as much astronomical detail as was possible in that age but imagined moon creatures that lived in caves. Modern historians of science fiction literature believe that Kepler's *Somnium* was really a vehicle to present his serious scientific ideas about the moon while attempting to avoid religious persecution. *Somnium* was published in 1634, four years after Kepler's death.

Kepler spent his later life as a professor of mathematics in Linz. He died in Regensburg on November 15, 1630, leaving a legacy that reaches across the centuries to the astronautics of the present day.

8.13 HISTORICAL NOTE: NEWTON AND THE LAW OF GRAVITATION

Newton's law of universal gravitation, Eq. (8.19), appears in every modern high school and college physics textbook; its existence is virtually taken for granted. Moreover, this equation is the very foundation for all modern astronautical calculations of motion through space, as discussed throughout this chapter. However, the disarming simplicity of Eq. (8.19) and its commonplace acceptance in classical physics belies the turmoil that swarmed about the concept of gravity before and during the seventeenth century, when Newton lived.

The earliest ideas on "gravity" were advanced by Aristotle during the period around 350 B.C. Believing that the four fundamental elements of the universe were earth, water, air, and fire, the aristotelian school held that everything in the universe had its appointed station and tended to return to this station if originally displaced. Objects made from "earth" held the lowest station, and thus heavy material objects would fall to the ground, seeking their proper status. In contrast, fire and air held a high station and would seek this status by rising toward the heavens. These ideas persisted until the age of Copernicus, when people began to look for more substantial explanations of gravity.

In 1600, the English scientist William Gilbert suggested that magnetism was the source of gravity and that the earth was nothing more than a gigantic lodestone. Kepler adopted these views, stating that gravity was "a mutual affection between cognate bodies tending towards union or conjunction, similar in kind to magnetism." Keplei used this idea in an attempt to prove his laws of planetary motion (see the previous section) but was not successful in obtaining a quantitative law for the force of gravity. About the same time, the French scientist and mathematician René Descartes (who introduced the cartesian coordinate system to the world of mathematics) proposed that gravity was the result of an astronomical fluid that was swirling in a vortex motion, pushing heavy objects toward the core of the vortex. Christian Huygens, a Dutch gentleman and amateur scientist, seemed to confirm Descartes's theory in the laboratory; he set up a whirlpool of water in a bowl and observed that pebbles "gravitated" to the center of the bowl.

Into this confused state of affairs was born Isaac Newton at Woolsthorpe near Grantham, Lincolnshire, England, on December 25, 1642. Newton's father died a few months before he was born and he was raised by his grandmother. His education ultimately led to studies at Trinity College, Cambridge University, in 1661, where he quickly showed his genius for mathematics. In 1666, he left Cambridge for his home in Woolsthorpe Manor to avoid the Great Plague of 1665–1666. It was here, at the fresh age of 24, that Newton made some of his discoveries and conclusions that were to revolutionize science and mathematics, not the least of which was the development of differential calculus. Also, Newton later maintained that during this stay in the country he deduced the law of centripetal force, i.e., that a body in circular motion experienced a radial force that varied inversely with the distance from the center. (In today's language, the centripetal acceleration due to circular motion is equal to V^2/r, as shown in all elementary physics books.) From this result applied to Kepler's third law, Newton further deduced that the force of gravity between two objects varied inversely as the square of the distance separating them, which led to the universal law of gravitation, as given by Eq. (8.19). However, Newton did not bother to publish immediately or otherwise announce his findings. The public was kept in the dark for another 30 years!

Throughout the history of science and engineering there are numerous examples of ideas whose "time had come" and which were conceived by several different people almost simultaneously. The same Christian Huygens made experiments with pendulums and circular moving bodies that led to his discovery of the law of centripetal force in 1673. With this, Robert Hooke (of Hooke's law fame), Christopher Wren (later to become an internationally famous architect), and Edmund Halley (of Halley's comet fame) all deduced the inverse-square law of gravity in 1679. Hooke wrote to Newton in the same year, telling him of the inverse-square discovery and asking Newton to use it to prove that a planet revolves in an elliptical orbit. Newton did not reply. In 1685, the problem was again posed to Newton, this time by Halley. Newton sent back such a proof. Halley was much impressed and strongly encouraged Newton to publish all his discoveries and thinking as soon as possible. This led to Newton's *Philosophiae Naturales Principia Mathematica*, the famous *Principia*, which has become the foundation of classical physics. It is interesting to note that the *Principia* was originally to be published by the Royal Society. But Hooke, who laid claim to the prior discovery of the inverse-square law and who was the curator of the Society, apparently discouraged such publication. Instead, the *Principia*, the most important scientific document to that time in history, was published at the personal expense of Halley.

Hooke again put forward his claim to the inverse-square law during a meeting of the Royal Society in 1693. Shortly thereafter, Newton had a nervous breakdown, which lasted about a year. After his recovery, Newton finally announced that he had made the basic discoveries of both the centripetal force law and the inverse-square law of gravitation back in 1666. Because of his high standing and reputation of that time, as well as subsequently, Newton's claim has been generally accepted through the present time. However, the record shows that we have only his word. Therefore, the claim by Robert Hooke is certainly legitimate, at least in spirit. Equation (8.19), which comes down to us as Newton's law of universal gravitation, could legitimately be labeled (at the very least) the "Newton-Hooke law."

Of course, this is not to detract from Newton himself, who was *the* giant of science in the seventeenth century. During his later years, Newton entered public life, becoming warden of the British Mint in 1696, advancing to the chief post of master in 1699. In this capacity, he made many important contributions during Britain's massive recoinage program of that time. In 1703, he was elected president of the Royal Society, a post he held for the next 25 years. During this period, Newton was embroiled in another controversy, this time with the German mathematician Gottfried von Liebniz over the claim of the discovery of calculus. Also, during these later years, Newton's imposing prestige and authority via the Royal Society apparently tended to squelch certain ideas put forward by younger scientists. Because of this, some historians of science hint that Newton may have hindered the progress of science during the first 30 years of the eighteenth century.

Newton died in Kensington on March 20, 1727. He is buried at a prominent location at Westminster Abbey. Without Newton, and without Kepler before him, this chapter on astronautics may never have been written.

8.14 HISTORICAL NOTE: LAGRANGE

In Sec. 8.3, a corollary to Newton's second law is introduced, namely, Lagrange's equation. Lagrange came after Newton. He was one of the small group of European scientists and mathematicians who worked to develop and augment newtonian (classical) physics during the eighteenth century; he was a contemporary of Laplace and a friend of Leonhard Euler.

Joseph L. Lagrange was born of French parents at Turin, Italy, on January 25, 1736. His father was an officer in the French army; hence it is no surprise that at the age of 19, Lagrange was appointed as professor of mathematics at the Turin Artillery School. Active in scientific thought, he helped to found the Turin

Academy of Sciences. In 1756, he wrote to Euler (see Sec. 4.20) with some original contributions to the calculus of variations. This helped to establish Lagrange's reputation. In fact, in 1766, he replaced Euler as director of the Berlin Academy at the invitation of Frederick II (Frederick the Great) of Prussia. For the next 20 years, Lagrange was extremely productive in the field of mechanics. His work was analytical, and he endeavored to reduce the many aspects of mechanics to a few general formulas. This is clearly reflected in the formalism discussed in Sec. 8.3. Lagrange's equations used in Sec. 8.3 were published in an important book by Lagrange entitled *Mécanique Analytique* in 1787. For these contributions, he is considered by some historians to be the greatest mathematician of the eighteenth century.

Lagrange moved to Paris in 1786. During the French Revolution, he was president of the committee for reforming weights and measures standards. At the time of his death in Paris on April 10, 1813, he was working on a revised version of his *Mécanique Analytique*.

8.15 HISTORICAL NOTE: UNMANNED SPACE FLIGHT

On the evening of October 4, 1957, the present author was a student of aeronautical engineering. The radio was on. Concentration on studies was suddenly interrupted by a news bulletin: the Soviet Union had just successfully launched the first artificial earth satellite in history. Labeled *Sputnik I* and shown in Figure 8.28, this 184-lb sphere circled the earth in an elliptical orbit, with apogee and perigee of 560 and 140 mi, respectively, and with a period of $1\frac{1}{2}$ h. The personal feeling of exhilaration that humanity had finally made the first great step toward space exploration was tempered by questions about the technical position of the United States in space flight. These feelings were to be reflected and amplified throughout America for weeks, months, and years to come. *Sputnik I* started a technological revolution which has influenced virtually all aspects of society, from education to business, from biology to philosophy. October 4, 1957, is a red-letter date in the history of humanity—the beginning of the space age.

Although the launching of *Sputnik I* came as a surprise to most of the general public, the technical community of the western world had been given some clear hints by Russian scientists. For example, on November 27, 1953, at the World Peace Council in Vienna, the Soviet academician A. N. Nesmeyanov stated that "science had reached such a stage that... the creation of an artificial satellite of the earth is a real possibility." Then, in April 1955, the U.S.S.R. Academy of Sciences announced the creation of the Permanent Interdepartmental Commission for Interplanetary Communications, with responsibility for developing artificial earth satellites for meteorological applications. In August of that year, the highly respected Russian scientist Leonid I. Sedov, at the Sixth International Astronautical Congress, in Copenhagen, said: "In my opinion, it will be possible to launch an artificial satellite of the Earth within the next two years, and there is the technological possibility of creating artificial satellites of various sizes and

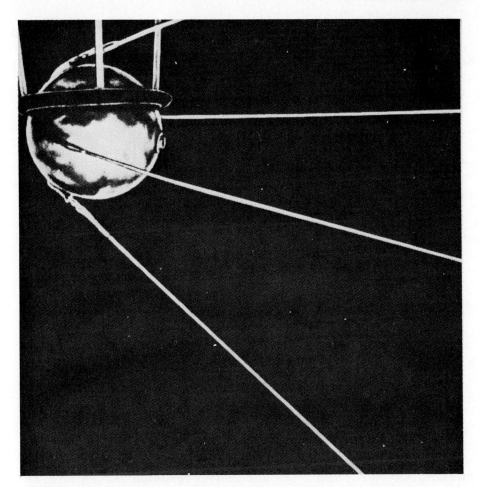

Figure 8.28 The first artificial earth satellite—*Sputnik I*—launched on October 4, 1957.

weights." Obviously, the Russian program kept to its schedule. Indeed, in June 1957, just four months before *Sputnik I*, the same A. N. Nesmeyanov blatantly stated that both the rocket launch vehicle and the satellite were ready and would be launched in a few months. Clear signs and clear words—yet the launching of *Sputnik I* still fell like a ton of bricks on the western world.

In 1957, the United States was not new to the idea of artificial satellites. Indeed, some far-sighted thinking and technical analyses on the prospects for launching such satellites were performed by the U.S. Navy and the U.S. Army Air Force beginning in 1945. Then, in May 1946 (just one year after Germany had been defeated in World War II), a Project RAND report entitled "Preliminary Design of an Experimental World-Circling Spaceship," was submitted to Wright Field, Dayton, Ohio. This report showed the feasibility of putting a 500-lb satellite in orbit at about 300 mi high. Moreover, it outlined how this could be accomplished in a five-year time scale! The authors of this report made some seriously prophetic statements, as follows:

> Although the crystal ball is cloudy, two things seem clear—1. A satellite vehicle with appropriate instrumentation can be expected to be one of the most potent scientific tools of the Twentieth Century.
>   2. The achievement of a satellite craft by the United States would inflame the imagination of mankind, and would probably produce repercussions in the world comparable to the explosion of the atomic bomb....

Then the authors go on to state:

> Since mastery of the elements is a reliable index of material progress, the nation which first makes significant achievements in space travel will be acknowledged as the world leader in both military and scientific techniques. To visualize the impact on the world, one can imagine the consternation and admiration that would be felt here if the U.S. were to discover suddenly that some other nation had already put up a successful satellite.

These were indeed prophetic words, written fully eleven years before *Sputnik I*.

The 1946 RAND report, along with several contemporary technical reports from the Jet Propulsion Laboratory at the California Institute of Technology, established some fundamental engineering principles and designs for rocket launch vehicles and satellites. However, these ideas were not seized upon by the U.S. Government. The period after World War II was one of shrinking defense budgets, and money was simply not available for such a space venture. Of probably more importance was the lack of a mission. What if a satellite were launched? What benefits would it bring, especially military benefits? Keep in mind this was in a period before miniaturized electronics and sophisticated sensing and telemetering equipment. Therefore, the first serious U.S. effort to establish a satellite program withered on the vine, and the idea lay essentially dormant for the next nine years.

Although upstaged by *Sputnik I*, the United States in 1957 finally did have an ongoing project to orbit an artificial satellite. On July 29, 1955, President Dwight D. Eisenhower announced that the United States would orbit a small earth satellite in conjunction with the International Geophysical Year. Making use of ten years of high-altitude sounding-rocket technology, which started with a number of captured German V-2 rockets, the United States established the Vanguard program, managed by the Office of Naval Research, in order to accomplish this goal. The Martin Company in Baltimore, Maryland, was chosen as the prime contractor. During the next two years, a rocket booster was designed and built to launch a small, 3-lb experimental satellite. By governmental edict, the Vanguard project was required not to draw upon or interface with the rapidly growing and high-priority ICBM program, which was developing large rocket engines for the military. Therefore, Dr. John P. Hagen, director of Project

Vangard, and his small team of scientists and engineers had to struggle almost as second-class citizens to design the Vangard rocket in an atmosphere of relatively low priority. (This is in sharp contrast to the Russian space program, which from the very beginning utilized and benefited from the Soviet military ICBM developments. Since Russian atomic warheads of that day were heavier than comparable U.S. devices, the Soviet Union had to develop more-powerful rocket boosters. Their space program correspondingly benefited, allowing *Sputnik I* and *II* to be the surprisingly large weights of 184 and 1120 lb, respectively.)

By October 1957, two Vangard rockets had been successfully tested at Cape Canaveral and the test program, which was aimed at putting a satellite into orbit before the end of 1958, was reasonably close to schedule. Then came *Sputnik I* on October 4. Not to be completely upstaged, the White House announced on October 11 that Project Vangard would launch a U.S. satellite "in the near future." Suddenly in the limelight of public attention, and now under intense political pressure, a third test rocket was successfully tested on October 23, carrying a 4000-lb dummy payload to an altitude of 109 mi and 335 mi downrange. Then, on December 6, 1957, in full view of the world's press, the first Vangard was prepared to put a small satellite into orbit. Unfortunately, the Vangard first-stage engine had its first (and last) failure of the program. With failing thrust, the rocket lifted a few feet off the launch pad and then fell back in a spectacular explosion. In Dr. Hagen's words: "Although we had three successful test launches in a row, the failure of TV-3 [the designation of that particular vehicle] was heard around the world."

Despite the original disadvantages of low priority, despite the emotional pressure after *Sputnik I*, and despite the inglorious failure of December 6, the Vangard project went on to be very successful. Vangards I, II, and III were put into orbit on March 17, 1958, February 17, 1959, and September 18, 1959, respectively, attributing to the fine efforts of Dr. Hagen and his group.

But Vangard I was not the first U.S. satellite. President Eisenhower's July 1955 announcement about U.S. plans to orbit a satellite was followed by much debate about whether or not military rocketry should be used. One proposal at the time was to make use of the rocket vehicles being developed at the Army's Redstone Arsenal at Huntsville, Alabama, under the technical direction of Dr. Wernher Von Braun. After the decision was made to go with the Vangard, the engineers at the Army Ballistic Missile Agency at Huntsville continued to propose a satellite program using the proven intermediate-range Jupiter C rocket. All such proposals were turned down. However, the picture changed after *Sputnik*. In later October 1957, Von Braun's group was given the green light to orbit a satellite—target date was January 30, 1958. A fourth stage was added to the Jupiter C rocket; this new configuration was labeled the Juno I. The target date was missed by only one day. On January 31, 1958, *Explorer I*, the first United States artificial satellite, was placed into orbit by Von Braun's team of scientists and engineers from Huntsville. The *Explorer I*, shown in Figure 8.29, weighed 18 lb, and its orbit had apogee and perigee of 957 and 212 mi, respectively; its period was 115 min. With the launchings of both *Sputnik I* and *Explorer I*, the two

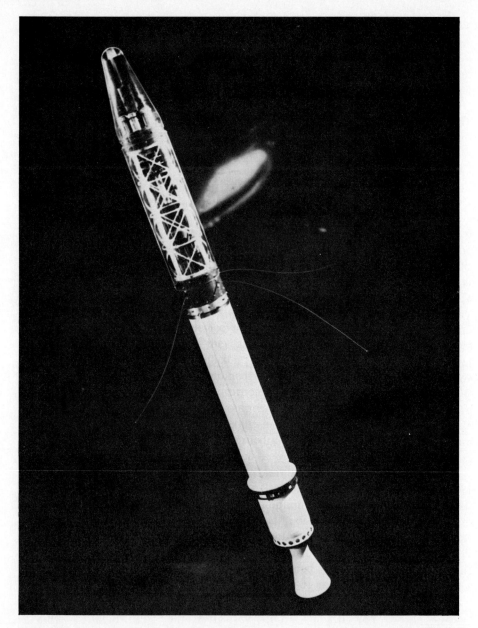

Figure 8.29 *Explorer I*, the first United States artificial earth satellite, launched on January 31, 1958.

technological giants in the world—the United States and the Soviet Union—were now in competition in the arena of space.

It is not the purpose here to give an exhaustive survey of space exploration. For an authoritative presentation, see the excellent book by Von Braun and Ordway, as well as others listed in the Bibliography at the end of this chapter.

8.16 HISTORICAL NOTE: MANNED SPACE FLIGHT

The previous section on unmanned space flight, the present section on manned space flight, and Sec. 9.12 on the early history of rocket engines are inexorably entwined—their division into three distinct sections in this book is purely artificial. Indeed, humanity's first imaginative thoughts about space flight involved the travel of human beings (not inanimate objects) to the moon. Later, during the technological revolution of the nineteenth and twentieth centuries, it was correctly reasoned that manned space travel would have to be preceded by unmanned attempts just to learn about the problems that might be encountered. Also during this period, the rocket engine was recognized as the only feasible mechanism for propulsion through the void of space. In fact, the three early pioneers of rocket engines—Tsiolkovsky, Goddard, and Oberth (see Sec. 9.12)—were inspired in their work by the incentive of space travel rather than the military applications that ultimately produced the first successful large rockets. Clearly, the histories of unmanned and manned space flight and rocketry are overlapping and in many cases are indistinguishable.

Manned space flight really has its roots in science fiction and reaches as far back as the second century A.D., when the Greek writer Lucian of Samosata conceived a trip to the moon. In this book *Vera Historia*, Lucian's ship is caught in a storm, lifted into the sky by the higher winds, and after seven days and seven nights, is quite accidentally blown to the moon. There he finds a land that is "cultivated and full of inhabitants." Lucian's work was followed by other science fiction fantasies over the ensuing centuries, including Kepler's *Somnium*, mentioned in Sec. 8.12. These science fiction stories served a useful purpose in fueling the imaginative minds of some people and spurring them to more-serious technological thought. Of particular note are books by Jules Verne and H. G. Wells in the nineteenth century, which were avidly read by many of the early rocket engineers. In particular, both Tsiolkovsky and Goddard avidly read Wells's *War of the Worlds* and Verne's *From the Earth to the Moon* and both have gone on record as being inspired by these works.

Considering that Wells and Verne wrote less than 100 years ago, and that just 40 years ago rockets were only the playthings of a few visionaries, it is astounding that manned space flight has now become a reality—and in the minds of the general public, a somewhat common reality. The ice was broken on April 12, 1961, when the Soviet Union orbited the 10,400-lb *Vostok I* spacecraft carrying Major Yuri A. Gagarin—the first human being to ride in space. Gagarin was a Russian air force major; his orbital flight lasted 1 h and 48 min, with an apogee of

203 mi. Upon reentry, the *Vostok* was slowed first by retro-rockets, and then by parachute, and came to rest on the solid ground somewhere deep within the interior of Russia. However, it is thought that, just before touchdown, Gagarin left the spacecraft and floated to earth with his own parachute. This reentry mode was followed by several other Russian astronauts during subsequent years. Unfortunately, Gagarin was later killed in an airplane crash on March 27, 1968.

Humanity was now on its way in space! Less than a year later, the first American in space, Marine Colonel John H. Glenn, Jr., was orbited on February 20, 1962. Executing three orbits in a Mercury capsule with an apogee and perigee of 162.7 and 100.3 mi, respectively, Glenn's flight lasted 4 h and 56 min from blast-off to touchdown. As with all subsequent U.S. manned spacecraft, Glenn rode the Mercury capsule all the way to the earth's surface, impacting at sea and being recovered by ship. Figure 8.30 shows a diagram of the single-seat Mercury space capsule and gives a clear picture of its size and shape relative to the astronaut himself. Glenn's successful flight in 1962 was a high point in Project Mercury, which was the United States' first manned space program. This project had its roots in an Air Force study entitled "Manned Ballistic Rocket Research System" initiated in March 1956—a full year and a half *before Sputnik I*. Within two years under this project, the Air Force, the NACA, and eleven private companies did much fundamental work on spacecraft design and life-support systems. After *Sputnik I*, and after the formation of NASA in 1958, this work was centralized within NASA and designated Project Mercury. Thus, when Gagarin went into orbit in 1961, the United States was not far behind.

Indeed, the U.S. manned space flight program was galvanized when President John F. Kennedy, in a speech before Congress on May 25, 1961, declared: "I

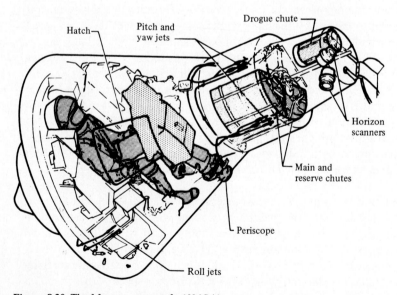

Figure 8.30 The Mercury spacecraft. (*NASA*.)

believe that this nation should commit itself to achieving the goal, before this decade is out, of landing [a person] on the Moon and returning him safely to Earth . . . ." In virtually a flash, the Apollo program was born. Over the next eight years, work on the Apollo manned lunar vehicle marshalled a substantial portion of the U.S. human and material aerospace resources. Then, almost like a page out of science fiction itself, at 4:18 P.M. (EDT) on July 20, 1969, a lunar descent vehicle named *Eagle*, carrying Neil A. Armstrong and Edwin E. Aldrin, Jr., came to rest on the moon's surface, with Michael Collins keeping watch in the Apollo Command Module orbiting above. President Kennedy's goal had been met; the dreams and aspirations of people over the centuries had been fullfilled; and the work of such minds as Copernicus, Kepler, Newton, and Lagrange had come to dramatic fruition.

The technical story of manned space flight is one of superhuman effort, fantastic advances in science and engineering, and unswerving dedication. It is still going on, albeit at a somewhat reduced frenzy after Apollo, and it will continue to progress as long as modern society exists. It is impossible to give justice to such a story in this short section; indeed, whole volumes have been written on this subject alone. Again, for a particularly authoritative and modern review, the reader is referred to the book by Von Braun and Ordway listed in the Bibliography.

8.17 CHAPTER SUMMARY

Some of the highlights of this chapter are summarized below:

1. The equation of the orbit or trajectory of a spacecraft under the influence of a central, inverse-square gravitational force field is

$$r = \frac{p}{1 + e\cos(\theta - C)} \tag{8.44}$$

where e is the eccentricity and C is the phase angle. If $e = 0$, the orbit is a circle; if $e < 1$, the orbit is an ellipse; if $e = 1$, the trajectory is a parabola; if $e > 1$, the trajectory is a hyperbola.

2. The eccentricity depends on the difference between kinetic and potential energies of the spacecraft, H.

$$e = \sqrt{1 + \frac{2h^2 H}{mk^4}} \tag{8.53}$$

3. Circular velocity is given by

$$V = \sqrt{\frac{k^2}{r}} \tag{8.57}$$

For earth satellites, circular, or orbital, velocity is 7.9 km/s, or approximately 26,000 ft/s (based on $r =$ earth's radius).

4. Escape velocity is given by

$$V = \sqrt{\frac{2k^2}{r}} \tag{8.58}$$

For escape from the earth, this velocity is 11.2 km/s, or approximately 36,000 ft/s.

5. Kepler's laws are: (1) a satellite describes an elliptical path around its center of attraction, (2) in equal times, the areas swept out by the radius vector of a satellite are the same, and (3) the periods of any two satellites about the same planet are related to their semimajor axes as

$$\left(\frac{\tau_1}{\tau_2}\right)^2 = \left(\frac{a_1}{a_2}\right)^3$$

6. The velocity variation of a ballistic reentry vehicle through the atmosphere is given by

$$\frac{V}{V_E} = e^{-\rho/2(m/C_D S)Z\sin\theta} \tag{8.87}$$

where ρ is a function of altitude, $m/C_D S$ is the ballistic parameter, θ is the entry angle, V_E is the initial entry velocity, and $Z = g_0/RT$. During reentry, the maximum deceleration is given by

$$\left|\frac{dV}{dt}\right|_{\max} = \frac{V_E^2 Z\sin\theta}{2e} \tag{8.101}$$

7. Reentry aerodynamic heating varies as the cube of the velocity:

$$\frac{dQ}{dt} = \tfrac{1}{2}\rho_\infty V_\infty^3 S C_H \tag{8.107}$$

To minimize aerodynamic heating, the vehicle should have a blunt nose.

BIBLIOGRAPHY

Allen, H. J., and Eggers, A. J., *A Study of the Motion and Aerodynamic Heating of Missiles Entering the Earth's Atmosphere at High Supersonic Speeds*, NACA TR 1381, 1958.

Anderson, J. D., Jr., "An Engineering Survey of Radiating Shock Layers," *AIAA Journal*, vol. 7, no. 9, Sept. 1969, pp. 1665–1675.

Chapman, D. R., *An Approximate Analytical Method for Studying Entry into Planetary Atmospheres*, NASA TR R-11, 1959.

Emme, E. M., *A History of Space Flight*, Holt, New York, 1965.

Hartman, E. P., *Adventures in Research: A History of Ames Research Center 1940–1965*, NASA SP-4302, 1970.

Nelson, W. C., and Loft, E. E., *Space Mechanics*, Prentice-Hall, Englewood Cliffs, NJ, 1962.

Von Braun, W. and Ordway, F. I., *History of Rocketry and Space Travel*, 3rd rev. ed., Crowell, New York, 1975.

PROBLEMS

8.1 At the end of a rocket launch of a space vehicle from earth, the burnout velocity is 13 km/s in a direction due south and 10° above the local horizontal. The burnout point is directly over the equator at an altitude of 400 mi above sea level. Calculate the trajectory of the space vehicle.

8.2 Calculate and compare the escape velocities from Venus, earth, Mars, and Jupiter, given the information tabulated below.

	Venus	Earth	Mars	Jupiter
k^2, m^3/s^2	3.24×10^{14}	3.96×10^{14}	4.27×10^{13}	1.27×10^{17}
r, m	6.16×10^6	6.39×10^6	3.39×10^6	7.14×10^7

8.3 The mass and radius of the earth's moon are 7.35×10^{22} kg and 1.74×10^6 m. Calculate the orbital and escape velocities from the moon.

8.4 It is known that the period of revolution of the earth about the sun is 365.3 days and that the semimajor axis of the earth's orbit is 1.495×10^{11} m. An astronomer notes that the period of a distant planet is 29.7 earth years. What is the semimajor axis of the distant planet's orbit? Check in a reference source (encyclopedia, etc.) what planet of the solar system this might be.

8.5 Assume that you wish to place in orbit a satellite that always remains directly above the same point on the earth's equator. What velocity and altitude must the satellite have at the instant of burnout of the rocket booster.

8.6 Consider a solid iron sphere reentering the earth's atmosphere at 8 km/s and at an angle of 30° below the local horizontal. The sphere diameter is 1.6 m. Calculate (a) the altitude at which maximum deceleration occurs, (b) the value of the maximum deceleration, and (c) the velocity at which the sphere would impact the earth's surface.

8.7 The aerodynamic heating rate of a given reentry vehicle at 200,000 ft traveling at a velocity of 27,000 ft/s is 100 Btu/(ft^2)(s). What is the heating rate if the velocity is 36,000 ft/s at the same altitude?

NINE

PROPULSION

We have sought power in the same fire which serves to keep the vessel aloft. The first which presented itself to our imagination is the power of reaction, which can be applied without any mechanism, and without expense: it consists solely in one or more openings in the vessel on the side opposite to that in which one wishes to be conveyed.

Joseph Montgolfier, 1783 — the first recorded technical statement in history on jet propulsion for a flight vehicle

I began to realize that there might be something after all to Newton's Laws.

Robert H. Goddard, 1902

9.1 INTRODUCTION

The old saying that "you cannot get something for nothing" is particularly true in engineering. For example, the previous chapters have discussed the aerodynamic generation of lift and drag, the performance, stability, and control of airplanes, and the motion of spacecraft. All this takes the expenditure of *power*, or *energy*, which is supplied by an engine or propulsive mechanism of some type. The study of *propulsion* is the subject of this chapter. Here, we will examine what makes an airplane or space vehicle go.

Throughout Chap. 1 the dominant role played by propulsion in the advancement of manned flight is clearly evident. George Cayley was concerned in 1799, and he equipped his airplane designs with paddles. Henson and Stringfellow did better by considering "airscrews" powered by steam engines, although they were unsuccessful in their efforts. In 1874, Felix Du Temple momentarily hopped off the ground in a machine powered by an obscure type of hot-air engine; he was followed by Mozhaiski in 1884 using a steam engine (see Figures 1.13 and 1.14). By the late nineteenth century, the early aeronautical engineers clearly recognized that successful manned flight depended upon the development of a lightweight but powerful engine. Fortunately, the advent of the first practical internal combustion engine in 1860 paved the way for such success. However, in spite of the rapid development of these gasoline-powered engines and their role in the early automobile industry, such people as Langley (Sec. 1.7) and the Wright brothers (Sec. 1.8) still were forced to design their own engines in order to obtain the high horsepower-to-weight ratio necessary for flight. Such internal combustion reciprocating engines driving a propeller ultimately proved to be a winning combination and were the only practical means of airplane propulsion up to World War II. In the process, such engines grew in horsepower from the 12-hp Wright-designed engine of 1903 to the 2200-hp radial engines of 1945, correspondingly pushing flight velocities from 28 to over 500 mi/h. Then, a revolution in propulsion occurred. Frank Whittle took out a patent in Britain in 1930 for a jet-propelled engine and worked ceaselessly on its development for a decade. In 1939, the German Heinkel He 178 airplane flew with a turbojet engine developed by Dr. Hans von Ohain. It was the first successful jet-propelled test vehicle. This led to the German Me 262 jet fighter late in World War II. Suddenly, jet engines became the dominant power plants for high-performance airplanes, pushing flight velocities up to the speed of sound in the 1950s, and beyond in the 1960s and 1970s. Today, the airplane industry rides on jet propulsion, and jet-propelled supersonic flight for both commercial and military airplanes is a regular occurrence. Meanwhile, another revolution of even more impact occurred: the advent of the successful rocket engine. Pioneered by Konstantin Tsiolkovsky (1857–1935) in Russia, Robert H. Goddard (1882–1945) in the United States, and Hermann Oberth (b. 1894) in Germany, the rocket engine first became operational in 1944 with the German V-2 missile. Being the only practical means of launching a vehicle into space, the rocket engine soon proved itself during the space age, allowing people to go to the moon and to probe the deep unknown regions of our solar system.

It is clear from the above brief historical sketch that propulsion has led the way for all major advancements in flight velocities. Propulsion is one of the major disciplines of aerospace engineering; therefore, in the following sections, some of the basic principles of propellers, reciprocating engines, turbojets, ramjets, and rockets will be examined. Such propulsion devices are highly aerodynamic in nature. Therefore, a firm understanding of the aerodynamic and thermodynamic fundamentals presented in Chaps. 4 and 5 will help you to grasp the propulsion concepts discussed in the present chapter.

9.2 THE PROPELLER

Airplane wings and propellers have something in common: they are both made up of airfoil sections designed to generate an aerodynamic force. The wing force provides lift to sustain the airplane in the air; the propeller force provides thrust to push the airplane through the air. A sketch of a simple three-blade propeller is

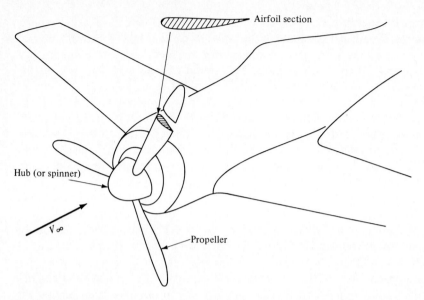

Figure 9.1 The airplane propeller, emphasizing that a propeller cross section is an airfoil shape.

given in Figure 9.1, illustrating that a cross section is indeed an airfoil shape. However, unlike a wing, where the chord lines of the airfoil sections are essentially all in the same direction, a propeller is twisted such that the chord line changes from almost parallel to V_∞ at the root, to almost perpendicular at the tip. This is illustrated in Figure 9.2, which shows a side view of the propeller, as well as two sectional views, one at the tip and the other at the root. This figure should be studied carefully. The angle between the chord line and the propeller's plane of rotation is defined as the *pitch angle β*. The distance from the root to a given section is r. Note that $\beta = \beta(r)$.

The airflow seen by a given propeller section is a combination of the airplane's forward motion and the rotation of the propeller itself. This is sketched in Figure 9.3*a*, where the airplane's relative wind is V_∞ and the speed of the blade section due to rotation of the propeller is $r\omega$. Here, ω denotes the angular velocity of the propeller in radians per second. Hence, *the relative wind seen by the propeller section is the vector sum of V_∞ and $r\omega$*, as shown in Figure 9.3*b*.

Clearly, if the chord line of the airfoil section is at an angle of attack α with respect to the local relative wind V, then lift and drag (perpendicular and parallel to V, respectively) are generated. In turn, as shown in Figure 9.4, the components of L and D in the direction of V_∞ produce a net *thrust T*:

$$T = L\cos\phi - D\sin\phi \tag{9.1}$$

where $\phi = \beta - \alpha$. This thrust, when summed over the entire length of the

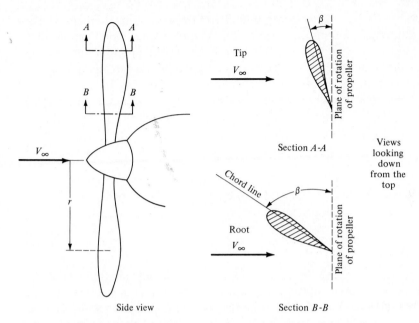

Figure 9.2 Illustration of propeller, showing variation of pitch along the blade.

propeller blades, yields the net thrust available (T_A, as defined in Chap. 6) which drives the airplane forward.

This simple picture is the essence of how a propeller works. However, the actual prediction of propeller performance is more complex. The propeller is analogous to a finite wing that has been twisted. Therefore, the aerodynamics of the propeller is influenced by the same induced flow due to tip vortices as was

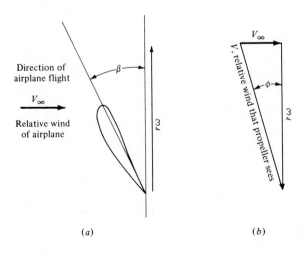

Figure 9.3 Velocity diagram for the flow velocity relative to the propeller.

described for the finite wing in Secs. 5.13 and 5.14. Moreover, due to the propeller twist and rotational motion, the aerodynamic theory is even more complicated. However, propeller theory has been extensively developed, and more details can be found in the books by Dommasch et al. and Glauret (see Bibliography at the end of this chapter). Such theory is beyond the scope of this book.

Instead, let us concentrate on the understanding of propeller efficiency η introduced in Sec. 6.6. From Eq. (6.30), the propeller efficiency is defined as

$$\eta = \frac{P_A}{P} \tag{9.2}$$

where P is the shaft brake power (the power delivered to the propeller by the shaft of the engine) and P_A is the power available from the propeller. As given in Eq. (6.31), $P_A = T_A V_\infty$. Hence, Eq. (9.2) becomes

$$\eta = \frac{T_A V_\infty}{P} \tag{9.3}$$

As explained above, T_A in Eq. (9.3) is basically an aerodynamic phenomenon which is dependent on the angle of attack α in Figure 9.4. In turn, α is dictated by the pitch angle β and ϕ, where ϕ itself depends on the magnitudes of V_∞ and $r\omega$. The angular velocity $\omega = 2\pi n$, where n is the number of propeller revolutions per second. Consequently, T_A must be a function of at least β, V_∞, and n. Finally, the thrust must also depend on the size of the propeller, characterized by the propeller diameter D. In turn, the propeller efficiency, from Eq. (9.3), must depend on β, V_∞, n, and D. Indeed theory and experiment both show that for a fixed pitch angle β, η is a function of the dimensionless quantity

$$J = \frac{V_\infty}{nD} \qquad \text{the advance ratio}$$

A typical variation of η with J for a fixed β is sketched in Figure 9.5; three curves are shown corresponding to three different values of pitch. Figure 9.5 is important; it is from such curves that η is obtained for an airplane performance analysis, as described in Chap. 6.

Examine Figure 9.5 more closely. Note that $\eta < 1$; this is because some of the power delivered by the shaft to the propeller is always lost, and hence $P_A < P$.

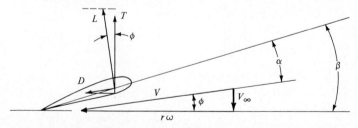

Figure 9.4 Generation of propeller thrust.

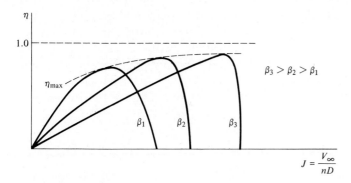

Figure 9.5 Propeller efficiency vs. advance ratio. Note that D denotes propeller diameter.

These losses occur due to several different effects. First imagine that you are standing in an open field. The air is still; it has no velocity. Then assume a propeller-driven vehicle goes zooming by you. After the propeller has passed, you will feel a stiff breeze moving in the direction opposite that of the vehicle. This breeze is part of the slipstream from the propeller; i.e., the air is set into both translational and rotational motion by the passage of the propeller. Consequently, you observe some translational and rotational *kinetic energy* of the air, where before there was none. This kinetic energy has come from part of the power delivered by the shaft to the propeller; it does no useful work and hence robs the propeller of some available power. In this fashion, the energy of the slipstream relative to the still air ahead of the vehicle is a source of power loss. Another source is frictional loss due to the skin friction and pressure drag (profile drag) on the propeller. Friction of any sort always reduces power. A third source is "compressibility" loss. The fastest-moving part of the propeller is the tip. For many high-performance engines, the propeller tip speeds result in a near-sonic relative wind. When this occurs, the same type of shock wave and boundary-layer separation losses which cause the drag-divergence increase for wings (see Sec. 5.10) now act to rob the propeller of available power. If the propeller tip speed is supersonic, η drops dramatically. This is the primary reason why propellers have not been used for transonic and supersonic airplanes. (After World War II, the NACA and other laboratories experimented with swept-back propellers, motivated by the success obtained with swept wings for high-speed flight, but nothing came of these efforts.) As a result of all the losses described above, the propeller efficiency is always less than unity.

Return again to Figure 9.5. Note that, for a fixed β, the efficiency is zero at $J = 0$, increases as J increases, goes through a maximum, and then rapidly decreases at higher J, finally again going to zero at some large finite value of J. Question: Why does η go to zero for the two different values of J? At the origin, the answer is simple. Consider a propeller with given values of n and D; hence, J depends only on V_∞. When $V_\infty = 0$, then $J = 0$. However, when $V_\infty = 0$, then

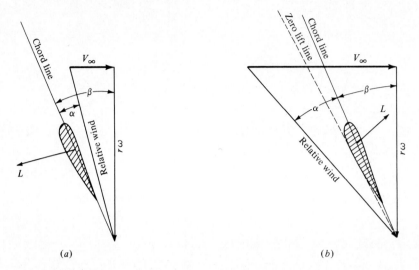

Figure 9.6 Explanation of the variation of propeller efficiency with advance ratio. (*a*) Velocity diagram for low V_∞. (*b*) Velocity diagram for high V_∞.

$P_A = T_A V_\infty = 0$; consequently, $\eta = P_A/P = 0$. Thus, propeller efficiency is zero at $J = 0$ because there is no motion of the airplane, hence no power available. At the other extreme, when V_∞, hence J, is made large, the propeller loses lift due to small angles of attack. This is seen in Figure 9.6. Consider a given propeller airfoil section at a distance r from the center. Assume ω, hence $r\omega$, remains constant. If V_∞ is small, the relative wind will be as shown in Figure 9.6*a*, where the airfoil section is at a reasonable angle of attack and hence produces a reasonable lift. Now, if V_∞ is increased, the relative wind approaches the chord line; hence α, and therefore lift coefficient, decreases. If the value of V_∞ is such that the relative wind corresponds to the zero-lift line, then the lift (hence thrust) is zero, and again $\eta = T_A V_\infty/P = 0$. In fact, if V_∞ is made even larger, the section will produce negative lift, hence reverse thrust, as shown in Figure 9.6*b*.

A consideration of the relative wind also explains why a propeller blade is twisted, with a large β at the root and a small β at the tip, as was first sketched in Figure 9.2. Near the root, r, and hence $r\omega$, is small. Thus, as shown in Figure 9.7*a*, β must be large to have a reasonable α. In contrast, near the tip, r, hence $r\omega$, is large. Thus, as shown in Figure 9.7*b*, β must be smaller in order to have a reasonable α.

Return again to Figure 9.5. All early airplanes before 1930 had *fixed-pitch propellers*, i.e., the values of β for all sections were geometrically fixed by the design and manufacture of the blades, and once the propeller was rigidly mounted on the engine shaft, the pilot could not change the blade angle. Thus, from the curves in Figure 9.5, maximum propeller efficiency could be obtained only at a specific value of the advance ratio J. At other flight velocities, the

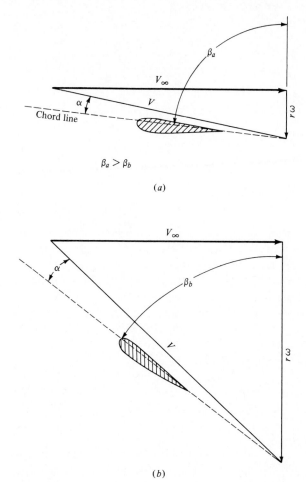

$\beta_a > \beta_b$

(a)

Figure 9.7 Difference in the relative wind along the propeller blade. (a) Near the root; (b) near the tip.

(b)

propeller always operated at efficiencies less than maximum. This characteristic severely limited airplane performance. Some improvement, although small, was attempted in 1916 at the Royal Aircraft Factory at Farnborough, England, by a design of a two-pitch propeller. However the ultimate solution was the *variable-pitch propeller*, patented in 1924 by Dr. H. S. Hele-Shaw and T. E. Beacham in England and first introduced into practical production in 1932 in the United States. The variable-pitch propeller is fixed to a mechanical mechanism in the hub; the mechanism rotates the entire blade about an axis along the length of the blade. In this fashion, the propeller pitch can be continuously varied to maintain maximum efficiency at all flight velocities. This can be visualized as riding along the peaks of the propeller efficiency curves in Figure 9.5, as shown by the dotted η_{max} line. A further development of this concept was the introduction in 1935 of the *constant-speed propeller*, which allowed the pitch angle to be varied continu-

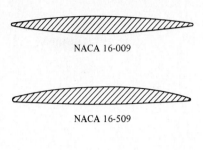

NACA 16-009

NACA 16-509

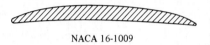

NACA 16-1009

Figure 9.8 Typical high-speed airfoil sections for propellers.

ously and automatically to maintain the proper torque on the engine such that the engine revolutions per minute were constant over the range of flight velocities. This is advantageous because the brake power output of aircraft piston engines is usually optimized at a given revolutions per minute. Nevertheless, the introduction of the variable-pitch and constant-speed propellers in the 1930s was one of the most important developments in the history of aeronautical engineering. As a result, values of η range from about 0.83 to 0.90 for most modern propellers.

A comment is in order concerning airfoil sections used for propellers. Early propellers from the World War I era typically utilized the RAF-6 airfoil; later, the venerable Clark Y shape was employed. During the late 1930s, some of the standard NACA sections were used. However, as aircraft speeds rapidly increased during World War II, special high-speed profiles were incorporated in propellers. The NACA developed a complete series, the 16 series, which found exclusive use in propellers. This series is different from the wing airfoil sections given in Appendix D; some typical shapes are sketched in Figure 9.8. These are thin profiles, designed to minimize the transonic flow effects near the propeller tips. They should be compared with the more conventional shapes in Appendix D.

9.3 THE RECIPROCATING ENGINE

For the first 50 years of successful manned flight, the internal combustion, reciprocating, gasoline-burning engine was the mainstay of aircraft propulsion. It is still used today in airplanes designed to fly at speeds below 300 mi/h, the range for the vast majority of light, private, general aviation aircraft (such as the hypothetical CP-1 in the examples of Chap. 6). A photograph of a typical internal combustion reciprocating engine is shown in Figure 9.9.

The basic operating principle of these engines is a piston moving back and forth (reciprocating) inside a cylinder, with values which open and close appropriately to let fresh fuel-air mixture in and burned exhaust gases out. The piston is connected to a shaft via a connecting rod which converts the reciprocat-

Figure 9.9 A large radial air-cooled internal combustion aircraft engine; the Pratt and Whitney Twin Wasp R-2000, produced from 1941 to 1959. (*Pratt and Whitney Aircraft, a Division of United Technologies.*)

ing motion of the piston to rotational motion of the shaft. A typical four-stroke cycle is illustrated in Figure 9.10. During the *intake* stroke (Figure 9.10*a*), the piston moves downward, the intake valve is open, and a fresh charge of gasoline-air mixture is drawn into the cylinder. This process is sketched on the *p-V* diagram (a plot of pressure vs. volume) in Figure 9.10*a*. Here, point 1 corresponds to the beginning of the stroke (where the piston is at the top, called top dead center), and point 2 corresponds to the end of the stroke (where the piston is at the bottom, called bottom dead center). The volume *V* is the total mixture volume between the top of the cylinder and the face of the piston. The intake stroke takes place at essentially constant pressure, and the total mass of fuel-air mixture inside the cylinder increases throughout the stroke. At the bottom of the intake stroke, the intake valve closes, and the *compression* stroke begins (Figure 9.10*b*). Here, the piston compresses the now constant mass of gas from a low pressure p_2

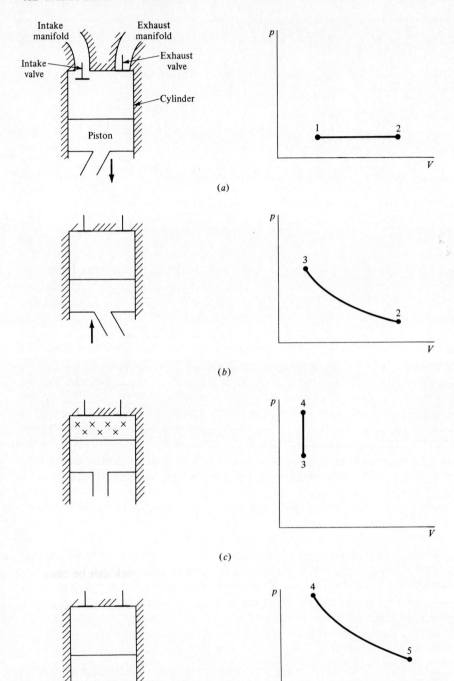

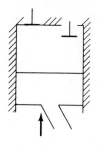

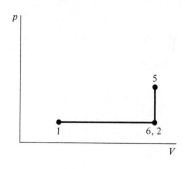

(e)

Figure 9.10 Elements of the four-stroke, internal combustion, reciprocating-engine cycle. (a) Intake stroke; (b) compression stroke; (c) constant-volume combustion; (d) power stroke; (e) exhaust stroke. Note that V denotes the gas volume in the cylinder.

to a higher pressure p_3, as shown in the accompanying p-V diagram. If frictional effects are ignored, the compression takes place isentropically (see Sec. 4.6) because no heat is added or taken away. At the top of the compression stroke, the mixture is ignited, usually by an electric spark. *Combustion* takes place rapidly, before the piston has moved any meaningful distance. Hence, for all practical purposes, the combustion process is one of *constant volume* (Figure 9.10c). Since energy is released, the temperature increases markedly; in turn, because the volume is constant, the equation of state, Eq. (2.7), dictates that pressure increases from p_3 to p_4. This high pressure exerted over the face of the piston generates a strong force which drives the piston downward on the *power stroke* (Figure 9.10d). Again, assuming that frictional and heat transfer effects are negligible, the gas inside the cylinder expands isentropically to the pressure p_5. At the bottom of the power stroke, the exhaust valve opens. The pressure inside the cylinder instantly adjusts to the exhaust manifold pressure p_6, which is usually about the same value as p_2. Then, during the *exhaust stroke*, Figure 9.10e, the piston pushes the burned gases out of the cylinder returning to conditions at point 1. Thus, the basic process of a conventional aircraft piston engine consists of a four-stroke cycle: intake, compression, power, and exhaust.

Due to the heat released during the constant-volume combustion, the cycle delivers a net amount of positive work to the shaft. This work can be calculated using the complete p-V diagram for the cycle, as sketched in Figure 9.11. Recall from Eq. (4.15) that the amount of work done on the gas due to a change in volume dV is $\delta w = -p\,dV$. In turn, the work done *by* the gas is simply

$$\delta w = p\,dV$$

For any part of the process, say during the power stroke, this is equivalent to the small sliver of area of height p and base dV, as shown in Figure 9.11. In turn, the work done by the gas on the piston during the whole power stroke is then

$$W_{\substack{\text{power}\\\text{stroke}}} = \int_{V_4}^{V_5} p\,dV \tag{9.4}$$

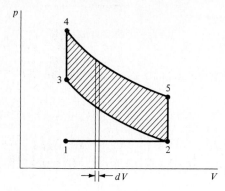

Figure 9.11 The complete four-stroke cycle for a spark-ignition internal combustion engine (the Otto cle).

This is given by the area under the curve from point 4 to point 5 in Figure 9.11. Analogously, the work done by the piston on the gas during the compression stroke is

$$\underset{\text{stroke}}{W_{\text{compression}}} = \int_{V_3}^{V_2} p \, dV \tag{9.5}$$

This is given by the area under the curve from point 2 to point 3. Consequently, the net work done during the complete cycle, W, is

$$W = \underset{\text{stroke}}{W_{\text{power}}} - \underset{\text{stroke}}{W_{\text{compression}}} \tag{9.6}$$

This is equal to the shaded area of the p-V diagram shown in Figure 9.11. Thus, we see the usefulness of p-V diagrams in analyzing thermodynamic processes in closed systems, namely, *the area bounded by the complete cycle on a p-V diagram is equal to the work done during the cycle.*

The power output of this arrangement is the *work done per unit time.* Consider the engine shaft rotating at n revolutions per second. The piston goes up and down once for each revolution of the shaft. Hence, the number of times the complete engine cycle is repeated in 1 s is $n/2$. The work output on each cycle is W, from Eq. (9.6). If the complete engine has N cylinders, then the power output of the engine is

$$\text{IP} = \frac{n}{2} NW \tag{9.7}$$

The symbol IP is used to signify *indicated power.* This is the power that is generated by the thermodynamic and combustion processes inside the engine. However, transmission of this power to the shaft takes place through mechanical linkages, which always generate frictional losses due to moving parts in contact. As a result, the power delivered to the shaft is less than IP. If the *shaft brake power* is P (see Sec. 6.6), then

$$P = \eta_{\text{mech}}(\text{IP}) \tag{9.8}$$

where η_{mech} is the mechanical efficiency which accounts for friction loss due to the moving engine parts. Then, from Eq. (6.30), the power available to propel the engine-propeller combination is

$$P_A = \eta\eta_{\text{mech}}(\text{IP}) \tag{9.9}$$

or, from Eq. (9.7),

$$\boxed{P_A = \eta\eta_{\text{mech}}\frac{n}{2}NW} \tag{9.10}$$

If rpm denotes the revolutions per minute of the engine, then $n = \text{rpm}/60$ and Eq. (9.10) becomes

$$P_A = \frac{\eta\eta_{\text{mech}}(\text{rpm})NW}{120} \tag{9.11}$$

Equation (9.11) proves the intuitively obvious fact that the *power available for a propeller-driven airplane is directly proportional to the engine rpm.*

The work per cycle, W, in Eq. (9.10) can be expressed in more detailed terms. Consider the piston shown in Figure 9.12. The length of the piston movement is called the *stroke s*; the diameter of the piston is called the *bore b*. The volume swept out by the piston is called the *displacement*, equal to $(\pi b^2/4)s$. Assume that a constant pressure p_e acts on the face of the piston during the power stroke; p_e is called the *mean effective pressure*. It is *not* the actual pressure acting on the piston, which in reality varies from p_4 to p_5 during the power stroke; rather, p_e is an artificially defined quantity which is related to the engine power output and which is an *average* representation of the actual pressure. Furthermore, assume that all the useful work is done on the power stroke. Thus, W is equal to the force

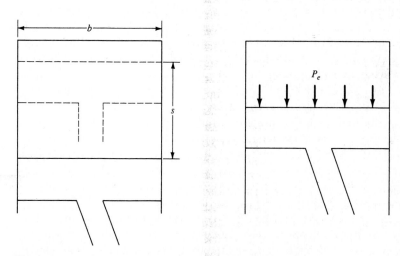

Figure 9.12 Illustration of bore, stroke, and mean effective pressure.

on the piston, $(\pi b^2/4)p_e$, times the distance through which the force moves, s; i.e.,

$$W = \frac{\pi b^2}{4} s p_e \qquad (9.12)$$

Combining Eqs. (9.11) and (9.12), we obtain

$$P_A = \eta \eta_{\text{mech}}(\text{rpm}) N \frac{\pi b^2}{4} \frac{s p_e}{120} \qquad (9.13)$$

The total displacement of the engine d is equal to the displacement of each cylinder times the number of cylinders:

$$d = \frac{\pi b^2}{4} s N \qquad (9.14)$$

Combining Eqs. (9.13) and (9.14) yields

$$\boxed{P_A = \frac{\eta \eta_{\text{mech}}(\text{rpm}) d p_e}{120}} \qquad (9.15)$$

Equation (9.15) indicates that power available is directly proportional to engine rpm, displacement, and mean effective pressure.

In Chap. 6, the altitude effect on P_A for a reciprocating engine–propeller combination was assumed to be governed by ambient density; i.e., P_A was assumed to be directly proportional to ρ_∞. More credence can now be added to this earlier assumption in light of the above discussion. For example, Eq. (9.15) shows that P_A is proportional to p_e. However, p_e is representative of the mass of air originally obtained at ambient conditions, then mixed with a small amount of fuel in the intake manifold, and then sucked into the cylinder during the intake stroke. If this mass of air is reduced by flying at higher altitudes where ρ_∞ is lower, then p_e will be correspondingly lower. In turn, from Eq. (9.15), P_A will be correspondingly reduced. Therefore, the assumption that $P_A \propto \rho_\infty$ is reasonable.

The reduction of P_A with altitude can be delayed if a *supercharger* is used on the engine. This is basically a pump, driven from the engine crankshaft (a geared supercharger) or driven by a small turbine mounted in the engine exhaust jet (a turbosupercharger). The supercharger compresses the incoming air before it reaches the intake manifold, increasing its density and thereby avoiding a loss in P_A at altitude. Early work on superchargers was performed in the 1920s by the NACA at Langley. This was important research, because an unsupercharged airplane of that day was limited to altitudes on the order of 20,000 ft or less. However, on May 18, 1929, Navy Lt. Apollo Soueck, flying an Apache airplane powered by a supercharged Pratt & Whitney Wasp engine, reached 39,140 ft, an altitude record for that time. Subsequently, a substantial portion of the NACA propulsion research was channeled into superchargers, which led to the high-performance engines used in military aircraft during World War II. For modern general aviation aircraft of today, supercharged engines are available as options on many designs and are fixed equipment on others.

A more extensive discussion of reciprocating, internal combustion engines is beyond the scope of this book. However, such engines are important to the general aviation industry. In addition, their importance to the automobile industry goes without saying, especially in light of the modern demands of efficiency and low pollutant emissions. Therefore, the interested reader is strongly encouraged to study deeper into the subject; for example, more details can be found in the book by Obert (see Bibliography at the end of this chapter).

Example 9.1 Consider a six-cylinder internal combustion engine with a stroke of 9.5 cm and a bore of 9 cm. The compression ratio is 10. [Note that the compression ratio in internal combustion (IC) engine terminology is defined as the volume of the gas in the cylinder when the piston is at bottom dead center divided by the volume of the gas when the piston is at top dead center.] The pressure and temperature in the intake manifold are 0.8 atm and 250 K, respectively. The fuel-to-air ratio of the mixture is 0.06 (by mass). The mechanical efficiency of the engine is 0.75. If the crankshaft is connected to a propeller with an efficiency of 0.83, calculate the power available from the engine-propeller combination for 3000 rpm.

SOLUTION Consider the ideal cycle as sketched in Figure 9.11. We want to calculate the work done per cycle in order ultimately to obtain the total power output. To do this, we first need to find $p_3, p_4, p_5, V_2 = V_5$, and $V_3 = V_4$. Since the compression stroke is isentropic, from Sec. 4.6,

$$\frac{p_3}{p_2} = \left(\frac{V_2}{V_3}\right)^\gamma = 10^{1.4} = 25.1$$

$$p_3 = 25.1(0.8) = 20.1 \text{ atm}$$

$$\frac{T_3}{T_2} = \left(\frac{V_2}{V_3}\right)^{\gamma-1} = (10)^{0.4} = 2.5$$

$$T_3 = 2.5(250) = 625 \text{ K}$$

Referring to Figure 9.11, we see the combustion process from point 3 to point 4 is at constant volume. The chemical energy release in 1 kg of gasoline is approximately 4.29×10^7 J. Hence, the heat released per kilogram of fuel-air mixture is (recalling the fuel-to-air ratio is 0.06)

$$q = \frac{(4.29 \times 10^7)(0.06)}{1.06} = 2.43 \times 10^6 \text{ J/kg}$$

From the first law of thermodynamics, Eq. (4.16), and from Eq. (4.23), for a constant-volume process:

$$\delta q = de + p\, dv = de + 0 = c_v\, dT$$

Hence

$$q = c_v(T_4 - T_3)$$

or

$$T_4 = \frac{q}{c_v} + T_3$$

The value of c_v can be obtained from Eq. (4.68), recalling from Example 4.5 that $c_p = 1008$ J/(kg)(K) for air. Assume that the specific heats and gas constant for the fuel-air mixture are approximated by the air values alone; this is reasonable because only a small amount of fuel is present in the mixture. Hence $c_v = c_p - R = 1008 - 288 = 720$ J/(kg)(K)

Thus

$$T_4 = \frac{q}{c_v} + T_3 = \frac{2.43 \times 10^6}{720} + 625 = 4000 \text{ K}$$

From the equation of state, noting that $V_4 = V_3$ and R is constant, we find $p_4/p_3 = T_4/T_3$.

Thus
$$p_4 = p_3 \frac{T_4}{T_3} = 20.1 \frac{4000}{625} = 128.6 \text{ atm}$$

For the power stroke, the process is isentropic. Hence

$$\frac{p_5}{p_4} = \left(\frac{V_4}{V_5}\right)^\gamma = \left(\frac{1}{10}\right)^{1.4} = 0.0398$$

$$p_5 = 128.6(0.0398) = 5.12 \text{ atm}$$

We now have enough thermodynamic information to calculate the work done per cycle. From Eq. (9.5)

$$W_{\substack{\text{compression} \\ \text{stroke}}} = \int_{V_3}^{V_2} p\,dV$$

For an isentropic process, $pV^\gamma = c$, where c is a constant. Thus $p = cV^{-\gamma}$ and

$$W_{\text{compression}} = c\int_{V_3}^{V_2} V^{-\gamma}\,dV = \frac{c}{1-\gamma}\left(V_2^{1-\gamma} - V_3^{1-\gamma}\right)$$

Since
$$c = p_2 V_2^\gamma = p_3 V_3^\gamma$$

$$W_{\text{compression}} = \frac{p_2 V_2 - p_3 V_3}{1-\gamma}$$

We need the volumes V_2 and V_3 to proceed further. Consider Figure 9.12. The stroke of the piston is 9.5 cm, and the compression ratio is 10. If x denotes the distance from the top of the cylinder to the piston top dead center position, then from the definition of compression ratio,

$$\frac{x+9.5}{x} = 10$$

$$x = 1.055 \text{ cm}$$

$$V_2 = \frac{\pi b^2}{4}(9.5 + 1.05) \qquad \text{where } b = \text{bore} = 9 \text{ cm}$$

$$= \pi(9)^2 \frac{9.5 + 1.05}{4} = 671.2 \text{ cm}^3$$

$$= 6.712 \times 10^{-4} \text{ m}^3 \qquad \text{(remember consistent units)}$$

$$V_3 = \frac{V_2}{10} = 0.6712 \times 10^{-4} \text{ m}^3$$

Thus
$$W_{\text{compression}} = \frac{p_2 V_2 - p_3 V_3}{1-\gamma}$$

$$= \frac{[0.8(6.712 \times 10^{-4}) - 20.1(0.6712 \times 10^4)]1.01 \times 10^5}{-0.4}$$

$$= 205 \text{ J}$$

Similarly, the work done by the power stroke from point 4 to point 5 (isentropic) is

$$W_{\substack{\text{power} \\ \text{stroke}}} = \int_{V_4}^{V_5} p\,dV$$

$$W_{\text{power}} = \frac{p_5 V_5 - p_4 V_4}{1-\gamma}$$

$$= \frac{[5.12(6.712 \times 10^{-4}) - 128.6(0.612 \times 10^{-4})]1.01 \times 10^5}{-0.4}$$

$$= 1312 \text{ J}$$

Finally, from Eq. (9.6), the net work per cycle is

$$W = W_{\text{power}} - W_{\text{compression}} = 1312 - 205 = 1107 \text{ J}$$

The total power available from the engine-propeller combination is, from Eq. (9.11),

$$P_A = \frac{1}{120} \eta \eta_{\text{mech}} (\text{rpm}) NW$$

$$= \frac{0.83(0.75)(3000)(6)(1107)}{120}$$

$$\boxed{P_A = 1.034 \times 10^5 \text{ J/s}}$$

From Sec. 6.6 A, we know that

$$1 \text{ hp} = 746 \text{ J/s}$$

Hence

$$\text{hp}_A = \frac{1.034 \times 10^5}{746} = \boxed{138.6 \text{ hp}}$$

Note: This example is rather long, with numerous calculations. However, it serves to illustrate many aspects of our discussion on IC engines, and the reader should examine it closely.

Example 9.2 For the engine in Example 9.1, calculate the mean effective pressure.

SOLUTION From Eq. (9.15)

$$P_A = \frac{1}{120} \eta \eta_{\text{mech}} (\text{rpm}) d p_e$$

where d is the displacement and p_e is the mean effective pressure. From Eq. (9.14),

$$d = \frac{\pi b^2}{4} sN$$

$$= \frac{\pi (9)^2 (9.5)(6)}{4} = 3626 \text{ cm}^3$$

$$= 3.626 \times 10^{-3} \text{ m}^3$$

Hence, from Eq. (9.15) and the results of Example 9.1,

$$103{,}366 = \frac{1}{120} (0.83)(0.75)(3000)(3.626 \times 10^{-3}) p_e$$

$$\boxed{p_e = 1.83 \times 10^6 \text{ N/m}^2 = 18.1 \text{ atm}}$$

9.4 JET PROPULSION—THE THRUST EQUATIONS

The previous two sections have discussed the production of thrust and power by a piston engine–propeller combination. Recall from Sec. 2.2 that the fundamental mechanisms by which nature communicates a force to a solid surface are by means of the surface pressure and shear stress distributions. The propeller is a case in point, where the net result of the pressure and shear stress distributions over the surface of the propeller blades yields an aerodynamic force, the thrust, which propels the vehicle forward. Also, an effect of this thrust on the propeller is an equal and opposite reaction which yields a force on the air itself, pushing it backward in the opposite direction of the propeller thrust; i.e., a change in momentum is imparted to the air by the propeller, and an alternate physical explanation of the production of thrust is that T is equal to the time rate of change of momentum of the airflow. For a propeller, this change in momentum is in the form of a large mass of air being given a small increase in velocity (about 10 m/s). However, keep in mind that the basic mechanism producing thrust is still the distribution of pressure and shear stress over the surface. Also, as in the case of lift produced by a wing, the thrust is primarily due to just the pressure distribution [see Eq. (9.1) and Figure 9.4]; the shear stress is predominately a drag-producing mechanism which affects the torque of the propeller.

These same principles carry over to jet propulsion. As sketched in Figure 9.13a, the jet engine is a device which takes in air at essentially the freestream velocity V_∞, heats it by combustion of fuel inside the duct, and then blasts the hot mixture of air and combustion products out the back end at a much higher velocity V_e. (Strictly speaking, the air velocity at the inlet to the engine is slightly larger than V_∞, but this is not important to the present discussion.) In contrast to a propeller, the jet engine creates a change in momentum of the gas by taking a small mass of air and giving it a large increase in velocity (hundreds of meters per second). By Newton's third law, the equal and opposite reaction produces a thrust. However, this reaction principle, which is commonly given as the basic mechanism for jet propulsion, is just an alternate explanation in the same vein as the discussion given above. The true fundamental source of the thrust of a jet engine is the net force produced by the pressure and shear stress distributions exerted over the surface of the engine. This is sketched in Figure 9.13b, which illustrates the distribution of pressure p_s over the internal surface of the engine duct, and the ambient pressure, essentially p_∞, over the external engine surface. Shear stress, which is generally secondary in comparison to the magnitude of the pressures, is ignored here. Examining Figure 9.13b, let x denote the flight direction. The thrust of the engine in this direction is equal to the x component of p_s integrated over the complete internal surface, plus that of p_∞ integrated over the complete external surface. In mathematical symbols,

$$T = \int (p_s \, dS)_x + \int (p_\infty \, dS)_x \qquad (9.16)$$

Since p_∞ is constant, the last term becomes

$$\int (p_\infty \, dS)_x = p_\infty \int (dS)_x = p_\infty (A_i - A_e) \qquad (9.17)$$

where A_i and A_e are the inlet and exit areas, respectively, of the duct, as defined in Figure 9.13b. In Eq. (9.17), the x component of the duct area, $\int (dS)_x$, is physically what you see by looking at the duct from the front, as shown in Figure 9.13c. The x component of surface area is geometrically the projected frontal area shown by the cross-hatched region in Figure 9.13c. Thus, substituting Eq. (9.17)

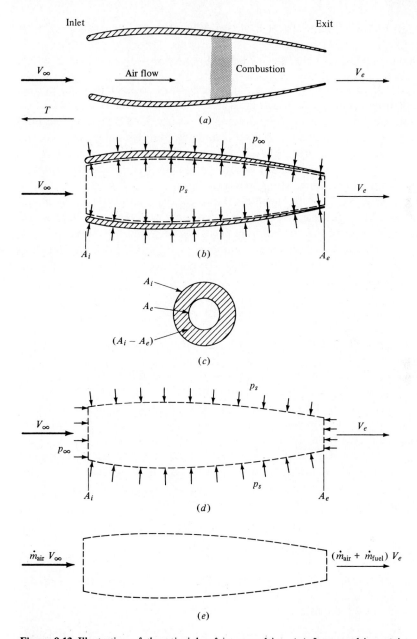

Figure 9.13 Illustration of the principle of jet propulsion. (*a*) Jet propulsion engine. (*b*) Surface pressure on inside and outside surfaces of duct. (*c*) Front view, illustrating inlet and exit areas. (*d*) Control volume for flow through duct. (*e*) Change in momentum of the flow through the engine.

into (9.16), we obtain for the thrust T of the jet engine

$$T = \int (p_s \, dS)_x + p_\infty (A_i - A_e) \tag{9.18}$$

The integral in Eq. (9.18) is not particularly easy to handle in its present form. Let us proceed to couch this integral in terms of the velocity and mass flow of gas through the duct. Consider the volume of gas bounded by the dashed lines in Figure 9.13b. This is called a *control volume* in aerodynamics. This control volume is sketched again in Figure 9.13d. The frontal area of the volume is A_i, on which p_∞ is exerted. The side of the control volume is the same as the internal area of the engine duct. Since the gas is exerting a pressure p_s on the duct as shown in Figure 9.13b, then by Newton's third law, the duct exerts an equal and opposite pressure p_s on the gas in the control volume as shown in Figure 9.13d. Finally, the rear area of the control volume is A_e, on which p_e is exerted. The pressure p_e is the gas static pressure at the exit of the duct. With the above in mind, and with Figure 9.13d in view, the x component of the force on the gas inside the control volume is

$$F = p_\infty A_i + \int (p_s \, dS)_x - p_e A_e \tag{9.19}$$

Now recall Newton's second law, namely, $F = ma$. This can also be written as $F = d(mV)/dt$; that is, the *force equals the time rate of change of momentum* (indeed, this is how Newton originally expressed his second law). Question: What is the time rate of change of momentum of the air flowing through the control volume? The answer can be obtained from Figure 9.13e. The mass flow of air (kg/s or slug/s) entering the duct is $\dot{m}_{air}$; its momentum is $\dot{m}_{air} V_\infty$. The mass flow of gas leaving the duct (remember that fuel has been added and burned inside) is $\dot{m}_{air} + \dot{m}_{fuel}$; its momentum is $(\dot{m}_{air} + \dot{m}_{fuel}) V_e$. Thus, the time rate of change of momentum of the airflow through the control volume is the difference between what comes out and what comes in: $(\dot{m}_{air} + \dot{m}_{fuel}) V_e - \dot{m}_{air} V_\infty$. From Newton's second law, this is equal to the force on the control volume,

$$F = (\dot{m}_{air} + \dot{m}_{fuel}) V_e - \dot{m}_{air} V_\infty \tag{9.20}$$

Combining Eqs. (9.19) and (9.20) yields

$$(\dot{m}_{air} + \dot{m}_{fuel}) V_e - \dot{m}_{air} V_\infty = p_\infty A_i + \int (p_s \, dS)_x - p_e A_e \tag{9.21}$$

Solving Eq. (9.21) for the integral term, we obtain

$$\int (p_s \, dS)_x = (\dot{m}_{air} + \dot{m}_{fuel}) V_e - \dot{m}_{air} V_\infty + p_e A_e - p_\infty A_i \tag{9.22}$$

We now have the integral in the original thrust equation Eq. (9.18), in terms of velocity and mass flow, as originally desired. The final result for the engine thrust is obtained by substituting Eq. (9.22) into Eq. (9.18):

$$T = (\dot{m}_{air} + \dot{m}_{fuel}) V_e - \dot{m}_{air} V_\infty + p_e A_e - p_\infty A_i + p_\infty (A_i - A_e) \tag{9.23}$$

The terms involving A_i cancel, and we have

$$T = (\dot{m}_{\text{air}} + \dot{m}_{\text{fuel}})V_e - \dot{m}_{\text{air}}V_\infty + (p_e - p_\infty)A_e \qquad (9.24)$$

Equation (9.24) is the fundamental *thrust equation* for jet propulsion. It is an important result and will be examined in more detail in subsequent sections. Keep in mind the reasoning that led to this result. First, the engine thrust was written down in purely mechanical terms, i.e., the thrust is due to the pressure distribution acting over the internal and external surfaces of the duct; this is the essence of Eq. (9.18). Then the internal pressure distribution acting over the internal surface was couched in terms of the change of momentum of the gas flowing through the duct; this is the essence of Eq. (9.22). Finally, the two lines of thought were combined to yield Eq. (9.24). You should reread the concepts presented in this section several times until you feel comfortable with the ideas and results. The above derivation of the thrust equation using the control volume concept is an example of a general method commonly used for the solution of many aerodynamic problems. You will see it again in more-advanced studies in aerodynamics and propulsion.

9.5 THE TURBOJET ENGINE

In 1944, the first operational jet fighter in the world was introduced by the German air force, the Me 262. By 1950, jet engines were the mainstay of all high-performance military aircraft, and by 1958 the commercial airlines were introducing the jet-powered Boeing 707 and McDonnell-Douglas DC-8. Today, the jet engine is the only practical propulsive mechanism for high-speed subsonic and supersonic flight. (Recall that our hypothetical CJ-1 in Chap. 6 was powered

Figure 9.14 The J52 turbojet engine. (*Pratt and Whitney Aircraft.*)

by two small jet engines.) A photograph of a typical turbojet engine is shown in Fig. 9.14.

The thrust of a turbojet engine is given directly by Eq. (9.24). The jet engine takes in a mass flow of cool air, $\dot{m}_{air}$, at velocity essentially equal to V_∞ and exhausts a mass flow of hot air and combustion products, $\dot{m}_{air} + \dot{m}_{fuel}$, at velocity V_e. This is illustrated in Figure 9.15. The mass of fuel added is usually small compared to the mass of air; $\dot{m}_{fuel}/\dot{m}_{air} \approx 0.05$. Thus, Eq. (9.24) can be simplified by neglecting $\dot{m}_{fuel}$:

$$T = \dot{m}_{air}(V_e - V_\infty) + (p_e - p_\infty)A_e \qquad (9.25)$$

Equation (9.25) explicitly shows that T can be increased by increasing $V_e - V_\infty$. Thus, the function of a jet engine is to exhaust the gas out the back end faster than it comes in through the front end. The conventional turbojet engine performs this function by inducting a mass of air through the *inlet* (location 1 in Figure 9.15). The flow is reduced to a low subsonic Mach number, $M \approx 0.2$ in a *diffuser* (point 1 to point 2 in Figure 9.15). This diffuser is directly analogous to the wind-tunnel diffusers discussed in Chap. 4. If V_∞ is subsonic, then the diffuser must increase the flow area to decelerate the flow, i.e., the diffuser is a *divergent duct* [see Eq. (4.83)]. If V_∞ is supersonic, the diffuser must be a convergent-divergent duct, and the decrease in flow velocity is accomplished partly through shock waves, as shown in Figure 9.15. For such supersonic inlets, a centerbody is sometimes employed to tailor the strength and location of the shock waves and to help form the convergent-divergent stream tube seen by the decelerating flow. In the diffusion process, the static pressure is increased from p_1 to p_2. After the diffuser, the flow is further compressed by a compressor (point 2 to point 3 in Figure 9.15) from p_2 to p_3. The compressor is usually a series of alternating rotating and stationary blades. The rotating sections are called *rotors*, and the stationary sections are *stators*. The rotor and stator blades are nothing more than airfoil sections which alternately speed up and slow down the flow; the work supplied by the compressor serves to increase the total pressure of the flow. The compressor sketched in Figure 9.15 allows the flow to pass essentially straight through the blades without any major deviation in direction; thus, such devices are called *axial flow compressors*. This is in contrast to *centrifugal flow compressors* used in some early jet engines, where the air was turned sometimes more than 90°. After leaving the compressor, fuel is injected into the airstream and burned at essentially constant pressure in the *combustor* (point 3 to point 4 in Figure 9.15), where the temperature is increased to about 2500°R. After combustion, the hot gas flows through the *turbine* (point 4 to point 5 in Figure 9.15). The turbine is a series of rotating blades (again, basically airfoil sections) which extract work from the flowing gas. This work is then transmitted from the turbine through a shaft to the compressor; i.e., the turbine drives the compressor. The flow through a turbine is an expansion process, and the pressure drops from p_4 to p_5. However, p_5 is still larger than the ambient pressure outside the engine. Thus, after leaving the turbine, the flow is expanded through a *nozzle* (point 5 to point

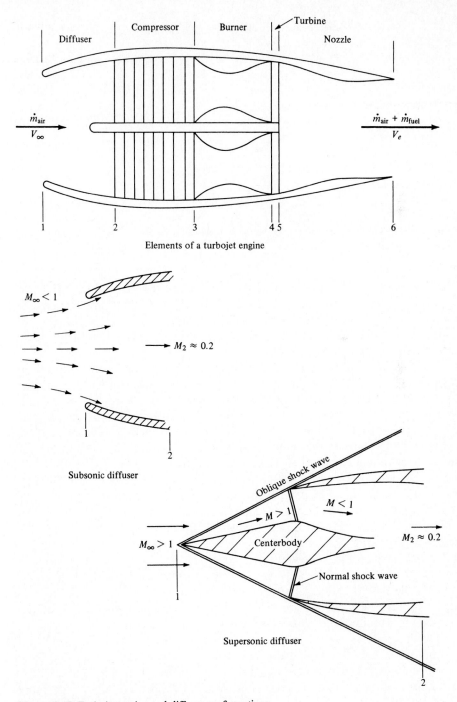

Figure 9.15 Turbojet engine and diffuser configurations.

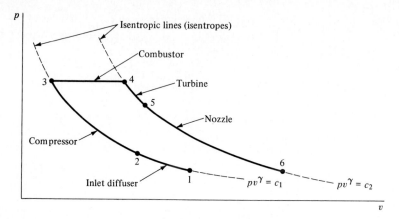

Figure 9.16 Pressure-specific volume diagram for an ideal turbojet.

6 in Figure 9.15) and is exhausted to the atmosphere at a high velocity V_e and at pressure $p_6 = p_e$. If the engine is designed for subsonic flight applications, the nozzle is usually convergent and V_e is subsonic, or at most sonic. However, if the engine is intended for supersonic aircraft, the exhaust nozzle is usually convergent-divergent and V_e is supersonic.

The thermodynamic process in an ideal turbojet engine is shown in the p-v diagram of Figure 9.16. The ideal process ignores the effects of friction and heat losses. Here, the air is isentropically compressed from p_1 to p_2 in the inlet diffuser, and the pressure is further isentropically increased to p_3 by the compressor. The process moves along the isentrope $pv^\gamma = c_1$, where c_1 is a constant (see Sec. 4.6). In the burner, the combustion process takes place at constant pressure (in contrast to the combustion process in an internal combustion reciprocating engine, which takes place at constant volume, as explained in Sec. 9.3). Since the temperature is increased by combustion and the pressure is constant, the equation of state, $pv = RT$, dictates that v must increase from v_3 to v_4 in the burner. Expansion through the turbine isentropically drops the pressure to p_5, and further isentropic expansion through the nozzle decreases the pressure to p_6. The turbine and nozzle expansions follow the isentrope $pv^\gamma = c_2$, where c_2 is a constant different from c_1. The ideal engine process further assumes that the nozzle expands the gas to ambient pressure, such that $p_e = p_6 = p_1 = p_\infty$. In the real engine process, of course, there will be frictional and heat losses; the diffuser, compressor, turbine, and nozzle processes will not be exactly isentropic, the combustion process is not precisely at constant pressure, and the nozzle exit pressure p_e will be something different from p_∞. However, the ideal turbojet shown in Figure 9.16 is a reasonable first approximation to the real case. The accounting of nonisentropic process in the engine is the subject for more-advanced studies of propulsion.

Return again to the turbojet engine thrust equation, Eq. (9.25). We are now in a position to understand some of the assumptions made in Chap. 6 concerning thrust available T_A for a turbojet. In our performance analysis of the CJ-1, we assumed that (1) thrust did not vary with V_∞ and (2) the altitude effect on thrust was simply proportional to ρ_∞. From the continuity equation, Eq. (4.2), applied at the inlet, we find that $\dot{m}_{\text{air}} = \rho_\infty A_i V_\infty$. Hence, as V_∞ increases, $\dot{m}_{\text{air}}$ increases. From Eq. (9.25), this tends to increase T. However, as V_∞ increases, the factor $V_e - V_\infty$ decreases. This tends to decrease T. The two effects tend to cancel each other, and the net result is a relatively constant thrust at subsonic speeds. With regard to altitude effects, $\dot{m}_{\text{air}} = \rho_\infty A_i V_\infty$ decreases proportionately with a decrease in ρ_∞; the factor $V_e - V_\infty$ is relatively unaffected. The term $(p_e - p_\infty) A_e$ in Eq. (9.25) is usually much smaller than $\dot{m}_{\text{air}}(V_e - V_\infty)$; hence even though p_e and p_∞ change with altitude, this pressure term will not have a major effect on T. Consequently, the primary consequence of altitude is to decrease ρ_∞, which proportionately decreases $\dot{m}_{\text{air}}$, which proportionately decreases T. Hence, our assumption in Chap. 6 that $T \propto \rho_\infty$ is reasonable.

Example 9.3 Consider a turbojet-powered airplane flying at a standard altitude of 30,000 ft at a velocity of 500 mi/h. The turbojet engine itself has inlet and exit areas of 7 ft^2 and 4.5 ft^2, respectively. The velocity and pressure of the exhaust gas at the exit are 1600 ft/s and 640 lb/ft^2, respectively. Calculate the thrust of the turbojet.

SOLUTION At a standard altitude of 30,000 ft, from Appendix B, $p_\infty = 629.66$ lb/ft^2 and $\rho_\infty = 8.9068 \times 10^{-4}$ slug/ft^3. The freestream velocity is $V_\infty = 500$ mi/h $= 500\,(88/60) = 733$ ft/s. Thus, the mass flow through the engine is

$$\dot{m}_{\text{air}} = \rho_\infty V_\infty A_i = (8.9068 \times 10^{-4})(733)(7)$$

$$= 4.57 \text{ slugs/s}$$

From Eq. (9.25), the thrust is

$$T = \dot{m}_{\text{air}}(V_e - V_\infty) + (p_e - p_\infty) A_e$$

$$= 4.57(1600 - 733) + (640 - 629.66)(4.5)$$

$$= 3962 + 46.5 = \boxed{4008.5 \text{ lb}}$$

9.6 THE TURBOFAN ENGINE

Section 9.4 established the relation between thrust and rate of change of momentum of a mass of air. In the turbojet engine (see Sec. 9.5), all of this mass flows through the engine itself, and all of it is accelerated to high velocity through the exhaust nozzle. Although this creates a large thrust, the *efficiency* of the process is

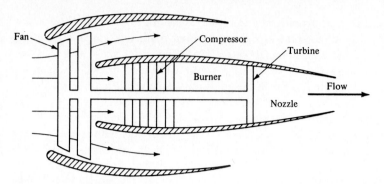

Figure 9.17 A turbofan engine.

adversely affected by the high exhaust velocities. Recall in Sec. 9.2 that one of the losses that reduces propeller efficiency is the kinetic energy remaining in the wake relative to the ambient air. In the case of the turbojet, the kinetic energy left in the jet exhaust is also a loss, and the high exhaust velocities produced by a jet engine just exacerbate the situation. This is why a piston engine–propeller combination is basically a more efficient device than a turbojet. (Remember, do not get efficiency and thrust confused—they are different things. A jet produces high thrust, but at at relatively low efficiency.) Therefore, the concepts of the pure turbojet and the propeller are combined in the *turbofan* engine. As sketched in Figure 9.17, a turbofan engine is a turbojet engine that has a large ducted fan mounted on the shaft ahead of the compressor. The turbine drives both the fan and compressor. The ducted fan accelerates a large mass of air that flows between the inner and outer shrouds; this unburned air then mixes with the jet exhaust downstream of the nozzle. The thrust of the turbofan is a combination of the thrust produced by the fan blades and jet from the exhaust nozzle. Consequently, the efficiency of a turbofan engine is better than that of a turbojet. This efficiency is denoted by the thrust-specific fuel consumption TSFC (see Sec. 6.12). For a typical turbojet, TSFC = 1.0 lb of fuel per pound of thrust per hour; for a typical turbofan, TSFC = 0.6 lb of fuel per pound of thrust per hour, a much better figure. This is why all modern commercial jet transports, such as the Boeing 747 and the McDonnell-Douglas DC-10, are equipped with turbofan engines. A photograph of a turbofan is given in Figure 9.18*a*, and a cutaway view of the same engine is shown in Figure 9.18*b*.

Of course, a further extension of this concept replaces the ducted fan and outer shroud with an out-and-out propeller, with the turbine driving both the compressor and the propeller. Such a combination is called a *turboprop*, where appoximately 85 percent of the thrust comes from the propeller and the remaining 15 percent comes from the jet exhaust. Turboprops are efficient power plants that have found application in the 300–500 mi/h range; one prime example is the Lockheed Electra transport of the 1950s.

(a)

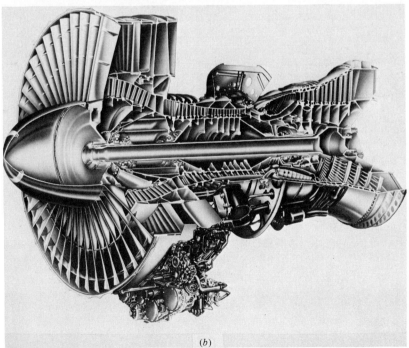

(b)

Figure 9.18 (*a*) The Pratt and Whitney JT9D turbofan engine. (*b*) A cutaway view. (*Pratt and Whitney Aircraft.*)

9.7 THE RAMJET ENGINE

Let us now move in the opposite direction from the last section. Instead of adding fans and propellers to a turbojet, let us get rid of *all* rotating machinery; i.e., consider the straight-through duct sketched in Figure 9.19, where air is inducted through the inlet at velocity V_∞, decelerated in the diffuser (point 1 to point 2), burned in a region where fuel is injected (point 2 to point 3), and then blasted out the exhaust nozzle at very high velocity V_e (point 3 to point 4). Such a simple device is called a *ramjet* engine. A cutaway drawing of a ramjet engine is shown in Figure 9.20. Because of their simplicity and high thrust, ramjets have always tickled the imagination of aerospace engineers. However, due to some serious drawbacks, they have not yet been used as a prime propulsive mechanism on a manned aircraft. On the other hand, they are used on numerous guided missiles, and they appear to be the best choice for future hypersonic airplanes. For this reason, let us examine ramjets more closely.

The ideal ramjet process is shown in the *p-v* diagram of Figure 9.21. All the compression from p_1 to p_2 takes place in the diffuser; i.e., a ramjet compresses the air by simply "ramming" through the atmosphere. Obviously, the compression ratio p_2/p_1 is a function of flight Mach number. In fact, to enhance combustion, the airflow entering the combustion zone is at a low subsonic Mach number; hence, assuming that $M_2 \approx 0$, then $p_2 \approx p_0$ (total pressure), and from Eq. (4.74),

$$\frac{p_2}{p_1} \approx \left(1 + \frac{\gamma - 1}{2} M_\infty^2\right)^{\gamma/(\gamma-1)} \tag{9.26}$$

The air decelerates isentropically in the diffuser, hence the compression from p_1 to p_2 follows the isentrope shown in Figure 9.21. Fuel is injected into the air at the end of the diffuser, and combustion takes place, stabilized by mechanical flame holders. This combustion is at constant pressure, hence the specific volume increases from v_2 to v_3. Then the hot, high-pressure gas is expanded isentropically through the exhaust nozzle, with the pressure dropping from p_3 to p_4.

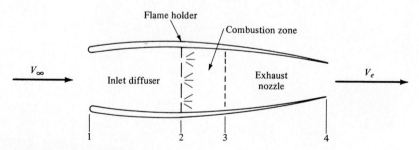

Figure 9.19 Ramjet engine.

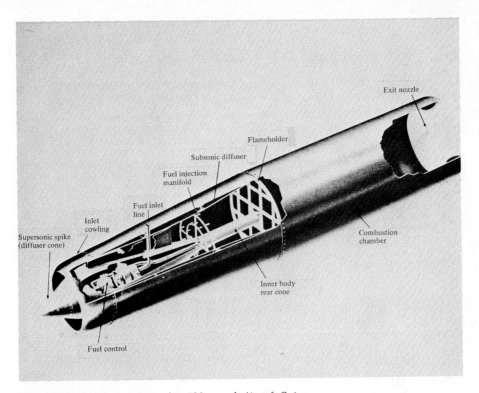

Figure 9.20 A typical ramjet engine. (*Marquardt Aircraft Co.*)

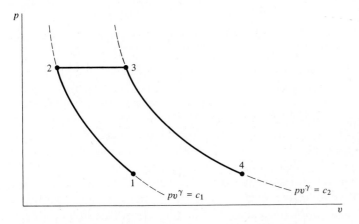

Figure 9.21 Pressure-specific volume diagram for an ideal ramjet.

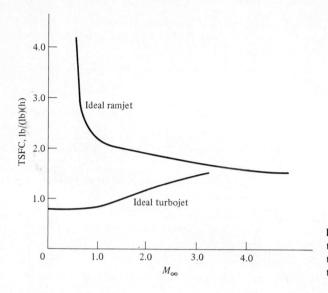

Figure 9.22 Comparison of thrust-specific fuel consumption for ideal ramjet and turbojet engines.

One disadvantage of a ramjet is immediately obvious from the above discussion, and especially from Eq. (9.26): in order to start and operate, the ramjet must already be in motion. Otherwise, there would be no compression in the diffuser; i.e., from Eq. (9.26), $p_2/p_1 = 1$ when $M_\infty = 0$. Therefore, all ramjet-powered vehicles must be launched by some independent mechanism (a catapult, or rockets) or must have a second engine of another type to develop enough flight speed to start the ramjet. At subsonic flight speeds, ramjets have another disadvantage. Although they produce high thrust, their subsonic efficiency is very low; typically TSFC $\approx$ 3 to 4 lb of fuel per pound of thrust per hour for ramjets at subsonic speeds. However, as shown in Figure 9.22, TSFC decreases to 2 or less at supersonic speeds.

Indeed, Figure 9.22 implicitly shows an advantage of ramjets for supersonic flight. At supersonic Mach numbers, TSFCs for turbojets and ramjets are somewhat comparable. Moreover, the curve for turbojets in Figure 9.22 is terminated at Mach 3 for a specific reason. In order to operate at higher Mach numbers, the turbojet must increase its combustion temperature. However, there is a material limitation. If the gas temperature leaving the turbojet combustor and entering the turbine is too hot, the turbine blades will melt. This is a real problem: the high-temperature material properties of the turbine blades limit the conventional turbojet to comparatively low to moderate supersonic Mach numbers. On the other hand, a ramjet has no turbine; therefore, its combustion temperatures can be much higher, and a ramjet can zip right into the high Mach number regime. Therefore, for sustained and efficient atmospheric flight at Mach numbers above 3 or 4, a ramjet is virtually the only choice, given our present-day technology.

Starting with Figure 9.19, we have described a conventional ramjet as a device which takes in the air at the inlet and diffuses it to a low subsonic Mach number before it enters the combustion zone. Consider this ramjet flying at $M_\infty = 6$. As a companion to Eq. (9.26), the temperature ratio T_2/T_1 can be estimated from Eq. (4.73) as

$$\frac{T_2}{T_1} \approx 1 + \frac{\gamma - 1}{2}M_\infty{}^2 \qquad (9.27)$$

(Note that the symbol T is used for both thrust and temperature; however, from the context, there should be no confusion.) If $M_\infty = 6$, Eq. (9.27) gives $T_2/T_1 \approx$ 7.9. If the ambient temperature $T_\infty = T_1 = 300$ K, then $T_2 = 2370$ K $= 4266°$R. At such high temperatures, the walls of the ramjet will tend to fail structurally. Thus, like turbojets, conventional ramjets are also limited by material problems, albeit at higher flight Mach numbers. Moreover, if the temperature of the air entering the combustor is too high, when the fuel is injected, it will be decomposed by the high temperatures rather than be burned; i.e., the fuel will absorb rather than release energy, and the engine will become a drag machine rather than a thrust-producing device. Clearly, for hypersonic flight at very high Mach numbers, something else must be done. This problem has led to the concept of a *supersonic combustion ramjet*, the *SCRAMjet*. Here, the flow entering the diffuser is at high mach number, say $M_1 = M_\infty = 6$. However, the diffuser decelerates the airflow only enough to obtain a reasonable pressure ratio p_2/p_1; the flow is still supersonic upon entering the combustor. Fuel is added to the supersonic stream, where "supersonic combustion" takes place. In this way, the flow field throughout the SCRAMjet is completely supersonic; in turn, the static temperature remains relatively low, and the material and decomposition problems associated with the conventional ramjet are circumvented. Therefore, the power plant for a hypersonic transport in the future will most likely be a SCRAMjet. Research on such devices is now in process. Indeed, SCRAMjet research constitutes the very frontier of propulsion research today. One such example is the SCRAMjet design concept pioneered by the NASA Langley Research Center since the mid-1960s. Intended for application on a hypersonic transport, the Langley SCRAMjet consists of a series of side-by-side modules blended with the underside of the airplane, as sketched in the upper right corner of Figure 9.23. The forward portion of the underside of the airplane acts as a compression surface; i.e., the air flowing over the bottom surface is compressed (pressure is increased) when it passes through the shock wave from the nose of the vehicle. The configuration of an individual module is shown in the middle of Figure 9.23. The compressed air from the bottom surface enters an inlet, where it is further compressed by additional shock waves from the leading edge of the inlet. This compressed air, still at supersonic velocity, subsequently flows over three struts, where H_2 is injected into the supersonic stream. A cross section of the struts is shown at the bottom left of Figure 9.23. Combustion takes place downstream of the struts. The burned gas mixture is then expanded through a nozzle at the rear of each module. The flow is further expanded over the smooth underbody at the rear of the

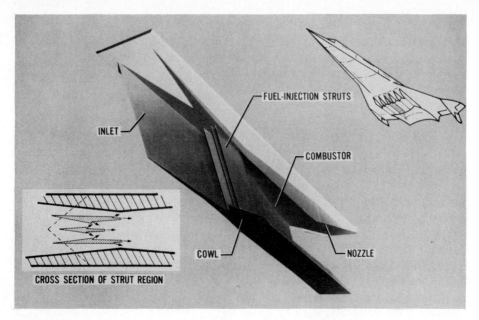

Figure 9.23 A concept for an airframe-integrated SCRAMjet engine, developed at NASA Langley Research Center. (*NASA.*)

airplane, which is intentionally contoured to act as an extension of the engine nozzles. For all practical purposes, the entire undersurface of the complete airplane represents the whole SCRAMjet engine—hence the concept is called an *airframe-integrated* SCRAMjet.

9.8 THE ROCKET ENGINE

With the launching of *Sputnik I* on October 4, 1957, and with the subsequent massive space programs of the United States and the Soviet Union, the rocket engine came of age. The rocket is the ultimate high-thrust propulsive mechanism. With it, people have gone to the moon, and space vehicles weighing many tons have been orbited about the earth or sent to other planets in the solar system. Moreover, rockets have been used on experimental aircraft; the rocket-powered Bell X-1 was the first manned airplane to break the sound barrier (see Sec. 5.22), and the rocket-powered North American X-15 was the first manned hypersonic aircraft (see Sec. 5.23). Finally, almost all types of guided missiles, starting with the German V-2 in World War II, have been rocket-powered. With this in mind, let us examine the characteristics of a rocket engine.

All the propulsion engines discussed in previous sections have been air-breathing: the piston engine, turbojet, ramjet—all depend on the combustion of fuel with air, where the air is obtained directly from the atmosphere. In contrast, as sketched in Figure 9.24, the rocket engine carries both its fuel and oxidizer and is completely independent of the atmosphere for its combustion. Thus, the rocket can operate in the vacuum of space, where obviously the air-breathing engines cannot. In Figure 9.24, fuel and oxidizer are sprayed into the combustion chamber, where they burn, creating a high-pressure, high-temperature mixture of combustion products. The mixture velocity is low, essentially zero. Therefore, the combustion chamber in a rocket engine is directly analogous to the reservoir of a supersonic wind tunnel (see Sec. 4.12). Hence, the temperature and pressure in the combustion chamber are the total values T_0 and p_0, respectively. Also directly analogous to a supersonic wind tunnel, the products of combustion expand to supersonic speeds through the convergent-divergent rocket nozzle, leaving with an exit velocity of V_e. This exit velocity is considerably higher than that for jet engines; hence, by comparison, rocket thrusts are higher, but efficiencies are lower. Figure 9.25 shows a typical rocket engine.

The thrust of a rocket engine is obtained from Eq. (9.24), where $\dot{m}_{\text{air}} = 0$ and $\dot{m}$ is the total mass flow of the products of combustion, $\dot{m} = \dot{m}_{\text{fuel}} + \dot{m}_{\text{oxidizer}}$. Hence, for a rocket engine,

$$\boxed{T = \dot{m}V_e + (p_e - p_\infty)A_e} \qquad (9.28)$$

The exit velocity V_e is readily obtained from the aerodynamic relations in Chap. 4. Write the energy equation, Eq. (4.41), between the combustion chamber

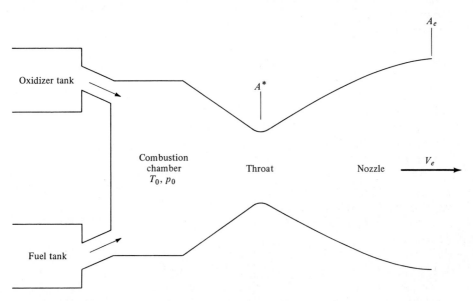

Figure 9.24 Schematic of a rocket engine.

Figure 9.25 The main rocket engine for the space shuttle. (*Rockwell International Corporation, Rocketdyne Corporation.*)

and the nozzle exit:

$$h_0 = h_e + \frac{V_e^2}{2} \tag{9.29}$$

$$c_p T_0 = c_p T_e + \frac{V_e^2}{2} \tag{9.30}$$

Solve Eq. (9.30) for V_e^2:

$$V_e^2 = 2c_p(T_0 - T_e) = 2c_p T_0 \left(1 - \frac{T_e}{T_0}\right) \tag{9.31}$$

The expansion through the nozzle is isentropic. Hence, from Eq. (4.36), $T_e/T_0 = (p_e/p_0)^{(\gamma-1)/\gamma}$. Also, from Eq. (4.69), $c_p = \gamma R/(\gamma - 1)$. Thus, Eq. (9.31) becomes

$$\boxed{V_e = \left\{\frac{2\gamma R T_0}{\gamma - 1}\left[1 - \left(\frac{p_e}{p_0}\right)^{(\gamma-1)/\gamma}\right]\right\}^{1/2}} \tag{9.32}$$

A comparative measure of the efficiency of different rocket engines can be obtained from the *specific impulse* I_{sp} defined as the thrust per unit weight flow at sea level:

$$I_{sp} \equiv \frac{T}{\dot{w}}$$

(9.33)

where $\dot{w} = g_0 \dot{m}$. (Recall that the weight is equal to the acceleration of gravity at sea level times the mass.) With this definition, the unit of I_{sp} in any consistent system of units is simply seconds. Furthermore, assume that the nozzle exit pressure is the same as the ambient pressure. Then, combining Eqs. (9.28) and (9.33), we get

$$I_{sp} = \frac{T}{\dot{w}} = \frac{T}{g_0 \dot{m}} = \frac{\dot{m} V_e}{g_0 \dot{m}} = \frac{V_e}{g_0}$$

(9.34)

Substitute Eq. (9.32) into Eq. (9.34), and note from chemistry that the specific gas constant R is equal to the universal gas constant $\bar{R}$ divided by the molecular weight $\bar{M}$; $R = \bar{R}/\bar{M}$.

$$I_{sp} = \frac{1}{g_0} \left\{ \frac{2\gamma R T_0}{\gamma - 1} \left[1 - \left(\frac{p_e}{p_0} \right)^{(\gamma - 1)/\gamma} \right] \right\}^{1/2}$$

(9.35)

Equation (9.35) is important. It tells what is necessary to have a high specific impulse; namely, the combustion temperature T_0 should be high and the molecular weight $\bar{M}$ should be low. The combustion temperature is primarily dictated by the chemistry of the oxidizer and fuel; a given combination, say oxygen and hydrogen, will burn at a specific T_0 called the adiabatic flame temperature, and this value of T_0 will be determined by the heat of reaction. The more highly reacting are the propellants, the higher is the T_0. Also, $\bar{M}$ is a function of the chemistry. If lightweight propellants are used, then $\bar{M}$ will be small. Therefore, outside of adjusting the oxidizer-to-fuel ratio (the O/F ratio), there is not much the engineer can do to change radically the I_{sp} for a given propellant combination. It depends primarily on the propellants themselves. However, Eq. (9.35) certainly tells us to choose a very energetic combination of lightweight propellants, as dramatized by the following tabulation.

Fuel-Oxidizer Combination	Adiabatic Flame Temperature, K	Average Molecular Weight of Combustion Products	I_{sp}, s
Kerosine-oxygen	3144	22	240
Hydrogen-oxygen	3517	16	360
Hydrogen-fluorine	4756	10	390

The kerosine-oxygen combination was used in the first stage of the Saturn 5 launch vehicle, which sent the Apollo astronauts to the moon; hydrogen-oxygen was used for the Saturn 5 second and third stages. However, the best combination is hydrogen-fluorine, which gives a specific impulse of 390 s, about the most we can expect from any propellant combination. Unfortunately, fluorine is extremely poisonous and corrosive and is therefore difficult to handle. Nevertheless, rocket engines using hydrogen-fluorine have been built.

Consider again the rocket engine schematic in Figure 9.24. We have discussed above that T_0 in the combustion chamber is essentially a function of the heat of reaction of the propellants, a chemical phenomenon. But what governs the chamber pressure p_0? The answer is basically the mass flow of propellants being pumped into the chamber from the fuel and oxidizer tanks, and the area of the nozzle throat, A^*. Moreover, we are in a position to prove this. From the continuity equation evaluated at the throat,

$$\dot{m} = \rho^* A^* V^* \tag{9.36}$$

where the superscript "*" denotes conditions at the throat. Recall from Chap. 4 that the velocity is sonic at the throat of a convergent-divergent supersonic nozzle; that is, $M^* = 1$. Thus, V^* is the speed of sound, obtained from Eq. (4.54) as

$$V^* = \sqrt{\gamma R T^*} \tag{9.37}$$

Also, from the equation of state,

$$\rho^* = \frac{p^*}{RT^*} \tag{9.38}$$

Substitute Eqs. (9.37) and (9.38) into (9.36):

$$\dot{m} = \frac{p^*}{RT^*} A^* \sqrt{\gamma R T^*} = \frac{p^* A^*}{\sqrt{RT^*}} \sqrt{\gamma} \tag{9.39}$$

Write Eqs. (4.73) and (4.74) between the combustion chamber and the throat:

$$\frac{T_0}{T^*} = 1 + \frac{\gamma - 1}{2} M^{*2} = 1 + \frac{\gamma - 1}{2} = \frac{\gamma + 1}{2} \tag{9.40}$$

$$\frac{p_0}{p^*} = \left(1 + \frac{\gamma - 1}{2} M^{*2}\right)^{\gamma/(\gamma - 1)} = \left(\frac{\gamma + 1}{2}\right)^{\gamma/(\gamma - 1)} \tag{9.41}$$

Substitute Eqs. (9.40) and (9.41) into (9.39):

$$\dot{m} = \sqrt{\frac{\gamma}{R}} A^* \left(\frac{\gamma + 1}{2}\right)^{-(\gamma+1)/2(\gamma-1)} \frac{p_0}{\sqrt{T_0}}$$

or

$$\dot{m} = \frac{p_0 A^*}{\sqrt{T_0}} \sqrt{\frac{\gamma}{R} \left(\frac{2}{\gamma + 1}\right)^{(\gamma+1)/(\gamma-1)}} \tag{9.42}$$

Equation (9.42) is important. It states that the mass flow through a nozzle that is choked, i.e., when sonic flow is present at the throat, is *directly proportional to p_0 and A^* and inversely proportional to the square root of T_0.* Moreover, Eq. (9.42) answers the previous question about how p_0 is governed in a rocket engine combustion chamber. For a given combination of propellants, T_0 is fixed by the chemistry. For a fixed nozzle design, A^* is a given value. Hence, from Eq. (9.42),

$$p_0 = (\text{const})(\dot{m})$$

If $\dot{m}$ is doubled, then p_0 is doubled, etc. In turn, since mass is conserved, $\dot{m}$ through the nozzle is precisely equal to $\dot{m}_{\text{fuel}} + \dot{m}_{\text{oxidizer}}$ being fed into the chamber from the propellant tanks. Therefore, we repeat again the conclusion that p_0 is governed by the mass flow of propellants being pumped into the chamber from the fuel and oxidizer tanks and the area of the nozzle throat.

Before we leave this discussion of rocket engines, we note the very restrictive assumption incorporated in such equations as Eqs. (9.32), (9.35), and (9.42), namely, that γ is constant. The real flow through a rocket engine is chemically reacting and is changing its chemical composition throughout the nozzle expansion. Consequently, γ is really a variable, and the above equations are not strictly valid. However, they are frequently used for preliminary design estimates of rocket performance, and γ is chosen as some constant mean value, usually between 1.2 and 1.3, depending on the propellants used. A more accurate solution of rocket nozzle flows taking into account the variable specific heats and changing composition must be made numerically and is beyond the scope of this book.

Example 9.4 Consider a rocket engine burning hydrogen and oxygen; the combustion chamber pressure and temperature are 25 atm and 3517 K, respectively. The area of the rocket nozzle throat is 0.1 m². The area of the exit is designed so that the exit pressure exactly equals ambient pressure at a standard altitude of 30 km. For the gas mixture, assume that $\gamma = 1.22$ and the molecular weight $\overline{M} = 16$. At a standard altitude of 30 km, calculate (a) specific impulse, (b) thrust, (c) area of the exit, and (d) flow Mach number at exit.

SOLUTION

(a) The universal gas constant, in SI units, is $\overline{R} = 8314$ J/(kg)(mole)(K). Hence, the specific gas constant is

$$R = \frac{\overline{R}}{\overline{M}} = \frac{8314}{16} = 519.6 \text{ J/(kg)(K)}$$

Thus, from Eq. (9.35)

$$I_{\text{sp}} = \frac{1}{g_0} \left\{ \frac{2\gamma \overline{R} T_0}{(\gamma - 1)\overline{M}} \left[1 - \left(\frac{p_e}{p_0} \right)^{(\gamma-1)/\gamma} \right] \right\}$$

$$= \frac{1}{9.8} \left\{ \frac{2(1.22)(8314)(3517)}{0.22(16)} \left[1 - \left(\frac{1.174 \times 10^{-2}}{25} \right)^{0.22/1.22} \right] \right\}$$

$$\boxed{I_{\text{sp}} = 397.9 \text{ s}}$$

Note that this value is slightly higher than the number tabulated in the previous discussion on specific impulse. The difference is that the tabulation gives I_{sp} for expansion to sea-level pressure, not the pressure at 30-km altitude as in this example.

(b) From Eq. (9.28),

$$T = \dot{m} V_e + (p_e - p_\infty) A_e$$

In this equation, at 30 km, $p_e = p_\infty$. Hence,

$$T = \dot{m} V_e \qquad \text{at 30-km altitude}$$

To obtain $\dot{m}$, use Eq. (9.42):

$$\dot{m} = \frac{p_0 A^*}{\sqrt{T_0}} \sqrt{\frac{\gamma}{R} \left(\frac{2}{\gamma + 1} \right)^{(\gamma + 1)/(\gamma - 1)}}$$

$$= \frac{25(1.01 \times 10^5)(0.1)}{\sqrt{3517}} \sqrt{\frac{1.22}{519.6} \left(\frac{2}{2.22} \right)^{2.22/0.22}}$$

$$= 121.9 \text{ kg/s}$$

To obtain V_e, recall that the nozzle flow is isentropic. Hence,

$$\frac{T_e}{T_0} = \left(\frac{p_e}{p_0} \right)^{(\gamma - 1)/\gamma}$$

$$T_e = 3517 \left(\frac{1.174 \times 10^{-2}}{25} \right)^{0.22/1.22}$$

$$= 3517(0.2517) = 885.3 \text{ K}$$

Also, from Eq. (4.69),

$$c_p = \frac{\gamma R}{\gamma - 1} = \frac{1.22(519.6)}{0.22} = 2881.4 \text{ J/(kg)(K)}$$

From the energy equation, Eq. (4.42),

$$c_p T_0 = c_p T_e + \frac{V_e^2}{2}$$

$$V_e = \sqrt{2 c_p (T_0 - T_e)} = \sqrt{2(2881.4)(3517 - 885.3)}$$

$$= 3894 \text{ m/s}$$

Thus, the thrust becomes

$$T = \dot{m} V_e = 121.9(3894) = \boxed{4.75 \times 10^5 \text{ N}}$$

Note that 1 N = 0.2247 lb. Hence

$$T = (4.75 \times 10^5)(0.2247) = \boxed{106{,}700 \text{ lb}}$$

(c) To obtain the exit area, use the continuity equation:

$$\dot{m} = \rho_e A_e V_e$$

To obtain the exit density, use the equation of state:

$$\rho_e = \frac{p_e}{R T_e} = \frac{1.1855 \times 10^3}{519.6(885.3)} = 2.577 \times 10^{-3} \text{ kg/m}^3$$

$$A_e = \frac{\dot{m}}{\rho_e V_e} = \frac{121.9}{(2.577 \times 10^{-3})(3894)}$$

$$\boxed{A_e = 12.14 \text{ m}^2}$$

(d) To obtain the exit Mach number,

$$a_e = \sqrt{\gamma R T_e} = \sqrt{1.22(519.6)(885.3)} = 749 \text{ m/s}$$

$$M_e = \frac{V_e}{a_e} = \frac{3894}{749} = \boxed{5.2}$$

9.9 HISTORICAL NOTE: EARLY PROPELLER DEVELOPMENT

The ancestry of the airplane propeller reaches as far back as the twelfth century, when windmills began to dot the landscape of western Europe. The blades of these windmills, which were essentially large wood and cloth paddles, extracted energy from the wind in order to power mechanical grinding mills. Only a small intellectual adjustment was necessary to think of this process in reverse—to power mechanically the rotating paddles in order to add energy to the air and produce thrust. Indeed, Leonardo da Vinci development a helical screw for a sixteenth century helicopter toy. Later, a year after the first successful balloon flight in 1783 (see Chap. 1), a hand-driven propeller was mounted to a balloon by J. P. Blanchard. This was the first propeller to be truly airborne, but it did not succeed as a practical propulsive device. Nevertheless, numerous other efforts to power hot-air balloons with hand-driven propellers followed, all unsuccessful. It was not until 1852 that a propeller connected to a steam engine was successfully employed in an airship. This combination, designed by Henri Giffard, allowed him to guide his airship over Paris at a top speed of 5 mi/h.

As mentioned in Chap. 1, the parent of the modern airplane, George Cayley, eschewed the propeller and instead put his faith mistakenly in oarlike paddles for propulsion. However, Henson's aerial steam carriage (see Figure 1.11) envisioned two pusher propellers for a driving force, and after that, propellers became the accepted propulsion concept for heavier-than-air vehicles. Concurrently, in a related fashion, the marine propeller was developed for use on steamships beginning in the early nineteenth century. Finally, toward the end of that century, the propeller was employed by Du Temple, Mozhaiski, Langley, and others in their faltering efforts to get off the ground (see Figures 1.13, 1.14, and 1.18, for example).

However, a close examination of these nineteenth century aircraft reveals that the propellers were crude, wide, paddlelike blades which reflected virtually no understanding of propeller aerodynamics. Their efficiencies must have been exceedingly low, which certainly contributed to the universal failure of these machines. Even marine propellers, which had been extensively developed by 1900 for steamships, were strictly empirical in their design and at best had efficiencies on the order of 50 percent. There existed no rational hydrodynamic or aerodynamic theory for propeller design at the turn of the century.

This was the situation when Wilbur and Orville returned from Kill Devil Hills in the fall of 1902, flushed with success after more than 1000 flights of their no. 3 glider (see Chap. 1) and ready to make the big step to a powered machine. Somewhat naively, Wilbur originally expected this step to be straightforward; the engine could be ordered from existing automobile companies, and the propeller could be easily designed from existing marine technology. Neither proved to be the case. After spending several days in Dayton libraries, Wilbur discovered that a theory for marine propellers did not exist and that even an appreciation for their true aerodynamic function had not been developed. So once again the Wright brothers, out of necessity, had to plunge into virgin engineering territory. Throughout the winter of 1902–1903, they wrestled with propeller concepts in order to provide accurate calculations for design. And once again they demonstrated that, without the benefit of a formal engineering education, they were the premier aeronautical engineers of history. For example, by early spring of 1903, they were the first to recognize that a propeller is basically a rotating wing, made up of airfoil sections that generated an aerodynamic force normal to the propeller's plane of rotation. Moreover, they made use of their wind-tunnel data, obtained the previous year on several hundred different airfoil shapes, and chose a suitably cambered shape for the propeller section. They reasoned the necessity for twisting the blade in order to account for the varying relative airflow velocity from the hub to the tip. Indeed, in Orville's words:

> It is hard to find even a point from which to start, for nothing about a propeller, or the medium in which it acts, stands still for a moment. The thrust depends upon the speed and the angle at which the blade strikes the air; the angle at which the blade strikes the air depends upon the speed at which the propeller is turning, the speed the machine is traveling forward, and the speed at which the air is slipping backward; the slip of the air backward depends upon the thrust exerted by the propeller and the amount of air acted upon. When any of these changes, it changes all the rest, as they are all interdependent upon one another. But these are only a few of the factors that must be considered....

By March of 1903, Wilbur had completed his theory to the extent that a propeller could be properly designed. Using a hatchet and drawknife, he carved two propellers out of laminated spruce and surfaced them with aluminum paint. Excited about their accomplishment, Orville wrote: "We had been unable to find anything of value in any of the works to which we had access, so we worked out a theory of our own on the subject, and soon discovered, as we usually do, that all the propellers built heretofore are *all wrong*, and then built a pair... based on our theory, which are *all right*!"

The propeller designed by the Wright brothers, principally by Wilbur, achieved the remarkably high efficiency of 70 percent and was instrumental in their successful flight on December 17, 1903, and in all flights thereafter. Moreover, their propellers remained the best in aviation for almost a decade. Indeed, until 1908, all other competitors clung to the older, paddlelike blades, both in the United States and in Europe. Then, when Wilbur made his first dramatic public flight on August 8, 1908, at Hunaundières, France, the impact of his highly efficient propeller on the European engineers was almost as great as that of the Wrights' control system, which allowed smoothly maneuverable flight.

As a result, subsequent airplanes in Europe and elsewhere adopted the type of aerodynamically designed propeller introduced by the Wrights.

Consequently, credit for the first properly designed propeller, along with the associated aerodynamic theory, must go to the Wright brothers. This fact is not often mentioned or widely recognized; however, this propeller research in 1903 represented a quantum jump in a vital area of aeronautical engineering, without which practical powered flight would have been substantially delayed.

The final early cornerstone in the engineering theory and design of airplane propellers was laid by William F. Durand about a decade after the Wright brothers' design was adopted. Durand was a charter member of the NACA and became its chairman in 1916 (see Sec. 2.6). Durand was also the head of the mechanical engineering department at Stanford University at that time, and during 1916–1917 he supervised the construction of a large wind tunnel on campus designed purely for the purpose of experimenting with propellers. Then, in 1917, he published NACA report no. 14 entitled "Experimental Research on Air Propellers." This report was the most extensive engineering publication on propellers to that date; it contained experimental data on numerous propellers of different blade shapes and airfoil sections. It is apparently the first technical report to give extensive plots of propeller efficiency vs. advance ratio. Hence, the type of efficiency curve sketched in Figure 9.5 dates back as far as 1917! Moreover, the values of maximum efficiency of most of Durand's model propellers were 75 to 80 percent, a creditable value for that point in history. It is interesting to note that, 60 years later, modern propeller efficiencies are not that much better, running between 85 and 90 percent. To Durand must also go the credit for the first dimensional analysis in propeller theory; in the same NACA reports, he shows by dimensional analysis that propeller efficiency must be a function of advance ratio, Reynolds number, and Mach number, and he uses these results to help correlate his experimental data. This early NACA report was an important milestone in the development of the airplane propeller. Indeed, a copy of the report itself is enshrined behind glass and is prominently displayed in the lobby of the new Durand Engineering Building on the Stanford campus.

9.10 HISTORICAL NOTE: EARLY DEVELOPMENT OF THE INTERNAL COMBUSTION ENGINE FOR AVIATION

The pivotal role of propulsion in the historical quest for powered flight was discussed in Chap. 1. The frustrating lack of a suitable prime mover was clearly stated as far back as 1852 by George Cayley, who wrote about his trials with a "governable parachute" (glider): "It need scarcely be further remarked, that were we in possession of a sufficiently light first mover to propel such vehicles by waftage, either on the screw principle or otherwise, with such power as to supply that force horizontally, which gravitation here supplies in the decent, mechanical aerial navigation would be at our command without further delay."

Indeed, Cayley devoted a great deal of thought to the propulsion problem. Before 1807, he had conceived the idea for a hot-air engine, in which air is drawn from the atmosphere, heated by passing it over a fire, and then expanded into a cylinder, doing work on a piston. This was to be an alternative to steam power. Considering his invention in a general sense, and not mentioning any possible application to flight, Cayley wrote in the October 1807 issue of Nicholson's *Journal* that "the steam engine has hitherto proved too weighty and cumbrous for most purposes of locomotion; whereas the expansion of air seems calculated to supply a mover free from these defects." Later, in 1843, Cayley summarized his work on aeronautical propulsion in a type of letter to the editor in *Mechanics Magazine*:

> The real question rests now, as it did before, on the possibility of providing a sufficient power with the requisite lightness. I have tried may different engines as first movers, expressly for this purpose [flight]. Gun powder is too dangerous, but would, at considerable expense, effect the purpose: but who would take the double risk of breaking their neck or being blown to atoms? Sir Humphrey Davy's plan of using solid carbonic acid, when again expanded by heat, proved a failure in the hands of our most ingenious engineer, Sir M. Isambard Brunel.
>
> As all these processes require nearly the same quantity of caloric to generate the same degree of power, I have for some time turned my own attention to the use, as a power, of common atmospheric air expanded by heat, and with considerable success. A five-horse engine of this sort was shown at work to Mr. Babbage, Mr. Rennie, and many other persons capable of testing its efficiency, about three years ago. The engine was only an experimental one, and had some defects, but each horse power was steadily obtained by the combustion of about $6\frac{1}{2}$ pounds of coke per hour, and this was the whole consumption of the engine, no water being required. Another engine of this kind, calculated to avoid the defects of the former one, is now constructing, and may possibly come in aid of balloon navigation—for which it was chiefly designed—or the present project, if no better means be at hand.

Thus, in keeping with his remarkable and pioneering thinking on all aspects of aviation, George Cayley stated the impracticality of steam power for flight and clearly experimented with some forerunners of the IC engine. However, his thoughts were lost to subsequent aeronautical engineers of the nineteenth century, who almost universally attempted steam-powered flight (see Chap. 1).

The development of IC engines gained momentum with Lenoir's two-cycle gas-burning engine in 1860. Then, in 1876, Nikolaus August Otto designed and built the first successful four-stroke IC engine, the same type of engine discussed in Sec. 9.3. Indeed, the thermodynamic cycle illustrated in Figure 9.11, consisting of isentropic compression and power strokes with constant-volume combustion, is called the *Otto cycle*. Although Otto worked in Germany, strangely enough, in 1877 he took out a U.S. patent on his engine. Otto's work was soon applied to land vehicle propulsion, heralding the birth of the automobile industry before 1900.

But automobiles and airplanes are obviously two different machines, and IC engines used in automobiles in 1900 were too heavy per horsepower for aeronautical use. One man who squarely faced this barrier was Samuel Pierpont Langley (see Sec. 1.7). He correctly recognized that the gasoline-burning IC engine was the appropriate power plant for an airplane. To power the newer versions of his Aerodromes, Langley contracted with Stephen M. Balzer of New York in 1898 for an engine of 12 hp weighing no more than 100 lb. Unfortunately, Balzer's delivered product, which was derived from the automobile engine, could produce only 8 hp. This was unacceptable, and Charles Manly, Langley's assistant, took the responsibility for a complete redesign of Balzer's engine in the laboratory of the Smithsonian Institution in Washington, D.C. The net result was a power plant, finished in 1902, which could produce 52.4 hp while weighing only 208 lb. This was a remarkable achievement; it was not bettered until the advent of "high-performance" aircraft toward the end of World War I, 16 years later. Moreover, Manly's engine was a major departure from existing automobile engines of the time. It was a radial engine, with five cylinders equally spaced in a circular pattern around a central crankshaft. It appears to be the first aircraft radial engine in history, and certainly the first successful one. Unfortunately, the failure of Langley's Aerodromes in 1903 obscured the quality of Manly's engine, although the engine was in no way responsible for these failures.

Five hundred miles to the west, in Dayton, Ohio, the Wright brothers also originally planned to depend on a standard automobile engine for the power plant for their Flyer. In the fall of 1902, after their stunning success with their no. 3 glider at Kill Devil Hills, the Wrights were rudely surprised to find that no automobile engine existed that was light enough to meet their requirement. Because Wilbur had taken the prime responsibility of developing a propeller (see Sec. 9.9) during this time, he assigned Orville the task of designing and building a suitable engine. It is interesting to note that Wilbur correctly considered the propeller to be a more serious problem than the engine. With the help of Charles Taylor, a mechanic who worked in the Wrights' bicycle shop, and using as a model the car engine of a Pope-Toledo (long since defunct), Orville expeditiously completed his engine design and construction in less than six weeks. In its first test in February 1903, the aluminum crankcase cracked. Two months later a local foundry finished casting a second case, and the engine was finally successfully tested in May.

The engine was a four-cylinder in-line design. It had only one speed, about 100 rpm, and could be stopped only by cutting off the supply of gasoline, which was fed to the cylinders by gravity. The engine produced 12 hp and weighed (without oil and fuel) about 100 lb. Although the Wrights' engine produced far less horsepower per pound of engine weight than Manly's design, it was nevertheless adequate for its purpose. The Wright brothers had little experience with IC engines before 1903, and their successful design is another testimonial to their unique engineering talents. In Orville's words: "Ignorant of what a motor this size ought to develop, we were greatly pleased with its performance. More experience showed us that we did not get one-half the power we should have had."

The Wright brothers' engine was obviously the first successful aircraft power plant to fly, by virtue of their history-making flight of December 17, 1903 (see Sec. 1.1). Subsequent development of the IC engine for airplanes came slowly. Indeed, nine years later, Captain H. B. Wild, speaking in Paris, gave the following pilot-oriented view of the aircraft engine:

> The comparatively crude and unreliable motor that we have at our disposal at the present time [1912] is no doubt the cause of many of the fatalities and accidents befalling the aeroplane. If one will look over the accessories attached to the aero engine of today, it will be noted that it is stripped clean of everything possible which would eliminate what he deems unnecessary parts in order to reduce the weight of the engine, and in doing so he often takes away the parts which help to strengthen the durability and reliability of the motor.

The eventual successful development of efficient, reliable, and long-endurance aircraft power plants is now a fact of history. However, it was accomplished only by an intensive and continuous engineering effort. Various reports on engine development—carburators, valves, radiators, etc.—perfuse the early NACA literature. Indeed, the recognition of the importance of propulsion was made clear in 1940 with the establishment of a complete laboratory for its research and development, the NACA Lewis Flight Propulsion Laboratory in Cleveland, Ohio.

The internal combustion reciprocating engine has now been supplanted by the gas turbine jet engine as the main form of aeronautical propulsion. However, IC engines are still the most appropriate choice for general aviation aircraft designed for speeds of 300 mi/h or less, and therefore their continued development and improvement will remain an important part of aerospace engineering.

9.11 HISTORICAL NOTE: INVENTORS OF THE EARLY JET ENGINES

By the late 1920s, the reciprocating-engine–propeller combination was so totally accepted as *the* means of airplane propulsion that other concepts were generally discounted. In particular, jet propulsion was viewed as technically infeasible. For example, the NACA reported in 1923 that jet propulsion was "impractical," but their studies were aimed at flight velocities of 250 mi/h or less, where jet propulsion is truly impractical. Eleven years later, the British Government still held a similar opinion.

Into this environment came Frank Whittle (now Sir Frank Whittle). Whittle is an Englishman, born on June 1, 1907, in Coventry. As a young boy, he was interested in aviation, and in 1923 he enlisted in the Royal Air Force. Showing much intelligence and promise, he soon earned a coveted student's slot at the RAF technical college at Cranwell. It was here that Whittle became interested in the possibilities of gas turbine engines for propelling airplanes. In 1928, he wrote a senior thesis at Cranwell entitled "Future Developments in Aircraft Design," in which he expounded the virtues of jet propulsion. It aroused little interest. Undaunted, Whittle went on to patent his design for a gas turbine engine in January 1930. For the next five years, in the face of polite but staunch disinterest, Whittle concentrated on his career in the RAF and did little with his ideas on jet propulsion. However, in 1935, with the help of a Cranwell classmate, a firm of bankers agreed to finance a private company named Power Jets Ltd. specifically to develop the Whittle jet engine. So in June 1935, Frank Whittle and a small group of colleagues plunged into the detailed design of what they thought would be the first jet engine in the world. Indeed, the engine was finished in less than two years and was started up on a test stand on April 12, 1937—the first jet engine in the world to successfully operate in a practical fashion.

However, it was not the first to fly. Quite independently, and completely without the knowledge of Whittle's work, Hans von Ohain in Germany developed a similar gas turbine engine. Working under the private support of the famous airplane designer Ernst Heinkel, von Ohain started his work in 1936. (As in the United States and England, the German government showed little initial interest in jet propulsion.) Three years after his work began, von Ohain's engine was mated with a specially designed Heinkel airplane. Then, on August 28, 1939, the He 178 (see Figure 9.26) became the first gas turbine–powered, jet-propelled airplane in history to fly. It was strictly an experimental aircraft, but von Ohain's engine of 838 lb of thrust pushed the He 178 to a maximum speed of 435 mi/h. Later, after the beginning of World War II, the German government reversed its lack of interest in jet propulsion, and soon Germany was to become the first nation in the world with operational military jet aircraft.

Meanwhile, in England, Whittle's success in operating a jet engine on a test stand finally overcame the Air Ministry's reluctance, and in 1938 a contract was let to Power Jets Ltd. to develop a revised power plant for installation in an airplane. Simultaneously, Gloster Aircraft received a contract to build a specially designed jet-propelled aircraft. Success was obtained when the Gloster E.28/39 airplane (see Figure 9.27) took off from Cranwell on May 15, 1941, the first airplane to fly with a Whittle jet engine. The engine produced 860 lb of thrust and

Figure 9.26 The German He 178—the first jet-propelled airplane in the world to fly successfully.

Figure 9.27 The Gloster E.28—the first British airplane to fly with jet propulsion.

powered the Gloster airplane to a maximum speed of 338 mi/h. The Gloster E.28/39 now occupies a distinguished berth in the Science Museum in London, hanging prominently from the top-floor ceiling of the massive brick building in South Kensington, London. The technology gained with the Whittle engine was quickly exported to the United States and eventually fostered the birth of the highly successful Lockheed P-80 Shooting Star, the first U.S. production-line jet airplane.

Today, after being knighted for his contributions to British aviation, Sir Frank Whittle lives the life of a retired RAF air commodore in Annapolis, Maryland, where he works and teaches on a part-time basis at the U.S. Naval Academy. On the other hand, Hans von Ohain continued in the propulsion business. Brought to the United States at the end of the World War II, von Ohain pursued a distinguished career at the Air Force's Aeronautical Research Laboratory at Wright-Patterson Air Force Base, Ohio, where he led a propulsion group doing research on advanced concepts. Indeed, the present author had the privilege of working for three years in the same laboratory with von Ohain and shared numerous invigorating conversations with this remarkable man. More recently, von Ohain became affiliated with the U.S. Air Force Aeropropulsion Laboratory at Wright Field, from which he retired in 1980.

9.12 HISTORICAL NOTE: EARLY HISTORY OF THE ROCKET ENGINE

"When it was lit, it made a noise that resembled thunder and extended 100 li [about 24 km]. The place where it fell was burned, and the fire extended more than 2000 feet.... These iron nozzles, the flying powder halberds that were hurled, were what the Mongols feared most." These words were written by Father Antonine Gaubil in 1739 in conjunction with his book on Genghis Khan; they describe how a Chinese town in 1232 successfully defended itself against 30,000 invading Mongols by means of rocket-propelled fire arrows. They are an example of the evidence used by most historians to show that rocketry was born and developed in the Orient many centuries ago. It is reasonably clear that the Chinese manufactured black powder at least as early as 600 A.D. and subsequently used this mixture of charcoal, sulfur, and saltpeter as a rocket propellant. Over the centuries, the rocket slowly spread to the West as a military weapon and was much improved as a barrage missile by Sir William Congreve in England in the early 1800s. (The "rocket's red glare" observed by Francis Scott Key in 1812 at Fort McHenry was produced by a Congreve rocket.) However, not until the end of the nineteenth and the beginning of the twentieth centuries was the rocket understood from a technical viewpoint and was its true engineering development begun.

The Soviet Union was first into space, both with an artificial satellite (*Sputnik I* on October 4, 1957) and with a human in orbit (Yuri Gagarin on April 12, 1961). Thus, in historical perspective, it is fitting that the first true rocket scientist was a Russian—Konstantin Eduardovitch Tsiolkovsky, born in September 1857 in the town of Izhevskoye. As a young student, he absorbed physics and mathematics and was tantalized by the idea of interplanetary space travel. In 1876, he became a school teacher in Borovsk, and in 1882 moved to the village of Kaluga. There in virtual obscurity, he worked on theories of space flight, hitting upon the idea of reactive propulsion in March 1883. Working without any institutional support, Tsiolkovsky gradually solved some of the theoretical problems of rocket engines. Figure 9.28 shows his design of a rocket, fueled with liquid hydrogen (H_2) and liquid oxygen (O_2), which was published in the Russian magazine *Science Survey* in 1903 (the same year as the Wright brothers' successful first powered airplane flight). The fact that Tsiolkovsky knew to use the high-specific-impulse combination of H_2-O_2 testifies to the sophistication of his rocket theory. Tsiolkovsky was neither an experimentalist (it took money that he did not have to develop a laboratory) nor an engineer. Therefore, he conducted no practical experiments nor generated any design data. Nevertheless, Tsiolkovsky was the first true rocket scientist, and he worked incessantly on his theories until his death on September 19, 1935, at the age of 78 years. In his later life, his contributions were finally recognized, and he became a member of the Socialist Academy (forerunner of the U.S.S.R. Academy of Science) in 1919, with a subsequent grant of a government pension.

At the turn of the century, progress in rocketry arrived in the United States in the form of Dr. Robert H. Goddard. Goddard was born at Worcester, Massachusetts, on October 5, 1882. His life had many parallels to Tsiolkovsky's: he too was an avid physicist and mathematician, he too was convinced that rockets were the key to space flight, and he too worked in virtual obscurity for most of his life. But there was one sharp difference. Whereas Tsiolkovsky's contributions were purely theoretical, Goddard successfully molded theory into practice and developed the world's first liquid-fueled rocket that worked.

Figure 9.28 Tsiolkovsky's rocket design of 1903, burning liquid hydrogen (H) and liquid oxygen (O).

Goddard was educated completely at Worcester, graduating from South High School in 1904, obtaining a bachelor's degree from Worcester Polytechnic Institute in 1908, and earning a doctorate in physics at Clark University in 1911. Subsequently, he became a professor of physics at Clark, where he began seriously to apply science and engineering to his childhood dreams of space flight. He too determined that liquid H_2 and O_2 would be very efficient rocket propellants, and he pursued these ideas during a leave of absence at Princeton University during 1912–1913. In July 1914, he was granted patents on rocket combustion chambers, nozzles, propellant feed systems, and multistage rockets. In 1917, he obtained a small grant ($5000) from the Smithsonian Institution in Washington, which permanently entrenched him in a career of rocketry. This grant led to one of the most historic documents of rocket engine history, a monograph entitled *A Method of Reaching Extreme Altitudes*, published as part of the Smithsonian Miscellaneous Collections in 1919. This book was a scholarly and authoritative exposition of rocket principles, but at that time few people seized upon Goddard's ideas.

Goddard increased his laboratory activities back at Worcester in the early 1920s. Here, after many tests and much engineering development, Goddard successfully launched the world's first liquid-fuel rocket on March 16, 1926. A picture of Goddard standing beside this rocket is shown in Figure 9.29. The vehicle was 10 ft long; the motor itself was at the very top (far above Goddard's head in Figure 9.29) and was fed liquid oxygen and gasoline though two long tubes which led from the propellant tanks at the rear of the vehicle (below Goddard's arm level in the figure). The conical nose on the fuel tanks was simply a deflector to protect the tanks from the rocket nozzle exhaust. This rocket reached a maximum speed of 60 mi/h and flew 184 ft. Although modest in performance, this flight on March 16, 1926, was to rocketry what the Wright brothers' December 17, 1903, flight was to aviation.

This work ultimately brought Goddard to the attention of Charles A. Lindbergh, who now had considerable stature due to his 1927 trans-Atlantic flight. Lindbergh was subsequently able to convince the Daniel Guggenheim Fund for the Promotion of Aeronautics to give Goddard a $50,000 grant to further pursue rocket engine development. Suddenly, Goddard's operation mag-

Figure 9.29 Robert H. Goddard and his first successful liquid-fuel rocket. This rocket made the world's first successful flight on March 16, 1926. (*National Air and Space Museum.*)

nified, and in 1930 he and his wife moved to a more suitable testing location near Roswell, New Mexico. Here, for the next 11 years, Goddard made bigger and better rockets, although still in an atmosphere of obscurity. The government was simply not interested in any form of jet propulsion research during the 1930s. Also, Goddard was somewhat from the same mold as the Wright brothers; he imposed a blanket of secrecy on his data for fear of others pirating his designs. However, at the beginning of World War II, the government's interest in Goddard's work turned from cold to hot, and his complete operation, personnel and facilities, was moved to the Naval Engineering Experiment Station at Annapolis, Maryland. Here, until July 1945, this group developed jet-assisted takeoff units for seaplanes and worked on a variable-thrust rocket engine.

On August 10, 1945, Dr. Robert H. Goddard died in Baltimore. Recognition for his contributions and realization of their importance to the development of modern rocketry came late. Indeed, only in the political heat of the post-*Sputnik* years did the United States really pay homage to Goddard. In 1959 he was honored by Congress; the same year he received the first Louis W. Hill Space Transportation Award of the Institute of Aeronautical Sciences (now the American Institute of Aeronautics and Astronautics). Also, on May 1, 1959, the new NASA Goddard Space Flight Center at Greenbelt, Maryland, was named in his honor. Finally, in 1960, the Guggenheim Foundation and Mrs. Goddard were given $1,000,000 by the government for use of hundreds of Goddard's patents.

During the 1930s, and completely independent of Goddard's operation, another small group in the United States developed rockets. This was the American Rocket Society (ARS), originally founded in March 1930 as the American Interplanetary Society and changing its name in 1934. This was a small group of scientists and engineers who believed in the eventual importance of rocketry. The society not only published technical papers, it also built and tested actual vehicles. Its first rocket, burning liquid oxygen and gasoline, was launched on May 14, 1933, at Staten Island, New York, and reached 250 ft. Following this, and up to World War II, the ARS was a public focal point for small rocket research and development, all without government support. After the beginning of the war, much of the ARS experimental activity was splintered and absorbed by other activities around the country. However, as an information dissemination society, the ARS continued until 1963, publishing the highly respected *ARS Journal*. Then, the American Rocket Society and the Institute of Aeronautical (by that time, Aerospace) Sciences were merged to form the present American Institute of Aeronautics and Astronautics.

As a brief example of how the threads of the history of flight are woven together, in 1941 members of the ARS formed a company, Reaction Motors, Inc., which went on to design and build the XLR-11 rocket engine. This engine powered the Bell X-1 and pilot Chuck Yeager to the first manned supersonic flight on October 14, 1947 (see Sec. 5.22 and Figure 5.54).

The early history of rocket engines forms a geographical triangle, with one vertex in Russia (Tsiolkovsky), another in the United States (Goddard), and the third in Germany. Representing this third vertex is Hermann Oberth,

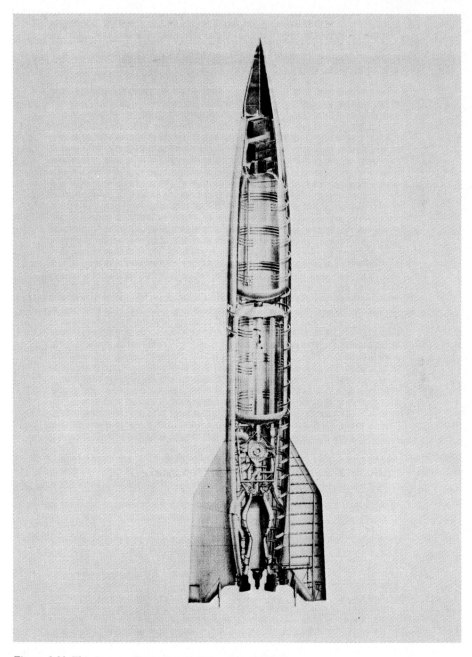

Figure 9.30 The German World War II V-2 rocket. (*NASA*.)

born in Transylvania on July 25, 1894, to later become a German citizen. Like Tsiolkovsky and Goddard before him, Oberth found inspiration in the novels of Jules Verne and began a mental search for a practical means of reaching the moon. During World War I, Oberth became interested in rockets, suggesting long-range liquid-fueled missiles to the German war department. In 1922, he combined these thoughts and suggested rockets for space flight. Oberth was at that time ignorant of the work of both Tsiolkovsky and Goddard. However, shortly thereafter, Goddard's work was mentioned in the German newspapers, and Oberth quickly wrote for a copy of the 1919 Smithsonian monograph. In 1923, Oberth published his own work on the theory of rocket engines, entitled *The Rocket into Planetary Space*. This was a rigorous technical text, and it laid the basis for the development of rockets in Germany. In order to foster Oberth's ideas, the German Society for Space Travel was formed in 1927 and began experimental work in 1929. (The American Rocket Society was subsequently patterned after the German society.) Oberth's ideas had a catalytic effect, especially on some of his students, such as Wernher Von Braun, and the 1930s found an almost explosive development of rocketry in Germany. This work, with Von Braun as the technical director, culminated in the development of the German V-2 rocket of World War II. Although an instrument of war, the V-2 was the first practical long-range rocket in history. A sketch of the V-2 is shown in Figure 9.30. Powered by liquid oxygen and alcohol, this rocket was 46.1 ft long, 65 in in diameter, and 27,000 lb in weight. It was the first vehicle made by humans to fly outside the sensible atmosphere (hence, in space), with altitudes above 50 mi and a range of 200 mi. The missile reached supersonic speeds during its flight within the atmosphere. During the closing phases of World War II, hundreds of production V-2s were captured by both Russian and U.S. forces and shipped back to their respective countries. As a result, all modern rockets can today trace their ancestry directly back to the V-2 and hence through Von Braun back to Hermann Oberth.

The development of modern rockets, culminating in the huge Saturn booster for the Apollo program, is a story in itself and is beyond the scope of this book. However, the early history sketched in this section is intended to add appreciation for the technical aspects of rocket engines discussed in Sec. 9.8. For an authoritative presentation on the history of modern rocketry, the reader is referred to the books by Von Braun and Ordway, and Emme (see Bibliography at the end of this chapter).

9.13 CHAPTER SUMMARY

A few important aspects of the chapter are itemized below:

1. The cross section of a propeller is an airfoil shape designed to produce an aerodynamic force in the direction of motion of the airplane, i.e., thrust. The efficiency of a propeller depends on the pitch angle and the advance ratio $J = V_\infty/nD$.

2. The four strokes of an Otto-cycle reciprocating internal combustion engine are intake, compression, power, and exhaust. Combustion takes place essentially at constant volume. The power generated by such an engine along with a propeller is the power available, expressed as

$$P_A = \eta \eta_{\text{mech}} \frac{n}{2} NW \tag{9.10}$$

where η = propeller efficiency, η_{mech} = mechanical efficiency, n = revolutions per second of the engine shaft, N = number of cylinders, and W = work produced during the complete four-stroke cycle. The power available can also be expressed as

$$P_A = \frac{\eta \eta_{\text{mech}} (\text{rpm}) d p_e}{120} \tag{9.15}$$

where rpm is the revolutions per minute of the engine shaft, d is the displacement, and p_e is the mean effective pressure.

3. The thrust equation for a jet propulsion device is

$$T = (\dot{m}_{\text{air}} + \dot{m}_{\text{fuel}}) V_e - \dot{m}_{\text{air}} V_\infty + (p_e - p_\infty) A_e \tag{9.24}$$

4. The turbojet engine process involves aerodynamic compression of the intake air in a diffuser, further compression in a rotating compressor, constant pressure combustion in the burner, expansion through a turbine which drives the compressor, and further expansion through an exhaust nozzle. In a turbofan engine, a large ducted fan is mounted on the shaft ahead of the compressor, which accelerates a large mass of auxiliary air outside the core of the engine itself, thus producing more thrust with higher efficiency. The ramjet engine has no rotating machinery and produces its thrust by means of aerodynamic compression in an inlet diffuser of the incoming air, burned at constant pressure in the combustor and exhausted through a nozzle.

5. The thrust for a rocket engine is

$$T = \dot{m} V_e + (p_e - p_\infty) A_e \tag{9.28}$$

A rocket carries its own fuel and oxidizer and is not dependent on atmospheric air for the generation of thrust.

6. The specific impulse is a direct measure of the efficiency of a rocket engine–propellant combination.

$$I_{\text{sp}} = \frac{T}{\dot{w}} = \frac{1}{g_0} \left\{ \frac{2\gamma R T_0}{\gamma - 1} \left[1 - \left(\frac{p_e}{p_0} \right)^{(\gamma-1)/\gamma} \right] \right\}^{1/2} \tag{9.35}$$

For a high specific impulse, the combustion temperature T_0 should be high and the molecular weight of the combustion gas should be low.

BIBLIOGRAPHY

Dommasch, D. O., Sherbey, S. S., and Connolly, T. F., *Airplane Aerodynamics*, 3d ed., Pitman, New York, 1961.

Emme, E. M., *A History of Space Flight*, Holt, New York, 1965.

Glauret, H., *The Elements of Aerofoil and Airscrew Theory*, Macmillan, New York, 1943.

Gray, G. W., *Frontiers of Flight*, Knopf, New York, 1948.

Hill, P. G., and Peterson, C. R., *Mechanics and Thermodynamics of Propulsion*, Addison-Wesley, Reading, MA, 1964.

Obert, E. F., *Internal Combustion Engines and Air Pollution*, Intext, New York, 1973.

Von Braun, W., and Ordway, F. I., *History of Rocketry and Space Travel*, 3d rev. ed., Crowell, New York, 1975.

Walsh, J. E., *One Day at Kitty Hawk*, Crowell, New York, 1975.

PROBLEMS

9.1 A reciprocating engine for light aircraft, modeled after the Avco Lycoming O-235 engine, has the following characteristics: bore = 11.1 cm, stroke = 9.84 cm, number of pistons = 4, compression ratio = 6.75, mechanical efficiency = 0.83. It is connected to a propeller with an efficiency of 0.85. If the fuel-to-air ratio is 0.06 and the pressure and temperature in the intake manifold are 1 atm and 285°, respectively, calculate the power available from the engine-propeller combination at 2800 rpm.

9.2 For the engine in Prob. 9.1, calculate the mean effective pressure.

9.3 Consider a turbojet mounted on a stationary test stand at sea level. The inlet and exit areas are the same, both equal to 0.45 m². The velocity, pressure, and temperature of the exhaust gas are 400 m/s, 1.0 atm, and 750 K, respectively. Calculate the static thrust of the engine. (Note: Static thrust of a jet engine is that thrust produced when the engine has no forward motion.)

9.4 Consider a turbojet-powered airplane flying at a standard altitude of 40,000 ft at a velocity of 530 mi/h. The turbojet engine has inlet and exit areas of 13 ft² and 10 ft², respectively. The velocity and pressure of the exhaust gas at the exit are 1500 ft/s and 450 lb/ft², respectively. Calculate the thrust of the turbojet.

9.5 Consider a turbojet in an airplane flying at standard sea level with a velocity of 800 ft/s. The pressure ratio across the compressor is 12.5 : 1. The fuel-to-air ratio (by mass) is 0.05. If the nozzle exhausts the flow to ambient pressure, calculate the gas temperature at the exit. (In solving this problem, assume the air in the diffuser is slowed to a very low velocity before entering the compressor. Also, assume the heat released per pound of fuel is 1.4×10^7 ft $\cdot$ lb/lb$_m$.)

9.6 A small ramjet engine is to be designed for a maximum thrust of 1000 lb at sea level at a velocity of 950 ft/s. If the exit velocity and pressure are 2000 ft/s and 1.0 atm, respectively, how large should the inlet be designed?

9.7 The mass flow through a rocket engine is 25 kg/s. If the exit area, velocity, and pressure are 2 m², 4000 m/s, and 2×10^4 N/m², respectively, calculate the thrust at a standard altitude of 50 km.

9.8 Consider a rocket engine where the combustion chamber pressure and temperature are 30 atm and 3756 K, respectively. The area of the rocket nozzle exit is 15 m² and is designed so that the exit pressure exactly equals ambient pressure at a standard altitude of 25 km. For the gas mixture, assume $\gamma = 1.18$ and the molecular weight is 20. At a standard altitude of 25 km, calculate: (a) specific impulse, (b) exit velocity, (c) mass flow, (d) thrust, and (e) throat area.

9.9 In a given rocket engine a mass flow of propellants equal to 87.6 lb$_m$/s is pumped into the combustion chamber, where the temperature after combustion is 6000°R. The combustion products have mixture values of $R = 2400$ ft $\cdot$ lb/(slug)(°R) and $\gamma = 1.21$. If the throat area is 0.5 ft², calculate the pressure in the combustion chamber.

STANDARD ATMOSPHERE, SI UNITS

Altitude		Temperature	Pressure	Density
h_G, m	h, m	T, K	p, N/m^2	ρ, kg/m^3
−5,000	−5,004	320.69	1.7761 + 5	1.9296 + 0
−4,900	−4,904	320.03	1.7587	1.9145
−4,800	−4,804	319.38	1.7400	1.8980
−4,700	−4,703	318.73	1.7215	1.8816
−4,600	−4,603	318.08	1.7031	1.8653
−4,500	−4,503	317.43	1.6848	1.8491
−4,400	−4,403	316.78	1.6667	1.8330
−4,300	−4,303	316.13	1.6488	1.8171
−4,200	−4,203	315.48	1.6311	1.8012
−4,100	−4,103	314.83	1.6134	1.7854
−4,000	−4,003	314.18	1.5960 + 5	1.7698 + 0
−3,900	−3,902	313.53	1.5787	1.7542
−3,800	−3,802	312.87	1.5615	1.7388
−3,700	−3,702	212.22	1.5445	1.7234
−3,600	−3,602	311.57	1.5277	1.7082
−3,500	−3,502	310.92	1.5110	1.6931
−3,400	−3,402	310.27	1.4945	1.6780
−3,300	−3,302	309.62	1.4781	1.6631
−3,200	−3,202	308.97	1.4618	1.6483
−3,100	−3,102	308.32	1.4457	1.6336
−3,000	−3,001	307.67	1.4297 + 5	1.6189 + 0
−2,900	−2,901	307.02	1.4139	1.6044
−2,800	−2,801	306.37	1.3982	1.5900
−2,700	−2,701	305.72	1.3827	1.5757
−2,600	−2,601	305.07	1.3673	1.5615

507

Altitude		Temperature	Pressure	Density
h_G, m	h, m	T, K	p, N/m^2	ρ, kg/m^3
−2,500	−2,501	304.42	1.3521	1.5473
−2,400	−2,401	303.77	1.3369	1.5333
−2,300	−2,301	303.12	1.3220	1.5194
−2,200	−2,201	302.46	1.3071	1.5056
−2,100	−2,101	301.81	1.2924	1.4918
−2,000	−2,001	301.16	1.2778 + 5	1.4782 + 0
−1,900	−1,901	300.51	1.2634	1.4646
−1,800	−1,801	299.86	1.2491	1.4512
−1,700	−1,700	299.21	1.2349	1.4379
−1,600	−1,600	298.56	1.2209	1.4246
−1,500	−1,500	297.91	1.2070	1.4114
−1,400	−1,400	297.26	1.1932	1.3984
−1,300	−1,300	296.61	1.1795	1.3854
−1,200	−1,200	295.96	1.1660	1.3725
−1,100	−1,100	295.31	1.1526	1.3597
−1,000	−1,000	294.66	1.1393 + 5	1.3470 + 0
− 900	− 900	294.01	1.1262	1.3344
− 800	− 800	293.36	1.1131	1.3219
− 700	− 700	292.71	1.1002	1.3095
− 600	− 600	292.06	1.0874	1.2972
− 500	− 500	291.41	1.0748	1.2849
− 400	− 400	290.76	1.0622	1.2728
− 300	− 300	290.11	1.0498	1.2607
− 200	− 200	289.46	1.0375	1.2487
− 100	− 100	288.81	1.0253	1.2368
0	0	288.16	1.01325 + 5	1.2250 + 0
100	100	287.51	1.0013	1.2133
200	200	286.86	9.8945 + 4	1.2017
300	300	286.21	9.7773	1.1901
400	400	285.56	9.6611	1.1787
500	500	284.91	9.5461	1.1673
600	600	284.26	9.4322	1.1560
700	700	283.61	9.3194	1.1448
800	800	282.96	9.2077	1.1337
900	900	282.31	9.0971	1.1226
1,000	1,000	281.66	8.9876 + 4	1.1117 + 0
1,100	1,100	281.01	8.8792	1.1008
1,200	1,200	280.36	8.7718	1.0900
1,300	1,300	279.71	8.6655	1.0793
1,400	1,400	279.06	8.5602	1.0687
1,500	1,500	278.41	8.4560	1.0581
1,600	1,600	277.76	8.3527	1.0476
1,700	1,700	277.11	8.2506	1.0373
1,800	1,799	276.46	8.1494	1.0269
1,900	1,899	275.81	8.0493	1.0167

Altitude		Temperature	Pressure	Density
h_G, m	h, m	T, K	p, N/m^2	ρ, kg/m^3
2,000	1,999	275.16	7.9501 + 4	1.0066 + 0
2,100	2,099	274.51	7.8520	9.9649 − 1
2,200	2,199	273.86	7.7548	9.8649
2,300	2,299	273.22	7.6586	9.7657
2,400	2,399	272.57	7.5634	9.6673
2,500	2,499	271.92	7.4692	9.5696
2,600	2,599	271.27	7.3759	9.4727
2,700	2,699	270.62	7.2835	9.3765
2,800	2,799	269.97	7.1921	9.2811
2,900	2,899	269.32	7.1016	9.1865
3,000	2,999	268.67	7.0121 + 4	9.0926 − 1
3,100	3,098	268.02	6.9235	8.9994
3,200	3,198	267.37	6.8357	8.9070
3,300	3,298	266.72	6.7489	8.8153
3,400	3,398	266.07	6.6630	8.7243
3,500	3,498	265.42	6.5780	8.6341
3,600	3,598	264.77	6.4939	8.5445
3,700	3,698	264.12	6.4106	8.4557
3,800	3,798	263.47	6.3282	8.3676
3,900	3,898	262.83	6.2467	8.2802
4,000	3,997	262.18	6.1660 + 4	8.1935 − 1
4,100	4,097	261.53	6.0862	8.1075
4,200	4,197	260.88	6.0072	8.0222
4,300	4,297	260.23	5.9290	7.9376
4,400	4,397	259.58	5.8517	7.8536
4,500	4,497	258.93	5.7752	7.7704
4,600	4,597	258.28	5.6995	7.6878
4,700	4,697	257.63	5.6247	7.6059
4,800	4,796	256.98	5.5506	7.5247
4,900	4,896	256.33	5.4773	7.4442
5,000	4,996	255.69	5.4048 + 4	7.3643 − 1
5,100	5,096	255.04	5.3331	7.2851
5,200	5,196	254.39	5.2621	7.2065
5,300	5,296	253.74	5.1920	7.1286
5,400	5,395	253.09	5.1226	7.0513
5,500	5,495	252.44	5.0539	6.9747
5,600	5,595	251.79	4.9860	6.8987
5,700	5,695	251.14	4.9188	6.8234
5,800	5,795	250.49	4.8524	6.7486
5,900	5,895	249.85	4.7867	6.6746
6,000	5,994	249.20	4.7217 + 4	6.6011 − 1
6,100	6,094	248.55	4.6575	6.5283
6,200	6,194	247.90	4.5939	6.4561
6,300	6,294	247.25	4.5311	6.3845
6,400	6,394	246.60	4.4690	6.3135
6,500	6,493	245.95	4.4075	6.2431

Altitude		Temperature	Pressure	Density
h_G, m	h, m	T, K	p, N/m^2	ρ, kg/m^3
6,600	6,593	245.30	4.3468	6.1733
6,700	6,693	244.66	4.2867	6.1041
6,800	6,793	244.01	4.2273	6.0356
6,900	6,893	243.36	4.1686	5.9676
7,000	6,992	242.71	4.1105 + 4	5.9002 − 1
7,100	7,092	242.06	4.0531	5.8334
7,200	7,192	241.41	3.9963	5.7671
7,300	7,292	240.76	3.9402	5.7015
7,400	7,391	240.12	3.8848	5.6364
7,500	7,491	239.47	3.8299	5.5719
7,600	7,591	238.82	3.7757	5.5080
7,700	7,691	238.17	3.7222	5.4446
7,800	7,790	237.52	3.6692	5.3818
7,900	7,890	236.87	3.6169	5.3195
8,000	7,990	236.23	3.5651 + 4	5.2578 − 1
8,100	8,090	235.58	3.5140	5.1967
8,200	8,189	234.93	3.4635	5.1361
8,300	8,289	234.28	3.4135	5.0760
8,400	8,389	233.63	3.3642	5.0165
8,500	8,489	232.98	3.3154	4.9575
8,600	8,588	232.34	3.2672	4.8991
8,700	8,688	231.69	3.2196	4.8412
8,800	8,788	231.04	3.1725	4.7838
8,900	8,888	230.39	3.1260	4.7269
9,000	8,987	229.74	3.0800 + 4	4.6706 − 1
9,100	9,087	229.09	3.0346	4.6148
9,200	9,187	228.45	2.9898	4.5595
9,300	9,286	227.80	2.9455	4.5047
9,400	9,386	227.15	2.9017	4.4504
9,500	9,486	226.50	2.8584	4.3966
9,600	9,586	225.85	2.8157	4.3433
9,700	9,685	225.21	2.7735	4.2905
9,800	9,785	224.56	2.7318	4.2382
9,900	9,885	223.91	2.6906	4.1864
10,000	9,984	223.26	2.6500 + 4	4.1351 − 1
10,100	10,084	222.61	2.6098	4.0842
10,200	10,184	221.97	2.5701	4.0339
10,300	10,283	221.32	2.5309	3.9840
10,400	10,383	220.67	2.4922	3.9346
10,500	10,483	220.02	2.4540	3.8857
10,600	10,582	219.37	2.4163	3.8372
10,700	10,682	218.73	2.3790	3.7892
10,800	10,782	218.08	2.3422	3.7417
10,900	10,881	217.43	2.3059	3.6946
11,000	10,981	216.78	2.2700 + 4	3.6480 − 1
11,100	11,081	216.66	2.2346	3.5932

Altitude		Temperature	Pressure	Density
h_G, m	h, m	T, K	p, N/m^2	ρ, kg/m^3
11,200	11,180	216.66	2.1997	3.5371
11,300	11,280	216.66	2.1654	3.4820
11,400	11,380	216.66	2.1317	3.4277
11,500	11,479	216.66	2.0985	3.3743
11,600	11,579	216.66	2.0657	3.3217
11,700	11,679	216.66	2.0335	3.2699
11,800	11,778	216.66	2.0018	3.2189
11,900	11,878	216.66	1.9706	3.1687
12,000	11,977	216.66	1.9399 + 4	3.1194 − 1
12,100	12,077	216.66	1.9097	3.0707
12,200	12,177	216.66	1.8799	3.0229
12,300	12,276	216.66	1.8506	2.9758
12,400	12,376	216.66	1.8218	2.9294
12,500	12,475	216.66	1.7934	2.8837
12,600	12,575	216.66	1.7654	2.8388
12,700	12,675	216.66	1.7379	2.7945
12,800	12,774	216.66	1.7108	2.7510
12,900	12,874	216.66	1.6842	2.7081
13,000	12,973	216.66	1.6579 + 4	2.6659 − 1
13,100	13,073	216.66	1.6321	2.6244
13,200	13,173	216.66	1.6067	2.5835
13,300	13,272	216.66	1.5816	2.5433
13,400	13,372	216.66	1.5570	2.5036
13,500	13,471	216.66	1.5327	2.4646
13,600	13,571	216.66	1.5089	2.4262
13,700	13,671	216.66	1.4854	2.3884
13,800	13,770	216.66	1.4622	2.3512
13,900	13,870	216.66	1.4394	2.3146
14,000	13,969	216.66	1.4170 + 4	2.2785 − 1
14,100	14,069	216.66	1.3950	2.2430
14,200	14,168	216.66	1.3732	2.2081
14,300	14,268	216.66	1.3518	2.1737
14,400	14,367	216.66	1.3308	2.1399
14,500	14,467	216.66	1.3101	2.1065
14,600	14,567	216.66	1.2896	2.0737
14,700	14,666	216.66	1.2696	2.0414
14,800	14,766	216.66	1.2498	2.0096
14,900	14,865	216.66	1.2303	1.9783
15,000	14,965	216.66	1.2112 + 4	1.9475 − 1
15,100	15,064	216.66	1.1923	1.9172
15,200	15,164	216.66	1.1737	1.8874
15,300	15,263	216.66	1.1555	1.8580
15,400	15,363	216.66	1.1375	1.8290
15,500	15,462	216.66	1.1198	1.8006
15,600	15,562	216.66	1.1023	1.7725
15,700	15,661	216.66	1.0852	1.7449
15,800	15,761	216.66	1.0683	1.7178
15,900	15,860	216.66	1.0516	1.6910

Altitude		Temperature	Pressure	Density
h_G, m	h, m	T, K	p, N/m^2	ρ, kg/m^3
16,000	15,960	216.66	1.0353 + 4	1.6647 − 1
16,100	16,059	216.66	1.0192	1.6388
16,200	16,159	216.66	1.0033	1.6133
16,300	16,258	216.66	9.8767 + 3	1.5882
16,400	16,358	216.66	9.7230	1.5634
16,500	16,457	216.66	9.5717	1.5391
16,600	16,557	216.66	9.4227	1.5151
16,700	16,656	216.66	9.2760	1.4916
16,800	16,756	216.66	9.1317	1.4683
16,900	16,855	216.66	8.9895	1.4455
17,000	16,955	216.66	8.8496 + 3	1.4230 − 1
17,100	17,054	216.66	8.7119	1.4009
17,200	17,154	216.66	8.5763	1.3791
17,300	17,253	216.66	8.4429	1.3576
17,400	17,353	216.66	8.3115	1.3365
17,500	17,452	216.66	8.1822	1.3157
17,600	17,551	216.66	8.0549	1.2952
17,700	17,651	216.66	7.9295	1.2751
17,800	17,750	216.66	7.8062	1.2552
17,900	17,850	216.66	7.6847	1.2357
18,000	17,949	216.66	7.5652 + 3	1.2165 − 1
18,100	18,049	216.66	7.4475	1.1975
18,200	18,148	216.66	7.3316	1.1789
18,300	18,247	216.66	7.2175	1.1606
18,400	18,347	216.66	7.1053	1.1425
18,500	18,446	216.66	6.9947	1.1247
18,600	18,546	216.66	6.8859	1.1072
18,700	18,645	216.66	6.7788	1.0900
18,800	18,745	216.66	6.6734	1.0731
18,900	18,844	216.66	6.5696	1.0564
19,000	18,943	216.66	6.4674 + 3	1.0399 − 1
19,100	19,043	216.66	6.3668	1.0238
19,200	19,142	216.66	6.2678	1.0079
19,300	19,242	216.66	6.1703	9.9218 − 2
19,400	19,341	216.66	6.0744	9.7675
19,500	19,440	216.66	5.9799	9.6156
19,600	19,540	216.66	5.8869	9.4661
19,700	19,639	216.66	5.7954	9.3189
19,800	19,739	216.66	5.7053	9.1740
19,900	19,838	216.66	5.6166	9.0313
20,000	19,937	216.66	5.5293 + 3	8.8909 − 2
20,200	20,136	216.66	5.3587	8.6166
20,400	20,335	216.66	5.1933	8.3508
20,600	20,533	216.66	5.0331	8.0931
20,800	20,732	216.66	4.8779	7.8435
21,000	20,931	216.66	4.7274	7.6015
21,200	21,130	216.66	4.5816	7.3671

Altitude		Temperature	Pressure	Density
h_G, m	h, m	T, K	p, N/m^2	ρ, kg/m^3
21,400	21,328	216.66	4.4403	7.1399
21,600	21,527	216.66	4.3034	6.9197
21,800	21,725	216.66	4.1706	6.7063
22,000	21,924	216.66	4.0420 + 3	6.4995 − 2
22,200	22,123	216.66	3.9174	6.2991
22,400	22,321	216.66	3.7966	6.1049
22,600	22,520	216.66	3.6796	5.9167
22,800	22,719	216.66	3.5661	5.7343
23,000	22,917	216.66	3.4562	5.5575
23,200	23,116	216.66	3.3497	5.3862
23,400	23,314	216.66	3.2464	5.2202
23,600	23,513	216.66	3.1464	5.0593
23,800	23,711	216.66	3.0494	4.9034
24,000	23,910	216.66	2.9554 + 3	4.7522 − 2
24,200	24,108	216.66	2.8644	4.6058
24,400	24,307	216.66	2.7761	4.4639
24,600	24,505	216.66	2.6906	4.3263
24,800	24,704	216.66	2.6077	4.1931
25,000	24,902	216.66	2.5273	4.0639
25,200	25,100	216.96	2.4495	3.9333
25,400	25,299	217.56	2.3742	3.8020
25,600	25,497	218.15	2.3015	3.6755
25,800	25,696	218.75	2.2312	3.5535
26,000	25,894	219.34	2.1632 + 3	3.4359 − 2
26,200	26,092	219.94	2.0975	3.3225
26,400	26,291	220.53	2.0339	3.2131
26,600	26,489	221.13	1.9725	3.1076
26,800	26,687	221.72	1.9130	3.0059
27,000	26,886	222.32	1.8555	2.9077
27,200	27,084	222.91	1.7999	2.8130
27,400	27,282	223.51	1.7461	2.7217
27,600	27,481	224.10	1.6940	2.6335
27,800	27,679	224.70	1.6437	2.5484
28,000	27,877	225.29	1.5949 + 3	2.4663 − 2
28,200	28,075	225.89	1.5477	2.3871
28,400	28,274	226.48	1.5021	2.3106
28,600	28,472	227.08	1.4579	2.2367
28,800	28,670	227.67	1.4151	2.1654
29,000	28,868	228.26	1.3737	2.0966
29,200	29,066	228.86	1.3336	2.0301
29,400	29,265	229.45	1.2948	1.9659
29,600	29,463	230.05	1.2572	1.9039
29,800	29,661	230.64	1.2208	1.8440
30,000	29,859	231.24	1.1855 + 3	1.7861 − 2
30,200	30,057	231.83	1.1514	1.7302

Altitude		Temperature	Pressure	Density
h_G, m	h, m	T, K	p, N/m^2	ρ, kg/m^3
30,400	30,255	232.43	1.1183	1.6762
30,600	30,453	233.02	1.0862	1.6240
30,800	30,651	233.61	1.0552	1.5735
31,000	30,850	234.21	1.0251	1.5278
31,200	31,048	234.80	9.9592 + 2	1.4777
31,400	31,246	235.40	9.6766	1.4321
31,600	31,444	235.99	9.4028	1.3881
31,800	31,642	236.59	9.1374	1.3455
32,000	31,840	237.18	8.8802 + 2	1.3044 − 2
32,200	32,038	237.77	8.6308	1.2646
32,400	32,236	238.37	8.3890	1.2261
32,600	32,434	238.96	8.1546	1.1889
32,800	32,632	239.55	7.9273	1.1529
33,000	32,830	240.15	7.7069	1.1180
33,200	33,028	240.74	7.4932	1.0844
33,400	33,225	214.34	7.2859	1.0518
33,600	33,423	241.93	7.0849	1.0202
33,800	33,621	242.52	6.8898	9.8972 − 3
34,000	33,819	243.12	6.7007 + 2	9.6020 − 3
34,200	34,017	243.71	6.5171	9.3162
34,400	34,215	244.30	6.3391	9.0396
34,600	34,413	244.90	6.1663	8.7720
34,800	34,611	245.49	5.9986	8.5128
35,000	34,808	246.09	5.8359	8.2620
35,200	35,006	246.68	5.6780	8.0191
35,400	35,204	247.27	5.5248	7.7839
35,600	35,402	247.87	5.3760	7.5562
35,800	35,600	248.46	5.2316	7.3357
36,000	35,797	249.05	5.0914 + 2	7.1221 − 3
36,200	35,995	249.65	4.9553	6.9152
36,400	36,193	250.24	4.8232	6.7149
36,600	36,390	250.83	4.6949	6.5208
36,800	36,588	251.42	4.5703	6.3328
37,000	36,786	252.02	4.4493	6.1506
37,200	36,984	252.61	4.3318	5.9741
37,400	37,181	253.20	4.2176	5.8030
37,600	37,379	253.80	4.1067	5.6373
37,800	37,577	254.39	3.9990	5.4767
38,000	37,774	254.98	3.8944 + 2	5.3210 − 3
38,200	37,972	255.58	3.7928	5.1701
38,400	38,169	256.17	3.6940	5.0238
38,600	38,367	256.76	3.5980	4.8820
38,800	38,565	257.35	3.5048	4.7445
39,000	38,762	257.95	3.4141	4.6112
39,200	38,960	258.54	3.3261	4.4819
39,400	39,157	259.13	3.2405	4.3566

Altitude		Temperature	Pressure	Density
h_G, m	h, m	T, K	p, N/m^2	ρ, kg/m^3
39,600	39,355	259.72	3.1572	4.2350
39,800	39,552	260.32	3.0764	4.1171
40,000	39,750	260.91	2.9977 + 2	4.0028 − 3
40,200	39,947	261.50	2.9213	3.8919
40,400	40,145	262.09	2.8470	3.7843
40,600	40,342	262.69	2.7747	3.6799
40,800	40,540	263.28	2.7044	3.5786
41,000	40,737	263.87	2.6361	3.4804
41,200	40,935	264.46	2.5696	3.3850
41,400	41,132	265.06	2.5050	3.2925
41,600	41,330	265.65	2.4421	3.2027
41,800	41,527	266.24	2.3810	3.1156
42,000	41,724	266.83	2.3215 + 2	3.0310 − 3
42,200	41,922	267.43	2.2636	2.9489
42,400	42,119	268.02	2.2073	2.8692
42,600	42,316	268.61	2.1525	2.7918
42,800	42,514	269.20	2.0992	2.7167
43,000	42,711	269.79	2.0474	2.6438
43,200	42,908	270.39	1.9969	2.5730
43,400	43,106	270.98	1.9478	2.5042
43,600	43,303	271.57	1.9000	2.4374
43,800	43,500	272.16	1.8535	2.3726
44,000	43,698	272.75	1.8082 + 2	2.3096 − 3
44,200	43,895	273.34	1.7641	2.2484
44,400	44,092	273.94	1.7212	2.1889
44,600	44,289	274.53	1.6794	2.1312
44,800	44,486	275.12	1.6387	2.0751
45,000	44,684	275.71	1.5991	2.0206
45,200	44,881	276.30	1.5606	1.9677
45,400	45,078	276.89	1.5230	1.9162
45,600	45,275	277.49	1.4865	1.8662
45,800	45,472	278.08	1.4508	1.8177
46,000	45,670	278.67	1.4162 + 2	1.7704 − 3
46,200	45,867	279.26	1.3824	1.7246
46,400	46,064	279.85	1.3495	1.6799
46,600	46,261	280.44	1.3174	1.6366
46,800	46,458	281.03	1.2862	1.5944
47,000	46,655	281.63	1.2558	1.5535
47,200	46,852	282.22	1.2261	1.5136
47,400	47,049	282.66	1.1973	1.4757
47,600	47,246	282.66	1.1691	1.4409
47,800	47,443	282.66	1.1416	1.4070
48,000	47,640	282.66	1.1147 + 2	1.3739 − 3
48,200	47,837	282.66	1.0885	1.3416
48,400	48,034	282.66	1.0629	1.3100

Altitude		Temperature	Pressure	Density
h_G, m	h, m	T, K	p, N/m^2	ρ, kg/m^3
48,600	48,231	282.66	1.0379	1.2792
48,800	48,428	282.66	1.0135	1.2491
49,000	48,625	282.66	9.8961 + 1	1.2197
49,200	48,822	282.66	9.6633	1.1910
49,400	49,019	282.66	9.4360	1.1630
49,600	49,216	282.66	9.2141	1.1357
49,800	49,413	282.66	8.9974	1.1089
50,000	49,610	282.66	8.7858 + 1	1.0829 − 3
50,500	50,102	282.66	8.2783	1.0203
51,000	50,594	282.66	7.8003	9.6140 − 4
51,500	51,086	282.66	7.3499	9.0589
52,000	51,578	282.66	6.9256	8.5360
52,500	52,070	282.66	6.5259	8.0433
53,000	52,562	282.66	6.1493	7.5791
53,500	53,053	282.42	5.7944	7.1478
54,000	53,545	280.21	5.4586	6.7867
54,500	54,037	277.99	5.1398	6.4412
55,000	54,528	275.78	4.8373 + 1	6.1108 − 4
55,500	55,020	273.57	4.5505	5.7949
56,000	55,511	271.36	4.2786	5.4931
56,500	56,002	269.15	4.0210	5.2047
57,000	56,493	266.94	3.7770	4.9293
57,500	56,985	264.73	3.5459	4.6664
58,000	57,476	262.52	3.3273	4.4156
58,500	57,967	260.31	3.1205	4.1763
59,000	58,457	258.10	2.9250	3.9482
59,500	58,948	255.89	2.7403	3.7307

STANDARD ATMOSPHERE, ENGLISH ENGINEERING UNITS

Altitude		Temperature	Pressure	Density
h_G, ft	h, ft	T, °R	p, lb/ft^2	ρ, slugs/ft^3
− 16,500	− 16,513	577.58	3.6588 + 3	3.6905 − 3
− 16,000	− 16,012	575.79	3.6641	3.7074
− 15,500	− 15,512	574.00	3.6048	3.6587
− 15,000	− 15,011	572.22	3.5462	3.6105
− 14,500	− 14,510	570.43	3.4884	3.5628
− 14,000	− 14,009	568.65	3.4314	3.5155
− 13,500	− 13,509	566.86	3.3752	3.4688
− 13,000	− 13,008	565.08	3.3197	3.4225
− 12,500	− 12,507	563.29	3.2649	3.3768
− 12,000	− 12,007	561.51	3.2109	3.3314
− 11,500	− 11,506	559.72	3.1576 + 3	3.2866 − 3
− 11,000	− 11,006	557.94	3.1050	3.2422
− 10,500	− 10,505	556.15	3.0532	3.1983
− 10,000	− 10,005	554.37	3.0020	3.1548
− 9,500	− 9,504	552.58	2.9516	3.1118
− 9,000	− 9,004	550.80	2.9018	3.0693
− 8,500	− 8,503	549.01	2.8527	3.0272
− 8,000	− 8,003	547.23	2.8043	2.9855
− 7,500	− 7,503	545.44	2.7566	2.9443
− 7,000	− 7,002	543.66	2.7095	2.9035
− 6,500	− 6,502	541.88	2.6631 + 3	2.8632 − 3
− 6,000	− 6,002	540.09	2.6174	2.8233
− 5,500	− 5,501	538.31	2.5722	2.7838
− 5,000	− 5,001	536.52	2.5277	2.7448

Altitude		Temperature	Pressure	Density
h_G, ft	h, ft	T, °R	p, lb/ft^2	ρ, slugs/ft^3
− 4,500	− 4,501	534.74	2.4839	2.7061
− 4,000	− 4,001	532.96	2.4406	2.6679
− 3,500	− 3,501	531.17	2.3980	2.6301
− 3,000	− 3,000	529.39	2.3560	2.5927
− 2,500	− 2,500	527.60	2.3146	2.5558
− 2,000	− 2,000	525.82	2.2737	2.5192
− 1,500	− 1,500	524.04	2.2335 + 3	2.4830 − 3
− 1,000	− 1,000	522.25	2.1938	2.4473
− 500	− 500	520.47	2.1547	2.4119
0	0	518.69	2.1162	2.3769
500	500	516.90	2.0783	2.3423
1,000	1,000	515.12	2.0409	2.3081
1,500	1,500	513.34	2.0040	2.2743
2,000	2,000	511.56	1.9677	2.2409
2,500	2,500	509.77	1.9319	2.2079
3,000	3,000	507.99	1.8967	2.1752
3,500	3,499	506.21	1.8619 + 3	2.1429 − 3
4,000	3,999	504.43	1.8277	2.1110
4,500	4,499	502.64	1.7941	2.0794
5,000	4,999	500.86	1.7609	2.0482
5,500	5,499	499.08	1.7282	2.0174
6,000	5,998	497.30	1.6960	1.9869
6,500	6,498	495.52	1.6643	1.9567
7,000	6,998	493.73	1.6331	1.9270
7,500	7,497	491.95	1.6023	1.8975
8,000	7,997	490.17	1.5721	1.8685
8,500	8,497	488.39	1.5423 + 3	1.8397 − 3
9,000	8,996	486.61	1.5129	1.8113
9,500	9,496	484.82	1.4840	1.7833
10,000	9,995	483.04	1.4556	1.7556
10,500	10,495	481.26	1.4276	1.7282
11,000	10,994	479.48	1.4000	1.7011
11,500	11,494	477.70	1.3729	1.6744
12,000	11,993	475.92	1.3462	1.6480
12,500	12,493	474.14	1.3200	1.6219
13,000	12,992	472.36	1.2941	1.5961
13,500	13,491	470.58	1.2687 + 3	1.5707 − 3
14,000	13,991	468.80	1.2436	1.5455
14,500	14,490	467.01	1.2190	1.5207
15,000	14,989	465.23	1.1948	1.4962
15,500	15,488	463.45	1.1709	1.4719
16,000	15,988	461.67	1.1475	1.4480
16,500	16,487	459.89	1.1244	1.4244
17,000	16,986	458.11	1.1017	1.4011

Altitude		Temperature	Pressure	Density
h_G, ft	h, ft	T, °R	p, lb/ft^2	ρ, slugs/ft^3
17,500	17,485	456.33	1.0794	1.3781
18,000	17,984	454.55	1.0575	1.3553
18,500	18,484	452.77	1.0359 + 3	1.3329 − 3
19,000	18,983	450.99	1.0147	1.3107
19,500	19,482	449.21	9.9379 + 2	1.2889
20,000	19,981	447.43	9.7327	1.2673
20,500	20,480	445.65	9.5309	1.2459
21,000	20,979	443.87	9.3326	1.2249
21,500	21,478	442.09	9.1376	1.2041
22,000	21,977	440.32	8.9459	1.1836
22,500	22,476	438.54	8.7576	1.1634
23,000	22,975	436.76	8.5724	1.1435
23,500	23,474	434.98	8.3905 + 2	1.1238 − 3
24,000	23,972	433.20	8.2116	1.1043
24,500	24,471	431.42	8.0359	1.0852
25,000	24,970	429.64	7.8633	1.0663
25,500	25,469	427.86	7.6937	1.0476
26,000	25,968	426.08	7.5271	1.0292
26,500	26,466	424.30	7.3634	1.0110
27,000	26,965	422.53	7.2026	9.9311 − 4
27,500	27,464	420.75	7.0447	9.7544
28,000	27,962	418.97	6.8896	9.5801
28,500	28,461	417.19	6.7373 + 2	9.4082 − 4
29,000	28,960	415.41	6.5877	9.2387
29,500	29,458	413.63	6.4408	9.0716
30,000	29,957	411.86	6.2966	8.9068
30,500	30,455	410.08	6.1551	8.7443
31,000	30,954	408.30	6.0161	8.5841
31,500	31,452	406.52	5.8797	8.4261
32,000	31,951	404.75	5.7458	8.2704
32,500	32,449	402.97	5.6144	8.1169
33,000	32,948	401.19	5.4854	7.9656
33,500	33,446	399.41	5.3589 + 2	7.8165 − 4
34,000	33,945	397.64	5.2347	7.6696
34,500	34,443	395.86	5.1129	7.5247
35,000	34,941	394.08	4.9934	7.3820
35,500	35,440	392.30	4.8762	7.2413
36,000	35,938	390.53	4.7612	7.1028
36,500	36,436	389.99	4.6486	6.9443
37,000	36,934	389.99	4.5386	6.7800
37,500	37,433	389.99	4.4312	6.6196
38,000	37,931	389.99	4.3263	6.4629
38,500	38,429	389.99	4.2240 + 2	6.3100 − 4
39,000	38,927	389.99	4.1241	6.1608
39,500	39,425	389.99	4.0265	6.0150

Altitude		Temperature	Pressure	Density
h_G, ft	h, ft	T, °R	p, lb/ft^2	ρ, slugs/ft^3
40,000	39,923	389.99	3.9312	5.8727
40,500	40,422	389.99	3.8382	5.7338
41,000	40,920	389.99	3.7475	5.5982
41,500	41,418	389.99	3.6588	5.4658
42,000	41,916	389.99	3.5723	5.3365
42,500	42,414	389.99	3.4878	5.2103
43,000	42,912	389.99	3.4053	5.0871
43,500	43,409	389.99	3.3248 + 2	4.9668 − 4
44,000	43,907	389.99	3.2462	4.8493
44,500	44,405	389.99	3.1694	4.7346
45,000	44,903	389.99	3.0945	4.6227
45,500	45,401	389.99	3.0213	4.5134
46,000	45,899	389.99	2.9499	4.4067
46,500	46,397	389.99	2.8801	4.3025
47,000	46,894	389.99	2.8120	4.2008
47,500	47,392	389.99	2.7456	4.1015
48,000	47,890	389.99	2.6807	4.0045
48,500	48,387	389.99	2.2173 + 2	3.9099 − 4
49,000	48,885	389.99	2.5554	3.8175
49,500	49,383	389.99	2.4950	3.7272
50,000	49,880	389.99	2.4361	3.6391
50,500	50,378	389.99	2.3785	3.5531
51,000	50,876	389.99	2.3223	3.4692
51,500	51,373	389.99	2.2674	3.3872
52,000	51,871	389.99	2.2138	3.3072
52,500	52,368	389.99	2.1615	3.2290
53,000	52,866	389.99	2.1105	3.1527
53,500	53,363	289.99	2.0606 + 2	3.0782 − 4
54,000	53,861	389.99	2.0119	3.0055
54,500	54,358	389.99	1.9644	2.9345
55,000	54,855	389.99	1.9180	2.8652
55,500	55,353	389.99	1.8727	2.7975
56,000	55,850	389.99	1.8284	2.7314
56,500	56,347	389.99	1.7853	2.6669
57,000	56,845	389.99	1.7431	2.6039
57,500	57,342	389.99	1.7019	2.5424
58,000	57,839	389.99	1.6617	2.4824
58,500	58,336	389.99	1.6225 + 2	2.4238 − 4
59,000	58,834	389.99	1.5842	2.3665
59,500	59,331	389.99	1.5468	2.3107
60,000	59,828	389.99	1.5103	2.2561
60,500	60,325	389.99	1.4746	2.2028
61,000	60,822	389.99	1.4398	2.1508
61,500	61,319	389.99	1.4058	2.1001
62,000	61,816	389.99	1.3726	2.0505
62,500	62,313	389.99	1.3402	2.0021
63,000	62,810	389.99	1.3086	1.9548

Altitude		Temperature	Pressure	Density
h_G, ft	h, ft	T, °R	p, lb/ft^2	ρ, slugs/ft^3
63,500	63,307	389.99	1.2777 + 2	1.9087 − 4
64,000	63,804	389.99	1.2475	1.8636
64,500	64,301	389.99	1.2181	1.8196
65,000	64,798	389.99	1.1893	1.7767
65,500	65,295	389.99	1.1613	1.7348
66,000	65,792	389.99	1.1339	1.6938
66,500	66,289	389.99	1.1071	1.6539
67,000	66,785	389.99	1.0810	1.6148
67,500	67,282	389.99	1.0555	1.5767
68,000	67,779	389.99	1.0306	1.5395
68,500	68,276	389.99	1.0063 + 2	1.5032 − 4
69,000	68,772	389.99	9.8253 + 1	1.4678
69,500	69,269	389.99	9.5935	1.4331
70,000	69,766	389.99	9.3672	1.3993
70,500	70,262	389.99	9.1462	1.3663
71,000	70,759	389.99	8.9305	1.3341
71,500	74,256	389.99	8.7199	1.3026
72,000	71,752	389.99	8.5142	1.2719
72,500	72,249	389.99	8.3134	1.2419
73,000	72,745	389.99	8.1174	1.2126
73,500	73,242	389.99	7.9259 + 1	1.1840 − 4
74,000	73,738	389.99	7.7390	1.1561
74,500	74,235	389.99	7.5566	1.1288
75,000	74,731	389.99	7.3784	1.1022
75,500	75,228	389.99	7.2044	1.0762
76,000	75,724	389.99	7.0346	1.0509
76,500	76,220	389.99	6.8687	1.0261
77,000	76,717	389.99	6.7068	1.0019
77,500	77,213	389.99	6.5487	9.7829 − 5
78,000	77,709	389.99	6.3944	9.5523
78,500	78,206	389.99	6.2437 + 1	9.3271 − 5
79,000	78,702	389.99	6.0965	9.1073
79,500	79,198	389.99	5.9528	8.8927
80,000	79,694	389.99	5.8125	8.6831
80,500	80,190	389.99	5.6755	8.4785
81,000	80,687	389.99	5.5418	8.2787
81,500	81,183	389.99	5.4112	8.0836
82,000	81,679	389.99	5.5837	7.8931
82,500	82,175	390.24	5.1592	7.7022
83,000	82,671	391.06	5.0979	7.5053
83,500	83,167	391.87	4.9196 + 1	7.3139 − 5
84,000	83,663	392.69	4.8044	7.1277
84,500	84,159	393.51	4.6921	6.9467
85,000	84,655	394.32	4.5827	6.7706
85,500	85,151	395.14	4.4760	6.5994
86,000	85,647	395.96	4.3721	6.4328
86,500	86,143	396.77	4.2707	6.2708

Altitude		Temperature	Pressure	Density
h_G, ft	h, ft	T, °R	p, lb/ft^2	ρ, slugs/ft^3
87,000	86,639	397.59	4.1719	6.1132
87,500	87,134	398.40	4.0757	5.9598
88,000	87,630	399.22	3.9818	5.8106
88,500	88,126	400.04	3.8902 + 1	5.6655 − 5
89,000	88,622	400.85	3.8010	5.5243
89,500	89,118	401.67	3.7140	5.3868
90,000	89,613	402.48	3.6292	5.2531
90,500	90,109	403.30	3.5464	5.1230
91,000	90,605	404.12	3.4657	4.9963
91,500	91,100	404.93	3.3870	4.8730
92,000	91,596	405.75	3.3103	4.7530
92,500	92,092	406.56	3.2354	4.6362
93,000	92,587	407.38	3.1624	4.5525
93,500	93,083	408.19	3.0912 + 1	4.4118 − 5
94,000	93,578	409.01	3.0217	4.3041
94,500	94,074	409.83	2.9539	4.1992
95,000	94,569	410.64	2.8878	4.0970
95,500	95,065	411.46	2.8233	3.9976
96,000	95,560	412.27	2.7604	3.9007
96,500	96,056	413.09	2.6989	3.8064
97,000	96,551	413.90	2.6390	3.7145
97,500	97,046	414.72	2.5805	3.6251
98,000	97,542	415.53	2.5234	3.5379
98,500	98,037	416.35	2.4677 + 1	3.4530 − 5
99,000	98,532	417.16	2.4134	3.3704
99,500	99,028	417.98	2.3603	3.2898
100,000	99,523	418.79	2.3085	3.2114
100,500	100,018	419.61	2.2580	3.1350
101,000	100,513	420.42	2.2086	3.0605
101,500	101,008	421.24	2.1604	2.9879
102,000	101,504	422.05	2.1134	2.9172
102,500	101,999	422.87	2.0675	2.8484
103,000	102,494	423.68	2.0226	2.7812
103,500	102,989	424.50	1.9789 + 1	2.7158 − 5
104,000	103,484	425.31	1.9361	2.6520
104,500	103,979	426.13	1.8944	2.5899
105,000	104,474	426.94	1.8536	2.5293
106,000	105,464	428.57	1.7749	2.4128
107,000	106,454	430.20	1.6999	2.3050
108,000	107,444	431.83	1.6282	2.1967
109,000	108,433	433.46	1.5599	2.0966
110,000	109,423	435.09	1.4947	2.0014
111,000	110,412	136.72	1.4324	1.9109
112,000	111,402	438.35	1.3730 + 1	1.8247 − 5
113,000	112,391	439.97	1.3162	1.7428
114,000	113,380	441.60	1.2620	1.6649

Altitude		Temperature	Pressure	Density
h_G, ft	h, ft	T, °R	p, lb/ft^2	ρ, slugs/ft^3
115,000	114,369	443.23	1.2102	1.5907
116,000	115,358	444.86	1.1607	1.5201
117,000	116,347	446.49	1.1134	1.4528
118,000	117,336	448.11	1.0682	1.3888
119,000	118,325	449.74	1.0250	1.3278
120,000	119,313	451.37	9.8372 + 0	1.2697
121,000	120,302	453.00	9.4422	1.2143
122,000	121,290	454.62	9.0645 + 0	1.1616 − 5
123,000	122,279	456.25	8.7032	1.1113
124,000	123,267	457.88	8.3575	1.0634
125,000	124,255	459.50	8.0267	1.0177
126,000	125,243	461.13	7.7102	9.7410 − 6
127,000	126,231	462.75	7.4072	9.3253
128,000	127,219	464.38	7.1172	8.9288
129,000	128,207	466.01	6.8395	8.5505
130,000	129,195	467.63	6.5735	8.1894
131,000	130,182	469.26	6.3188	7.8449
132,000	131,170	470.88	6.0748 + 0	7.5159 − 6
133,000	132,157	472.51	5.8411	7.2019
134,000	133,145	474.13	5.6171	6.9020
135,000	134,132	475.76	5.4025	6.6156
136,000	135,119	477.38	5.1967	6.3420
137,000	136,106	479.01	4.9995	6.0806
138,000	137,093	480.63	4.8104	5.8309
139,000	138,080	482.26	4.6291	5.5922
140,000	139,066	483.88	4.4552	5.3640
141,000	140,053	485.50	4.2884	5.1460
142,000	141,040	487.13	4.1284 + 0	4.9374 − 6
143,000	142,026	488.75	3.9749	4.7380
144,000	143,013	490.38	3.8276	4.5473
145,000	143,999	492.00	3.6862	4.3649
146,000	144,985	493.62	3.5505	4.1904
147,000	145,971	495.24	3.4202	4.0234
148,000	146,957	496.87	3.2951	3.8636
149,000	147,943	498.49	3.1750	3.7106
150,000	148,929	500.11	3.0597	3.5642
151,000	149,915	501.74	2.9489	3.4241
152,000	150,900	503.36	2.8424 + 0	3.2898 − 6
153,000	151,886	504.98	2.7402	3.1613
154,000	152,871	506.60	2.6419	3.0382
155,000	153,856	508.22	2.5475	2.9202
156,000	154,842	508.79	2.4566	2.8130
157,000	155,827	508.79	2.3691	2.7127
158,000	156,812	508.79	2.2846	2.6160
159,000	157,797	508.79	2.2032	2.5228
160,000	158,782	508.79	2.1247	2.4329
161,000	159,797	508.79	2.0490	2.3462

SYMBOLS AND CONVERSION FACTORS

SYMBOLS

meter, m
kilogram, kg
second, s
kelvin, K
foot, ft
pound force, lb or lb_f
pound mass, lb_m
degree rankine, °R
newton, N
atmosphere, atm

CONVERSION FACTORS

1 ft = 0.3048 m
1 slug = 14.594 kg
1 slug = 32.2 lb_m
1 lb_m = 0.4536 kg
1 lb = 4.448 N
1 atm = 2116 lb/ft^2 = 1.01×10^5 N/m^2
1 K = 1.8°R

AIRFOIL DATA

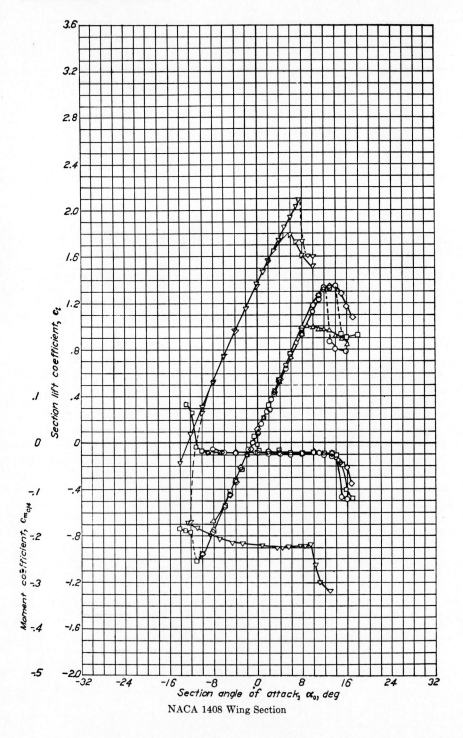

NACA 1408 Wing Section

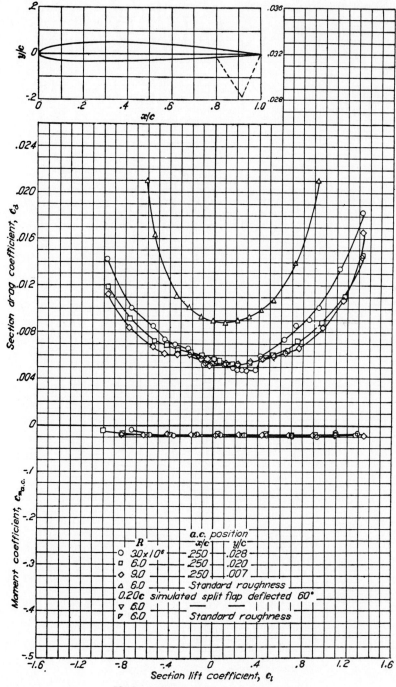

NACA 1408 Wing Section (*Continued*)

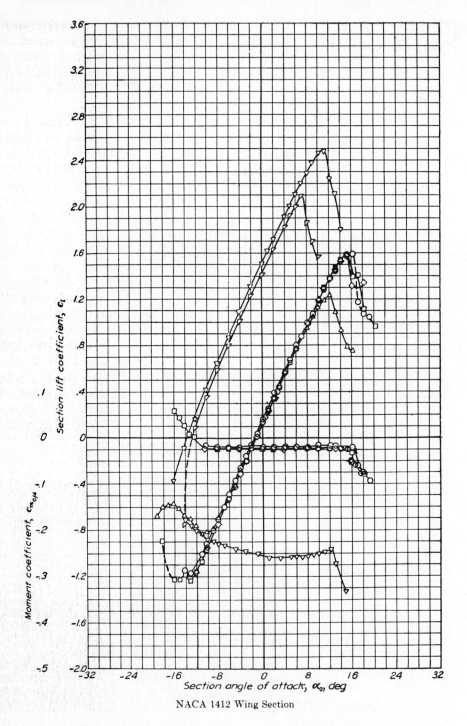

NACA 1412 Wing Section

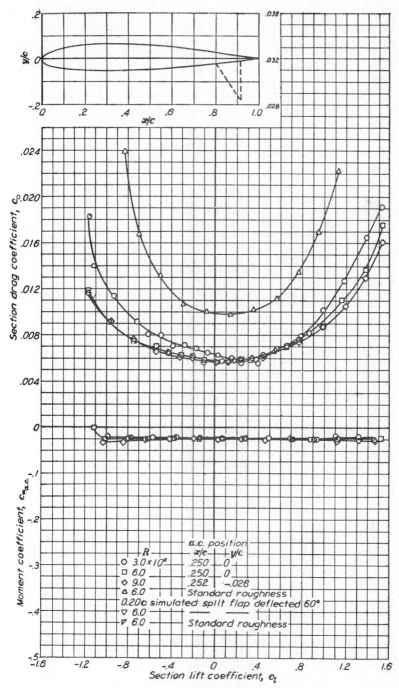

NACA 1412 Wing Section (*Continued*)

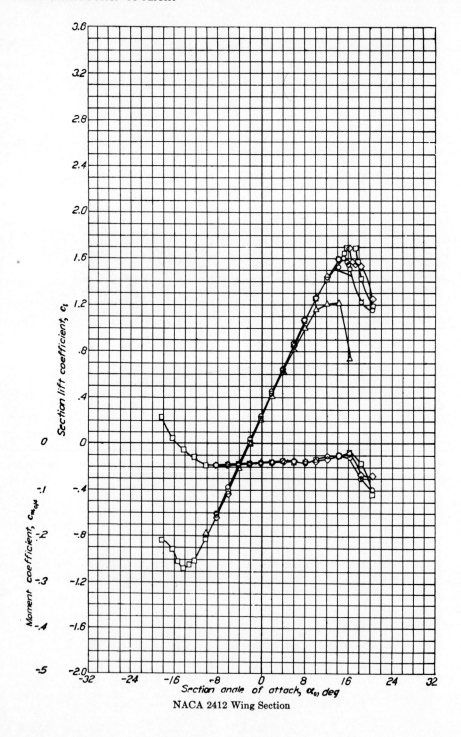

NACA 2412 Wing Section

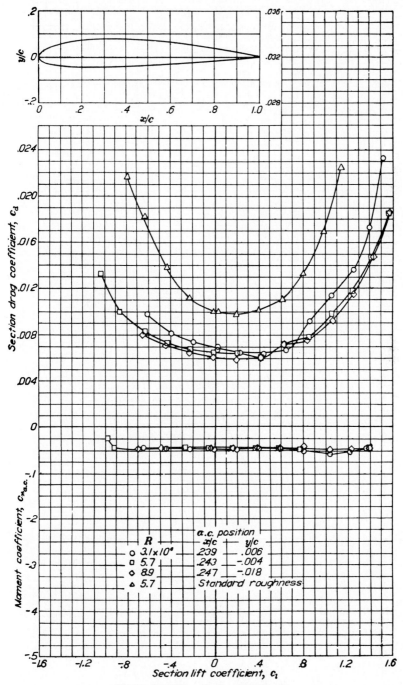

NACA 2412 Wing Section (*Continued*)

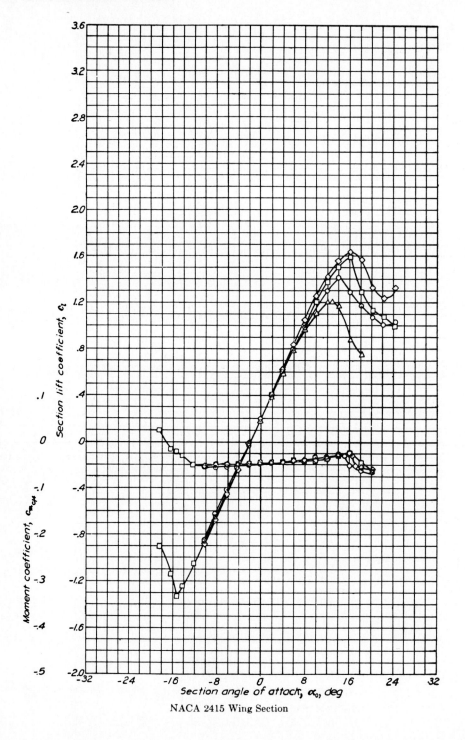

NACA 2415 Wing Section

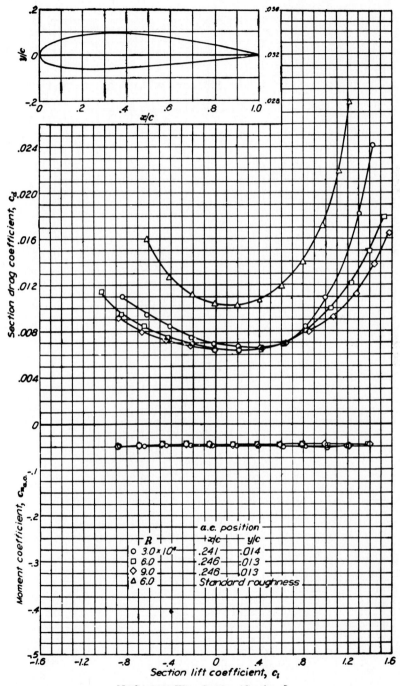

NACA 2415 Wing Section (*Continued*)

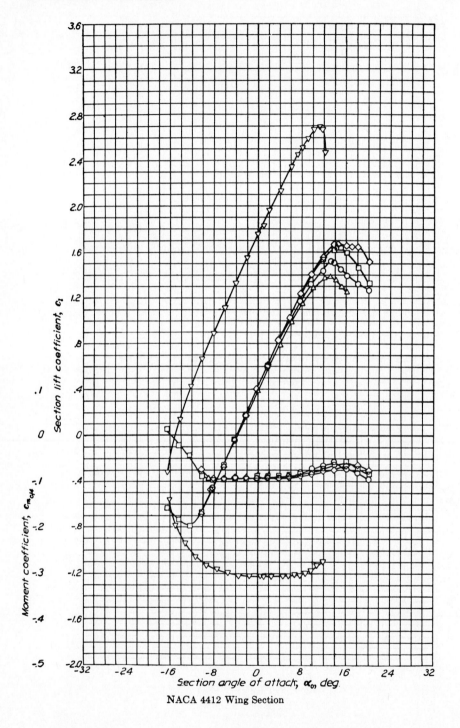

NACA 4412 Wing Section

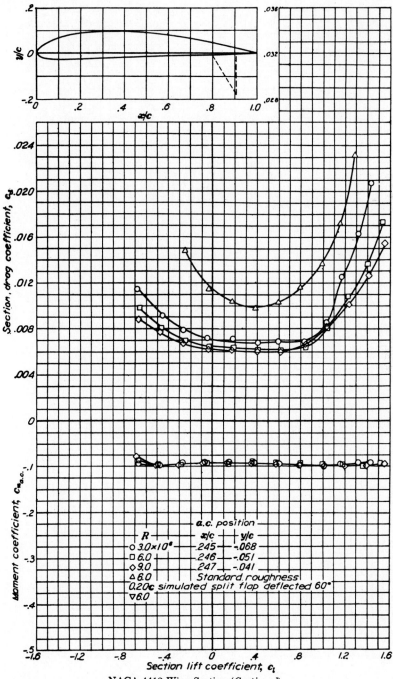

NACA 4412 Wing Section (*Continued*)

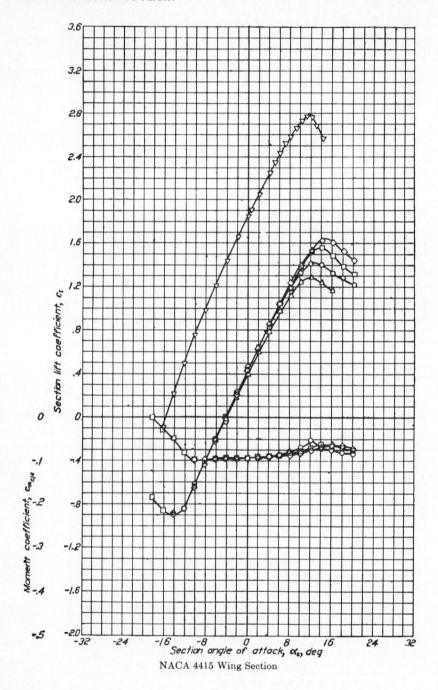

NACA 4415 Wing Section

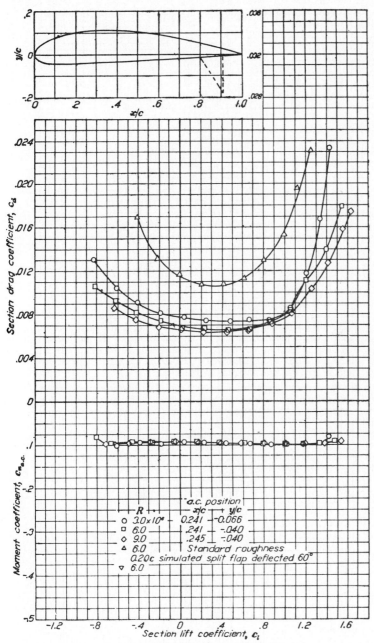

NACA 4415 Wing Section (*Continued*)

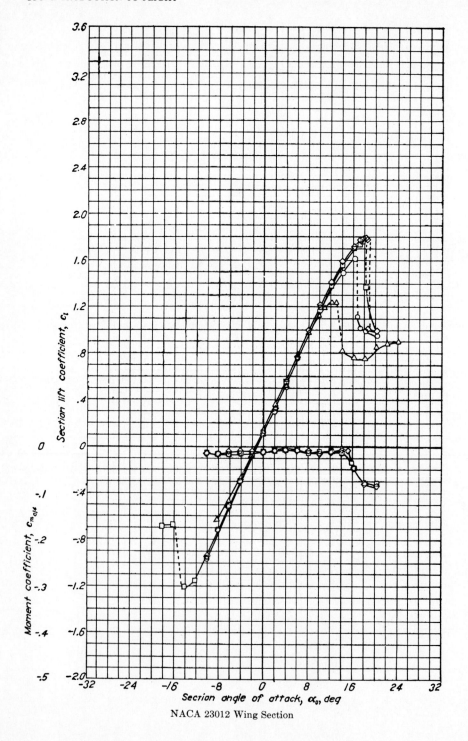

NACA 23012 Wing Section

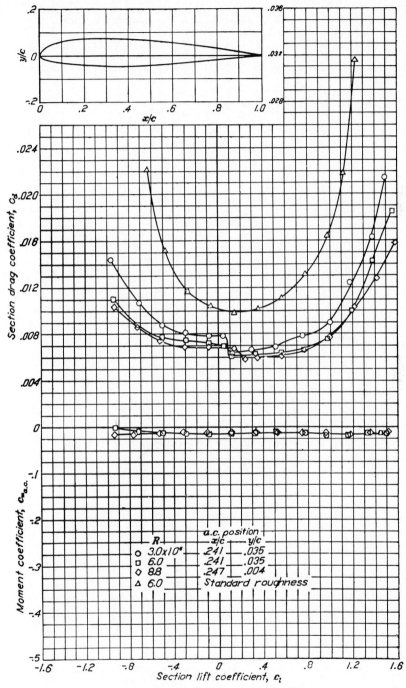

NACA 23012 Wing Section (*Continued*)

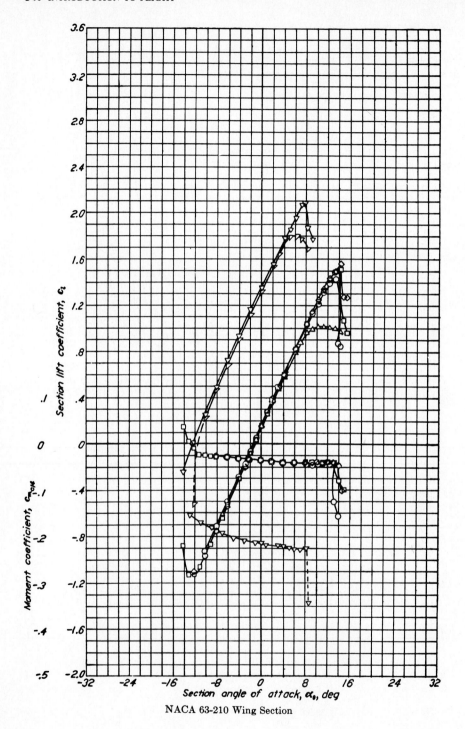

NACA 63-210 Wing Section

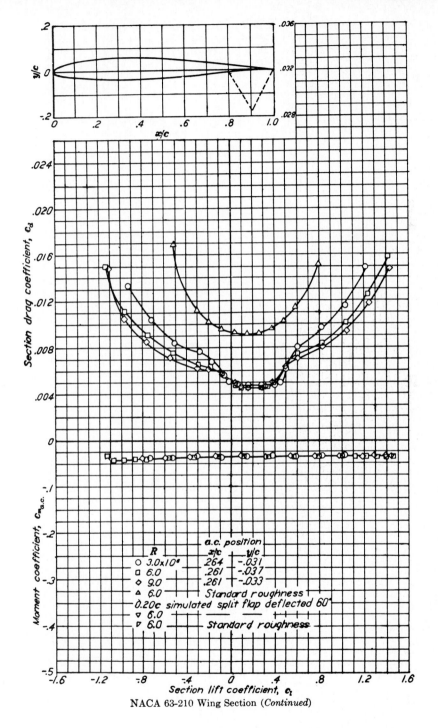

NACA 63-210 Wing Section (*Continued*)

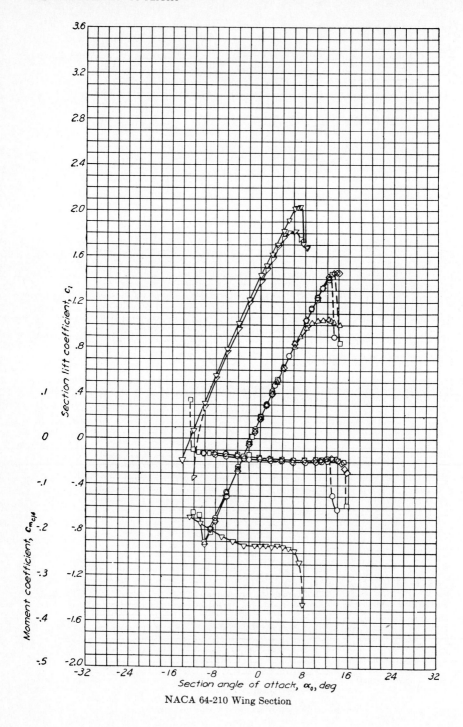

NACA 64-210 Wing Section

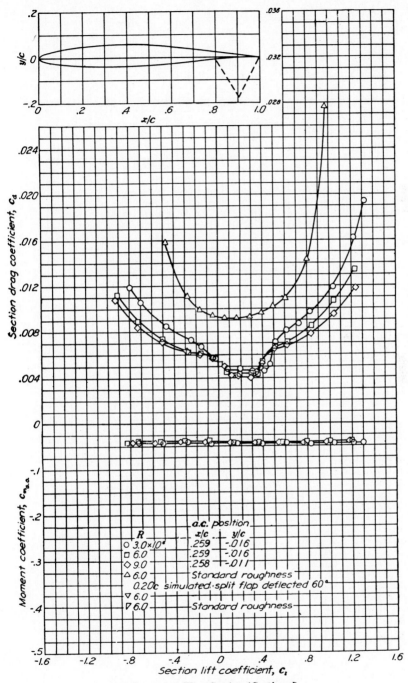

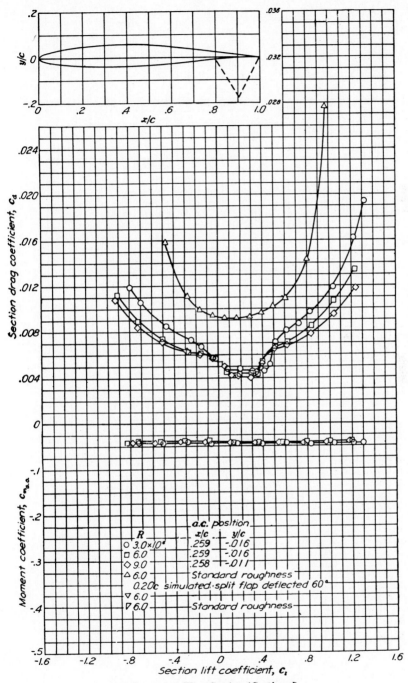

	R	x/c	y/c
○	3.0×10^6	.259	-.016
□	6.0	.259	-.016
◇	9.0	.258	-.011
△	6.0	Standard roughness	

0.20c simulated split flap deflected 60°

▽	6.0	Standard roughness
▽	6.0	

NACA 64-210 Wing Section (*Continued*)

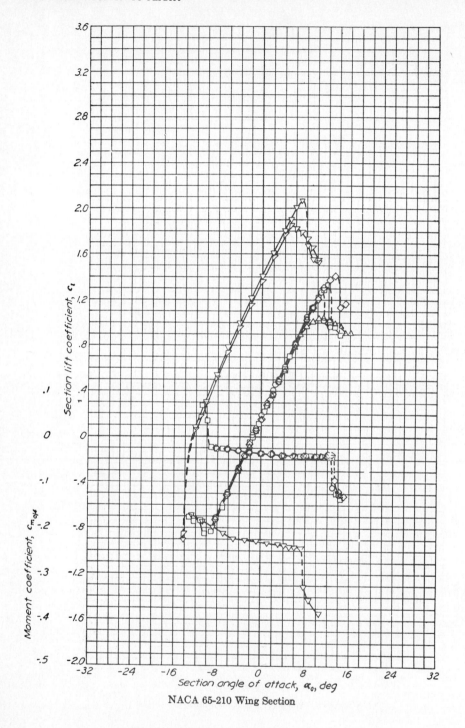

NACA 65-210 Wing Section

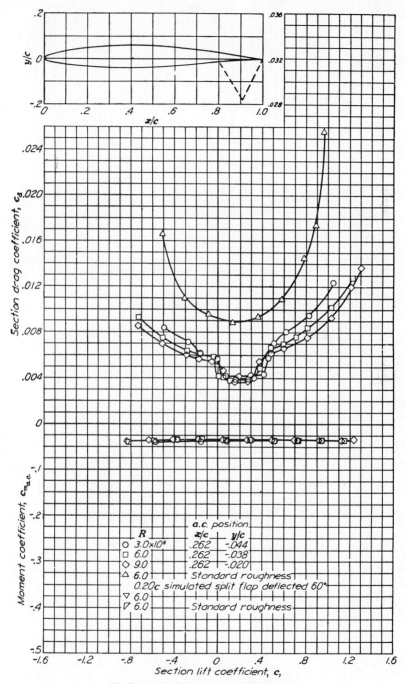

NACA 65–210 Wing Section (*Continued*)

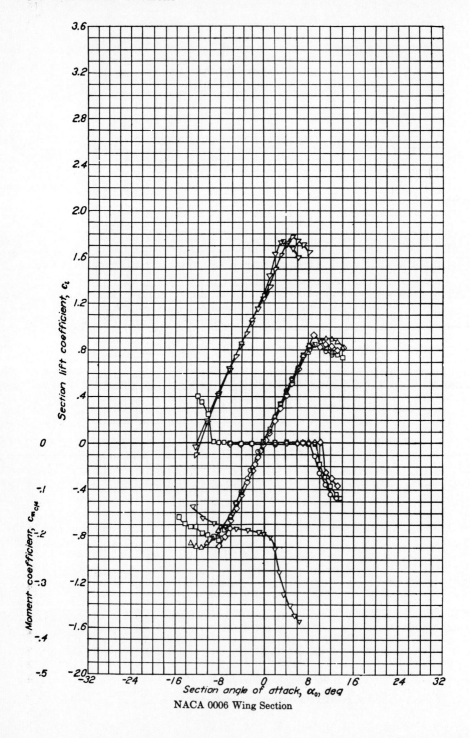

NACA 0006 Wing Section

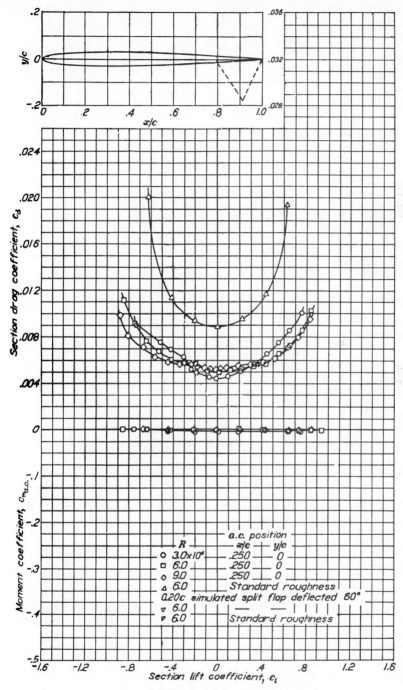

a.c. position

	R	x/c	y/c
○	3.0x10⁶	.250	0
□	6.0	.250	0
◇	9.0	.250	0
△	6.0	Standard roughness	

0.20c simulated split flap deflected 60°

▽	6.0	
▽	6.0	Standard roughness

NACA 0006 Wing Section (*Continued*)

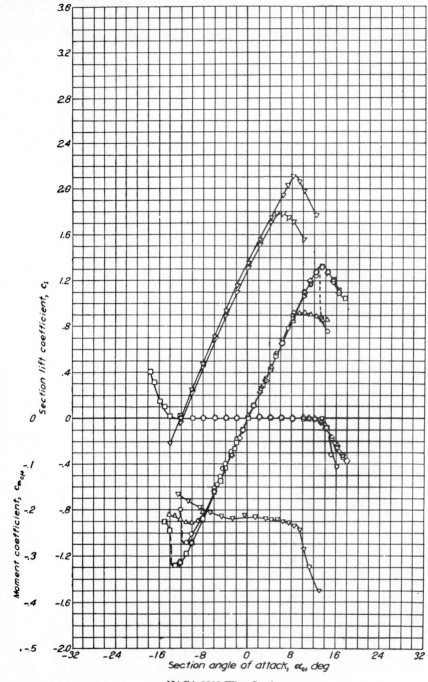

NACA 0009 Wing Section

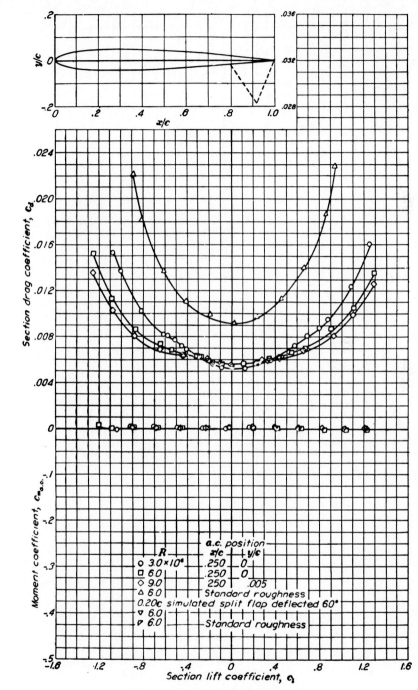

NACA 0009 Wing Section (*Continued*)

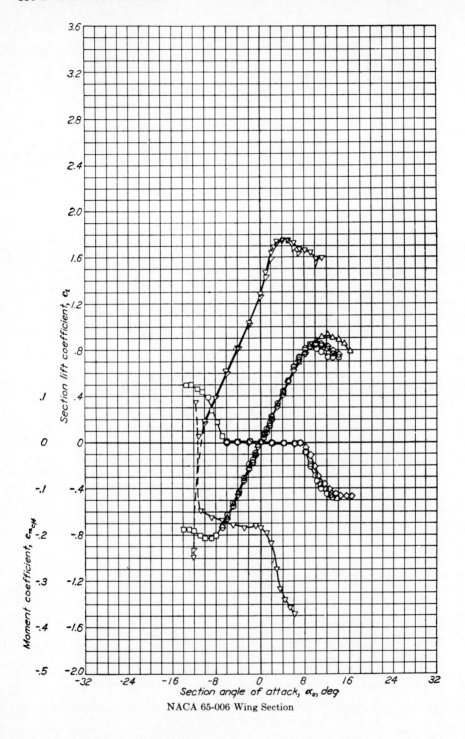

NACA 65-006 Wing Section

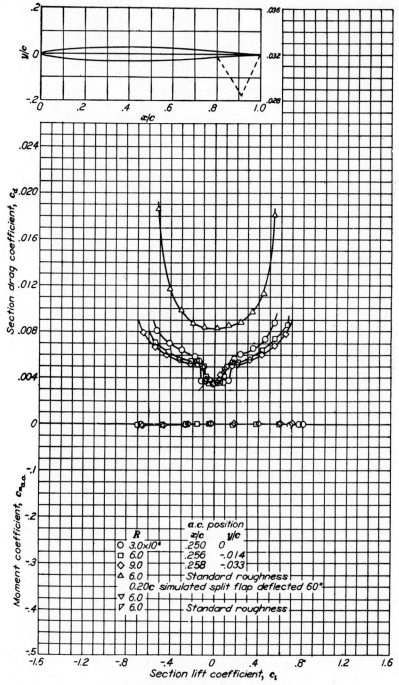

NACA 65-006 Wing Section (*Continued*)

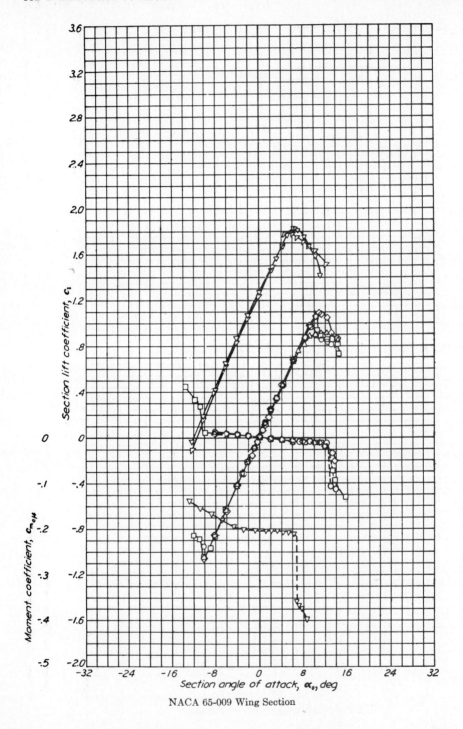

NACA 65-009 Wing Section

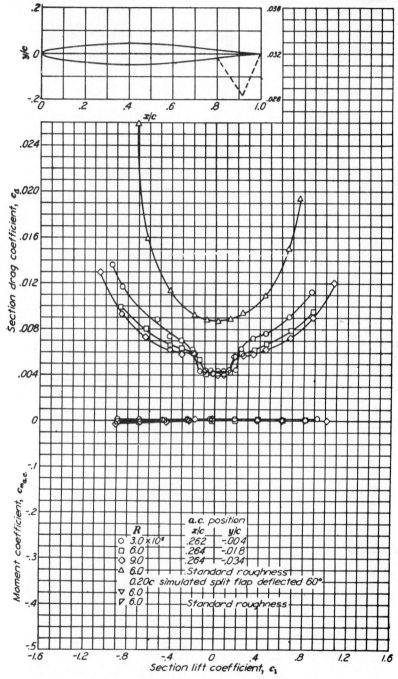

NACA 65-009 Wing Section (*Continued*)

INDEX